T. £19.45

Benchmark Papers
in Geology

Series Editor: Rhodes W. Fairbridge
Columbia University

PUBLISHED VOLUMES

Benchmark Papers
in Geology / 31

A BENCHMARK® Books Series

PALEOBIOGEOGRAPHY

Edited by

CHARLES A. ROSS
Western Washington State College

Dowden, Hutchinson
& Ross, Inc.

STROUDSBURG, PENNSYLVANIA

Distributed by

**HALSTED
PRESS**

A Division of
John Wiley & Sons, Inc.

LIBRARY OF CONGRESS CATALOGING IN PUBLICATION DATA

Main entry under title:
Paleobiogeography
  (Benchmark papers in geology / 31)
  Includes indexes.
    1. Paleobiogeography—Addresses, essays, lectures.
I. Ross, Charles Alexander.
QE721.2.P24P34   560'.9   76-12969
ISBN: 0-87933-226-3

Exclusive Distributor: **Halsted Press**
A Division of John Wiley & Sons, Inc.
ISBN: 0-470-15135-8

ACKNOWLEDGMENTS
AND PERMISSIONS

ACKNOWLEDGMENTS

AMERICAN MUSEUM OF NATURAL HISTORY—*Bulletin of the American Museum of Natural History*
Fossil Floras of the Southern Hemisphere and Their Phytogeographical Significance
The Mesozoic Tetrapods of South America; Discussion

AUSTRALIAN AND NEW ZEALAND ASSOCIATION FOR THE ADVANCEMENT OF SCIENCE—
Report of the Hobart Meetings
Gondwanaland: A Problem in Palaeogeography
The Problem of Sub-Antarctic Plant Distribution

CARNEGIE INSTITUTION OF WASHINGTON—*Carnegie Institution of Washington Publications*
Climates of Geologic Time (No. 192)
Evolution of Desert Vegetation in Western North America (No. 590)

ECONOMIC GEOLOGY PUBLISHING COMPANY—*The American Geologist*
On the Faunal Provinces of the Middle Devonic of America and the Devonic Coral Sub-provinces of Russia, with Two Paleogeographic Maps

GEOLOGICAL SOCIETY OF AMERICA—*Bulletin of the Geological Society of America*
Affinities and Origin of the Antillean Mammals
Correlation and Chronology in Geology on the Basis of Paleogeography
Paleogeographic Significance of the Cenozoic Floras of Equatorial America and the Adjacent Regions
Relations Between the Mesozoic Floras of North and South America
The Relations of the American and European Echinoid Faunas

PACIFIC SCIENCE ASSOCIATION
Proceedings of the Pan-Pacific Science Congress, Australia, The Australian National Research Council Meetings, Melbourne, 1923
The Paleogeography of Permian Time in Relation to the Geography of Earlier and Later Periods
Proceedings of the 6th Pacific Science Congress, Berkeley and San Francisco, 1939
Antarctica as a Faunal Migration Route

UNIVERSITY OF CHICAGO PRESS—*Journal of Geology*
Farewell Lecture by Professor Eduard Suess on Resigning His Professorship

PROFESSOR DR. W. ZEIL—*Geologischen Rundschau*
Sakmarian Geography

Acknowledgments and Permissions

PERMISSIONS

The following papers have been reprinted with the permission of the authors and copyright holders.

AMERICAN ASSOCIATION FOR THE ADVANCEMENT OF SCIENCE—*Science*
    Plate Tectonics and Australasian Paleobiogeography

AMERICAN GEOPHYSICAL UNION—*Journal of Geophysical Research*
    Fossil Floras Suggest Stable, Not Drifting, Continents; Discussion

DEUTSCHE AKADEMIE DER NATURFORSCHER—*Leopoldina*
    The Making of Paleogeographic Maps

DUKE UNIVERSITY PRESS for THE ECOLOGICAL SOCIETY OF AMERICA—*Ecological Monographs*
    Geology and Plant Distribution
    Tertiary Centers and Migration Routes

GEOLOGICAL SOCIETY OF LONDON—*Quarterly Journal of the Geological Society of London*
    On the Age and Correlations of the Plant-Bearing Series of India, and the Former Existence of an Indo-Oceanic Continent

TIMES MIRROR MAGAZINES, INC.—*The Popular Science Monthly*
    Biologic Principles of Paleogeography
    Biologic Principles of Paleogeography

UNIVERSITY OF CHICAGO PRESS—*Journal of Geology*
    Pennsylvanian Floral Zones and Floral Provinces

SERIES EDITOR'S PREFACE

The philosophy behind the "Benchmark Papers in Geology" is one of collection, sifting, and rediffusion. Scientific literature today is so vast, so dispersed, and, in the case of old papers, so inaccessible for readers not in the immediate neighborhood of major libraries that much valuable information has been ignored by default. It has become just so difficult, or so time consuming, to search out the key papers in any basic area of research that one can hardly blame a busy man for skimping on some of his "homework."

This series of volumes has been devised, therefore, to make a practical contribution to this critical problem. The geologist, perhaps even more than any other scientist, often suffers from twin difficulties—isolation from central library resources and immensely diffused sources of material. New colleges and industrial libraries simply cannot afford to purchase complete runs of all the world's earth science literature. Specialists simply cannot locate reprints or copies of all their principal reference materials. So it is that we are now making a concerted effort to gather into single volumes the critical material needed to reconstruct the background of any and every major topic of our discipline.

We are interpreting "geology" in its broadest sense: the fundamental science of the planet Earth, its materials, its history, and its dynamics. Because of training and experience in "earthy" materials, we also take in astrogeology, the corresponding aspect of the planetary sciences. Besides the classical core disciplines such as mineralogy, petrology, structure, geomorphology, paleontology, and stratigraphy, we embrace the newer fields of geophysics and geochemistry, applied also to oceanography, geochronology, and paleoecology. We recognize the work of the mining geologists, the petroleum geologists, the hydrologists, the engineering and environmental geologists. Each specialist needs his working library. We are endeavoring to make his task a little easier.

Each volume in the series contains an Introduction prepared by a specialist (the volume editor)—a "state of the art" opening or a summary of the object and content of the volume. The articles, usually some twenty to fifty reproduced either in their entirety or in significant extracts, are selected in an attempt to cover the field, from the key papers of the last century to fairly recent work. Where the original works are in foreign languages, we have endeavored to locate or commission translations. Geologists, because of their global subject, are often acutely aware of the oneness of our world. The selections cannot, therefore, be restricted to any one country, and whenever possible an attempt is made to scan the world literature.

To each article, or group of kindred articles, some sort of "highlight commentary" is usually supplied by the volume editor. This commentary should serve to bring that article into historical perspective and to emphasize its particular role in the growth of the field. References, or citations, wherever possible, will be reproduced in their entirety—for by this means the observant reader can assess the background material available to that particular author, or, if he wishes, he, too, can double check the earlier sources.

A "benchmark," in surveyor's terminology, is an established point on the ground, recorded on our maps. It is usually anything that is a vantage point, from a modest hill to a mountain peak. From the historical viewpoint, these benchmarks are the bricks of our scientific edifice.

RHODES W. FAIRBRIDGE

PREFACE

Rapid advances in a particular field of science may overshadow and partially obscure the past principles, hypotheses, theories, and assumptions that governed the early growth of the field on which these new interpretations rest. In the late 1960s renewed interest in paleobiogeography generated a surprising number of symposia and collected volumes with promises of more to come. This renewed interest in paleobiogeography is resulting in reinterpretation of earlier fossil distributions, new data, and the use of plate tectonic models to reconstruct biogeographic provinces. The new tectonic concepts of sea-floor spreading and plate tectonics are strikingly different from previous earth crustal models, and commonly the earlier work on which the principles of paleobiogeography were founded is little mentioned. Although the tectonic models may change, the history of paleobiogeography reveals that methods of fossil analyses and interpretation are still strongly rooted to nearly a hundred years of previous studies and debates.

In reviewing the pertinent literature for this volume, I have been struck by the diversity of topics commonly covered in articles having paleobiogeography as the central part of their theme. Because the number of pages available for reprinting are limited, many articles have been shortened and many that dealt principally with physical aspects of paleogeography have been omitted. Because our understanding of paleobiogeography is closely related to our concepts and understanding of physical paleogeography, I have included in the Introduction some notes and references about the physical aspects and in the reproduced material have concentrated on studies in which fossils were the focal point. The recent advances in sea-floor spreading and reinterpretations of fossil distributions have been well covered in several recent compendia and symposia, so this concept and its paleobiogeographic implications are illustrated in this volume by only one paper that investigates a particularly puzzling phytogeographic problem. This collection of papers is concerned mostly with the interval from about 1875 to 1960 and outlines the understanding of paleobiogeographic problems in view of the geological, geophysical, climatic, and evolutionary assumptions during that time.

Preface

This is the time of the great debates between several schools of thought on the permanency of continents and continental drift.

The papers are arranged into five divisions. The first aims at establishing the general perspectives of the subject. The following division centers around the recognition of distinctive parts of the biota and how they are dispersed. Next comes a group of papers that generally investigates phyletic lineages and their origins, and then follows a group that looks into the permanency of oceans and continents. The final set shows examples of how paleobiogeography has been used and interpreted.

In preparing this volume I particularly thank June R. P. Ross, Department of Biology, Western Washington State College, for discussion and help in selecting these articles, and Patricia E. Hamilton, Department of Geology, Western Washington State College, for typing various drafts of the editor's text.

CHARLES A. ROSS

CONTENTS

Contents

Contents

CONTENTS BY AUTHOR

PALEOBIOGEOGRAPHY

INTRODUCTION

In developing our present ideas about paleobiogeography, we rely a great deal on other branches of science for supporting data and theory. Paleobiogeographic analysis involves as much, if not more, synthesis with related sciences as evolutionary analysis. For this reason many changes in existing theories or additional new hypotheses in paleobiogeography commonly must wait for new developments in other sciences before they can be tested and accepted. This has been a slow process because ecology, evolution, systematics, sedimentation, stratigraphy, tectonics, geophysics, oceanography, and many other fields are presently major contributors to the framework on which paleobiogeography must be constructed.

The significant influence that each of these other fields has had in forming the physical and biologic framework is clearly shown by the different steps through which paleobiogeography has progressed. Paleontological and paleobotanical work moved ahead only following advancement in evolution and ecology and the gradual increase in the data generated by systematics. Of particular interest to paleobiogeography are Darwin and Wallace's contributions in formulating the theory of evolution. Darwin's *The Origin of Species* (1859) brought together into a single conceptual framework a great amount of biological data that had been slowly and carefully collected during the first half of the nineteenth century, and related these data to a historical, or geological, context. Wallace's contributions in zoogeography were clearly more descriptive. He demonstrated that faunas in particular parts of the world are at present isolated from one another and, through long evolutionary processes, had been isolated for a long period of time. Both of these biological syntheses greatly expanded investigative imagination and stimulated further delving into needed details to support these syntheses. At that time evolutionary theory lacked a convincing mechanism for inheritance. Zoogeographical theory lacked a plausible control for dispersal and isolation of terrestrial faunas and, as it has eventuated, it also lacked a valid paleogeographic history.

Both Darwin and Wallace were strongly influenced by similarities and differences in the geographical distribution of animals. They recognized that similar (or related) species occur in the same or adjacent geographical areas although each species might be adapted to contrasting ecological niches. Thus the species comprising the floras and faunas of the pampas of Argentina are more closely related to those of the tropical rain forests of Brazil than to the grasslands or tropical rain forests of either Africa or Australia. Darwin (1859) reasoned that species had to have ancestors, and that similar species arose from common ancestral species that also had geographic (and ecological) limitations. In this way, paleobiogeography and species dispersals were introduced as crucial links in the evolution of the biota. This comprehensive view rapidly put an end to the active speculation of the mid-1850s about "centers of creation," where these centers were located, and how many centers had existed.

Darwin also recognized that ranges of organisms constantly shift and have not had static boundaries in the geologic past, or even in the memory of a generation, and that changes in climates result in shifts of biotic ranges. These shifts of biota may cause geographic isolation and the evolution of distinct species in the disjunct ranges. Darwin was also interested in the geographic history of biotic dispersals and the means by which dispersal took place. He understood the general changes that a lowering of sea level or orogenic movements of the earth's crust would cause in the connection of an isthmus or the flooding of terrestrial connections; he saw little reason to support the views of Edward Forbes that called for extensive continental masses across present ocean floors to connect continents. Darwin argued that almost all oceanic islands were volcanic (and basaltic in composition) and that granitic material was not present; it was therefore difficult to imagine foundered continents in the middle of the present oceans. He was a better geologist than many have given him credit for; yet both he and James Dwight Dana believed that the present continents and oceans were of great antiquity and fixed in position. Darwin, however, was willing to hedge the point by qualifying his time reference to within the period of the present species. Darwin's prime interest in zoogeography was to find supporting data to fit into his general theory of evolution. He emphasized ecological and Cenozoic historical aspects of zoogeography and developed concepts and theories based on the general rather than systematic observations that he and others made.

After 1858, Alfred Wallace (1875) became more interested in organizing, identifying, and delineating the limits of zoogeographic units; he gradually built up the concepts for zoogeographical realms, provinces, and subprovinces that are still used today. Between them, Darwin and Wallace laid the foundations for our present biological methods in biogeography and paleobiogeography.

Some of the most striking changes in paleobiogeographic concepts have occurred in the area of physical geographic reconstruction. In his early geology text, Lyell (1830–1833) stated that every part of the land had once been beneath the sea and that every part of the oceans had once been land. Although the first part of the statement was true and enough of the land surface had been studied to warrant the statement, the second part was at most an assumption. Knowledge of the ocean basins was too meager, however, to disprove the latter, and together the two parts appeared to complement each other quite logically.

In North America, Dana (1856) proposed the hypothesis that continents have always been continents (even if they were sometimes flooded by seas) and implied, therefore, that ocean basins had always been ocean basins. His interesting hypotheses on the plan of development of the geologic history of North America related a shrinking crust and the relative size of an ocean basin to the magnitude of deformation of linear mountain belts on continental coasts. His statements on the relationship of various mountain ranges to oceans are entirely geographical and descriptive, and his conclusion relating size of ocean basin to intensity of mountain building is based on the hypothesis of a cooling earth with a shrinking crust and not on the geologic histories of these orogenic belts. In *Manual of Geology* (1863) Dana clearly sets forth his ideas concerning the permanence and fixity of continents and ocean basins.

Nearly all North American geologists, geophysicists, and paleontologists believed that Dana had established firm proof for his several related hypotheses, and thus arrived at the conclusion that continents were proved to be locked in their respective geographic positions. So firmly were Dana's hypotheses held that R. T. Chamberlin (1924), using the latest geophysical and earth model data, went to great length to elaborate on Dana's hypotheses and later hypotheses concerned with mountain-building, and to attack the "switching around of continents," as exemplified by Emile Haug's (1900) reconstruction of the world in "Les géosynclinaux et les aires continentales." In concluding his fine summary of crustal tectonic theory, as viewed by most North American geologists at that time, Chamberlin (1924, p. 574) says, "in explaining the peculiarities of life-distribution, much more may be left to the traveling capacities of the plants and animals themselves." This is what was done by most North American paleontologists and neontologists until the mid-1960s. (See also Willis, 1929, for a synthesis of continental genesis.)

Other aspects of the problems of paleobiogeography were contributed mainly by European geologists. Suess (1885, 1888, 1892) was strongly influenced by the similarities of fossils between South America, Africa, and India, and in his *Antlitz der Erde* placed considerable emphasis on the relations of these continents. He gave the name Gondwana to this continent, which he connected by large land masses in what are now the South Atlantic and Indian oceans. Thus a third hypothesis was set forth: continents could founder to form ocean basins (Suess, 1893). Although this concept was originally adopted in principle, if not in detail, by Charles Schuchert, the total problems with foundering continents were not brought to light until the 1930s when geophysical data, particularly gravity studies, demonstrated that continental sial and oceanic sima crusts were consistently different in density and in thickness.

A number of paleobiogeographical syntheses in the early part of this century (Arldt, 1907, 1919–1922; Haug, 1900; von Huene, 1929; Nopcsa, 1934) and structural paleogeographic syntheses (Stille, 1924; Kossmat, 1936) generated considerable interest, and a journal, *Zoogeographica* was started in Jena in the early 1930s. A renewed interest in the 1960s encouraged the start of another journal, *Palaeoecology, Palaeogeography and Palaeoclimatology.*

Charles Schuchert was the strongest influence in the field of paleogeography in North Ameica in the first third of the twentieth century. He was instrumental in translating the works of Suess, von Huene, and Stille, thus making their ideas available to a wide audience in North America, and he frequently organized and encouraged studies in paleogeography and paleobiogeography. Among his many original ideas was the hypothesis of "borderlands" along the oceanward edge of geosynclines, which he believed were needed to supply sources for the great mass of sediment in geosynclines. His contributions are numerous and lengthy and only a few can be reproduced in this volume. A listing of his bibliography (Dunbar, 1943) is in itself informative, as Schuchert was an active contributor until his death at 85 in 1942. Schuchert (1910), as we shall see in several papers reprinted in this volume, had established his criteria for constructing paleogeographic maps by 1910 or earlier. Although he changed his emphasis from one aspect to another for different geologic periods, he steadfastly maintained that the Precambrian core of North America was permanently fixed and that all historical geology of later periods was peripheral to this core. After the source of most of the clastic sediments in the Appalachian geosyncline was shown to have originated to the east of the present position of the geosyncline, Schuchert patched up his paleogeography (and violated Dana's basic premise) by postulating a land area, Appalachia, which subsequently foundered in the North Atlantic. Having created Appalachia, it was only a minor step to add Llanoria and Cascadia and many other borderlands. On the other hand, the South Atlantic and Indian Ocean, which Suess and others had considered parts of a foundered Gondwana continent, were examined by Schuchert (1932) and Willis (1932); instead of a foundered continent, they proposed an isthmian connection between these continents. This interpretation was partly the result of new information about the South Atlantic ocean floor, but in most part was a reaction to Wegener's (1912, 1915, and later editions) ideas of continental drift, about which Schuchert (1929) and others in North America initially were curious. In the 1930s, geophysical data supported a nondrifting interpretation because they showed that the Atlantic and other ocean basins were made up of oceanic crust. Except for a few islands in the southwest Pacific and in the Indian oceans, continental crust was associated only with continents.

F. B. Taylor (1910, 1923, 1928a, 1928b), H. B. Baker (1911), and Alfred Wegener (1912) each independently developed and presented ideas of continental drift. Taylor approached the problem mainly through structural geology, volcanology, and geophysics, and he was principally interested in the Tertiary. In a number of details Taylor's hypothesis disagreed with Wegener's (1915 and later), but these were tangential to the basic concepts. Baker (1911) postulated a single continent, similar to Pangaea, but his rates of displacement motion and his catastrophic approach to the interpretation were unacceptable to most geologists.

It was Wegener's work that caused the tempest in Europe and

North America, for here was a meteorologist and geophysicist who was using geology, paleontology, tectonics, paleoclimatology, and geophysical data to support a hypothesis that inherently was illogical in the light of previously accepted dogma; he rejected both the permanence of continents and ocean basins and also crustal shrinkage as the cause of fold belts.

Several symposia were organized to discuss continental drift. One of the earliest in North America was organized for the American Association of Petroleum Geologists by Waterschoot van der Gracht in 1926 (published 1928). In general, reaction was negatively inclined toward the Wegener's hypothesis, and a great number of objections were raised to various parts of it. Wegener's book passed through four editions, and each contained new data with modified details and concepts. For many years only Wegener's (1924) third edition was translated into English, which was unfortunate, because in the fourth edition (1929; translated in 1967) he accepted continental shelf margins for his reconstructions and leaned toward the concepts suggested by Joly (1925) and Holmes (1931) of subcrustal convection currents generated by radioactivity as the driving mechanism. Wegener died in Greenland on an exploration trip the next year (1930).

DuToit from South Africa took up the ideas of Wegener and set about looking for direct evidence to compare the geology of Africa with South America (1927). (Schuchert's 1928 review and the discussion by DuToit, 1929, indicate the types of misunderstandings and conflicting opinion that existed.) DuToit's book *Our Wandering Continents* (1937) was a comprehensive and updated synthesis of the continental drift theory. There were many unanswered problems, many apparent contradictions, and still too few answers to provide a mechanism.

From the late 1920s until the mid-1960s the debate about drifting continents lay dormant in North America. In many of the Gondwana continents the hypothesis was widely adopted, and also by some European geologists. With few exceptions most northern hemisphere English-speaking geologists rejected it. The topic was discussed as a possible explanation to various distributional patterns in plants or animals, as in the symposium *Origin and Development of Natural Floristic Areas with Special Reference to North America* [Ecol. Monog., 17 (1947)] and a symposium held in 1949 and published as *The Problem of Land Connections Across the South Atlantic, with Special Reference to the Mesozoic* (Mayr, 1952). In neither of these symposia did continental drift receive much general acceptance. One reason is that emphasis was placed on angiosperms and mammals which only evolved in the middle Mesozoic and reached their present dispersals in the later part of the Cretaceous or in the Cenozoic.

Another interesting symposium, dealing with geological and structural features of the earth's crust, was organized by S. W. Carey (1958a) in Tasmania in 1956. From this appeared an extensive analysis by Carey (1958b) that greatly expanded the theme of structural geologic analysis initiated by Taylor nearly fifty years earlier. In the same symposium, E. Irving (1958) summarized the implications of the data obtained by new geophysical techniques in rock magnetism. The paleomagnetic poles appeared to wander through geologic time and with little apparent direct relationship between different continents, but when fitted to Wegener's or DuToit's reconstruction of continental drift geography, the paleomagnetic poles for the different continents agreed well for Carboniferous, Permian, and later times. (See also reviews by Cox and Doell, 1960; Irving, 1964; Blackett et al., 1965; and Runcorn, 1962.)

During the early 1960s a major scientific effort was mounted to learn more about the geology of the oceans. The results of this effort are still appearing, but what has been learned already has entirely altered our understanding of ocean basins and has given rise to an unanticipated mechanism for ocean basin dynamics: sea-floor spreading and the resulting hypotheses of plate tectonics. A large proportion of these data is geophysical, particularly with regards to magnetic striping of the ocean floor and heat-flow studies; but a significant portion is paleontologic and shows that the basalts of the ocean floors become progressively older away from the actively forming mid-oceanic ridges. Indeed, the ocean basins are not permanent in the sense of Dana, the oldest preserved ocean crust being probably no older than early or middle Mesozoic. With the sea-floor structure explained, a new theory of plate "mobilism" became structurally feasible. With this, the castle of "fixism" had been stormed and the gates flung open to "mobilists" and "drifters." (See also Bullard, 1969; Heirtzler et al., 1968; Hess, 1962; Hurley, 1968; Phinney, 1968; Dietz and Holden, 1970; Johnson and Smith, 1970; Dewey and Bird, 1970; Kay, 1969; Tarling and Runcorn, 1972; Bird and Isacks, 1972; Cox, 1973; and Le Pichon et al., 1973, for more details of tectonic, geologic and geophysic aspects of this hypothesis.)

Paleontologists, who had struggled to explain Paleozoic and Mesozoic faunal provinces or distributions using present-day geography, began reexamining nearly all their earlier ideas of faunal dispersals and evolutionary events. The American Philosophical Society held a symposium with the theme "Gondwanaland revisited" (Piel et al., 1968) in which Romer (1968) reexamined the South American Gondwana vertebrate fossil record. This record is further examined in Keast et al. (1972). In addition, Middlemiss et al. (1971) examined faunal provinces in space and time, Hughes (1971) studied organisms and continents through time, Hallam (1973a, 1973b) organized an atlas of paleobiogeography, and Ross (1974) organized a symposium on provinces and provinciality. A subsection of the 24th International Geological Congress in Montreal (1972) was devoted to plate tectonic theory and continental drift, and other subsections examined the sedimentologic and paleontologic evidence for or against the relative movements of continental blocks. In addition, geologists, mostly from the Gondwana countries, have organized a number of conferences to study the stratigraphy and paleontology of the Gondwana System. Thus far these conferences have met in Uruguay (1967), South Africa (1970), and Australia (1973).

Recent developments in sea-floor spreading and plate tectonics add appreciably to the understanding of processes and possibilities that geologists, paleontologists, and neontologists have available for their use in paleobiogeography. As the reevaluation of fossil distributions continues, we should be able to obtain a substantially improved understanding of the evolutionary and geographic history of life.

REFERENCES

Arldt, T. (1907). Die Entwicklung der Kontinente und ihrer Lebewelt, ein Beitrag zur vergleichenden Erdgeschichte. W. Engelmann, Leipzig, 729 p.

———. (1919–1922). Handbuch der Palaeogeographie. Borntraeger, Leipzig: 1 (1919), 679 p., 2 (1922), 967 p.

Baker, H. B. (1911). The origin of the moon. Detroit Free Press, April 23.

Blackett, P. M. S., E. Bullard, and S. K. Runcorn, eds. (1965). A symposium on continental drift. Phil. Trans. Roy. Soc. London, 258 (1088) 1–323.

Bird, J. M., and B. Isacks, eds. (1972). Plate tectonics. Am. Geophys. Union, Washington, D. C., 453 p.

Bullard, E. (1969). The origin of the oceans. Sci. Am., 211, 66–75.

Carey, S. W., conv. (1958a). Continental drift, a symposium. Geol. Dept., Univ. Tasmania, Hobart, 363 p.

———. (1958b). A tectonic approach to continental drift, in S. W. Carey, conv., Continental drift, a symposium. Geol. Dept., Univ. Tasmania, Hobart, p. 177–355.

Chamberlin, R. T. (1924). The significance of the framework of the continents. Jour. Geol., 32, 545–574.

Cox, A., ed. (1973). Plate tectonics and geomagnetic reversals. W. H. Freeman, San Francisco, xiv + 702 p.

_____, and R. R. Doell (1960). Review of paleomagnetism. Bull. Geol. Soc. America, 71, 645–768.

Dana, J. D. (1856). On the plan of development in the geologic history of North America. Am. Jour. Sci., 2nd ser., 22, 335–349.

_____. (1863) Manual of geology. Philadelphia, 812 p.

Darwin, Charles (1859). The origin of species by means of natural selection or the preservation of favoured races in the struggle for life. Murray, London, ix + 490. (Reprinted 1964 with introduction by E. Mayr, Harvard Univ. Press, Cambridge, Mass., xxvi + ix + 502 p.)

Dewey, J. F., and J. M. Bird (1970). Lithosphere plate–continental margin tectonics and evolution of the Appalachian Orogen. Bull. Geol. Soc. America, 81, 1031–1060.

Dietz, R. S., and J. C. Holden (1970). Reconstruction of Pangaea: breakup and dispersion of continents, Permian to present. Jour. Geophys. Res., 75, 4939–4956.

Dunbar, C. O. (1943). Memorial to Charles Schuchert. Proc. Geol. Soc. America, 1943, 217–240.

DuToit, A. L. (1927). A geological comparison of South America with South Africa. Carnegie Inst. Wash., Publ. 381, 157 p.

_____. (1929). The continental displacement hypothesis as viewed by DuToit. Am. Jour. Sci., 5th ser., 17, 179–183.

_____. (1937). Our wandering continents. Oliver & Boyd, Edinburgh, xiii + 366 p. (Reprinted 1957, Hafner, New York.)

Hallam, A. (1973a). A revolution in the earth sciences. Oxford Univ. Press, Oxford, 127 p.

_____, ed. (1973b). Atlas of palaeobiogeography. Elsevier, Amsterdam, 531 p.

Haug, E. (1900). Les Géosynclinaux et les aires continentales. Bull. Soc. Geol. France, Ser. 3, 28, 617–711.

Heirtzler, J. R., G. O. Dickson, E. M. Herron, W. C. Pitman III, and X. Le Pichon (1968). Marine magnetic anomalies, geomagnetic field reversals, and motions of the ocean floor and continents. Jour. Geophys. Res., 73, 2119–2136.

Hess, H. H. (1962). History of the ocean basins, in A. E. J. Engel, H. L. James, and B. F. Leonard, eds., Petrological studies: a volume in honor of A. F. Buddington. Geol. Soc. America, New York, p. 599–620.

Holmes, Arthur (1931). Radioactivity and earth movements. Trans. Geol. Soc. Glasgow, xviii, p. 559–606.

Hughes, N. F., ed. (1971). Organisms and continents through time. Palaeont. Assoc., Spec. Pap. 12; System. Assoc. Publ. 9, London, vi + 334 p.

Hurley, P. M. (1968). The confirmation of continental drift. Sci. Am., 218, 52–62.

Irving, Edward (1958). Rock magnetism: a new approach to the problems of polar wandering and continental drift., in S. W. Carey, conv., Continental drift, a symposium. Geol. Dept., Univ. Tasmania, Hobart, p. 24–61.

————. (1964). Paleomagnetism and its application to geological and geophysical problems. Wiley, New York, xvi + 399 p.

Johnson, H., and B. L. Smith, eds. (1970). The megatectonics of continents and oceans: Rutgers Univ. Press, New Brunswick, N. J., xii + 282 p.

Joly, John (1925). The surface-history of the earth. Clarendon Press, Oxford, 192 p.

Kay, G. M., ed. (1969). North Atlantic—geology and continental drift. Am. Assoc. Petrol. Geol., Mem. 12, 212–233.

Keast, A., F. C. Erk, and B. Glass, eds. (1972). Evolution, mammals, and southern continents. State Univ. New York Press, Albany, N. Y., 543 p.

Kossmat, Franz (1936). Paläogeographie und Tektonic. Borntraeger, Berlin, xxiii + 413 p. (See Charles Schuchert, 1937, The evolving face of the earth, Am. Jour. Sci., 5th ser., 33, 308–311.)

Le Pichon, X., J. Francheteau, and J. Bonnin (1973). Plate Tectonics. Elsevier, Amsterdam, xiv + 300 p.

Lyell, Charles (1830–1833). Principles of geology. Murray, London (v. 1, 1830; v. 2, 1832; v. 3, 1833); 4th ed. (1872) D. Appleton, New York., v. 1, xx + 671 p., v. 2, xix + 652 p.

Mayr, E., ed. (1952). The problem of land connections across the South Atlantic, with special reference to the Mesozoic. Bull. Am. Mus. Nat. Hist., 99(3), 79–258.

Middlemiss, F. A., P. F. Rawson, and G. Newall, eds. (1971). Faunal provinces in space and time. Geol. Jour., Spec. Issue 4, Seel House Press, Liverpool, 236 p.

Nopcsa, F. (1934). The influence of geological and climatological factors on the distribution of non-marine fossil reptiles and Stegocephalia. Quart. Jour. Geol. Soc. London, 90, 76–140.

Phinney, R. A., ed. (1968). History of the earth's crust. Princeton Univ. Press, Princeton, N.J., viii + 244 p.

Piel, G., et al. (1968). Gondwanaland revisited: new evidence for continental drift. A symposium. Proc. Am. Philos. Soc., 112(5), 307–353.

Romer, A. S., (1968). Fossils and Gondwanaland. Proc. Am. Philos. Soc., 112(5), 335–343.

Ross, C. A., ed. (1974). Paleogeographic provinces and provinciality. Soc. Econ. Paleont. Mineral., Spec. Publ. 21, 233 p.

Runcorn, S. K., ed. (1962). Continental drift. Intl. Geophys. Ser. v. 3, Academic Press, New York, 338 p.

Schuchert, Charles (1910). Biologic principles of paleogeography. Popular Sci. Monthly, 76, 591–600.

————. (1919). Paleogeography of North America. Bull. Geol. Soc. America, 20, 427–606.

————. (1928). The continental displacement hypothesis as viewed by DuToit. Am. Jour. Sci., 5th ser., 16, 266–274.

————. (1929). The hypothesis of continental drift. Smithsonian Inst. Ann. Rept. 1928, p. 249–282.

————. (1932). Gondwana land bridges. Bull. Geol. Soc. America, 43, 875–915.

Stille, Hans (1924). Grundfragen der vergleichenden Tektonik. Borntraeger, Berlin, vii + 443 p. (See Charles Schuchert, 1926, Stille's analysis and synthesis of the mountain structures of the earth, Am. Jour. Sci., 5th ser., 12, 277–292.)

Suess, E. (1885). Das Antlitz der Erde, v. 1, pt. 2, Die Gebirge der Erde. F. Tempsky, Vienna, p. 237–778.

———. (1888). Das Antlitz der Erde, v. 2, pt. 3, Die Meere der Erde. F. Tempsky, Vienna, 704 p.

———. (1893). Are great ocean depths permanent? Natural Sci., 2, 180–187.

Tarling, D. H., and S. K. Runcorn, eds. (1972). Implications of continental drift to the earth sciences. Academic Press, New York, v. 1, xvi + 624 p; v. 2, xvi + 560 p.

Taylor, F. B. (1910). Bearing of the Tertiary mountain belt on the origin of the earth's plan. Bull. Geol. Soc. America, 21, 179–226.

———. (1923). The lateral migration of land masses. Proc. Washington Acad. Sci., 13, 445–447.

———. (1928a). North America and Asia, a comparison in Tertiary diastrophism. Bull. Geol. Soc. America, 39, 985–1000.

———. (1928b). Sliding continents and tidal and rotational forces, *in* Waterschoot van der Gracht et al., The theory of continental drift, a symposium. Am. Assoc. Petrol. Geol., Tulsa, Okla., p. 158–177.

von Huene, F. (1929). Los Saurisquios y Ornitisquios del Cretáceo Argentino. Mus. LaPlata Anal. 3(2a), 196 p., 44 pl., 91 fig. (See C. Schuchert, 1930, Cretaceous and Cenozoic continental connections according to von Huene. Am. Jour. Sci., 5th ser., 19, 55–66.)

Wallace, A. R. (1875). The geographical distribution of animals. 2 vol. Harper, New York, xxii + 503 p., xi + 553 p. (Reprinted 1962, Hafner, New York.)

Waterschoot van der Gracht, W. A. J. M. (1928). The problem of continental drift, *in* W. A. J. M. Waterschoot van der Gracht et al., Theory of continental drift, a symposium. Am. Assoc. Petrol. Geol., Tulsa, Okla., p. 1–75, 197–226.

———, et al. (1928). Theory of continental drift, a symposium. Am. Assoc. Petrol. Geol., Tulsa, Okla., x + 240 p.

Wegener, A. (1912). Die Entstehung der Kontinente. Peterm. Mitt., p. 185–195, 253–256, 305–309.

———. (1915). Die Entstehung der Kontinente und Ozeane. Braunschweig, 94 p.; 2nd ed. (1920), 135 p.; 3rd ed. (1922), 144 p.; 4th ed. (1929), 231 p.

———. (1924). The origin of continents and oceans. (English transl. of 3rd ed.) Methuen, London, 212 p.

———. (1967). The origin of continents and oceans. (English transl. of 4th ed.) Methuen, London, 248 p.

Willis, Bailey (1929). Continental genesis. Bull. Geol. Soc. America, 40, 281–336.

———. (1932). Isthmian links. Bull. Geol. Soc. America, 43, 917–952.

Part I

PALEOBIOGEOGRAPHIC
PERSPECTIVES

Editor's Comments
on Paper 1

1   **SUESS**
    *Farewell Lecture by Professor Eduard Suess on Resigning
    His Professorship*

Eduard Suess was one of those remarkable gentlemen of European geology who greatly influenced the thought and trend of geological inquiries for many generations. He was eighty-three when he passed away in 1914 after spending most of his adult life in Vienna as an academic in the university, in the city council, and as a member of the Austrian parliament.

In the 1850s he began comprehensive studies of fossil Paleozoic and Mesozoic Brachiopoda. Vienna in the early 1860s had difficulty with an impure water supply, and the resultant diseases, particularly typhoid fever, were a serious problem. Suess strongly advocated a 110-kilometer aqueduct system to bring pure water from the Alps to the city. This was eventually constructed, and, as a result, Suess was so well thought of that he was elected and re-elected to the Austrian parliament for thirty years. He was an enthusiastic teacher and retired at the age of seventy. His international reputation was thoroughly established by his great geological synthesis *Das Antlitz der Erde* (1885–1909) (*The Face of the Earth*), which was translated into both English and French. This set of volumes is the only attempt thus far to summarize the regional geology of the globe. In it, Suess first established the basic ideas of crustal plates, continental shields, and global orogenic belts, as well as eustacy and worldwide stratigraphic cycles. The direct influence of this man in geological thought was monumental, in part because of his personal warmth and in part because of the range, perception, and stimulation of his geological investigations. Neumayr, Mojsisovics, Fuchs, Waagen, and Penck were among his many students and from comments by persons who had known him, such as Geikie and de Margeire, Suess was obviously greatly respected and admired.

Paper 1 is a translation of Eduard Suess's farewell lecture when he resigned as Professor of Geology at the University of Vienna. It needs little additional comment except to say that his message is as clear, fresh, and farsighted today as it was in 1901.

REFERENCES

Schuchert, C. (1914). Eduard Suess: Science, n.s., 39, 933–935.
Suess, E. (1885, 1888, 1901, 1909). Das Antlitz der Erde. G. Freytag, Leipzig. [Translated as The face of the earth, 5 vol. (1904, 1906, 1908, 1909, 1924), Clarendon Press, Oxford.]

1

Reprinted from *Jour. Geol.*, **12**, 264–274 (1904)

FAREWELL LECTURE BY PROFESSOR EDUARD SUESS ON RESIGNING HIS PROFESSORSHIP.[1]

In the last lecture we occupied ourselves with the structure of South America. We saw that the earlier volcanic occurrences are restricted entirely to the Cordillera of the Andes, but that in the course of their appearance there are long interruptions.

We have therefore arrived at the close of our hasty survey of the earth's entire surface, and today we will review the events which have been set forth during the last two semesters. The present lecture, moreover, also closes my active life as a professor, and I stand at the end of a career of teaching at this university, which I have been permitted to enjoy for eighty-eight semesters. Before I take up the short summary mentioned, I believe it suitable to say a few words in regard to the changes which our science has undergone during this long period.

My collegiate work as lecturer on general paleontology was begun October 7, 1857—two years before the appearance of Darwin's book, *The Origin of Species.*

It is well known that in the eighteenth century prominent thinkers, as Leibnitz, Herder, and others, properly recognized the connection and unity of all organic life. But, at the beginning of the nineteenth century, Cuvier, essentially by means of the fossils of the chalk of Montmartre, was able to present the surprising evidence that there had lived on the earth genera of animals which today are wholly extinct, and that similar changes have again and again occurred in the animal kingdom. He thus concluded that there had been repeated revolutions. In this he was followed by the great majority of inquirers, and at that time—the year 1857—everyone was completely under the influence of Cuvier's views. Personally, a paper by Edward Forbes, on the influence of the glacial period on migrations, had a great effect on me; the article merits reading even to this day.

[1] Given July 13, 1901, in the Geological Lecture Hall, Vienna University; taken stenographically by Mr. H. Beck. For the original lecture, in German, see *Mitth. Pal. u. Geol. Inst.*, Universität Wien, 1902, pp. 1–8. Translator, Charles Schuchert.

After the appearance of Darwin's book, there occurred a great and general change of view in all branches of biology. In fact, outside the great discoveries of Copernicus and Galileo, there cannot be cited another example having so deep an influence on the general opinions of naturalists. Darwin was not the first to conceive and pronounce upon the unity of all life; but that he was able to produce stronger proofs and to direct the trend of thought constitutes his undying fame.

In the field of paleontology the consummation of this change did not, of course, go on so simply and, at least with us, so entirely in accordance with the views of Darwin as one is apt to imagine. The Darwinian theory of the variability of species was essentially based on selection and related appearances. Paleontology, however, teaches otherwise. It teaches that the terminology for single divisions of the stratified terranes, characterized by their fossil remains, finds application over the entire earth. Therefore from time to time there must have occurred, in some way, general changes affecting the entire physical condition of the world. Nor is there seen a perpetual and continuous changing of organic beings, as would be the case through the constant influence of selection. On the contrary, there are entire groups of animals appearing and disappearing. Darwin sought to explain this by means of gaps in our knowledge, but today it is known that these supposed gaps possess too great a horizontal extension.

There now arises the thought that the changes in the outer conditions of life have a controlling influence. I may here state that on this question there was some correspondence between Darwin and our widely mourned Neumayr, and that Darwin in no wise took a dissenting stand against the objections. In this connection it is most remarkable that the great and general knowledge of paleontology which I have just indicated, should apparently have made upon so great a mind as Darwin's less of an impression than those small lines of variation noticed in certain fossil fresh-water snails, as, for example, in *Valvata* or *Paludina*.

Here and there conditions are combined which permit somewhat closer analyses of the relations of this subject. This for instance, is the case in the superposition of the Tertiary land faunas of Europe, more particularly of Vienna. Here one recognizes the following: Living beings are dependent, on the one hand, on certain outer physical circumstances, as climate, moisture, etc.; on the other hand, they also are mutually and socially dependent upon one another. Every living province—or, as it is usually expressed, every zoölogical province—forms, as it were, an economical unity in which for so many flesh-eaters there must be so many plant-feeding food animals; for so many plant-feeders, so many food plants; honey-sucking Lepidoptera presuppose flowers; for insect-feeding song-birds a certain number of small insects are necessary, etc. The disturbance of one member of this unity can possibly destroy the balance of the whole.

According to all appearances, such disturbances have occurred from time to time in land faunas, and they may have been of very diverse kinds. Then again an entire fauna is seen to vanish over all Europe, or over a still greater region, and a new fauna comes in to take its place. This new fauna nevertheless always has a more or less strongly vicarious relationship to its predecessor; it is clearly a variation of the former, probably in the main a resulting adaptation to changed conditions; and even if the sequence of strata were completely unknown, one could readily discern which was the first, the second, or the third fauna.

Besides this, the numerous phylogenetic lines which unite nearly all the great groups of fossil animals; or the unity in the developmental nature of single organs, as the extremities; or the general superposition of gills and lungs; or the rows of striking harmonies that exist between the development of certain groups of animals, and of single individuals of these groups—all indicate with certainty the correctness of the Darwinian basal idea, namely, the unity of life.

Stratigraphic geology and paleontology show that the evolution of organic life was probably never completely interrupted, but that it did not go on in a uniform manner. Disturbances have occurred. The struggle for existence continues; yet it is only of secondary importance. Single very old types, as *Hatteria* (*Sphenodon*), have continued to maintain themselves to our day with but slight changes.

Allow me now to speak of a few tectonic questions.

When I began my collegiate work, there prevailed, especially in

Germany, the idea that mountain chains were built symmetrically; one group of oldest rocks formed the lifted or lifting axis, and upon each side were arranged younger rocks in parallel zones. Thus, you will still find in my own writing on the substructure of Vienna, in the year 1862, a presentation of the Alps as symmetrical mountains.

Of course, this idea did not prevail without objections. At nearly every gathering of German naturalists at that time, the old Bergrath Dücker arose to protest against it. No one listened to him. With Schimper it was the same. The authority of Leopold von Buch, who expressed himself for symmetrical construction, remained unshaken. Then Leopold von Buch died. Upon this primary question of modern geology you will find no explanation for the origin of mountains in the leading text-books of that time, as, for instance, Lyell's justly celebrated *Principles of Geology*.

For the investigation of this problem no part of Europe was more advantageously situated than Austria. There the land is arrayed before us in unusual variety. Hardly anywhere in Europe are tectonic contrasts so plainly presented—contrasts between the Bohemian Mass and the Alps, between the portion of Russian table-land beneath the Galician plain and the Carpathians, the peculiar connection of Alps and Carpathians, the continuance of the Turkestan depression over the Aral Sea into the depression of the Danube and to Vienna, and much besides. In the year 1857 the idea was still often maintained that the deposits found in the eastern Alps did not occur at all outside of the Alps, so great were the difficulties which the application of the accepted stratigraphic divisions of England and south Germany bore to the strange occurrences in the Alps themselves.

Soon, however, it was recognized that in the Bohemian Mass the stratigraphic sequence was far less complete than in the adjoining regions of the Alps, and that in Bohemia particularly there is an extraordinary interruption of marine deposits extending upward into the Middle Cretaceous, whereas in the Alps all these great epochs are represented by marine strata. This same transgression of the Middle and Upper Cretaceous shows again in Galicia, then far into Russia, on the other side of the French Central Plateau, on the Spanish Meseta, in large parts of the Sahara, in the valley of the Mississippi, and northward over this region to the vicinity of the Arctic Sea, in

Brazil, finally on the shores of central and southern Africa, in east India; and, in fact, over such extraordinarily vast regions that it became impossible longer to explain such transgressions of the sea, according to the older views of Lyell, by means of the elevation and depression of continents.

Through this and similar observations the newer idea has recently come into prominence that some general change must have occurred either in the shape of the hydrosphere or in its entire volume. It was seen that by the forming of a new oceanic depth, due to sinking, a certain amount of the hydrosphere was drawn off into the new depression, and that at the same time there appeared to be a general land elevation, or, more correctly, there must have resulted a general sinking of the beach lines. The older view of the numerous oscillations of the continents has also given way more and more to the teachings of marine transgressions, and through the denudation of continents, a more exact examination into the actual mountain movements has become possible.

If one were to assert that the Alps are folded, but that the Bohemian Mass is not, and that because of this there has resulted a damming up, then this assertion would not be exact. The Bohemian Mass is also folded, and there is at present no known portion of the earth's surface of which at least the archaic base is not folded. The difference, however, consists in this, that the folding has ended early at certain places; at others it has continued into a later or very late time, and possibly has also continued with a change in the ground-plan.

In this respect central Europe shows a quite peculiar arrangement. The oldest folding is seen in the gneiss of the western Hebrides. Younger and of pre-Devonic age are the folds of the Caledonians, which can be traced down to Ireland. On these, farther south, are ranged the Armoricanian and Varischian folds, which embrace south-western England, Normandy and Brittany, the Central Plateau, the mountains of the Rhine, and the Bohemian Mass, inclusive of the Sudetes. Its principal folding was accomplished before the close of Carboniferous time, but minor movements of various kinds have followed. The Alps and Carpathians even underwent decided folding in the Miocene. Each part has moved northward toward the preceding, or toward the horsts, in which the earlier member was dissolved by sinking, and thus Europe has resulted through a succession of younger and younger folds.

Meanwhile, more and more light came regarding the strange development which certain Mesozoic deposits, particularly the Triassic of the Alps, show when compared to the north-lying lands, as Würtemberg or Franconia. The observations in Asiatic highlands, especially in the Himalayas, taught that this type of Triassic development has a very wide distribution toward the east; and it even became possible to prove that directly across present Asia, from the existing European Mediterranean to the Sunda Islands, there once extended a continuous sea. This sea has, as you know, received the name Tethys. The old continent along its southern side was named Gondwana Land, and that on its northern side, Angara Land. The present Mediterranean is a remnant of Tethys.

This Mediterranean, however, consists of a series of areas of diverse construction, and we have had opportunity to convince ourselves that, since Middle Tertiary time, first a portion was separated, as, for instance, the Danube plain, then a portion was added, as the Ægean Sea.

The progress of geological research during the last ten years, however, has been so extremely great that a far more extensive knowledge of the seas has become possible. They are of different kinds. We examine a world-map, and thereby, in accordance with oft-repeated warning, seek to guard against the deception which the distortion of Mercator's projection so easily produces. We see that, with the exception of the two Chinese rivers, Yang-tse-kiang and Hoang-ho, hardly another great stream finds its way to the Pacific Ocean. All waters of the continents flow toward the Atlantic or Indian Ocean. Many years ago the Russian General von Tillo drew on a little map the watershed of the earth, and showed how surprisingly small an amount of fresh water the Pacific receives.

These two oceanic areas differ also in a feature of far greater importance. At the beginning of these lectures I noted the remarkable fact that from the mouth of the Ganges eastward to Cape Horn the continents are bounded ocean-ward by long arcuate mountain ranges, all of which appear to be moving toward the Pacific Ocean.

21

When, however, one follows the coast from the mouth of the Ganges westward, and again to Cape Horn, totally different conditions are met. Disregarding the bending of the mountains at Gibraltar and vicinity, which the American Cordilleras in the Antilles also show— at both places, as you know, folded mountain chains do approach the Atlantic area, but they bend backward as if held back by some secret force—one sees encircling the Atlantic and the Indian Oceans only similar amorphous coast lines, namely, such as are in no wise predicated by the structure of the lands. Therefore we have distinguished a Pacific and an Atlantic type of coast.

We can go still farther. In whatever direction one proceeds from the land to the Pacific, an unfolding sequence of marine series is seen. If one goes from the wide Archean areas of South America, on which lie horizontal Paleozoic sediments, toward the west, in the Andes are found marine beds of the Jura, the Lower Cretaceous, also the Middle and Upper Cretaceous. It is the same if one goes from the old Laurentian Mass in Canada westward toward the sea. This is also the case in Japan, etc. From the foregoing we may conclude that the Pacific is of very ancient origin, and that it has existed for an extraordinarily long time.

With the other oceans it is different. When one nears the Indian Ocean, horizontally disposed marine beds are met with, not folded strata as in the Pacific area. These, however, do not begin with the Trias, but in east Africa as in western Australia start with the Middle Jura, and in Madagascar with the Middle Lias. Similarly, on the shores of the Atlantic Ocean horizontal non-folded strata are found, and these, in west Africa as in North America and Brazil, begin with the Middle and Upper Cretaceous. From this we conclude that the Pacific Ocean is older, the Indian Ocean younger, and the Atlantic Ocean essentially still younger.

I have mentioned yet another ocean, Tethys, which in Mesozoic times lay across present Asia, and whose remnants constitute our Mediterranean. The entire area of Tethys is laid in folds, and from the Pacific Ocean to the Caucasus throughout these folds are also moving southward; their margins in the south are overthrusted; the entire province of the sea is crushed from the north, and even remnants of the old southern foreland—the Gondwana Land, or the

Indian peninsula—are included within this folding. You have heard that Kinchinjinga and its neighbors, the highest peaks of the earth, though within the folds of the Himalayas, still have, so far as known from their foothills, the stratigraphic sequence of Gondwana Land.

We will now take a glance at the distribution of the lines of folding on the earth's surface. In the region of Lake Baikal lies an extensive, somewhat crescentically arranged mass of very ancient Archean rocks. It is folded, with a nearly northeast strike in the east and a northwest strike in the west, and the folds are of pre-Cambrian age. This old strike locus or vertex embraces Sabaikalia, northern Mongolia, and the East Sajan. Farther northwest there is developed another, younger vertex, or a second center of folding—the Altai. From this second younger locus proceeds an extraordinarily great system of bow-shaped folds, which, in an almost incomprehensible manner, embraces the entire Northern Hemisphere. The Altai encircle the old vertex, and its bows repeat themselves in the east from Japan and Kamschatka to the Bonin Islands. Toward the west they form the broad ranges of the Tian-shan and Bei-shan. Their southeastern branches appear in the bows of Burmah. In front of them to the south lie the marginal bows of the Himalayas— the Iranic; and farther along, the Tauric-Dinaric bows. They press over the Caucasus to Europe, and form here the two previously mentioned chains of folds.

These two chains of folds are themselves preserved in different ways. The one, older, embracing the Varischian and Armoricanian folds, is first discernible in Mähren. It reaches the Atlantic Ocean in southwestern Ireland and Brittany, and disappears as a Rias coast. Years ago, however, Marcel Bertrand called attention to the fact that such a broad and mighty mountain system—on the Atlantic coast it is as broad as the bows of the Himalayas—could not possibly suddenly end here, but that in all probability it is continued to the other side of the ocean in the Rias coast of Newfoundland. As you have heard, Marcel Bertrand accordingly continued the Armoricanian primary lines directly across the ocean to the Appalachians.

Of the Appalachians, however, it has been learned in recent years that they are far longer than was formerly believed. They form a bow which is not, as in the Asiatic and European chains, folded toward the convex side, but toward the concave side, first westerly, then northerly, and continues west of the Mississippi into the Washita Mountains.

The second or younger type, the Altai, strikes with decided flexing, narrowed through older horsts, from the Balkans to the Carpathians and the Alps, and at Gibraltar the latter join those bows of the western Mediterranean that are completely reversed.

Let us return once more to North America. As we have heard, the American term as Laurentia the wide Archean area which embraces the region of the Hudson Bay, middle Canada, and the central part of the United States. The Appalachians to the east and south of this mass, as we have seen, have a concave strike, are folded toward Laurentia, and vanish in the Washita hills. West of Laurentia, also, it is similar. It could have been shown that the Cordillera, whose connection with northern Asia has of course not yet been established, is, on its eastern side, in Canada, also folded toward Laurentia. It, too, bends toward the south with a more and more concave strike; continuing through Mexico, it is folded to the northeast, and then part of its folds finally turn toward Cuba and in the direction of the Antilles.

Thus on both sides is North America encircled by concave-striking chains of folds. It is as if the folds extended away from Asia and toward Laurentia. This entire grand phenomenon may be illustrated by a comparison. By the eruption of Krakatoa the oceans were moved; long waves proceeded from the place of eruption, traveled around the entire earth, and met themselves on the other side of the sphere. This is merely a comparison, not an explanation.

In the Southern Hemisphere the state of things is wholly different. For some time it has been known that in East India and South Africa, during Permian and Trias time, there flourished identical land floras—the Gondwana floras. Accordingly, it is concluded that these two continents were once united, and the area was named Gondwana Land. Later such floras were also found in Australia; then in the Argentine Republic. Thus it spread around the south. But the conclusion drawn from this as to the continuity of so great a continent was shattered by the circumstance that not only the characterizing plants of Lower Gondwana, but, in addition, the South African occurrences of associated animals, were also found in the Permian deposits of Perm in north Russia.

What then results is an exceedingly similar distribution of land plants and land animals of that time, and a great continent in the south; yet immediate proof of its continuity is lacking.

In fact, only on the Pacific margins of this supposed or actually united continent is it found that folding has taken place, and, indeed, in the east of Australia and the west of South America; while the intermediate Atlantic and Indian coasts are without younger folds. It is true that more recently, folding of pre-Carboniferous time has been described in South Africa, but in general the entire area between the western South American Cordilleras and the eastern Australian Cordilleras appears dead and unmovable. This is in contradistinction to the great diversity in movements of the Northern Hemisphere.

In general, these are the chains which we have sought to follow in detail in the course of these two semesters. The attempt toward a geometric arrangement of the mountain chains, which recently has been undertaken by distinguished specialists, finds, I fear, but little confirmation in actual occurrences. The tectonic lines that are met in nature tend generally at most to follow straight lines only in fissures or faults. The foldings, however, maintain themselves more like long waves, and they give way to the older horsts. This is seen more clearly in the youngest Alps, or that branch of the Altai trending toward Europe; the bows of the Banda Islands are similar.

I should now like to say a little about the conditions of life upon the earth. We have already spoken of the wide distribution of the land faunas and land floras of Lower Gondwana. Earlier types of Carboniferous land floras had spread themselves from the Arctic region to South Africa. The Culm flora is known in Europe, Mongolia, and Australia. Still more noteworthy is the fact that in the basalt streams of western Greenland there are interbedded plant layers of Lower and Middle Cretaceous, as well as of Tertiary times, and that during all this period there lived in this Arctic region first ferns and then leaf-bearing trees. In a word, in west Greenland are seen occurrences of different times which throughout cannot be brought into harmony with the climatic conditions of the glacial period nor with those of the present; thus this entire younger epoch appears as an exception. One gets the impression that not at all times did there exist the present diversity of climate, and also that the diversity of life was not at all times a varied one. The great Indian land fauna of today, with its tigers and elephants, can be considered as an independent unity, but here and there it is accompanied by older Malayan remnants which increase the diversity.

Gentlemen, as you see by this attempted survey, I can point out only some of the various directions in which our studies may be continued, and there exist so many hundreds and hundreds of questions that all, even the keenest ambition, will find the-portals open and may hope for satisfaction. New discoveries are in prospect for all conscientious inquirers.

[*Editor's Note:* Suess then gives tribute to several of his students in particular and to all of them in general.]

Editor's Comments
on Paper 2

2   SCHUCHERT
     The Making of Paleogeographic Maps

Charles Schuchert was one of the masters of North American paleogeography, and Paper 2 serves as an introduction to the breadth, scope, and the interwoven fabric of physical paleogeography and paleobiogeography. The first two sections carefully, but broadly, outline the history of geological and paleontological thought and their influence in paleogeography. The ideas behind this paper were a long time in being assembled, formulated, and set down in print, and it may be considered the culmination of nearly thirty years of active concentration by Schuchert on the subject.

The third section of his article deals with difficulties in analyzing paleogeography and discusses a number of problems that are still largely unresolved half a century later. Among his references are several volumes that are still classics in the field.

The methods of paleogeography occupy the remainder of the article. The first four of these Schuchert considered to be virtually absolute natural laws: uniformitarianism, permanency of continental masses and ocean basins, cyclic nature of geologic phenomena (i.e., similar evolution of features in geosynclines), and an increase in surface water with time. Of these four, the second and third became increasingly disputed during the 1950s. Even the first and fourth are no longer taken as axiomatic. Schuchert develops his ideas of the need for "borderlands" in trying to explain the sedimentation and diastrophism of the continental margins. The "borderland" concept, which he firmly held to in his own thinking and in his textbook, *Historical Geology*, throughout all its many editions, became firmly ingrained in the minds of at least three generations of North American geologists. So ingrained was this concept that authors of competing historical geology textbooks, who might have been doubtful of the existence of "Appalachia," "Llanoria," or "Cascadia," commonly were careful to obscure these supposed areas on their maps in decorative clouds or conveniently placed legends!

Areal stratigraphy and faunal zonation are only briefly treated; however, inasmuch as these were two topics on which Schuchert personally relied heavily in his own reconstructions, we will examine his technique later. Similarly, superposition and sedimentary petrology are only briefly mentioned. Schuchert treated paleontology as a unifying element in palaeogeography, and this philosophy sets the theme for many of the subsequent articles.

2

Reprinted from *Leopoldina (Leipzig),* **4** (Amerikaband),
116–125 (1929)

The Making of Paleogeographic Maps.

C. Schuchert, M. A. N.

Professor Emeritus of Paleontology and Historical Geology,
Yale University, New Haven, Conn.

The writer has often been asked what the principles are that guide him in the making of paleogeographic maps, and he has frequently answered this question in lectures to his graduate students and others. Teachers of Historical Geology, however, have expressed a desire to have such information placed before them in printed form, and this paper has been written to meet their wishes.

Geography and paleogeography. — The geography of to-day can only be understood in the light of the geologic past, and on the other hand, paleogeography can only be deciphered when guided by a knowledge of the causation of the changes in the geography of the present. Geography has been defined as the science which describes the earth's surface, including the distribution of the living things upon it at the present time, and more especially of man and his cultural works. Paleogeography, on the contrary, treats of the succession of geographies of the past, i. e., the prehistoric geographies as known to Paleontology and Geology. It seeks not only to describe and picture the shapes and interrelations of the various marine areas and the lands, but as well to discern, on the latter, the leading physiographic forms; the volcanic, desert, and glacial areas; the major drainage lines; and, throughout time, the paleoclimatology. For the seas and oceans, the connections are worked out on the basis of the fossils entombed in the sediments (paleobiogeography), which likewise indicate something about the temperature, depth, and salinity of the water. In other words, paleogeography is the synthesis of the several earth sciences based on the record of the rocks of the earth's surface.

The geographer's methods of observation are direct, for he measures and describes what he sees, but the geographic and organic relations of any given geologic time are discerned by the paleogeographer mainly by indirect methods; he sees no actual seas and oceans, no hills, valleys, mountains, or rivers, no actual living things of any past time, yet the paleogeographic record of all these is hidden in the rocks and fossils, and can be read by him as easily as the antiquarian or archeologist deciphers, in his unearthed relics, forgotten civilizations and languages, or wholly unknown human cultures.

Since it has taken the geographers more than two thousand years to bring their science to its present high stand as depicted on modern maps and described in modern books, it follows naturally that the present knowledge of paleogeographers must be sketchy rather than detailed, and in a way comparable to the geography of the ancients. This conclusion becomes all the more apparent when we realize that the sciences of Geology and Paleontology began as such but a little more than a century ago, and that large parts of some of the continents still remain almost unknown geologically. The paleogeographers therefore picture to us only what geologists and paleontologists know. As yet the geographies of the past are all much generalized and must long remain more or less inaccurate; nevertheless we are learning rapidly how to picture on maps the paleogeographic values ascertained by the geologists and paleontologists.

It was natural that the term "paleogeography" should be readily suggested by such older usages as "ancient geography" and "geologic geography", and so we find T. Sterry Hunt coining the word for the first time in 1872 in his article entitled "The Paleogeography of the North American Continent" (Jour. Amer. Geog., New York, IV, p. 416). He says:

"The student of organic fossils constructs from their history the sciences of *paleophytology* and *paleozoology;* and we may also, from the records of the attendant physical changes, construct what may be appropriately named *paleogeography,* or, the geographical history of these ancient geological periods."

The term, however, did not come into wide usage until subsequent to 1896.

History. — Now let us look a little into the history of paleogeography. It appears that the French geologists were the first to take up this study, and De Lapparent the first to disseminate it widely. Élie de Beaumont, in his geologic lectures at the College of France, used to outline to his students, as early as 1829, the relation of the lands and seas as shown by certain of the better known formations in the center of Europe. His first published map appeared in 1833, but it was mainly his students who spread the knowledge in printed maps. In America, the first to show paleogeographic maps was the Swiss geologist, Arnold Guyot, in his Lowell lectures of 1848, but apparently the first such maps to be published in this country were the three included in James Dwight Dana's epochal work of 1863, the "Manual of Geology". Of world paleogeography, Jules Marcou published the first map in France in 1866, but it appears to have attracted little attention, the most celebrated of the early attempts along world lines being the widely known map published by the Viennese paleontologist Neumayr in 1883, and based on ammonite distribution. In 1894 Karpinsky issued seventeen maps relating to the historical geology of western Russia, which proved, as had nothing before, the periodicity of continental flooding by the oceans. Even earlier, Jukes-Browne in his "Building of the British Isles", 1888, showed fifteen maps, and Canu in 1896 put out an atlas with fifty-seven maps relating to France and Belgium. Then in 1910 Schuchert in his "Paleogeography of North America" published fifty maps. The book that has had the greatest influence in this branch of Geology, however, is De Lapparent's text-book, "Traité de Géologie", the fourth edition of which (1900) had twenty-one maps of France, thirty of Europe, and twenty-two of the

30

world; the fifth edition, of 1906, had twenty-five of France, thirty-four of Europe, and twenty-three of the world, besides ten from other authors.

Since 1833 there have appeared something like 1000 different paleogeographic maps, and of this number about 450 relate to North America. Nevertheless, paleogeography is still in its infancy; practically all of the maps embrace too much geologic time, the ancient shore lines are much generalized and are drawn in sweeping curves unlike modern strands with their islands, bays, and headlands, the ancient lands are usually drawn as almost featureless, and only rarely do the maps indicate the areas of probable mountains, the volcanoes, and the drainage. Moreover, no one has attempted to picture the geographies back of the Cambrian — a vast time almost devoid of fossils. This means that all of our attempts are limited to the shorter latter half of geologic time, which alone has an abundance of fossils in its formations.

Paleogeographic difficulties. — It was stated above that all paleogeographic maps embrace too much geologic time. Hence none can give an exact geographic picture of any one time in the sense that maps showing the present geography do, since the latter embrace the conditions of our time as seen during the life of an individual. To get at the significance of these statements, however, we must analyze them still further.

Geology used to say that the age of the earth is of the order of about one hundred million years, but ever since the rate of disintegration of radium-bearing minerals became known, it has been clear that geologic time is ten to fifteen times longer. As more than one half of this time lies back of the Cambrian, we have at least five hundred million years of geologic time since this period, with an abundance of fossils to help us to decipher the succession of paleogeographies. Accordingly, if we make one map for each million years, we should eventually have five hundred, and yet of no continent are there at hand as many as seventy-five successive maps, while of the world there are fewer than thirty-five for all the time since the beginning of the Cambrian. Even one million years is surely too long a time to give a true picture of the paleogeographic conditions of a given time, and it seems safe to predict that some day paleontologists will aim at making more than five hundred maps for all North America and two or three times as many more depicting the spread of faunules over limited areas.

All paleogeography is drawn over the geographic base of the present, a method of illustration started by the first paleogeographer, Élie de Beaumont, and followed by everyone since. At present it is the only correct method, because we can not understand the ancient, but tentative geographies except on the base of the present one; maps drawn by plotting the present position of the fossils and then taking away the present continental outlines, as the writer has done on more than one occasion, become meaningless. And yet, as everyone knows, the modern base of the continental areas is a greatly foreshortened one in all the places where the roots of the ancient mountains and the peaks of the modern ones occur, and consequently the geography of the present gives a more or less erroneous picture of the actual size and shape of the ancient seas in the areas older than the times of mountain making. These crushed areas are variably wide, in the Appalachians now around 150 miles, while the Cordilleras of western North America are usually not less than 500 and not more than 1000 miles across. Previous to the Permian,

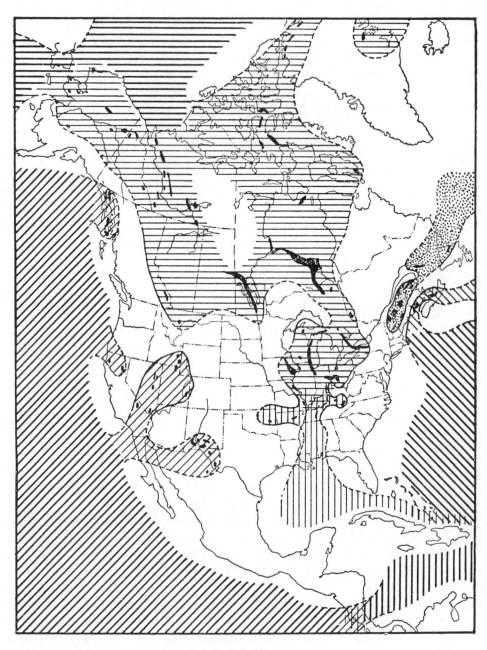

however, the Appalachian area must have been between 100 and 200 miles wider, while that of the Cordilleras must have shrunk something like 300 to 500 miles since the Devonian. It is these folded and overthrust areas that eventually will

32

have to be smoothed out in order to give true areal relationships to the paleogeographies older than the times of orogeny. This matter is discussed at greater
length by Dacqué (1915, pp. 370—375). For the present, however, this need not
be done excepting for special cases, since now it is more important that we first
agree upon what we are showing in the various paleogeographic maps; as yet our
maps are far from harmonious.

Another difficulty in drawing paleogeographic maps, and especially for world
conditions, is the type of projection. For the world, Mercator's projection is commonly used. Here, however, the enlargement toward the poles becomes ever
greater and accordingly Arctic seaways are unduly enlarged. Lapparent at first used
the Mercator projection, but with the fifth edition of his text-book he changed to
a globe or stereographic projection. Another good type is the four-pointed-star
projection radiating from the north pole, as drawn by Steinhäuser. There are still
other good types, but for small areas and even for single continents, almost any
projection has no distortions that are greatly misleading.

Books on paleogeography. — The best aids for paleogeography in the literature
are the following: The text-book of E. Dacqué, "Grundlagen und Methoden der
Paläogeographie", Gustav Fischer, 1915; the two volumes of Th. Arldt, "Handbuch
der Paläogeographie", Bornträger, 1919—1922, which are the most comprehensive
of all, especially from the zoogeographic side; for North America, Schuchert's
"Paleogeography of North America" (Bull. Geol. Soc. America, vol. 20, 1910,
pp. 427—606); for maps of France, Europe, and the world, A. de Lapparent's
"Traité de Géologie", 5th ed., 3 vols., Masson et Cie., 1906.

Methods of Paleogeography.

Turning now to the question of more direct methods, the paleogeographer is guided
by at least nine factors or principles, five of which are of more immediate application
in the making of his maps, and the others rather of a theoretic nature. We will first
take up the more theoretic principles, and then go on to those directly concerned
with the making of maps.

The paleogeographer, as has been said above, makes use of all the knowledge of
the earth's surficial formations and their internal structures, plus the knowledge
of the fossils and the climates indicated by these. The ancient geographies were automatically recorded in the rocks of the earth's surface by the geologic processes, and
this silent but significant information is obtained chiefly from the distribution
of the sedimentary formations and their fossils. It is therefore the ever changing
fossils, the variable rock nature of the formations, and the geologic structures of
the crust that furnish the direct evidence for the deciphering of the ancient geographies. In all of this the paleogeographer is steadied by four theories:

1. *Uniformitarianism.* — The laws of nature have always operated more or less as
we see them doing now, although not necessarily with their present intensity; hence
in our work of deciphering we are guided by the results we see to-day. In other
words, paleogeographers, like geologists, are conformists or uniformitarians, followers

of a doctrine first propounded by Hutton more than a century ago, and later put into working order by Lyell.

2. *Permanency of the Earth's Greater Facial Features.* — The first question that a paleogeographer asks himself is: Have the continental masses and the oceanic basins always been as we see them to-day, and have they always held their present relations to one another? This question is fundamental in paleogeography, for if there has been a repeated or irregular interchange of these areas, the discerning of the ancient geographies would be most difficult if not impossible. By "repeated interchange" is meant the changing of oceanic basins to dry land and the sinking of continents at different times to below two miles of depth and their rising again into dry land. If, on the other hand, there has been more or less permanency in the position of the continents and oceanic basins, then the work of the paleogeographer is much simplified.

Because most of the continents are deeply buried beneath strata laid down by marine waters, the older geologists naturally came to the conclusion that the deep oceans had often interchanged places with the continents. They did not then know that the fossils first, and then the sediments, would also indicate depth of water, and temperature as well. Accordingly, Sir Charles Lyell, whose text-books had wide influence, taught that "every part of the land had once been beneath the sea, and every part of the oceans had once been land". We agree with Lyell that, at one time or another, every part of the lands has been beneath *shallow* seas, but as for the rest of his dictum, all geologists now hold that there is not the slightest evidence present on the lands to show that the continents have ever been submerged beneath great depths of oceanic water. And that every part of the oceans once has been land, there is not the slightest direct evidence to prove. Geophysical and geological studies, on the contrary, all show that the oceans of to-day have been deep-water basins almost since their origin, or at least since before Cambrian time, that they have always remained in essentially their present places, and finally that they have apparently grown somewhat deeper, and somewhat larger as well, at the expense of the continents during the geologic ages. These conclusions are at once demonstrated by the fact that the crust beneath the oceans is of rocks about three per cent heavier than those of the continents, and therefore on the basis of the law of isostasy which has to do with the balance of masses, there can be no interchange between them as to their relief. On the other hand, the lands rarely have any abyssal or oceanic deposits, and even those known to be present near the margins of the continents do not cover more than one per cent of the land surfaces.

In regard to the possibility that the continents may have been displaced, or have slid about horizontally, Geology appears to be coming to hold that this may be true to a limited extent; and that something of a compensatory nature has gone on is demonstrated by the fold-mountains. On the other hand, the distribution of the best known faunas of the Paleozoic and Mesozoic shows clearly that the continents since at least the Cambrian must have been about the present distances apart, for if they had been much nearer together the number of species common to two continents would be far greater than it is known to be, i. e., rarely exceeding five per cent.

We now know that the North American continent has been more or less widely flooded by the oceans at least seventeen times, and that these shallow-water floods varied in extent between 154,000 and 4,000,000 square miles. At times there were vast inland seas, many of them far greater in areal extent than the present Hudson Bay. These facts show how constantly the geography of the geologic past has been changing. To-day we are living in the first phase of a new cycle and, a new era, and the oceans have already begun their spread over the continents, as seen in Hudson Bay, the North and Baltic seas, and the more or less wide extension of the shallow seas over the shelves that border all continents.

Here we may digress a little to say that by *seas* is meant those oceanic waters that spread with limited depths over the continents. These waters in paleogeography are called *epicontinental seas;* when such are restricted to the margins of the continents, they are called *shelf seas;* when they spill more or less widely over the interior basin areas, they are *epeiric seas;* while the long and narrow ones immediately on the inner side of the borderlands are the *geosynclinal seas.* All of these basins are thought to have had shallow waters, as a rule under 500 feet of depth. Therefore their deposits are *littoral* when near shore, and farther away are *neritic;* both series of deposits are, accordingly, within the zone of wave action and sunlight penetration, with greatest abundance of bottom life. It is the rarest exception that the epicontinental seas have deeper waters approaching those of the bathyal regions of the oceans.

Finally, it should be said that the epicontinental seas do not necessarily everywhere deposit sediments continuously, because in certain areas the currents and wave action are so strong as not to permit the silt and muds to accumulate and become cemented. Where tidal action is strong, as in the littoral regions and over the submerged "ridges", and in certain places in the oceanic depths, the bottoms are swept clean of all the finer sediments and these are piled up elsewhere, on the "banks" or in the deeper offshore waters where the currents and wave actions cease.

3. *Cyclic Nature of Geologic Phenomena.* — It is now a demonstrated fact that the oceans have repeatedly transgressed the continents and as often have left these areas. Usually, one such completed movement takes place in the duration of a single period, but some of the periods, as now delimited, have two and possibly three floodings. Again, minor mountain makings (= disturbances) often appear during the latter half or last third of the periods, with greater ones (revolutions) closing the eras. In all of this we see that crustal compensation takes place periodically in all continents, with the result that the geologic processes of the lands proceed from rejuvenated youth to maturity, from high lands to low lands, and in many continents from small to greater areas of dry lands, and from small to vast floodings.

4. *Increase in Volume of Water with Time.* — We now postulate that the volume of water on the surface of the earth has increased with geologic time. This modern view is opposed to that of the older geologists, who held that there was at first a universal ocean over the earth's surface, and that it originally had far more water than is now visible, since great quantities had soaked into the cooling earth subsequent to its astral condition. This last conclusion is only partially true, the amount

so absorbed being relatively small. It is known, through deep mining and oil well borings, that the stratified rocks tend to become drier with depth. Furthermore, we know that most volcanos, when in action, are delivering to the earth's surface new waters (juvenile waters), which have become freed by the rock differentiations and chemical changes deep within the crust. The volcanoes were most active in the earliest history of the earth, and therefore the oceanic waters increased in volume then far more quickly than subsequently. The writer believes that probably three fourths of the present oceanic waters were in existence at the close of the first half of the earth's history (Proterozoic), and that the remaining twenty-five per cent has been added since. However, this postulate of an increasing hydrosphere is not known to have markedly affected the ancient geographies.

5. *Diastrophism.* — Now we must consider another condition of the earth's surface that has an important bearing on the making of paleogeographic maps, namely, diastrophism. Under diastrophism are comprehended all the movements of the earth's surface, and this means that the oceanic bottoms, as well as the dry lands, have moved; in consequence, the oceanic level in relation to the lands must have been in repeated movement up and down. Geologists now clearly understand that the stratigraphic record is discontinuous, that it is much interrupted by "breaks", and that the local sequences are very variable. What any geologist sees is therefore the local records, and these fragments Historical Geology pieces together to make one more or less continuous story. Everywhere the geologic sequence is full of these "breaks" or "intervals" of time, when no local record was being made other than that of erosion, which of course means removal of record. These breaks or intervals are as a rule not general and the fossils show that the recording goes on now here and now elsewhere, and that it is continuous probably only in the ocean basins that are forever inaccessible to geologists.

The continents are spoken of as the positive or rising elements of the earth's surface, because the sum of their crustal movements is positive or upward. It is well known, however, that parts of them are more positive than others, i. e. certain areas have a local tendency to rise more or less frequently and to variable heights. The most striking of these positive regions are the marginal areas of the continents, called *borderlands*. These have for long times a decided, but pulsatory, tendency to rise and renew their highland form, and are therefore among the most striking of the geanticlines. On the other hand, the interiors of the continents have far less dominant, smaller, and lower subpositive areas, which reappear time and again as domes and depressed geanticlines. For these periodically rising areas the paleogeographer should ever be on the lookout, since they indicate to him the position of the dry lands, hence also the locations of shore lines and the sources of sediments.

It is, however, not an easy matter to locate with exactness the actual positions of the shore lines; in paleogeographic maps in which the scale is small, and especially those seen in octavo books, the solidly drawn lines, which indicate the best known shore lines, are probably nearly always in error anywhere from 5 to 25 miles, and where the lines are broken they may be 50 miles out of place or altogether wrong. The presence of sandstones passing into conglomerates and the thinning out of formations against the positive areas are among the surest indications of nearness to shore lines.

Geology now affirms that geosynclinal areas, once uplifted into fold-mountains, have a tendency to be re-elevated time and again, and therefore to remain as continuous lands. It is rare to find places where the oceans have spread completely across the site of a Paleozoic mountain range, or one of younger age, but it is not so rare to find that Paleozoic or younger seas have spread completely across mountains of Proterozoic or Archeozoic time. A most wonderful transgression of this nature is recorded in the Grand Canyon of the Colorado River, where Cambrian seas have spread horizontal strata across the upturned and worn-down roots of mountains that were uplifted at different times during the Archeozoic and Proterozoic.

In all the continents there are more or less large areas that have the most ancient rocks known to geologists, and upon whose flanks lie younger and younger strata as one proceeds away from them. These are known as the nuclei or shields of the continents, the parts that were elevated earliest into mountainous areas and welded into very resistant masses. Since their final folding and peneplanation, they have been rarely transgressed by the seas, and hence are the longest enduring lands, which guide the paleogeographer in his mapping of the seaways about them.

6. Areal Stratigraphy. — The units in paleogeography are the geologic formations and faunal zones, and as the first essential in mapping is to determine the places of the transgressing seas, it follows that this cartography is most concerned with the sediments of marine deposition. Accordingly, the "continental deposits" must be clearly distinguished from those of the seas, and this is done chiefly through the entombed fossils, though fresh-water deposits are often devoid of such life records and they are usually lacking in those made by desert conditions. The present geographic distribution of the formations is the first basis for deciphering the geographies of the past. These formations are masses of rock, either stratified or unstratified, of one kind or of various kinds, formed in the seas and oceans or on the lands, with or without fossils, and composed either of sediments or igneous materials. Just how many formations there are in North America since the beginning of the Cambrian is unknown, but the well known and more or less widely spread ones probably exceed several hundred. Some day there will be more than 5oo faunal zones, but so far the writer has been able to gather material for only about 125 maps covering the whole of North America.

One of the greatest drawbacks in making paleogeographic maps is the areas where the formations at the surface go beneath younger ones and do not reappear on the other side of the basin of deposition. Even more deterrent are the great lava fields, as for instance those of the Columbia and Snake rivers, which cover at least 15o,ooo square miles of older strata; or the regions of great bathyliths, such as the one of the Coast Range of British Columbia, where the older formations have been swallowed up by the igneous masses or have been eroded away since their rising. In all these areas one is dependent in his studies upon the distribution of the formations in adjacent regions with synchronous faunas.

7. Superposition. — The geologic age of a formation is determined first through superposition. It naturally follows, if in a given area there are several superposed formations which have not been disturbed through crustal deformation, that the lowest one must be older than those above. Also that when they are cut by igneous rocks these crystalline masses must be younger than the formations through which they rose when molten.

8. Petrology. — The petrologic factor, or the nature of the stratified rocks, tells whether they were formed in water or by the winds, in seas and oceans or in rivers and lakes, or on dry lands under pluvial or arid climates. Also the probable depth of water in the seas, the amount of agitation, and whether the rocks were formed near to, or far from, the shores. The environment automatically impresses itself to a considerable extent upon the accumulating sediments, just as it brings about most of the evolution of all living things.

9. Paleontology. — Fossils are the basis of most geologic chronology, because all life is in the constant state of change comprehended under the term organic evolution. The individuals change into different species or even genera, and the forms of an assemblage die out or migrate elsewhere or are added to by the introduction of new ones of local origin or by immigrants. In other words, the organic assemblages constitute a series of vital records automatically made by nature, and the order of their appearance in time is checked by the law of stratigraphic superposition. Therefore any fauna or flora indicates more or less accurately the geologic time of its existence.

Fossils, furthermore, tell much about the environment in which they lived, whether in the seas or oceans, on lowlands or highlands, in swamps, rivers, or lakes, and whether the climates were cold, temperate, or tropical. As examples may be cited, for normally marine seas, an abundance of bryozoans, brachiopods, echinoids, crinoids, and cephalopods; warm seas have the greatest variety of invertebrates, along with reef-building corals and thick-shelled and cemented bivalves (Chamacea and especially Rudistacea), large foraminifers and an abundance of shelled cephalopods; near-shore seas have Ostracea and barnacles; cold marine waters lack reef corals, large foraminifers, cemented bivalves, oysters, and have small faunas though the individuals are very abundant. In all of this, however, the paleontologist is guided by the distribution of life as it is at present, and by the Hutton-Lyell principle that at all times the laws of nature have operated uniformly and about as we see them to-day.

38

Editor's Comments
on Paper 3

3   **KNOWLTON**
     Biologic Principles of Paleogeography

In North America, Frank Hall Knowlton is known as a geologist
for the United States Geological Survey, custodian of Mesozoic
plants at the United States National Museum, and pioneer in North
American paleofloral analysis, including paleofloral distributions
(paleophytogeography). In Paper 3 he states the principles of paleofloral analysis in clearly formulated concepts.

Knowlton has studied the fossil floras of Alaska, Yellowstone
National Park, John Day Basin in Oregon, Florissant lakebeds in
Colorado, the Laramie flora of the Denver Basin, the Green River
Formation in Utah and Wyoming, the Latah Formation in Washing-
ton and Idaho, and the Denver Formation in Colorado. Further-
more, he assembled a catalogue of the North American Mesozoic
and Cenozoic plants in the U.S. National Museum (1919) and wrote
Plants of the Past (1927), an introductory textbook.

The biologic principles of paleogeography was the theme of a
symposium held by the Paleontology Society at their annual meeting
in 1909 and many aspects of the subject were discussed and debated.
The contributed papers were then published in *Popular Science
Monthly* (volume 79, 1910). The collection of papers outlined the
state of the art of paleobiogeography at that time. Knowlton's contribution, along with Charles Schuchert's essay with the identical
title (reproduced as Paper 5), which was also part of this symposium,
are probably the clearest statements of their understanding of the
problems. Knowlton is particularly careful to emphasize the need
for reliable identification of species, for land connections between
identical floras, and for an understanding of the relation of plant
distributions, climates, and climatic changes.

REFERENCES

Knowlton, F. H. (1919). A catalogue of the Mesozoic and Cenozoic plants of North America. U.S. Geol. Surv. Bull. 696.

―――. (1927). Plants of the past. Princeton Univ. Press, Princeton, N.J., xix + 275 p.

3

BIOLOGIC PRINCIPLES OF PALEOGEOGRAPHY

By Dr. F. H. KNOWLTON

U. S. GEOLOGICAL SURVEY

CONSIDERING the breadth and intricacy of the subject assigned me, and the limited time that can be given to its consideration, it has seemed best to me to restrict my remarks to two or three of the obviously more important phases of the problem.

Aside from the study of the rock-masses themselves—which are often difficult of interpretation—reliance for an interpretation of paleogeography must be placed in the former life found entombed, and of the two biologic elements, plants undoubtedly hold a very high—probably the highest—place.

In making use of plants in the study of paleogeography we may first consider distribution. If we find two fossil floras identical or similar in all essential or important details, we feel justified in regarding them for all practical geologic purposes as contemporaneous. In order that we may be certain that the two floras are identical, they must be composed of types that are readily identifiable, that is, forms so well characterized that they may be easily and certainly recognized. As examples of such floral elements mention may be made of many ferns and fern allies, most cycads, conifers and peculiar, well-marked or characteristic dicotyledons. Having settled the contemporaneity of the floras, inquiry may next be made as to the probable manner in which the separated or isolated areas were reached by these floras. Here again we must carefully consider the character of the flora and the means for its natural dispersal. The living flora, and for that matter probably the floras from at least the beginning of the Tertiary progressively to the present time, has developed in many ways means for the comparatively rapid and wide-spread dissemination of their reproductive parts (seeds, etc.). For example, a large percentage of the members of the dominant living family of seed-plants—the Compositæ—have developed seeds with an attachment of soft, fluffy hairs which serve to float them in the air, often to great distances. In many other living groups there are similar, or at least as effective, devices for dissemination, but as we go back in time adaptations calculated to be of aid in distribution grow less and less, and soon even seeds of any kind are unknown, or known but imperfectly, and reproduction is normally by means of spores, that is, reproductive bodies in which there is no embryo already formed when they leave the parent plant. It is obvious that plants that are reproduced by seeds, in which there is both an embryo and a supply of food for use during germination, must possess a decided advantage over those reproduced by means of spores.

41

In the groups of spore-bearing plants ordinarily found fossil, the spores are not known to have developed any particular devices for their wide dissemination, such as flotation in air, attachment to animals, etc. They are produced in vast quantities, and depend upon a few reaching situations favorable for successful germination. Their vitality is also of apparently exceedingly limited duration, and it is doubtful if they could long survive immersion in salt water.

The bearing of the above digression is apparent. Given a fossil flora made up of ferns or fern allies, exclusive of what are known to belong to the cycadofilices, and when such flora is found in two or more separated areas, we are justified, in my opinion, in arguing a practically continuous land connection. They were incapable of crossing very wide reaches of open water, particularly salt water. Fresh-water streams have been to some extent avenues of distribution, but many fossil floras—and living floras as well—are too widely spread to be explained by this means. When, as is usually the case, identical floras occupying different areas are mixed floras, the bearing on the means of reaching the various areas is more complicated. An example may better serve to bring this out. Thus, the Jurassic flora is practically world-wide in its distribution, ranging from Franz Josef Land, 82° N., to Louis Philippe Land, 63° S. It is composed of ferns, fern-allies, cycads and conifers, a large percentage being true ferns. The probability of a close land connection argued on the basis of the true ferns, has already been alluded to. The cycads—the Jurassic is called the age of cycads —were abundant in individuals and numerous in forms. On the basis of our knowledge of living types, it may be stated that cycad seeds germinate immediately on falling from the cone without any necessary resting period. They are not known to retain their vitality for a longer period than three years, and usually but two years. They sink promptly in fresh water and as the stony coat is easily penetrated by water, they either germinate or rot at once. In salt water they will probably sink and decay even more quickly. Therefore, the probability of their being transported for any great distance over open water is reduced to a minimum. The conifers of the Jurassic were reproduced by seeds. They belong to types not known to enjoy any special means for transportation, nor is it probable they could better withstand fresh- or salt-water immersion than the cycads. All classes of vegetation present in the Jurassic, therefore, argue for a practically continuous land connection.

In considering the bearing of any flora on the paleogeographic problem the process is similar to that outlined above. That is, an analysis of the composition of the flora, a study of the means of natural dissemination which includes duration of vitality, and finally a judgment as to its probable means or avenues of transportation, involving a land connection or otherwise.

A word may be said as to the presence of land plants in marine deposits. That the trunks of trees may float for a considerable time and to great distances is undeniably possible, but unfortunately the study of fossil wood has not yet reached that degree of refinement in most cases that will permit of its general use, and reliance in identification must be placed largely in foliar and reproductive organs. The delicate fronds of ferns, leaf-clad branchlets of conifers and the leaves of seed-bearing plants are incapable of long withstanding the immersion and wave action of salt waters. In my judgment, therefore, the presence of fronds, leaves and similar organs in marine deposits argues very near-by land.

The only other point I shall consider is the bearing of plants on the interpretation of *climate*. Since it is generally acknowledged that plants furnish the most reliable data for this phase of the subject, an inquiry as to the kinds of plants that have been found most valuable in this connection may be of interest. Obviously our interpretation of the probable conditions under which the plants of past geological ages grew, must be on a basis of a knowledge of present conditions found to obtain for similar or closely related groups. That we may occasionally err in this is possible, especially if reliance is based on too few forms, but when all the various elements of a flora are considered, the results are thought to be within a close approximation of the truth. Thus, since *Artocarpus*—the bread-fruit tree—only grows at the present day within 20° of the equator, it follows that when *Artocarpus* is found fossil in Greenland, 72° N., the conditions at the time it flourished there must have been tropical or subtropical, and this conclusion is confirmed by the tree ferns and cycads associated with it. Palms can not flourish with a temperature below 40°: a fossil flora, rich in palms of well-defined types, could hardly have grown under very much cooler conditions. Tree-ferns are practically confined to within 30° of the equator and a temperature of approximately 60°. A fossil flora, such, for example, as the Triassic of Virginia, that contains numbers of tree-ferns, must have grown under tropical or subtropical conditions. A fossil flora rich in types, the living representatives of which can withstand a temperature of — 40° to — 60°, or even lower, must have been at least cool-temperate. Cycads are now found only within 30° of the tropics: a rich cycad flora argues then for a tropical or at least a subtropical climate.

Examples of this kind could be multiplied almost indefinitely. In interpreting geological climate selection is made so far as possible of the plants or groups of plants, that are confined at the present day within relatively narrow limits of temperature, be this high, medium or low.

Editor's Comments
on Paper 4

4 JUST
Geology and Plant Distribution

Theodore Just's paper went a long way toward broadening the perspectives of phytoecologists, taxonomists, and paleobotanists in North America. His message clearly stated that the distribution patterns of plants are a result of a series of evolutionary, climatic, and geologic events and that one aspect is not complete without the others. In many regards, Just's message is as timely for the present-day expanded interest in paleobiogeography and the theory of plate tectonics as it was in 1947, and, of course, the basis for analysis is similar for fossil and living groups as well as terrestrial plants.

In this Paper 4 Just examines with typical European thoroughness the hotly debated topics of paleobiogeography of the late 1940s and achieves a coherence and perspective that welds the distribution of plants into a historical framework. Just first examines the questions of land bridges, the permanence of oceans and continents, and continental drift, and then reviews geologic rhythms and processes as related to evolution, concepts of origin, and ecology of areas of origin; finally, he investigates centers of origin and dispersal, all in a concise seven pages. In later sections of this volume, several of these concepts will be considered in more detail; Paper 4 is an introduction to these concepts.

4

Reprinted from *Ecol. Monogr.*, **17**(2), 129–137 (1947), with permission of
the publisher, Duke University Press, Durham, N.C.

GEOLOGY AND PLANT DISTRIBUTION

Theodor Just

INTRODUCTION

It has long been recognized that geological factors are much more important causes of plant distribution than are others. Thus no additional proof of this conclusion reached by De Candolle is needed here. However, the tremendous increase in our geological and botanical knowledge since the days of the early plant geographers has widened our horizon and enlarged our retrospect. With it came also many new problems awaiting their solution often before the existing ones could be solved. Thus the pertinent literature is extensive, covering as it does many borderline fields, controversial and frequently contradictory. This situation is not the outgrowth of lack of mutual understanding on the part of geologists, paleobotanists, taxonomists and others but largely the result of incomplete records and of the difficulties encountered in collecting, identifying and interpreting the necessary data. Although the whole story has not yet been worked out, this goal appears to be closer at hand than it was only some time ago.

PALEOGEOGRAPHY AND PALEOCLIMATOLOGY

If the distribution and extent of land masses and oceans throughout geological history were known in greater detail, problems of plant distribution would offer fewer difficulties. Actually this is still one of the crucial problems confronting geology, paleontology and biogeography. Various theories have been proposed and discussed with considerable vigor but none has so far resolved all difficulties. Disregarding certain variants, the main geological theories bearing on problems of plant distribution can be grouped in the following order.

LAND BRIDGES

Former land connections of varying extent and duration between continents have been postulated. These or some former continental masses have subsided leaving the present distribution of continents and oceans. Examples are the Austro-melanesian continent of F. Sarasin, Ihering's Archhelenis, Atlantis, etc. (Högbom 1941, Joleaud 1939, Schröter 1932, Schuchert 1932).

Although the number and extent of land bridges proposed to date can not be recounted here, it is generally conceded that former land connections did exist both in the northern and southern hemispheres (Bucher 1933, Copeland 1940, Dacqué 1932, Florin 1940, Lindsey 1940, Netolitzky 1933, Reinig 1937, Setchell 1935, Seward 1933, Skottsberg 1940, du Toit 1940). The needs of animal and plant geographers however vary sufficiently with their respective interests and so do the land bridges assumed by them. Land as such may not always be the deciding factor

in the migration of floras, for Holttum (1940) has demonstrated the effectiveness of a rather uniform climate as a barrier.

PERMANENCE OF CONTINENTS AND OCEANS

This theory is widely accepted among geologists. Its most notable biological adherents are Darwin and Wallace. The fundamental assumption of it is that the continental land masses and oceanic basins have remained in their present position throughout geological history, though obviously with different outlines and physiographic structure. (Bucher 1933, Holmes 1944).

CONTINENTAL DRIFT

The most disputed theories of the origin of continents and oceans are now collectively referred to as continental drift, as their common feature is the assumption of a horizontal drift of entire continental masses, regardless of the particular configuration attributed to each continent at any particular geological period (Daly 1942, Gutenberg 1939, Haddock 1936, Holland 1937, Holmes 1944, Joly 1930, Lee 1939, Longwell 1944a & b, Rastall 1929 & 1946, Schiller 1942, Tiercy 1945, du Toit 1937, Waterschoot van der Gracht 1928). According to Wulff (1943) the "permanence of the relative area of land and sea taken as a whole" is a fundamental feature of the well known theory of Wegener and may be carried over to other forms given it by various authors. For some 30 years the discussion has gone on concerning the possibility of such large scale phenomena and seems to have centered largely around the forces necessary to explain such vast changes in the earth's crust. In the opinion of Zeuner (1946) the idea of continental drift is "increasingly acceptable to geologists" who are beginning to recognize the need for horizontal movements of some sort though their intensity and rate are still subject to real scrutiny, as they would have to be about 10 to 100 times larger than the vertical ones so far readily accepted in geology.

Rastall (1946) makes the following comment: "This present writer believes that the rejection of continental drift in some form involves difficulties quite as great as its acceptance. The usual argument against it, of course, is that adequate forces are unknown. But everybody believes in the folding of mountain chains: the Alps show that it did happen, and that quite recently. But as yet we have no clear conception of what the forces were that did it. The cases surely are comparable, and it is illogical to swallow one whole and to make such a fuss about the other."

The theory of continental drift has been regarded as a fairy tale (Willis 1944), or vigorously supported with data recruited from all fields related to this complex problem. It is clear by now that the whole question is exceedingly involved and that it can not be resolved in relatively simple and appealing terms. Thus in the opinion of some geologists (Daly 1942, Staub 1944) no major orogenies can take place without continental drift. Or, in Daly's words, "it appears that the mountain chains of the lands are byproducts of the horizontal displacement of whole continents over the earth's body, and that this horizontal motion of the crust is possible only because the subcrustal layer is nearly or quite as weak as water. Furthermore, . . . , the weak substratum is worldcircling, beneath ocean as well as continent. Perhaps, too, this conclusion may yet give a basis for solving a supreme puzzle, namely, the concentration of the sial, the earth's lighter rock, in the continental sectors, leaving the oceanic sectors without any continuous cover of sialic rock." Similarly Staub (1944) rejects Wegener's particular version of continental drift but believes that the idea of moving continents is correct. According to this author, displacement of solid continental blocks is the fundamental geological phenomenon as it involves movements of the entire lithosphere. Orogenies are only partial results of this whole process though they may extend through entire geological periods after passing through prolonged preliminary phases, viz., the Alps and other mountain ranges. This conclusion is essentially the same as that reached by Longwell (1944) who says: "in the genetic study of major earth-features I can not believe we have arrived at a stage that permits discarding the method of multiple hypothesis."

Staub's summary (1944) provides a good example of such an hypothesis. It reads as follows: "Neither contraction alone, as classical geology assumed, nor equatorial drift or a primarily western drift, as Wegener thought, nor magmatic movements alone, as suggested by Otto Ampferer 40 years ago in opposition to the theory of contraction, evoke the complicated mechanism of crustal movements and of tectogenesis on earth. Rather all of these factors are integrated, complement each other, follow or release each other throughout geological history. As a result, contraction of the earth is still going on. Equatorial drift away from the poles directs the migrating masses as on other celestial bodies; magmatic movements contribute to these movements tending toward isostasy; the decomposition of radioactive substances, in addition to isostatic processes and further contraction of the planet, continually revives the otherwise slackening movements. Therefore we know now:

"Neither the theory of contraction as such nor continental movements caused solely by equatorial drift nor magmatic movements can independently account for crustal movements; rather all of these factors are integrated into a grand interplay of forces producing the tectonic phenomena of the earth. . . .'"

SHIFTS OF CLIMATIC BELTS

Intimately associated with problems of paleogeography are those relating to the climates of the past (Brockmann-Jerosch 1914, Brooks 1926, Dacqué 1932, Eckardt 1925, Kerner 1930, Kubart 1929, Seward'

1933). Fossils have been found far out of range of occurrence of their nearest living relatives (Berry 1920, 1930, 1945, Gothan 1924, et al.). There is also other evidence which speaks for different climates in various geological periods and their different distribution on the globe. Some periods, particularly the Jurassic, seem to be characterized by greater uniformity of climate (Gothan 1924), whereas the paleobotanical evidence indicates a certain zonation for others. The first sign of such is found in the Permo-Carboniferous floras of Asia studied by Halle (1937).

A shifting of the climatic belts is commonly assumed, particularly in conjunction with continental drift and its accompanying movements of the poles and the earth's axis. For example, Simpson et al. (1930) concluded that the "most cogent argument that has been offered in favor of continental drift" is the fact that "widespread ice caps reached to low latitudes in the southern continents." Halle (1937) on the other hand has difficulty in the interpretation of the eastern Asiatic Permo-Carboniferous floras in the light of continental drift. Similarly, other paleobotanists encounter difficulties with continental drift as a basis for the explanation of the distribution of fossil plants (Høeg 1937, Gothan, et al.). It should be pointed out however that many paleobotanists are favorable to continental drift (Hirmer 1938, 1942, Mägdefrau 1942, Sahni 1936, Zimmermann 1930, et al.) or at least expectant of future verification by geophysical research (Seward 1933, 1939).

Umbgrove (1946) has recently reviewed various theories on polar displacement and shown that calculations by Milankovitch used in support of Köppen-Wegener's findings are in error. Other interpretations and details given by him are inconclusive and of no immediate bearing in this connection. Berry (1945) adds a note of caution and even challenges meteorologists "gifted with sufficient knowledge" to discuss the effects in the late Paleozoic of a wide Mediterranean Sea (Tethys) and an enormous southern continent. Finally, Zeuner (1946) made some rough estimates of the supposed movement of the equator relative to shifting climatic zones. He found that since the Tertiary the equator would have had to move at a rate of 4500 km. in 50 million years, or 9 cm. per year, from Spain southward in order to reach its present position. Seward (1933) has raised the question of the reliability of plants as "thermometers of the ages." The great care needed in the study of fossil plants has been repeatedly demonstrated and examples are given elsewhere in this paper. In short, our knowledge of the climates of the past is far from being complete and greatly in need of additional records and interpretation.

RHYTHMS OF GEOLOGICAL PHENOMENA

Although the forces responsible for many geological phenomena are still unknown or poorly understood, there is considerable evidence of their periodic operation and rhythmic occurrence (Grabau 1940, Seward 1933, Umbgrove 1942). This is at least one of the fundamental assumptions of Haarmann's oscillation theory of diastrophism (see Longwell 1930). Umbgrove (1942) gives a more detailed account of the many ways in which this periodicity in earth history is demonstrated. He distinguishes four major periods, each extending about 250 million years, and some 25 minor periods of diastrophism and concludes that "diastrophism is going on all the time somewhere in the world, which seems reasonable if we consider that the ultimate causes, whatever they may be, are probably always in operation." The possibility of continental drift as seen by Wegener and du Toit is rejected by Umgrove in view of the thin sial layer found on the floor of the Atlantic and Indian oceans which would prevent sialic continents from moving through it. But Rastall promptly points out that "icebergs can plough through pack-ice." Ice ages also seem to follow this major periodicity of 250 million years, the period between the Karroo Ice Age of the late Carboniferous and Pleistocene, and are roughly contemporaneous with major orogenies. Umbgrove claims that these cycles are linked with cosmic causes, including the rotation of the galaxy. While these explanations may well be far removed from the immediate considerations of the plant geographer, it is equally true that repeatedly recourse has been made to such causes as possible evolutionary factors as will be shown below.

Thus a certain periodicity of geological phenomena seems well established. But it is not at all clear what bearing it had on the evolution of organic life beyond the repeated speculation of the effect of the main orogenies on accelerating the rate of evolutionary processes and the slowing down of the latter during more quiet geological periods (Rubtzov 1945).

GEOLOGICAL PROCESSES AS EVOLUTIONARY FACTORS

The notion that geological processes are effective evolutionary factors has appeared often, ranging from the outmoded cataclysmic theory of Cuvier to various recent versions. Schindewolf (1937) in his review of the more important contributions to this problem concludes that neither macro- nor mega-evolution could be attributed directly to the action of geological phenomena though local physiographic changes may readily be involved in micro-evolution or speciation. Thus new habitats are frequently provided by tectonic phenomena, though they can hardly be said to induce evolution on a major scale. While some indirect effects of tectonic phenomena like changes in climatic zonation, opening and closing of migration routes, etc., may be far-reaching, they are not the real causes of the disappearance of major groups of wide distribution. It seems that other, possibly internal, causes are here involved.

Similarly, volcanic phenomena are limited in occurrence and often even more local in character. The classic case of the eruption of the Krakatau has shown this clearly, for the island was invaded by the old flora and fauna and not by new types. Thus the old Buffonian concept postulating direct effects of geological processes on the evolution of plants and animals is untenable though the remote possibility of local micro-evolutionary effects remains.

In Schindewolf's opinion, cosmic causes with world-wide effects on every ecological niche would be required to induce evolution in many groups at the same time and without their pursuing any special evolutionary course. Changes in the intensity and nature of light and heat rays have been suggested in this connection, for they are believed to bring about periods of increased mutability and effect profound changes of germ plasm. Events of this sort seem to have taken place at the beginning of the Permian and the close of the Cretaceous.

Rubtzov (1945) on the other hand attributes acceleration of evolutionary processes as indicated at the boundaries of various geological periods to the profound effects of orogenetic processes, whereas quiet periods are generally marked by slow evolution. Similarly, areas subjected to great geographical changes are said to show the result of more rapid evolution, viz., both sides of the Atlantic are supposedly characterized by a greater abundance of phylogenetically higher groups while eastern Asia is now a refuge of many old and elsewhere extinct forms. By comparison, Berry (1945) questions the possibility of the sudden change in the fossil floras from Paleozoic to Mesozoic and offers as an alternative interpretation that this change was "a gradual transition foreshortened by our imperfect knowledge into seeming suddenness."

ORIGIN AND TYPES OF AREAS

Plant distribution is usually recorded by the various forms of areas occupied by plants such as relic, vicarious, discontinuous, progressive (expanding) and contracting (retrogressive) areas (see Wulff 1943, Cain 1944, Hayek 1926). In this context the relic and discontinuous areas are of greatest interest as their interpretation involves major geological events and theories as well as paleoclimatological considerations. Various botanists (Campbell 1943, 1944, Hirmer 1938, 1942, Irmscher 1922, 1929, Koch 1924-1933, Studt 1926, Herzog 1926, Mägdefrau 1942 and others) have accumulated voluminous data from living and fossil plant distribution in support of continental drift. The last to join these ranks is Hutchinson (1946) who states his case clearly: "As a botanist who has studied the distribution of plants for many years, particularly African plants, I am a firm believer in Wegener's ideas." However, not all botanists are convinced of the validity of continental drift as the sole explanation of discontinuous areas and have voiced strong objections (Diels 1928, 1934, Fernald 1944, Merrill 1943, Schuster 1931, Suessenguth 1938, Berry and others). As is to be expected, often the same group of plants is used to illustrate both sides of the controversy, viz., the Cycadaceae (Koch 1925 vs. Schuster 1931).

The greatest difficulty in interpretation is offered by the largest discontinuous areas known, e.g., the so-called austral or subantarctic disjuncts. Schröter

(1932) gives five possible explanations proposed at different times and from different points of view. The whole question is linked, if not identical, with the problem of the bipolar distribution and origin of modern floras (Campbell 1943, 1944, Du Rietz 1940, Florin 1940). Hutchinson (1946) for instance regards the families of flowering plants and smaller groups peculiar to South Africa and some to Australia as well as *"austral types evolved independently in the Southern Hemisphere."* Similarly, Florin (1940) favors "from the Permian onwards" two major centers of development of modern conifers, one of which is definitely southern and has left no trace of representatives in the northern hemisphere.

The critical point in most of these discussions is the possibility and probability of the following four processes, monotopy, monophylesis, polytopy and polyphylesis. As it is impossible to review here the entire controversy (see also Cain 1944 and Schröter 1932), the following remarks must be limited to the strong suggestion made by Suessenguth (1938) regarding the origin of the large areas occupied by entire plant families. This author is of the opinion that certain areas can be explained best by assuming the possibility of a polytopic as well as polyphyletic origin, provided the original stock was sufficiently widely distributed. According to this view, several or even many parallel lines are evolving more or less simultaneously in several species or genera. This mode of origin would do away with the problem of large scale migrations and the invariably long spans of time required for such. Unfortunately, not a single case is presented by Suessenguth to illustrate this view. Rather he offers immediately as a possible objection to this mode of origin the discontinuous areas occupied by such groups as the Cycadaceae, Gnetaceae, Balanophoraceae, etc. all of which are treated as contracting areas. Conversely, plant families characterized by expanding areas are said to have reached their maximum extent if the family is supposedly phylogenetically younger, viz., Gramineae, Cyperaceae, Umbelliferae, Labiatae, Compositae etc. Here, apparently, fossil records would aid greatly in determining the relative phylogenetic age of the major angiosperm groups, as it can not be stated categorically that the Compositae for instance may not be as old as many other old angiosperm stocks.

MODERN FLORAS AND THEIR ANALYSES

Detailed analyses of known fossil (Goodspeed *et al.* 1936, Gothan 1937, Hirmer 1935-1942, Seward 1933) and living floras have yielded striking, though not always conclusive results. Customarily floras are divided into elements, either geographical, genetical or otherwise, depending on the particular needs or reasons for study. Excellent examples are the careful studies of the British flora by Matthews (1946), the South African flora by Weimarck (1941), the flora of Asia Minor by Schwarz (1937), the flora of the Pacific islands and adjacent Asia by Andrews (1940), Hu (1940), Kanehira (1940), Lam (1938, 1940),

Merrill (1926), Ridley (1937), Setchell (1935), Skottsberg (1940) and Tardien-Blot (1940).

As it is impossible to treat all methods employed in the analysis of floras, only representative examples are given here.

STATISTICAL METHODS

In his detailed analysis of the Post-Pleistocene flora of the eastern Baltic area Kupffer (1930) used various statistical criteria. He characterized the flora of this area in terms of its floristic dissimilarity (ratio of species limited to this area *versus* all species occurring there), its similarity (number of species in common with one or more areas of comparison), and its floristic gradient (expressed in terms of distance within or outside the area studied). Other statistical methods are reviewed by Cain (1944).

Following his intensive studies of the distribution of the insect order Homoptera Metcalf (1946) found "a correlation between zoogeographic regions and taxonomic groups." Apparently no group of plants has ever been analyzed from this point of view.

The elaborate task of dividing the world into regions on the basis of the number of species of flowering plants represented in each was attempted by Wulff (1937). Fully cognizant of the multitude of possible errors in delimiting species etc., Wulff established five classes differing in the number of species represented in areas referred to these. They are: 1) none to 500 species, 2) 500 to 2,000 species, 3) 2,000 to 3,000 species, 4) 3,000 to 7,000 or 8,000 species, and 5) more than 8,000 species. From the map prepared on the basis of his analyses Wulff was able to draw some general conclusions. Generally the number of species increases from the polar regions toward the equator with the general increase in temperature in that direction, although subtropical steppen and desert areas have fairly low numbers of species in view of their great aridity. In Europe and North Africa the number of species decreases for the same reason from W to E, whereas in temperate Asia the reverse is true, namely, from E to W. Mountainous areas are generally characterized by larger numbers of species in view of the greater humidity, physiographic changes in relief and soil, limited competion and the presence of Tertiary relics. The greatest climatic changes since the Tertiary have taken place in the polar and temperate areas accompanied by a complete change of the plant world represented in the fossil record and found there now. By comparison, the tropics and subtropics retained almost unchanged ecological conditions, accompanied by a virtually uninterrupted evolution of their floras. This is especially true of southeastern Asia and South America whose floras may range from 20,000 to 45,000 species and are thus the richest known floras of the world. Tropical Africa apparently witnessed fewer changes and its flora is therefore poorer in species ranging from 10,000 to 15,000 species. The Mediterranean area, southeastern and southwestern North America were characterized by more favorable climatic conditions and therefore retained a large part of the Tertiary flora.

ECOLOGICAL METHODS

These have been widely employed by paleobotanists and others though inherent dangers of misinterpretation are always present (Berry 1945, Cain 1944, Chaney in Goodspeed 1936, Brockmann-Jerosch 1914, Gothan 1924, Seward 1933, Weiss 1925). In an effort to avoid erroneous interpretations made on the basis of limited comparisons, Bews (1927) analyzed all known fossil angiosperms ecologically and statistically. He compared entire groups of fossils with modern types of vegetation regarding leaf size, margin, texture, etc. Hansen (1930) has shown that the groups and genera of greatest fossil age are now represented only by phanerophytes, whereas genera of Miocene or Pleistocene age are today represented by an increasing percentage of chamaephytes, hemicryptophytes, and therophytes. Any disharmony between climatic periodicity and phasic development in plants in a given area is regarded by Scharfetter (1922) and Fritzsche (1936) as proof of the origin of these plants in other areas. Wulff (1943) is of the opinion that the evidence in this connection is inconclusive as it omits the possibility of change *in situ*.

HISTORICAL (GENETICAL) METHODS

Long ago Blytt (1882) stated that "present day vegetation reflects, as in a mirror, the geological history of a country, different groupings of species being the expression of different stages in that history." Since then the science of areography (Cain 1944) or phytochorology (Reinig 1937, 1938, Schwarz 1937) has been developed and its aims and methods variously defined. For example, Reinig (1939) recognizes two major principles of chorology, namely, constant dependence on habitat, and monotopous origin of systematic forms. Working on an entire flora, viz., that of Asia Minor, Schwarz (1937) used various new and old concepts and demonstrated their applicability as far as possible. These concepts are: primary, secondary, relic and migrant floras, relic areas (refuges), centers of origin, and two major phases of evolution, viz., progressive and retrograde. Relic floras are relatively young with reference to their ecological environment, though largely composed of fairly old stocks; the species may be of rather recent origin, since speciation goes on continuously; the areas however are old. Migrant floras on the other hand are the youngest floras and are composed of species of divergent relationships and distribution. Usually the appearance of a migrant flora disturbs the sociological equilibrium bringing about definite deterioration of ecological conditions. In this case migrant floras may be regarded as retrograde. Only after the sociological equilibrium has been re-established, can the migrant flora begin to function as a new (secondary) center of origin. Centers of origin are either primary or secondary. Primary centers are characterized by an abundance of related though well marked species and of forms intermediate between those isolated in the relic areas. Here the ecological conditions are relatively stable and speciation is thus solely the expression of the available genetic material and of continued mutations and recombination as well as dispersal. Primary floras are made up of very old stock though composed of relatively young species, the product of almost uninterrupted speciation. The areas in turn are very old. Gene filters appear where relic or primary floras come in contact. Here geographical clines appear, which are followed by the breakdown of the species and recombination of their entities, and often by new evolutionary outbursts. Gradually gene filters either absorb entirely the old floras or become wholly or partly isolated and thus begin to function as secondary centers of origin. These in turn are populated by secondary floras composed of species of heterogeneous origin and relationship. Migrant floras indicate that contact has been established between relic areas, poor in biotypes, and centers of origin which are comparatively rich in biotypes, and that migration has started in the direction of least resistance, e.g., toward relic areas. Thus migrant floras are the youngest floras, as indicated above. Each of these phases may be regarded as a stage in the development of a modern flora, very much like geological strata mark subsequent periods of earth history. Thus any area may eventually be characterized by its "chorogenetic spectrum" rather than as a mere geographic entity. At present no one flora represents all of these possible stages but is likely to be the result of several.

CENTERS OF ORIGIN AND DISPERSAL

As Cain (1944) has formulated the important criteria for indication of center of origin, this discussion can be restricted to the general problems connected with the origin of vascular floras. As far as our modern floras are concerned, two main theories are to be considered.

Generally Heer (1868) is given credit for proposing the theory of the polar origin of our modern floras, particularly its most widely accepted form, viz., that of the monopolar (boreal) origin of floras in the northern hemisphere. It is also known as the monoboreal relic hypothesis (of Thiselton-Dyer, Lydekker et al., see Schröter 1932). According to it, the northern hemisphere is the center of origin of all floras and faunas whence they migrated southward in successive waves and left their relic members in the southern hemisphere. The broad latitudinal distribution of boreal organisms (holarctic) is said to account for their migration southward into the three southern continents and their analogous development there. Considerable paleobotanical evidence has been accumulated that apparently serves to strengthen this widely held theory (Engel 1943, Harris 1937, Hirmer 1942, Koch 1924-1933a, Seward 1939, Stebbins 1940). The alternate possibility of a bipolar origin, previously referred to, is gaining ground, at least for certain groups and periods (Du Rietz 1940, Florin 1940). The extreme view of the bipolar origin of temperate floras held by Campbell (1943, 1944) is based on du Toit's assumption of two primordial continents, Laurasia and Gondwana. Hutchinson (1946) also assumes an austral origin of the families and groups confined to the southern hemisphere.

The earliest history of known vascular (pteridophyte) floras is too poorly known to permit any specific statement regarding their origin, as the entire period from the appearance of the pteridophytes in the Upper Silurian to the end of the Lower Carboniferous is marked by the great uniformity of its plant life (Hirmer 1938). With the Upper Carboniferous, however, and prior to the beginning of the Permian, fairly distinct floral regions can be recognized, at least in eastern Asia (Halle 1937). These regions are: a) Euramerican region: comprising Europe and North America, b) Angara region: extending from the rivers Dwina and Petschora in the west to the Kusnezk and Minnussinsk coal basins, c) Cathaysia region: comprising Shansi and Korea; the most typical plant life of this region occurred in the Upper Carboniferous and Lower Permian; an Upper Carboniferous (Stephanian) flora of Sumatra may also belong here, and d) Gondwana region; including central and southern South America, Antarctica, southern and central Africa, India south of the Himalayas, Australia with Tasmania and New Zealand. The famous Gondwana flora was essentially confined to the Gondwana region from the end of the Carboniferous throughout the Permian. The only connections during this time were apparently those with the Angara region. With the beginning of the Triassic Gondwana elements entered the Euramerican region and conversely Euramerican elements migrated southward. These contrasts of the northern Arcto-Carboniferous flora and the Glossopteris flora of the Gondwana region can not be explained by continental drift in the opinion of Halle, though Hirmer thinks so.

At the end of the Triassic and prior to the beginning of the Jurassic plant life is once more characterized by considerable uniformity. Throughout most of the Mesozoic the floras of the southern continents are quite uniform, a fact regarded by Hirmer as evidence in support of continental drift. Signs of latitudinal zonation of the subtropical flora, extending into the Arctic, make their appearance during the Upper Cretaceous and Lower Tertiary. These generalizations could be substantiated by including evidence drawn from the distribution of fossil gymnosperms.

Studies by Seward and Conway (1935a & b) of the Cretaceous strata of western Greenland have yielded some striking results. Here extensive beds bearing remarkable plant fossils are found, but are difficult to separate into the usual stratigraphic units. The most significant result seems to be the fact that many plants belonging to the preceding Jurassic and Wealden formations are accompanied by the first known flowering plants. Thus the first angiosperms appear in a vegetation otherwise composed of Mesozoic ferns and gymnosperms. The uppermost (Patoot) strata contain more Dicotyledones than do the lowest (Kome). It seems that the Cretaceous plants of western Greenland with their ancient character constitute remnants of the vegetation known from the

Rhaetic and Liassic deposits of eastern Greenland (Harris 1937, Seward & Conway 1935), although no direct connection between them is known. Conspicuous among the old Jurassic and Wealden types of almost cosmopolitan distribution are the Gleicheniaceae, Matoniaceae, Dipteridaceae, Cycadophyta s.l., and Coniferae.

Seward and Conway believe that the angiosperms originated in the Arctic and that the geographic and climatological conditions of Greenland as well as the geological ones during the Cretaceous were conducive to the origin and rapid evolution of the angiosperms. Greenland was then probably part of a much larger continent covered with vegetation for the most part and, on the whole, less disturbed or threatened by the Cenomanian transgression than areas south of it. Proximity to the pole is regarded as productive of new types due to increased activity during the northern summer and a prolonged rest period during long winters with higher temperatures than are known from there today. From this center the angiosperms are believed to have spread southward in Eurasia and America.

Comparison of these floras with others of comparable age and composition is difficult indeed, for the Greenland floras are the only known mixed floras of this kind, containing Mesozoic ferns and gymnosperms as well as the earliest angiosperms. The latter are quite similar to modern types. Because of this mixed character, the Cretaceous floras of Greenland cannot be assigned to Lower and Upper Cretaceous strata as is customary elsewhere. The slight resemblance of the floras of the Potomac formations of the coastal plains of the eastern United States is attributable to their poor representation of Mesozoic types and the appearance of angiosperms in the uppermost strata, whereas the Greenland beds contain angiosperms even in the lowest strata. Likewise other American Cretaceous floras lack older pteridophyte and gymnospermous types but consist mainly of angiosperms (New Jersey, Gulf States, Dakotas). Approximately contemporaneous Eurasian floras too are not as rich as the Greenland floras.

The climate of that time was apparently more or less subtropical though many deciduous trees were represented. In Seward's opinion a small movement of the earth's crust could account for this climate even if the pole remained stationary. At the same time it must be assumed that different oceanic conditions prevailed permitting warm currents to reach the shores of Greenland. For instance, an increase of eight degrees in winter and of four degrees in summer today would provide Norway at a latitude of 65 degrees N with a subtropical climate such as is found on the northern island of New Zealand. Moreover, the climate of the area occupied at present by any plant does not always indicate fully the plant's ecological requirements. The genus *Platanus*, though definitely more southern in distribution, can withstand the coldest winter in Europe as demonstrated by transplants. *Ginkgo*, *Sciadopitys* and other genera behave likewise.

Hollick's extensive studies (*fide* Hirmer 1942) of the Upper Cretaceous and Eocene floras of Alaska disclosed that the Mesozoic types contained resemble those found in Bohemia and Sachalin and thus fit in general the concept of the Arctic origin of modern floras. These floras too are subtropical in character, and show a close relationship to the modern floras of North America, Europe and East Asia including even Australasia.

The great expansion of the angiosperms took place during the early Tertiary. This event is described in one of Seward's (1939) addresses and may fittingly be quoted here: "The fossil flora of Mull represents an early phase of what may be called the modern type of vegetation, which overspread the world in the later stages of the Cretaceous period and has persisted with few major modifications until now. Evolution seems to have been characterized by bursts of production when new and successful types exercised a transforming influence; and these periods of exceptional creative activity were separated by periods of relative stability. The early Tertiary floras belong to a stage when a new order had become well established and an older order had passed its prime. The one great difference that emerges from comparison of the Mull flora and the existing European floras is not a difference in the components of the world forests but a contrast in the geographical positions occupied by the various genera in the northern hemisphere; for the most part a western home has been exchanged for a home in the Far East."

But with all Berry (1945) seems to be right in his appraisal of our knowledge when he says: "In spite of the great volume of descriptive literature of modern paleobotany and the light which it has shed on geological problems and plant history, it has been a history of modern families and not of the founders of dynasties."

SUMMARY

1. Geologists are not agreed on the interpretation of the origin and development of continental masses and oceans. A modified form of the theory of continental drift may prove acceptable if shown to be connected with diastrophism.

2. Many data pertaining to fossil and living plant distribution, especially large discontinuous areas, have been interpreted in the light of continental drift. Other data can not be construed as evidence in support of it.

3. Available information concerning the earliest known angiosperm records from the Cretaceous of Greenland indicates an Arctic center of origin, the existence of a subtropical flora of wide distribution composed of many old stocks and their subsequent southward migration. These facts can be used as evidence in favor of the theory of the monoboreal (northern) origin of modern angiosperm floras.

4. A bipolar evolution of certain stocks, notably of the conifers, appears quite probable, at least of their modern representatives.

5. Evolution has apparently been characterized "by bursts of production" that were followed "by periods of relative stability."

6. The composition of modern floras is exceedingly complex due to the interaction of many known and unknown processes such as climatic changes and migrations, as the members vary in age and origin. Only detailed analysis of all fossil and living members can disclose the true composition and history of modern floras.

7. With few exceptions the old doctrine of historical plant geography is apparently still valid: areal continuity presupposes genetic continuity.

LITERATURE CITED

Andrews, E. C. 1940. Origin of the Pacific Insular Floras. Sixth Pac. Sci. Congr. Proc. **4**: 613-620.

Berry, E. W. 1920. Paleobotany: A Sketch of the Origin and Evolution of Floras. Smithson. Inst. Ann. Rpt. **1918**: 289-407.
1930. The Past Climate of the North Polar Region. Smithson. Inst. Misc. Collect. **82(6)**: 29 pp.
1945. The Origin of Land Plant and Four Other Papers. J. Hopkins Univ. Studies Geol. **14**.

Bews, J. W. 1927. Studies in the Ecological Evolution of the Angiosperms. New Phytol. Reprint **16**.

Brockmann-Jerosch, H. 1914. Zwei Grundfragen der Paläophytogeographie. Englers Bot. Jahrb. **50** (Suy.): 249-267.

Brooks, C. E. P. 1926. Climate through the Ages. London.

Bucher, W. H. 1933. The Deformation of the Earth's Crust. Princeton.

Cain, S. A. 1944. Foundations of Plant Geography. New York.

Campbell, D. H. 1943. Continental Drift and Plant Distribution. Privately printed.
1944a. Living fossils. Science **100(2592)**: 179-181.
1944b. Relations of the Temperate Floras of North and South America. Calif. Acad. Sci. Proc. **4**. ser. **25(2)**: 139-146.

Copeland, E. B. 1940. Antarctica as the Source of Existing Ferns. Sixth Pac. Sci. Congr. Proc. **4**: 625-627.

Dacqué, E. 1932. Paläogeographie und Paläoklimatologie. Handw. Naturw. **7**: 609-628. 2 ed. Jena.

Daly, R. A. 1942. The Floor of the Ocean. Chapel Hill.

Diels, L. 1928. Kontinentalverschiebung und Pflanzengeographie. Ber. Deut. Bot. Gesell. **46**: (49)-(58).
1934. Die Flora Australiens und Wegeners Verschiebungstheorie. Sitzber. Preuss. Akad. Wiss., Phys.math. Kl., **31/33**: 533-545.

Du Rietz, E. G. 1940. Problems of Bipolar Plant Distribution. Acta Phytogeog. Suecica **13**: 215-282.

Eckardt, W. R. 1925. Die klimatischen Verhältnisse der geologischen Vergangenheit im Lichte von Alfred Wegeners Hypothese der Kontinentalverschiebungen. Naturw. **13**: 77-83.

Engel, A. 1943. Mécanisme et historique des migrations forestières de l'époche tertiaire à nos jours. Avec une Note sur les conditions géographiques et climatiques aux époques tertiaires et quaternaires, par Elie Gagnebin (pp. 187-217). Mém. Soc. Vaud. Sci. Nat. **7(8)**: 167-218.

Fernald, M. F. 1944. Continental Drift and Plant Distribution. Rhodora **46(546)**: 249-251.

Florin, R. 1940. The Tertiary Fossil Conifers of South
Chile and their Phytogeographical Significance with
a Review of the Fossil Conifers of Southern Lands.
Kgl. Svenska Vetensk.-Akad. Handl., 3 ser. **19(2):**
107 pp., 6 pls.

Fritzsche, G. 1936. Klimarhythmik und Vegetations-
rhythmik. Bioklim. Beibl. **3:** 119-123.

Goodspeed, T. H. (Editor). 1936. Essays in Geobotany
in Honor of William Albert Setchell. University of
California Press.

Gothan, W. 1924. Palaeobiologische Betrachtungen
über die fossile Pflanzenwelt. Fortschr. der Geol. u.
Paläont. **8:** 178 pp.
1937. Paläobotanik. In O. H. Schindewolf, Fortschr.
der Paläon. **1:** 345-374.

Grabau, A. W. 1940. The Rhythm of the Ages. Peking.

Gutenberg, B. [Editor]. 1939. Internal Constitution of
the Earth. New York.

Haddock, M. H. 1936. The Wandering of the Continents. Discovery **17(195):** 67-70.

Halle, T. G. 1937. The Relation between the Late
Palaeozoic Floras of Eastern and Northern Asia.
Comp. Rend. 2. Cong. Avanc. Etud. Stratigr. Carboni
fére Heerlen **1:** 237-248.

Hansen, H. M. 1930. En Undersøgelse over de Raun-
kiaerske Livsformers Palaeontologie. Copenhagen.

Harris, T. M. 1937. The Fossil Flora of Scoresby Sound
East Greenland. Part 5: Stratigraphic Relations of
the Plant Beds. Meddel. om Grønland **112(2):** 114 pp.

Hayek, A. 1926. Allgemeine Pflanzengeographie. Berlin.

Herzog, T. 1926. Geographie der Moose. Jena.

Hirmer, Max. 1935-1940. Paläobotanik. In F. von
Wettstein, Fortschr. der Bot. **4:** 95-126; **5:** 72-103;
7: 71-124; **8:** 91-130; **9:** 409-470. Berlin.
1938. Geographie und zeitliche Verbreitung der fossilen Pteridophyten. In F. Verdoorn, Manual of
Pteridology: 474-495. Hague.
1942. Die Forschungsergebnisse der Paläobotanik auf
dem Gebiet der Känophytischen Floren. Ein Sam-
melbericht über die Erscheinungen der Jahre 1936-
1941. Englers Bot. Jahrb. **72(3/4):** 347-563, pl. 8-20.

Høeg, O. A. 1937. Plant Fossils and Paleogeographical
Problems. Compt. Rend. 2. Cong. Avanc. Etud. Stratigr. Carbonifére Heerlen **1:** 291-311.

Högbom, A. G. 1941. Die Atlantisliteratur unserer Zeit.
Betrachtungen eines Geologen. Geol. Inst. Univ. Upsala Bul. **28:** 17-78.

Holland, T. H. 1937. The Permanence of Oceanic Basins
and Continental Masses. Huxley Memorial Lecture.
London.

Holmes, A. 1944. Principles of Physical Geology. London.

Holttum, R. E. 1940. The Uniform Climate of Malaya
as a Barrier to Plant Migration. Sixth Pac. Sci. Cong.
Proc. **4:** 669-671.

Hu, Hsen-Hsu. 1940. Constituents of the Flora of
Yunnan. Sixth Pac. Sci. Cong. Proc. **4:** 641-653.

Hutchinson, J. 1946. A Botanist in Southern Africa.
London.

Irmscher, E. 1922. Pflanzenverbreitung und Entwicklung
der Kontinente. Studien zur genetischen Pflanzengeographie. Mitt. Inst. Allg. Bot. Hamburg **5:** 15-
235, pl. 1-12.
1929. Idem. II. Teil. Weitere Beiträge zur genetischen

Pflanzengeographie unter besonderer Berücksichtigung
der Laubmoose. Loc. cit. **8:** 169-374, pl. 11-26.

Joleaud, L. 1939. Atlas de Paléobiogéographie. Paris.

Joly, J. 1930. The Surface History of the Earth.

Kanehira, R. 1940. On the Phytogeography of Micro-
nesia. Sixth Pac. Sci. Cong. Proc. **4:** 595-611.

Kerner-Marilaun, F. 1930. Paläoklimatologie. Berlin.

Koch, F. 1924. Über die rezente und fossile Verbreitung
der Koniferen im Lichte neuerer geologischer Theo-
rien. Mitt. Deut. Dendrol. Gesell. **1924:** 81-99.
1925. Die Cycadeen im Lichte der Wegenerschen Kon-
tinent- und Polwanderungstheorie. Loc. cit. **1925:** 67-
74.
1927. Zur Frage der fossilen und rezenten Verbreitung
der Koniferen. Loc. cit. **38:** 182-184.
1930. In den Wäldern Guatemalas. Mit Bemerkungen
über Florenwanderungen. Loc. cit. **42:** 337-347.
1931. Die Entwicklung und Verbreitung der Kontinente und ihrer höheren pflanzlichen und tierischen
Bewohner. Braunschweig.
1933. Die Bedeutung der Wegenerschen Theorie für
die Dendrologie. Mitt. Deut. Dendrol. Gesell. **45:** 184-
193.
1933a. Ueber das entwicklungsgeschichtliche Alter
der Bäume und ihre geographische Verbreitung in Vergangenheit und Gegenwart. Loc. cit. **45:** 194-199.

Kubart, B. 1929. Das Problem der tertiären Nordpolar-
floren. Ber. Deut. Bot. Gesell. **46:** 392-402.

Kupffer, K. R. 1930. Die pflanzengeographische Bedeutung des Ostbaltischen Gebietes. Rep. spec. nov. regn.
veget. Beih. **61:** 1-31.

Lam, H. J. 1938. Studies in Phylogeny. I. and II.
Blumea **3:** 114-158.
1940. Some Notes on the Distribution of the Sapotaceae of the Pacific Region. Sixth Pac. Sci. Cong.
Proc. **4:** 673-683.

Lee, J. S. 1939. Continental Drift. Geol. Mag. **76(7):**
289-293.

Lindsey, A. A. 1940. Biology and Biogeography of the
Antarctic and Subantarctic Pacific. Sixth Pac. Sci.
Cong. Proc. **2:** 715-721.

Longwell, C. R. 1930. The "Oscillation Theory" of Diastrophism. Amer. Jour. Sci. **219:** 217-220.
1944a. Some Thoughts on the Evidence for Continental Drift. Amer. Jour. Sci. **242(4):** 218-231.
1944b. Further Discussion of Continental Drift.
Amer. Jour. Sci. **242(9):** 514-515.

Mägdefrau, K. 1942. Paläobiologie der Pflanzen. Jena.

Matthews, J. R. 1937. Geographical Relationships of the
British Flora. Jour. Ecology **25(1):** 1-90.
1946. Plant Life in Britain: Its Origin and Distribution. Jour. Roy. Hort. Soc. **71(8):** 225-239; **71(9):**
259-273.

Merrill, E. D. 1926. Correlation of the Indicated Biologic Alliances of the Philippines with the Geologic
History of Malaysia. Reprinted in Merilleana, Chron.
Bot. **10(3/4):** 216-236, 1946.
1943. Foreword to Wulff, E. V. 1943. An Introduction
to Historical Plant Geography. Waltham, Mass.

Metcalf, Z. P. 1946. The Center of Origin Theory.
Elisha Mitchell Sci. Soc. Jour. **62(2):** 149-175, pl. 23-
41.

Netolitzky, F. 1933. Eine neue Hypothese zur Erklär-
ung der zirkumpolaren Verbreitung von Pflanzen und
Tieren. Fac. Sti. Cernauti Bul. **6(1/2):** 135-137.

Rastall, R. H. 1929. On Continental Drift and Cognate Subjects. Geol. Mag. **66:** 447-456.
1946. Review of Umbgrove, J. H. F. 1942. The Pulse of the Earth. Geol. Mag. **83(1):** 46-47.

Reinig, W. F. 1937. Die Holarktis. Ein Beitrag zur diluvialen und alluvialen Geschichte der zirkumpolaren Faunen- und Florengebiete. Jena.
1938. Elimination und Selektion. Eine Untersuchung über Merkmalsprogressionen bei Tieren und Pflanzen auf genetisch- und historisch-chorologischer Grundlage. Jena.
1939. Die Evolutionsmechanismen, erläutert an den Hummeln. Zool. Anz. Sup. **12:** 170-206.

Ridley, H. N. 1937. Origin of the Flora of the Malay Peninsula. Blumea Sup. **1:** 183-192.

Rubtzov, I. A. 1945. The Inequality of Rates of Evolution. Jour. Gen. Biol. **6(6):** 411-441. [English summary: 439-441.]

Sahni, B. 1936. Wegener's Theory of Continental Drift in the Light of Paleobotanical Evidence. Ind. Bot. Soc. Jour. **15:** 319-332.

Scharfetter, R. 1922. Klimarhytmik, Vegetationsrhytmik and Formationsrhytmik. Studien zur Bestimmung der Heimat der Pflanzen. Osterr. Bot. Ztschr. **71(7/9):** 153-171.

Schiller, Walther. 1942. Las Antiguas Montañas de la Provincia de Buenos Aires ¿Qué comprueban en favor o en contra de la hipótesis de Wegener? Notas Mus. La Plata, Geologia **7(22):** 247-252.

Schindewolf, O. H. 1937. Geologisches Geschehen und Organische Entwicklung. Geol. Inst. Univ. Upsala Bul. **27:** 166-188.

Schröter, C. 1932. Genetische Pflanzengeographie (Epiontologie). Handw. Naturw. **4:** 1002-1044. 2. ed. Jena.

Schuchert, C. 1932a. Permian Floral Provinces and their Interrelations. Amer. Jour. Sci. **24:** 405-415.
1932b. Gondwana Land Bridges. Bull. Geol. Soc. Amer. **43:** 875-915.

Setchell, W. A. 1935. Pacific Insular Floras and Pacific Paleogeography. Amer. Nat. **69:** 289-310.

Schuster, J. 1931. Ueber das Verhältnis der systematischen Gliederung, der geographischen Verbreitung und der paläontologischen Entwicklung der Cycadaceen. Englers Bot. Jahrb. **64 (1/2):** 165-260, pl. 4-11.

Schwarz, O. 1938. Phytochorologie als Wissenschaft, am Beispiele der vorderasiatischen Flora. Rep. spec. nov. regn. veget. Beih. C: 178-228.

Seward, A. C. 1933. Plant Life through the Ages. Cambridge.
1935. Selections from the Story of Plant Migration Revealed by Fossils. Sci. Prog. **30(118):** 193-217.
1939. The Western Isles through the Mists of Ages. Rep. British Assoc. Adv. Sci. **1939(1):** 11-29.

Seward, A. C. & V. Conway. 1935a. Fossil Plants from Kingigtok and Kagdlungnak, West Greenland. Meddel. om Grønland **93(5):** 41, 5 pls.
1935b. Additional Cretaceous Plants from Western

Greenland. Kgl. Svenska Vetensk. Akad. Handl. **3.** ser. **15(3):** 41, 6 pls.

Simpson, G. C. et al. 1930. Discussion on Geological Climates. Roy. Soc. Proc. London **106(B):** 299-317.

Skottsberg, C. 1940. The Flora of the Hawaiian Islands and the History of the Pacific Basin. Sixth Pac. Sci. Cong. Proc. **4:** 685-707.

Staub, R. 1944. Die Gebirgsbildung im Rahmen der Erdgeschichte. Verh. Schweiz. Naturf. Gesell. **1944:** 25-48.

Stebbins, G. L., Jr. 1940. Additional Evidence for a Holarctic Dispersal of Flowering Plants in the Mesozoic Era. Sixth Pac. Sci. Cong. Proc. **3:** 649-660.

Studt, W. 1926. Die heutige und frühere Verbreitung der Koniferen und die Geschichte ihrer Arealgestaltung. Mitt. Inst. Allg. Bot. Hamburg **6:** 167-307.

Suesenguth, K. 1938. Neue Ziele der Botanik. Munich.

Tardien-Blot, M. L. 1940. Etude Phytogéographique des Fougère d'Indochine. Sixth Pac. Sci. Cong. Proc. **4:** 579-593.

Tiercy, G. 1945. Réflexions sur la théorie des translations continentales. Verh. Naturf. Gesell. Basel **56(2):** 531-543.

Toit, A. L. du. 1937. Our Wandering Continents. An Hypothesis of Continental Drifting. London.
1940. Observations on the Evolution of the Pacific Ocean. Sixth Pac. Sci. Cong. Proc. **1:** 175-183.

Umbgrove, J. H. F. 1942. The Pulse of the Earth. 2nd Ed. Hague. 1947.
1946. Recent Theories on Polar Displacement. Amer. Jour. Sci. **244(2):** 105-113.

Waterschoot van der Gracht, W. A. T. M. et al. 1928. Theory of Continental Drift. A Symposium on the Origin and Movement of Land Masses both Inter-Continental and Intra-Continental, as Proposed by A. Wegener. Tulsa.

Weimarck, H. 1941. Phytogeographical Groups, Centres and Intervals within the Cape Flora. A Contribution to the History of the Cape Element seen against Climatic Changes. Lunds Univ. Årsskr. N. F. Avd. 2, **37(5):** 143 pp.

Weiss, F. E. 1925. Plant Structure and Environment with Special Reference to Fossil Plants. Jour. Ecology **13:** 301-313.

Willis, Bailey. 1944. Continental Drift, ein Märchen. Amer. Jour. Sci. **242(9):** 509-513.

Wulff, E. V. 1934. Essay at Dividing the World into Phytogeographical Regions According to the Numerical Distribution of Species. Appl. Bot., Gen. & Plantbr., Bul. Ser. 1, No. 2: 315-354.
1943. An Introduction to Historical Plant Geography. Waltham.

Zeuner, F. E. 1946. Dating the Past, an Introduction to Geochronology. London.

Zimmermann, W. 1930. Phylogenie der Pflanzen. Jena.
1938. Vererbung "erworbener Eigenschaften" und Auslese. Jena.

Editor's Comments
on Paper 5

5   **SCHUCHERT**
    Biologic Principles of Paleogeography

Paper 5 is from the same symposium as Paper 3 (Knowlton, 1910) and is of particular interest to invertebrate paleontologists and biostratigraphers. It is one of the first of a long series of papers in which Schuchert explained how he organized thoughts and data into paleogeographic maps for each geologic period and epoch. This 1910 paper fills in the gap left by the brief, almost cursory, mention of invertebrate paleontology, biostratigraphy, and sedimentary features that Schuchert made in Paper 2, a later (1929) discussion on paleogeographic map making. This 1910 statement by Schuchert probably outlines his philosophy of paleogeographic analysis and map construction from about 1904 until the early 1920s, when his attention became more and more focused on orogenies (revolutions), geologic rhythms, and glaciations. In 1904, at the age of forty-six, he joined the faculty at Yale University after a series of careers, including that of a paleontologist at the U.S. National Museum. At Yale he initially had difficulty conveying the wonderous world of paleontology and the concepts of paleontology and stratigraphy to his students; he searched for a way to organize the great wealth of detailed information that was being analyzed and hit upon the plan of constructing paleogeographic maps as a teaching aid. This started his continuing interest in paleogeography and formed the framework for his textbooks in historical geology. This then is Charles Schuchert in 1910, having wrestled with the problems of paleogeography for a number of years and having the experience of using paleogeographic maps to convey ideas.

5

BIOLOGIC PRINCIPLES OF PALEOGEOGRAPHY

Charles Schuchert
Yale University

IN deciphering the ancient geography as to the position of the marine waters and the land masses, we as pioneers in this work must be controlled primarily by the known fossilized life and secondarily by the character and place of deposition of the geologic formations. This record is most extensive and best preserved in the deposits of the continental and the littoral region along the continental shelves of the oceanic areas. Back of these two principles, however, there is another that eventually will become the primary guiding factor. It is the principle of diastrophism—one seeking to explain the causes for the periodic movements of the lithosphere.

In our study of the ancient seas with their sediments and entombed life we have safe guidance in the phenomena of the present. Ludwig in 1886 estimated the species of animals then known to naturalists as upwards of 312,000, and in 1905 Stiles thought this great total had increased to about 470,000 forms. Of this sum fully 60 per cent. are insects, and of the remainder, the writer concludes that about 25 per cent., or 115,000 species, live in the sea, and 71,000 have their habitat on the land or in the waters of the land. Of the 115,000 kinds of known animals inhabiting the seas nearly 70 per cent. are Cœlenterata, Echinodermata, Molluscoidea and Mollusca, the types of organisms most often found by the stratigrapher and on which he is largely dependent in deciphering the ancient geography.

Let us now examine into the number of available fossil forms made known by the paleontologists. As early as 1868, Bigsby in his "Thesaurus Siluricus" listed 8,897 species from the strata beneath the Devonic, and in his "Thesaurus Devonico-Carboniferous" of 1878, he further enumerated about 5,600 Devonic and 8,700 Carbonic forms. In 1889 Neumayr concluded that there were then known about 10,000 Jurassic species. We may therefore estimate that the paleontologists of to-day have access to at least 100,000 species of fossils. Their numbers in the geologic scale are about as follows: Cambric 2,000, Ordovicic 8,000, Siluric 8,000, Devonic 9,000, Lower Carbonic 7,000, Upper Carbonic 8,000, Permic 4,000, Triassic 6,000, Jurassic 15,000, Cretacic 10,000 and Tertiary 25,000. The end of species-making is not at all in sight, and the day will come when paleontologists will deal with ten times as many species as are now known.

Stiles tells us that zoologists know but from 10 to 20 per cent. of the living forms, and there should therefore be from 3,760,000 to 4,700,000 different kinds of animals alive to-day, ranging from the protozoa to man. Now let us compare the abundance of living animals with those of the geologic ages, and especially with the Jurassic period, of which life we have probably a better knowledge than of any time back of the Tertiary. The European Jurassic has long been divided into 33 zones (Buckman hints at a probable 100), and if we hold that each one of these times had only one quarter as many species as in the lowest estimate of the present world, there must have lived during the entire Jurassic something like 31,000,000 kinds of animals. Yet paleontologists have described not more than 15,000 Jurassic forms. The great imperfection of the extinct life record is thus forcibly brought to our attention, and we learn from these estimates that for each kind of animal preserved in the rocks more than 2,000 other kinds are utterly blotted out of the geologic record.

Much of this more apparent than real imperfection, however, is due to the vast number of insect species now living—animals that must have been comparatively few in the Jurassic, due in the main to the absence of flowering plants. From these figures, however, we must not conclude that the geologic record is equally imperfect throughout; for the paleontologist studying marine fossils well knows that he can not, as a rule, hope to study other than those kinds of animals that have hard and calcareous or siliceous external or internal skeletons. Of such there may be in the present seas about 250,000 kinds, of which about 25,000 have been named. Therefore on this basis we can say that the student of Jurassic faunas knows 1 species in every 54 of shelled animals that lived during this period.

This admittedly great imperfection of the life record needs to be further explained so that the reader will not arrive at the erroneous conclusion that modern stratigraphy rests upon very insecure foundations. The stratigrapher in determining the age of a given deposit, and in the identification of it from place to place and from country to country, and even across the great oceans, deals in his work not with quantity of species, but with comparatively small numbers of constantly recurring hard parts of certain species that are more often of marine than of land origin. Many of these forms have but local value but others have spread thousands of miles, and some of the long enduring species range over the greater part of the earth. Some of the best guide fossils in the Paleozoic are the brachiopods because they are present in nearly all the strata of this era. The writer in 1897 listed 1,859 forms then known from these rocks of North America. Of these about 28 per cent., or 537 species, had great geographic distribution. 117 species are found in the Rocky Mountain area, the Mississippi valley and the Appalachian region, and of these 36 are also known to occur in foreign countries. The number of species common to North America and other continents, however, is 121. It is upon faunal assemblages of this quantity and nature that the stratigrapher relies most in deciphering the former extent of the continental seas.

In the making of paleogeographic maps or in the determination of geologic time, using fossils as the essential basis, we have guidance in those of marine faunas, and the floras and faunas of the land and its fresh waters. Of these widely differing realms or habitats we now know that the fossils of the marine faunas are the more reliable not only because there are so many more of them than of the land dwellers, but more especially because their geologic succession is far more complete. The conditions of preservation, that is, appropriate burial in sediments, are always at hand in marine waters, but on the land entombment occurs only exceptionally, whereas the life of fresh waters is very meager and almost unchanging during geologic time. Then, too, marine life is " less affected by meteorologic factors, and more dependent upon conditions which affect the whole hydrosphere rather than small areas of it. The struggle for life is less intense, the food supply generally more adequate, enemies less vigorous, and dangerous fluctuations of temperature far less frequent, in the sea than on land. The same features make the land fauna more clearly indicative of minor divisions of the scale, and of the progress of organic evolution in the general region concerned; while less conclusive as to the contemporaneity of widely separated though analogous faunas."[1]

In regard to the probable geographic position of the shore lines we rarely have safe guidance in the fossils, and for this depend on the nature of the deposits. Greatest dependence is placed upon the geographic position of sandstones and especially on conglomerates to indicate the probable former shores. Limestones of uniform character and wide distribution are indicative of greater distance from land. Shallowness of the continental seas is proved by a rapid change in the character of the sediments both laterally and vertically, and by the oolite and dolomite deposits. Intraformational conglomerates, coral reefs, ripple marks, and shrinkage cracking furnish further evidence to the same conclusion. Storm waves are known to plough the present sea-bottom to depths of 160 feet. Calcareous muds are now forming in tropical and subtropical waters at sea-level around coral reefs, and elsewhere in these latitudes at depths from 200 to 600 meters. It is probable that all of the ancient great limestone deposits are of warm waters, and, if so, are an additional aid in discerning the geologic times and regions of milder climates.

Phosphatic concretions form in the littoral region where the temperature changes are rapid, as off the coast of the New England states, and periodically cause much destruction of the individual life. The carcasses decompose at the bottom of the sea, making nuclei for the accretion of phosphate of lime, and because of the irregular periodicity of accumulation come to be arranged in definite stratigraphic zones. Old Red sandstone fishes are also usually found in clay nodules but abundantly only in limited zones (Scaumenac, Canada and Wildungen, Germany). Have these also been killed by rapid changes in the temperature of these waters? In any event the fish-bearing beds are always found near the shore lines of Devonic seas.

Scour of sea bottom is met with in the present seas where great streams of water are forced through narrow passages, as the Gulf Stream in the Floridian area; or where such streams impinge against the continental shelf, as north of Cape Hatteras, or flow across submerged barriers " a few miles broad," as the Wyville-Thomson ridge connecting the British and Faeroese plateaus (Johnstone, 1908, 31). Strong currents preventing sedimentation also occur in long and narrow bays, as that of Fundy, where the undertow caused by the very high tides of this region sweeps the bottom clean. These exceptional and, after all is said, rather local occurrences can not be the explanation for the many known breaks in the geological sedimentary record, the disconformities of stratigraphers. These breaks are at times as extensive as the North American continent (post Utica break), and are usually

---

[1] Dall, *Jour. Geol.*, 1909, 494.

of very wide extent. Scour of the bottom by the currents of the ancient continental seas will not explain away the presence of these truly land times, but it is to be sought in the oscillatory nature of the seas of all time which is probably caused by the periodic unrest of the earth's crust due to earth shrinkage. We agree with Suess that " Every grain of sand which sinks to the bottom of the sea expels, to however trifling a degree, the ocean from its bed," and every movement of the sea-bottoms and the periodic down fracturing of the horsts causes the strand lines to tremble in and out, be they of a positive or transgressive or of a negative or land-making character.

The ancient marine life had similar zoogeographic arrangement to that of the present. It can be grouped into local faunas and these combined into subprovinces, provinces and realms. Their distribution is governed primarily by the presence or absence of land barriers, and secondarily by temperature and latitude. In the present seas temperature is one of the main factors controlling the distribution of the species, but during the geologic ages the climate was, as a rule, far more uniform than now, as we are living under the influence of polar ice caps and a passing glacial period, or possibly even an Interglacial period.

The faunas with which the stratigraphic paleontologist works appear in many instances as suddenly introduced biotas. Our collaborators of half a century ago explained them as Special Creations, but since their time we have learned that the suddenly appearing faunas are not such in reality but only seem to appear rather quickly due to the slowness of sedimentary accumulation. Ulrich estimates that the American Paleozoic has less than 100 mapable units or formations, each with a duration of probably not less than 175,000 years. Accordingly, each foot of average sedimentary rock has taken not less than 833 years to accumulate. Our knowledge regarding the average rate of sedimentary marine accumulation is, however, as yet very insecure, and to make this clear some of the remarks made by Sollas, President of the Geological Society of London (1910), will be quoted. He was led to make these remarks after the reading of a paper by Buckman correlating the Jurassic sections of South Dorset. He said, " The correlation of thin seams with thick deposits was a matter of great importance. . . . It might afford some hints as to the order of magnitude of the scale of time. If we assumed that one foot of sediment might accumulate in a century, in an area of maximum deposition, then in the case of the seam two inches thick, which was represented by 250 feet in the Cotteswolds, the rate of formation would be less probably than 1 foot in 150,000 years." What Ulrich's estimate of time necessary for the accumulation of one foot of average sediment means to migratory faunas may be illustrated by the spreading of *Littorina littorea*. In the last century this edible European gastropod was introduced at Halifax, Nova Scotia, and in 50 years attained the Delaware Bay and north to Labrador. Taking this dispersion as the basis for calculating faunal migrations, we learn that they may spread 500 miles, while one sixteenth of an inch of average sediment is depositing, or 8,000 miles during the time of one foot of sedimentary accumulation. If, therefore, Paleozoic faunas migrated " only one fiftieth as fast as this living shell, then we may reasonably assert essential contemporaneity for stratigraphic correlations extending entirely across the continent." We have here an explanation for the apparently sudden distribution of the Ordovicic brachiopod *Rhynchotrema capax,* that everywhere holds an identical geologic horizon from Anticosti to the Big Horns and from El Paso, Texas, to Arctic Alaska. *Spirifer hungerfordi* spreads during the first half of Upper Devonic time from the Urals to Iowa, and another brachiopod, *Stringocephalus burtoni,* migrates during the last third of Middle Devonic time from western Europe to Manitoba.

The life of the present seas extends from the strand-line to the deepest abyss, but by far the greatest quantity and variety lives in the upper sunlight, photic or diaphanous region. Photographically the light of the sun is detectable in exceptionally clear-water tropical seas to a depth of about 2,000 feet, but Johnstone places the average depth for all waters at 650 feet, beyond which there is more or less of total darkness, the aphotic realm.

Sunlight is the first essential for the existence of life. Where it penetrates, there plant life is possible, and this life is the substratum on which all animal life is ultimately dependent for food. Near the surface of the sea lives the plankton, sometimes referred to as the " pastures of the sea " and compared with the " grass of the fields." Most of this plankton consists of diatoms that at present are by far more prolific in the cooler polar waters. At times of greatest abundance in Kiel Bay as many as 200 of these " jewels of the plant world " are contained in a drop of water, and in the Antarctic seas there is an area of ten and one half million square miles where diatom ooze is accumulating. They are the principal food supply for most of the sessile benthos, or bottom life, among which the mollusca and brachiopods are of the greatest importance in paleogeography.

Geologic deposits rich in diatoms are sometimes regarded as those of the deep sea, at least as of deeper waters than those of continental seas. The English Carbonic deposits, rich in diatoms, have a fauna whose species are all of the shallow water kinds. The vast Miocene diatom deposits of California, described by Arnold, have living bottom types of foraminifera that, according to Bagg, do not indicate a depth of over 500 fathoms.

From the present distribution of marine life we learn that the greatest bulk of invertebrates are restricted to the bottom of the shallow seas within the depth to which sunlight readily penetrates, that is, a depth on the average not over 600 feet. The value of this observation to the paleogeographer and the student of fossil marine life lies in the confirmation of paleontologists that continental seas are shallow seas, to the bottom of which in most places sunlight permeates. These seas are to be compared with the littoral regions of the present oceans, and they are the areas that are most exposed to climatic and physical changes, due to their proximity to the atmosphere and the lands. The life of these waters is, therefore, subject to an environment that is more or less changeable, and one of the basic causes underlying organic change. It is the invertebrates of the littoral and shallow seas that the paleontologist studies.

In the tropical and subtropical shallow seas one meets with the greatest variety of life and with the brighter colored and more ornamental shelled animals, but we are much surprised when told that the greatest number of individuals occur in the colder shallow waters of the temperate and polar regions. Johnstone states, " There is little doubt that the distribution of life in the sea is exactly opposite to that on the land. The greatest fisheries are those of the temperate and arctic seas. . . . Nowhere are sea birds so numerous as in polar waters. The benthic fauna and flora are also most luxuriant." The Bay of Naples has a " richly varied, but (in mass) a scanty fauna and flora," and " at the very least the amount of life in polar seas is not less than in the tropics."[2]

Marine life is also more prolific near river mouths of the temperate zones, probably because of the great quantities of dissolved " salts of nitrous and nitric acid and ammonia, and other substances which are the ultimate food-stuffs of the plankton." Just outside of the estuary of the Mersey in Lancashire there were " not less than twenty, and not more than two hundred animals varying in size from an amphipod (one fourth inch long) to a plaice (eight to ten inches long) on every square meter of bottom " (Johnstone, 1909: 149, 176, 195–6). Finally the quantity of life in the shallow waters of the sea is not directly governed by favorable habitat, such as shallow sunlight waters in constant circulation and of equable temperature, but seems to be primarily controlled by the amount of the minimal food elements. Sea-water may be regarded as a dilute food-solution having the essential materials on which life is dependent. Of these nitrogen and the compounds of silica and phosphoric acid are present in the smallest amount. Johnstone tells us that " The density of the marine plants will therefore fluctuate according to the proportions of these indispensable food-stuffs " (234). " It is only the protophyta among the

[2] " Life in the Sea," 1908, 201–205.

plankton which can utilize the CO_2 and the nitric acid compounds, and so we see that upon these rest the greater part of the task of elaborating the dissolved food-stuff of the sea" (239).

Undoubtedly much of the land-derived nitrogen, estimated at 38 million tons per annum, is used up in the shallow areas by the plants. We therefore arrive at the conclusion that shallow seas bordering naked, cold, or arid lands should have the smallest amount of life, and that those of temperate regions adjacent to low lands under pluvial climates should have the greatest number of individuals. This conclusion, however, may be decidedly altered by the oceanic currents in that they distribute far and wide the salts of the sea.

These factors also suggest that during "critical periods" the faunas should be least abundant and varied, and that at the times of extreme base levels and sea transgressions they ought to be at their maximum development. These suggestions are borne out by the small Cambric, Permic and earliest Eocene faunas and the large cosmopolitan biotas of the Siluric, Jurassic and Oligocene times.

Sessile algæ are not common on muddy or sandy grounds, and these areas in the present seas have been compared with the desert areas of the lands. That muddy grounds are now nearly devoid of algous growth has particular significance in stratigraphy, because in the geologic column at many levels and in nearly all regions occur black shale formations that are not only devoid of plant fragments but are also usually very poor in fossils of the sessile benthos. When the latter are present it is seen that they are usually thin-shelled and small forms, or are types of organisms that live in the upper sunlight realm and are either of the swimming plankton or the floating nekton. As examples of such deposits may be cited the widely distributed Utica formation of the Ordovicic extending from southern Ohio to Lake Huron and east to Montreal, and the Genesee (Devonic) of New York. In these cases what appears to be of the sessile benthos is thought to belong to the nekton attached to floating seaweeds or other floating objects, and eventually all of the life of the nekton and the plankton sinks to the bottom of the sea. Therefore the carbonaceous matter of the black shales may be of algous origin like that of the New York Genesee, but it is far more probable that it is largely of animal origin, as the crude petroleum of such deposits usually has the optical properties of animal oil and especially those of fish oil.[3] Plants may be torn from rocky bottoms of the shallow areas by the action of the storms and then carried by the currents into eddying areas like the present Sargossa Sea, which has among its algæ a very characteristic assemblage of animals. It is probable, however, that black shales having wide distribution were more often the deposits in closed arms of the sea (cul

[3] Dalton, *Economic Geology*, 1909, 627.

de sacs), or when of small areal extent, as the result of fillings of holes in the sea bottom. In all such places there is defective circulation and lack of oxygen resulting in foul asphixiating bottoms.

These are the " halistas " of Walther and the " dead grounds " of Johnstone. To-day such are the Black Sea and the Bay of Kiel, where sulphur bacteria abound in greatest profusion. These decompose the dead organisms that rain from the photic region into such suffocating areas, or the carcasses which are drawn there by the slow undertow from the higher ground. These bacteria in the transforming process deposit in their cells sulphur that ultimately combines with the iron that is present and replaces the calcareous skeletons of invertebrates by iron pyrite or marcasite. In this way are formed the wonderfully interesting pseudomorphs of *Triarthrus becki,* the Utica trilobite preserving the entire ventral limbs, and of the other well preserved but small invertebrates from the Coal Measures black shale of Danville, Illinois.

Brackish-water and especially deep-sea shelled animals tend to have thin shells, while increase of salinity tends towards the thickening and roughening of the calcareous shells. It is a well known fact that in the dolomite-depositing continental seas like that of the Guelph (Siluric), all of the molluscs have ponderous thick shells. These have been interpreted as reef-living species but actual reefs in the Guelph are unknown. The molluscs are often common but corals are represented by but a few species. Similar conditions are known to occur in other dolomite faunas. Further, the Guelph was of a time of decided progressive emergence and restrictional seas under an arid climate, and therefore the waters must have been abnormally salty.

Rivers constantly discharge into the sea great quantities of plant material, but as a rule little of it other than the wood is swept far out to sea. At present the rivers of northern Siberia float into the sea vast numbers of logs that drift with the currents to Spitzbergen, East and West Greenland and Arctic America. This wide dispersal of wood by the sea is met with only in the cold regions, whereas in tropical waters the wood is rapidly decomposed. Single leaves are rarely transported far from their place of origin, and when of good preservation in geologic deposits, give decisive evidence of the nearness of the shore. On the other hand, tough palm leaves have been seen in the sea 70 miles from land and rafts of leaves are often met with 200 or more miles beyond the mouths of the Kongo and the Amazon. Proximity to shore is also indicated by the presence in marine faunas of land molluscs, insects and bones of land vertebrates.

With tillites now known in the Lower Huronian of Canada, in the Lower Cambric of northern Norway, China, South Africa and Australia, and in the Permic of India, South Africa, Australia and Brazil, we observe the recurrence of glacial climates. The Siluric and Devonic coral reefs occurring in Arctic regions, the sponge, coral and bryozoa reefs in the Jurassic of northern Europe, the rudistid and other cemented pelecypods in reefs of wide distribution in the Cretaceous, and the almost world-wide distribution of the Nummulitidæ (north of Siberia) in the late Eocene and Oligocene point as clearly to warm waters and mild polar climates. Further the widely distributed Carbonic foraminifers of the family Fusulinidæ that swarmed in temperate and tropical regions are unknown to Arctic and Antarctic regions. In other words, long before we have a fossil record the earth had climatic zones, and for long periods the climate was mild to warm, punctuated by shorter intervals of cold to mild climates.

The volume of sea water to-day is very great, but we must ask ourselves: Has this quantity always been such or was it even greater, as some geologists still hold? We no longer agree with Laplace and Dana that the earth passed through an astral stage, but rather agree with Chamberlin that it always has had a more or less cold exterior. Through volcanic activity much juvenile water from the interior of the earth was extruded in geologic time and was added to the vadose waters of the surface. Suess states that "the body of the earth has given forth its oceans and is in the middle phase of its gas liberations." Accordingly, the Paleozoic oceans must have been quantitatively smaller than those of the present, and the gradual increase in the volume of vadose waters has been accommodated by the periodic increase of oceanic depth.

We also agree with Walther that the oceans of Paleozoic and earlier time did not have the great abyssal depths they now have. The accentuated deepening of the permanent oceanic basins did not begin until the Triassic, for in none of the great depths of the present oceans are found traces of Paleozoic organisms, and all here are of Mesozoic or Tertiary origin. In the shallow regions, however, are still found a few Paleozoic testaceous-bearing genera of brachiopods, tubicular annelids, pelecypods, gastropods, *Nautilus,* and *Limulus.* The deepening of the Pacific, the Indian, and especially the Atlantic oceans has been at the expense of the lands or horsts, for the ancient continents, Gondwana and Laurentia, have each towards the close of the Mesozoic been broken into several masses. We may therefore speak of permanent oceans, and transgressed, fractured, and partially down faulted, continents or horsts.

These are some of the factors that control the making of some of the modern paleogeographic maps.

Part II

DISPERSALS AND TAXONOMIC RESEMBLANCE OF FAUNAS AND FLORAS

Editor's Comments
on Papers 6 and 7

6   **GREGORY**
    The Relations of the American and European Echinoid Faunas

7   **SCHUCHERT**
    On the Faunal Provinces of the Middle Devonic of America and the Devonic Coral Sub-provinces of Russia, with Two Paleogeographic Maps

Papers 6 and 7 are early comparisons of some North American marine invertebrate faunas with those of Europe (which at that time were much better known). In Paper 6, Gregory points to the problems that phylogenists and taxonomists have when assuming one particular set of isolating mechanisms in contrast to another set. Thus, in the minds of the taxonomists, the availability of a dispersal route, or the apparent lack of a dispersal route, may have had a conscious (or subconscious) influence on how they constructed their taxonomies. One extension of this thinking was the commonly used criterion of distance between fossil collections as a basis for recognizing a new species—500 miles being a frequent yardstick! Hopefully, we now have a better appreciation of species ranges and morphologic (and ecologic) clines.

Paper 7 is an early discussion by Schuchert comparing the Devonian coral subprovinces in Russia with assemblages of Devonian fossils from different parts of North America. It is a good summary of Lebedev's 1902 article about the Russian fauna, which was written in Russian but supplied with an extended German abstract (Schuchert's first language). The discussion of the North American Devonian fauna is less clearly presented, and in this volume only a precis of pages 144–148 and 151–162 of the original article is included. This seems to be Schuchert's first attempt at a paleobiogeographic analysis of the marine invertebrate fauna of a geologic period and indicates the exploratory nature of such endeavors at that time. It is significant that relatively few marine invertebrate paleontologists undertook this type of study until nearly thirty years later.

6

Reprinted from *Bull. Geol. Soc. America*, **3**, 101–108 (1891)

THE RELATIONS OF THE AMERICAN AND EUROPEAN ECHINOID FAUNAS

J. W. Gregory, F.G.S., F.Z.S.
The British Museum of Natural History

Contents.

INTRODUCTION.

Probably every paleontologist who lives on the western border of the great galearctic province occasionally chafes against the limitation which the Atlantic places upon our knowledge of the origin or derivation of successive fossil faunas. In many cases researches on the paleontology of central and eastern Europe have given the desired information as to the origin of a British or western European fauna; but in other cases groups of genera and species appear suddenly in a certain zone and as suddenly disappear. The probabilities in such cases are in favor of the migration of these forms from some western area. If the species in question possessed a great range, either in depth or of latitude, they present no especial difficulty; if their bathymetrical distribution was or appears to have been great, they may have come directly eastward; if they were spread over a wide area or were boreal forms, they may have worked their way around the shallow waters of the northern margins of the Atlantic. But there are cases that cannot be thus easily explained. The genera in question may be shallow water and tropical forms to which the deep and cold abysses of the Atlantic would present as insuperable an obstacle as an actual land barrier. If, as seems most probable, these forms did come from the west, how did they cross such a barrier, or was it in existence at that time? To solve the difficulties presented by such cases, many geologists have sought to give a scientific basis to the legends of the fabled Atlantis, and have called a new world into existence in the mid-Atlantic to explain the difficulties of paleozoological distribution in the old world: but, on the other hand, a school composed mainly of zoologists have adopted a more aquatic attitude by accepting the theory of the permanence of oceans and continents, which leaves these difficulties unexplained. Certain physical arguments have been adduced in support of this view, but they do not seem of any great value, and the whole question seems to turn on zoological, and especially on paleontological distribution. If the Atlantic has been permanently a deep ocean basin no such littoral tropical forms could have entered Europe from the west except during periods when the arctic area enjoyed a temperate climate, and a theory which postulates a series of such warm periods would be unsatisfactory even if there were not evidence in some cases against the " northwest passage."

The question is one of some importance to workers in most departments of paleontology. The phylogenist who accepts the theory of the permanence of oceans and continents is likely to train the branches of his phylogenetic tree along very different lines from those that would be preferred by one who admitted the possibility

67

of occasional direct intercourse between the southern palearctic and nearctic faunas. To the geologists and paleontologists who try to trace the origin and migrations of extinct faunas and their evidence as to the physiography of the past, the question is also of primary importance.

The evidence that would be most conclusive now, of course, lies buried beneath the Atlantic, and the paleontologist has to turn to America to see whether he can trace among its fossils the origin of any of the constituents of the old world faunas, and, if so, to see if he can discover when they entered the European area and by what route they traveled.

Any comparison of the European and American faunas that might be made with this end in view must be conducted with greater care than it would be possible for any one paleontologist to give to the whole of the evidence. A mere examination of lists of species is quite inadequate. Hence probably more reliable data can be gained from the detailed study of one group than from an attempt to handle all the available evidence; at least, this is all the present writer can attempt. The echinoidea offer especial advantages: the bathymetrical range of the species is fairly restricted; the deep-sea forms are very easily distinguished; the adults at least, and in some cases the young, are practically non-migratory; the echinoids are mostly tropical or temperate in habitat; they occur in abundance from the Carboniferous to the present; and, finally, as their classification rests upon the hard parts, their affinities can be more definitely decided than in the cases of most other classes. Hence in this paper attention is restricted to the echinoidea. It must, however, be admitted that conclusions based on one class alone are likely to be modified when the evidence of all the other groups is worked out. The final conclusion will probably be the mean of the results given by the independent study of the different divisions of the animal kingdom.

THE CARBONIFEROUS FAUNAS.

Neglecting the problematical Silurian and the rare Devonian echinoidea as giving no adequate data for comparison, it is with the Carboniferous system that the species become sufficiently numerous to form definite faunas.

In Mr. S. A. Miller's useful "Catalogue of North American Paleozoic fossils" we find a fairly long list of Carboniferous echinoidea. Deducting one or two synonyms, the list stands as 41 species and 10 genera, to which must be added several new species recently described and several undescribed forms that occur in the American museums. Of this fauna of about 50 species, not one representative occurs in Europe. It is true that 20 of these belong to the genus *Archæocidaris*, and most of them have been based on spines and isolated plates; and that while the discovery of better material would probably reduce the number of species, it might at the same time demonstrate the identity of some of them with European forms; but at present I feel bound to admit that I have seen no evidence of the existence of any one Carboniferous echinoid on both sides of the Atlantic. The comparison of the genera is still more valuable and brings out a great difference between the two faunas. Of the ten American genera only three occur in Europe, viz, *Archæocidaris*, *Palæchinus*, and *Perischodomus*.* The other seven genera are peculiar to

---

* *Eocidaris* may seem an additional genus, but the European species referred to it really belong to *Cidaris*, and the name has been abandoned as a synonym. The specimen described by Vanuxem as *Eocidaris drydenensis* proves to belong to a very different genus. The type is now in the New York State museum at Albany.

North America. In the same way three of the six European Carboniferous genera are peculiar to the Eurasian area. The difference between the two faunas is thus extremely marked, and clearly shows that there was no close connection between the echinoids of the two areas. The absence from Europe of the great family of the *Melonitidæ* is especially striking.

Permian-Jurassic Faunas.

After the Carboniferous system the next fauna of any special value is in the Cretaceous. The Permian of both continents yields a few species, but not sufficient for any definite comparison. The paucity of species in the American Jurassic is also disappointing, as the European echinoids of this age are so exceptionally well known. Descriptions of several species by Professor Clark are now passing through the press and serve to encourage the hope that more may be discovered. As yet, however, the few species known are not sufficient for comparison with the European faunas.

The Cretaceous Faunas.

The Cretaceous system yields much evidence which has been admirably summarized by Professor W. B. Clark in a " Revision of the Cretaceous echinoidea of North America," * issued as a preliminary notice to his forthcoming monograph. In this he enumerates 43 species belonging to 19 genera; in addition to this are the 7 species described by M. Cotteau from Mexico, including representatives of two other genera; some new species found by Professor Clark; and a species of *Linthia* in the museum of the Boston Natural History Society, which, so far as one can judge from the brief diagnosis of *Linthia tumidula*, appears to be new. There are also several more species from South America and the West Indies; the former, however, closely resemble the Mexican species, and the latter are a rather isolated group and may be neglected.† The Cretaceous echinoids of the mainland of North America may therefore be estimated at about 55 species, distributed among 25 genera.‡

If this fauna be examined as a whole it presents a very familiar facies to a European echinologist. Only one genus occurs that is not also found in Europe, while several species are common European forms; but if we separate them into their successive faunas we find one interesting point brought out—*i. e.*, that the members of the earlier faunas agree more closely with the trans-Atlantic species than do those of the upper beds, such as of the Yellow limestone of New Jersey. This is especially well shown by the small fauna described by M. Cotteau from Mexico. This yields six good species, of which three are characteristic of the European lower Cretaceous (Aptien and Urgonien), viz, *Diplopodia malbosi, Salenia prestensis*, and *Pseudocidaris saussurei*. The *Enallaster texanus*, moreover, is not unlike some European species, and only the form upon which the late Professor Duncan founded the genus *Lanieria* is quite distinct. The identification of these species rests on the authority of M. Cotteau; his opinion is of especial weight, as the general impression

---

* Johns Hopkins Univ. Circ. no. 86, 1891.

† The best known of the South American species is the *Enallaster karstein* from Ecuador, described by M. de Loriol. An examination of the type of *Spatangus columbianus*, Lea, now in the museum of the Academy of Natural Sciences in Philadelphia, shows that they are identical, and it must therefore be known as *Enallaster columbianus* (Lea).

‡ The following is the list of those recognized in addition to those mentioned in Professor Clark's " Revision: " *Stereocidaris, Diplopodia, Coptosoma, Lanieria*, and *Cardiaster*.

of his work seems to be that he is inclined to limit specific variation within much narrower limits than do many workers on the echinoids. In the larger faunas from the upper Cretaceous, as in that from New Jersey, the whole of the species are peculiar to America, and in most cases the species are quite distinct from their European representatives. The abundance and variety of the species of *Cassidulus* is the most striking feature in this upper Cretaceous fauna, and they are all quite distinct from the European species. Dr. Clark does not admit one species as occurring in the eastern hemisphere (excluding, of course, those described by M. Cotteau), and, so far as I have been able to examine the American collections, I am inclined to agree with him except, possibly, in the case of *Holaster simplex*, Shum. (*H. comanchesi*, Marc.), from the Comanche series of Fort Worth, Texas. There are two good specimens of this species in the American Museum of Natural History, New York. These seem to be indistinguishable from the European *H. lævis* (De Luc), a very variable species in which several well characterized varieties are recognized. The same variations seem to occur in the American forms, and one of the two is our *H. lævis*, var. *trecensis*, the other being *H. lævis*, var. *planus*. Other species from the Comanche series are very different from the European ones—*e. g.*, the *Goniopygus zitteli*, Clark, and *Holectypus planatus*, Roemer. The latter is an interesting species, as its ornamentation rather resembles that of the Jurassic forms. The resurrection of the fifth genital pore is also noteworthy, as it happens in Europe in some allied genera of the same age.

Hence in the American Cretaceous echinoidea we find the relations to their European representatives to indicate that the two faunas were very closely allied in the lowest Cretaceous, but that in later periods of this age the two faunas developed on independent lines. The evidence of this system is of especial value, as in Europe there is practically a complete series of echinoid faunas from the Valangian to the Danian, and thus the difference between these and the upper American faunas cannot be ascribed to differences of age. The New Jersey Middle marl fauna must be not only homotaxial but synchronous with some of the echinoids between the Gault and the upper Chalk.

EOCENE AND OLIGOCENE FAUNAS.

A list of the paleogene echinoids from the United States, copied from existing literature, would give but a poor idea of the composition of this fauna or of its affinities. The whole group is in urgent need of revision, and it certainly does not seem a sparse one. Thus, the collection of the American Museum of Natural History includes species of *Sarsella*, *Euspatangus*, and *Breynella*,* none of which have been previously recorded from America. The Smithsonian Institution collections also add the genera *Cidaris* and *Echinarachnius*, and the Academy of Natural Sciences the genus *Monostychia*.

The most striking feature in the echinoid faunas of these two systems is the predominance of the group of flat clypeastroidea, belonging to the genera *Mortonia*, *Periarchus*, *Echinanthus* (Leske non Breynius), *Scutella* and *Echinarachnius*, and of the numerous species of *Cassidulus* and *Pygorhynchus*. The great series of spatangoids found in the European Eocenes are hardly represented. The abundance of the two last genera mentioned is of interest, as they were common forms in the Ameri-

---

*The *Echinanthus* of MM. de Loriol and Cotteau, but not of Alexander Agassiz and other American authors. See a discussion of this question in a paper, now in the press, by the present writer, on the Maltese echinoids, in the Trans. Roy. Soc. Edinb.

can Cretaceous. It therefore appears that the gradual differentiation of the echinoids of the two areas, which commenced in the Cretaceous, had gone on until the faunas appear strikingly different.

Until a detailed revision of the American Eocene species has been undertaken it is perhaps not advisable to carry the comparison further; but the following notes on the synonyms of a few of the species appear necessary in order to render intelligible the use of some of the above-quoted generic terms. This is especially necessary in the case of the genus *Mortonia* and its allies. This genus was founded by Desor in his "Synopsis des echinides fossiles." The diagnosis was well drawn, obviously from specimens. The only species given was named *M. rogersi*, and a reference given to Dr. Samuel Morton's figure of *Scutella rogersi*. This was unfortunate, as Morton's species is a true clypeastroid, with twinned margins, and belongs to the genus *Echinanthus* (Leske non Breynius). The species which Desor actually described was the *Scutella quinquefaria* of Say. Desor's mistake has led to great confusion, and the names are applied very differently in different American collections. In many cases *Mortonia* is regarded as synonymous with *Periarchus*, but this genus seems worthy of recognition. The type species is *S. altus*, Conrad, but I have not been able to see the type of this species. The common species, *S. pileus-sinensis*, is, however, a good example. The names, therefore, accepted by the writer for this group are:

Mortonia rogersi, Desor non Morton.
Echinanthus quinquefaria (Say).
Periarchus altus (Conrad).

Another thin, flat form, in which a change of nomenclature seems necessary, is the *Sismondia marginalis*, Conrad. The type of this is in the Academy of Natural Sciences, and with its smaller ally, *S. plana*, Conrad, must be transferred to *Monostychia*.

THE MIOCENE FAUNAS.

The Miocene echinoid fauna of the mainland of America is numerically smaller than that of the Eocene and Oligocene, but it gains considerably in size if the West Indian species be included. Most of the echinoidea described by Ravenel and Tuomey from South Carolina, and referred by them to the Pliocene, must also be referred to the Miocene. On the other hand, some species from the western states usually referred to this system seem to be Pliocene or Pleistocene, and are the common living species; thus some of the specimens labelled *Scutella striatula*, Rem., really belong to the living *Echinarachnius excentricus*. Some of the species referred to the West Indian Miocene seem also to be of later date, such as the *Rhynchopygus guadaloupensis*, Mich., a synonym of *R. caribbaearum*.

Taking, then, the Miocene echinoid fauna with these additions and restrictions, we find it to present a remarkable resemblance to the Miocene echinoids of the Mediterranean basin. This resemblance is established (1) by the presence of several species common to the two faunas—*e. g.*, *Cidaris melitensis*, *Schizaster parkinsoni*, and *Schizaster scillæ*; (2) by the fact that other genera are represented by closely allied species, as in the case of the Maltese and Jamaican species of *Heteroclypeus*; and (3) by the presence in both of genera with a very restricted distribution—*e. g.*, *Agassizia*.

Professor Alexander Agassiz, in his interesting account of the origin and affinities of the long existing West Indian echinoid fauna, has argued that the fact that so many of the genera are represented by equivalent species on the two sides of Centra America is clear proof of the former connection between the waters of the Antillean and Panamaic regions; but the resemblance between the echinoidea of these two provinces seems to be less close than is that between the Mediterranean and West Indian Miocene. No one species of echinoid is common to both shores of Central America, and the representative species are often more distinct than those of the two Miocene faunas. Hence if Professor Agassiz is justified in his conclusion of the common origin of the Antillean and Panamaic echinoidea, then so also must the Antillean and West Indian Miocene faunas have been derived from a common source. And just as it is considered to prove in the one case a depression of Central America which brought the waters of the Pacific and the Caribbean into connection, so in the other case we must assume a period of elevation which produced a band of shallow sea across the mid-Atlantic. Whether it be assumed that the fauna originated in the Mediterranean and migrated to the West Indies, or vice versa, or whether it developed in some area in the Atlantic now deeply submerged, this shallow water connection is essential.

But there are two explanations that might be proposed that could not involve any such complete opposition to the theory of the permanence of the ocean basins. It might be urged (1) that the common element in the two faunas worked its way around from the one area to the other along the shallow northern shores of the Atlantic; or (2) that the connection was established by the free-swimming larval forms. But we are not without evidence against both of these hypotheses. If we follow the Echinoid fauna of the Helvetian (middle Miocene) from its typical development in Egypt, Malta, Sicily, and Italy toward the north we find at the most northerly area in Brittany that though a considerable series of echinoids remain, the group of species and genera which ally the Mediterranean to the West Indian fauna has completely disappeared. It is just the same in America; the Miocene of South Carolina has yielded none of the same group, which is replaced by species of *Mellita, Encope, Echinocardium*, etc. This fauna has resemblances to the West Indian, but it is by an element not typically represented in the Mediterranean. Thus, on both sides of the Atlantic the evidence seems fairly conclusive that the migration did not follow the northern route. But we are fortunately not compelled to rely on negative evidence alone. In the Azores, in Madeira, and in the Grand Canary there are Miocene beds which have yielded a small echinoid fauna; in each case the species when determinable are found to be those characteristic of or close allies to the Mediterranean Miocene; in some cases the species are represented by the same varieties. This is, of course, proof only of the original extension of the Mediterranean fauna as far west as the Azores, but this is a very considerable step across the Atlantic; and some West Indian forms, as *Temnechinus*, occur elsewhere only at the Azores, and thus serve to show the completion of the bridge.

In regard to the second hypothesis explaining the connection by the free-swimming larvæ it may be objected that the chances of so delicate an organism as a pluteus surviving the journey across the Atlantic must be somewhat remote, and the species would have no chance of establishing itself unless a number of the plutei arrived simultaneously at a suitable locality. I do not remember that the Challenger surface nets ever collected any plutei of a littoral species in mid-ocean. But here again we are fortunately not left to decide on mere probabilities such as these. Many living echinoidea are now known to be viviparous and to have no free-swimming stage. Now *Schizaster parkinsoni* has in a very marked degree all the characters of a viviparous form, while *Schizaster scellæ* was probably the same. The occurrence therefore of these species in both the Mediterranean and Antillean faunas is quite sufficient of itself to demonstrate the inadequacy of any explanation based on the passage of the pluteal forms; some of the forms that crossed the Atlantic had an abbreviated development without any pluteal stage.

THE PLIOCENE FAUNAS.

In the Pliocene period the echinoidea are scarcer and less well known than in the Miocene, and now that most of the species described by Ravenel and Tuomey have been transferred to the earlier division no very definite fauna is left. In fact on the mainland there are only a few recent species, such as *Mellita sexforis* from Carolina and *Echinarachnius excentricus* (syn. *Scutella striatula*, Rem. non Marc. de Serres) from the Pacific slope. The collections of the Academy of Natural Sciences of Philadelphia and the Smithsonian Institution also contain some specimens of the living *Echinanthus reticulatus*, Linn. sp. (*sensu* Lovén; the *Echinanthus*—or *Clypeaster—rosaceus*, Auct.) from Coloosahatchie, Florida. These, however, seem to be all recent species, whereas in the European Pliocene but few living species are represented. The few echinoids from beds of this age in the United States have no particular affinities with the European ones.

There are, however, two species of echinoidea from deposits in the West Indies that may be referable to this age, and which cannot be overlooked, as they have important bearing on questions of physical geography. They are *Cystechinus crassus*, Greg., and *Asterostoma*, n. sp., both from the Radiolarian marls of Barbados. The geological bearing of the discovery of such a typically deep-sea genus as *Cystechinus* was referred to at the time of its description, but it has gained considerably in interest by the recent work of Professor Agassiz. At the time of the discovery of the Barbados specimen the genus was only known from the Antarctic and the China sea. It has now, however, been dredged by Professor Agassiz in deep water off the western coast of Central America, but the species is so far known only by the few remarks made about it by Professor Agassiz in his preliminary report on the results of the cruise; yet as far as we can judge from these it is closely allied. The species of *Asterostoma* is of interest from the light it throws on the age of the beds in Cuba, from which the original specimens of this genus were derived, from their resemblance to *Echinocorys* (*Ananchytes*). M. Cotteau referred them to the Cretaceous, but the discovery of this Barbadian specimen renders it highly probable that they should be transferred to the upper Cenozoic.

The paucity of American Pliocene echinoidea is to be regretted, as those of this age in Europe have been in most cases carefully collected and monographed. With the few Pliocene echinoids from America they have nothing in common; but as the writer has pointed out in a recent "Revision of the British fossil Cenozoic echinoidea," those of the English Crag have many affinities with the existing fauna of the West Indies. The Crag echinoids number 22 species, and may be divided into two groups: (1) the common northern European forms, or species closely allied to these; and (2) a group of genera represented together elsewhere only in the West Indian area. Thus, in the English Crag there are species of *Temnechinus*, *Agassizia*, *Rhynchopygus*, and *Echinolampas*, of which the nearest allies are Caribbean species. Now, these are all either tropical or littoral forms, and it is of interest to note that they do not occur elsewhere among the European Pliocene deposits. The fauna which agrees best with that of the English Crag (excluding the few patches of Pliocene sand in northern France) is that of Belgium. This, however, contains but two British species, though as a rule the species are allied; the main difference consists in the presence of some Mediterranean species and the absence of the four genera of the western group. The richest of the Belgian beds is the Diestian, which is older than our Coralline Crag. This, therefore, suggests that the "western group," as we may call the second element in the Crag fauna, did not reach Europe until post-Diestian times, and thus did not penetrate so far east as Belgium.

In this case the same suggestions as to the possible northern migration or the floating across of the larvæ might be made, and there is less evidence on the subject than in the Miocene. The only well-known species of *Temnechinus* from the Crag (*T. woodi*, Ag.) was probably viviparous, and it may be that the West Indian species is so also; otherwise there is no evidence to directly disprove this second hypothesis. As there is no known European Pliocene fauna north of the Crag, and as the Pliocene series from the American mainland is also very scanty, there is no such means of disproving the northern extension of these tropical or subtropical forms; but had this happened we might have expected a much greater mingling of the faunas of different zones of latitude than has happened. The echinoidea of the European shore agree more closely with those of the corresponding isotherms on the American side than with the faunas north and south of them. The presence of *Temnechinus maculatus* at the Azores as well as in the West Indies also further suggests that the connection was established somewhere in the mid-Atlantic.

SUMMARY OF CONCLUSIONS.

A brief comparison of the successive echinoid faunas of Europe and America has thus been attempted, and it may be advisable briefly to summarize the conclusions arrived at.

In the Carboniferous period there was an almost complete difference between the two faunas, whereas in the succeeding Urgonien and Aptien the two faunas are almost identical. But the Cretaceous period was marked by a gradual differentiation; species ceased to be common to the two areas, and the representative forms became more distinct. In the Eocene and Oligocene the same independent evolution seems to have gone on; the American fauna was rich in species of *Cassidulus* and *Pygorhynchus*, genera also common in the Cretaceous beds of the same continent, and the faunas were more distinct than were the Cretaceous. During the Miocene there was again a change: a fresh connection was established that enabled the echinoidea of corresponding latitudes in the new and the old worlds to commingle; and later still, in the Pliocene, there is evidence to show the introduction into the European area of some American echinoids. The possibilities of this connection across the Atlantic by free-swimming larvæ or by the adults having worked around the northern margin have been examined and evidence adduced against them, and one case is quoted in which the dissimilarities of fauna cannot be explained as due to difference of age.

It is therefore urged that the comparison of the succession of the echinoid faunas of Europe and America present a series of phenomena wholly incompatible with the theory of the permanence of the great ocean basins.

Remarks were made upon the topic of the paper by Mr. L. C. Johnson.

7

Reprinted from *Am. Geologist*, **32**, 137–143, 149–150 (Sept. 1903)

ON THE FAUNAL PROVINCES OF THE MIDDLE DEVONIC OF AMERICA AND THE DEVONIC CORAL SUB-PROVINCES OF RUSSIA, WITH TWO PALEOGRAPHIC MAPS.

Charles Schuchert

U.S. National Museum

The recent work on the corals of Russia by Lebedew* gives Americans their first opportunity for more or less detailed comparisons among the coral faunas of that Empire and those of Europe and America. In this paper will be given (1) an abstract of Lebedew's conclusions, (2) some observations by the present writer on the affinity of the coral subprovinces of America and Europe, and (3) extended remarks on the "American" and "Eurasiatic" provinces of North America.

Devonic corals are often of very wide geographic distribution and from this fact have considerable value as indicators for the intercommunication of seas and oceanic provinces. It seems almost certain that many of the world-widely distributed species are loosely defined and contain more than one form, and that many of the genera also lack precision. This, however, cannot militate against the general conclusions here drawn, since it still remains a fact that the forms so recognized are closely allied, if not identical.

Lebedew finds that the Devonic corals of Russia occurring in ten regions can be grouped into three subprovinces each having a more or less uniform faunal development. These are:

1. *The West European type,* embracing the areas of West Europe, Poland, Transkaukasia and Petchora-land.

2. *The Central Russian type,* embracing the area of the main Devonian field of Russia and Northwest and Central Russia.

3. *The Ural-Altai type,* including the Urals (also the Mugadzhar Mts.), Altai. West Siberia and Turkestan.

WEST EUROPEAN TYPE.

Poland.—In this country occur fifty-seven species of Devonic corals. Only eight of these are identical with or are repre-

---

* Die bedeutung der Korallen in den Devonischen ablagerungen Russlands [in Russian with an extended abstract in German]. By N. LEBEDEW. (*Mem. du Comite Geologique*, xvii, No. 2, 1902, pp, i-x, 1-130 in Russian, 137-180 in German, pls. I-V).

sented by closely related forms in Northwest and Central Russia, and five of these are very widely distributed forms.

Of the twenty-eight species found in the upper zones of the Middle and the entire Upper Devonic, but eight occur in Northwest and Central Russia. However, when comparison is made with Germany, "one finds that all the forms found in the various zones of Poland also occur in Germany." This holds true even to varieties.

The developmental aspect of the corals is also in agreement with the balance of the fauna, showing that the Northwestern and Central Russian Devonic is a distinct subprovince from that of West Europe. Lebedew therefore concludes that these areas were not directly connected. Such characteristic West European or Poland forms as *Cyathophyllum quadrigeminum* Goldfuss, the genus *Spongophyllum,* and *Calceola sandalina* are absent in Northwest and Central Russia.

Of the fifty-seven species of corals of Poland, one finds that but ten are analogous with American forms, and further, that the species characteristic of one area are absent in the other. A coral of special interest in this connection is *Calceola sandalina* which is absent in America but is known in the Urals and even in the Altai.

"The character of the coral faunas of the Devonian of Poland is typically West European in its essential characters as: (1) the scarcity of corals in the Lower Devonian, (2) the preponderance of massive forms of *Favosites* in the Lower, and branching species in the Middle and Upper Devonian, (3) the great abundance of operculate corals of the families *Calceolidae* and *Cystiphyllidae* in the lower zones of the Middle Devonian, (4) the great horizontal distribution of *Calceola sandalina* in the Middle Devonian, (5) the reappearance of *Acervularia* and *Phillipsastraea* in the upper zones of the Middle and lower beds of the Upper Devonian, and (6) the absence of corals in the upper horizons of the Upper Devonian." (p. 159.)

Transkaukasia.—But one coral horizon has been clearly determined although there are more. This is the zone of *Calceola sandalina.* Besides this species it has *Favosites, Alveolites, Heliolites, Cyathophyllum, Endophyllum, Mesophyllum, Phillipsastraea* and *Cystiphyllum.* "Nearly all of the forms are common also in the Devonian of West Europe." (p. 175.)

Petchora-land.—Here the Middle Devonic has eleven species and the Upper Devonic twelve, or together seventeen forms. The faunal aspect is that of West Europe and not that of the Urals. This is shown by the occurrence in both regions of rare forms, as *Cyathophyllum minus* and *C. kunthi.* These species have much stratigraphic significance in Germany. A more decisive connection between the coral faunas of Petchora-land and West Europe is shown by the presence of *Campophyllum* in the Upper Devonic; also for the northern region of Russia but not for the Urals where another species of the genus is found. Further faunal connection is shown by the presence of *Acervularia* and *Phillipsastraea. Calophyllum,* always rare in species, is represented by one form and is further proof "that there was close connection between the Devonic basin of West Europe and that of Petchora-land." The next nearest affinity is with West Siberia and the Altai.

CENTRAL RUSSIAN TYPE.

The main Devonic field of European Russia.—This region is poor in corals, though certain species make considerable reefs in the upper zone of the Middle Devonic. Here nine species are known and, as is to be expected, five of these also occur in Central Russia. These are *Aulopora serpens, A. tubaeformis, Favosites cristatus. Cyathophyllum caespitosum* and *C. hexagonum.* The other species are *Aulopora orthocerata, Chaetetes intricatus, Favosites reticulatus* and *Strombodes.*

The characterizing feature of the coral faunas of North and Central Russia is the absence of *Coenites, Striatopora, Phillipsastraea, Acervularia* and *Endophyllum.*

Central Russia.—Here twenty-seven species are known. These are mainly from two horizons, *i.e.,* fifteen species from the upper beds of the Middle Devonic and the same number from the uppermost zone of the Upper Devonic. The latter is strongly tinged with Carboniferous species as ten of the fifteen also pass upward.

Of the fifteen Middle Devonic species, twelve also occur in West Europe but only five in Poland and these are cosmopolitan forms. This area has *Favosites* 1. *Chaetetes* 1, *Syringoporidae* 5, *Cyathophyllidae* 8.

Note the absence here of *Cystiphyllum, Acervularia* and *Phillipsastraea,* forms widely distributed in the Middle and lowest Upper Devonic.

There was no direct connection between West Europe and Central Russia. One American species occurs here.

Northwest and North Russia.—In this region but nine species are known in the upper part of the Middle Devonic and the lower zone of the Upper Devonic, but in places they form reefs of considerable extent. These belong to the families *Favositidae, Chaetetidae, Syringoporidae, Cyathophyllidae* and *Cystiphyllidae.*

The corals of this region have a decided relationship with those of Central Russia. Compared with those of the Timan region there is a great difference and the conclusion is warranted that no direct connection existed between these areas during Devonic time.

THE URAL-ALTAI TYPE.

Urals.—Here we have sixty-six species distributed over a great area extending from north to south. The *Favositidae* and *Cyathophyllidae* are most abundant. The great coral horizons are in the lower and upper zones of the Lower Devonic, and the middle and upper beds of the Middle Devonic.

The lowest Lower Devonic is characterized by *Favosites* (some Silurian species) and *Cystiphyllidae.* The upper zone of the Lower Devonic has such characteristic species as *Cyathophyllum caespitosum, C. ceratites,* etc.

Comparing the corals of the Urals with those of West Europe and America, it is seen that the lowest Lower Devonic species are of the West European type and that in the upper zones of the Lower Devonic the first distinctly American species are met, as *Favosites forbesi* and *Syringopora hisingeri.* The distinctly American species increase in number in the Middle Devonic where occur *Favosites placenta, F. nitellus, Syringopora perelegans, S. nobilis, S. tabulata* and *Emmonsia hemispherica.* In the Upper Devonic, corals are rare and the data for comparison are not satisfactory.

"We recognize that the Devonian corals of the Urals have in general a great resemblance to those of West Europe, although here also appear strange elements as local and American forms. These occurrences seem to show that between the Devonian of the Urals and that of North America there was a less decided connection than between the former and West Europe." (p. 167.)

Lebedew's extensive tables show that in the Middle and Upper Devonic faunas of the Urals there are no *Acervularia, Endophyllum, Mesophyllum, Spongophyllum* or *Diphypnyllum* and but one *Phillipsastraea,* which occurs in the basal Middle Devonic. In this respect these Ural faunas have the characteristics of the Central Russian type, and it is strange that Lebedew does not call attention to this decided affinity.

Mugadzhar Mts.—These mountains are directly south of the Urals and terminate towards the Aral sea. But ten species are known here. In the upper Middle Devonic occur *Alveolites* 1, *Cyathophyllum* 2, *Acervularia* 2, *Phillipsastraea* 3. The author remarks that the faunal aspect is rather that of the Altai and West Siberia than that of the Urals.

Turkestan.—The corals of this area are too few in number and are otherwise not sufficiently significant to indicate faunal relationship with the other areas.

Altai and West Siberia.—Here are known sixty-six species. Of these "twenty-five are common to the Urals, eight are identical with or very closely related to American species, while the greater part of the remainder, with the exception of local forms, recur in West Europe. In general, the corals of the Altai and West Siberia resemble those of the Urals (and the Mugadzhar Mts.) with respect to (1) relationship, (2) the preponderance of West European forms, and (3) the noticeable recurrence of American species. In connection with this it is to be noted that here, as there, the American species are found in the same homotaxial horizons: they concentrate themselves in the main in the adjoining horizons of the Lower and Middle Devonian, while the lower zone of the Middle and the entire Upper Devonian are free of them." (p. 172.) The absence of American species in these formations he thinks is probably due to differing geological horizons in the two areas.

Decidedly characteristic for the Urals (including the Mugadzhar Mts.), Petchora-land, and the Altai (with West Siberia), appear to be certain corals that occur only in these regions. Such are *Alveolites goldfussi* Billings, *Heliolites porosus* Goldfuss, *Cyathophyllum hypocrateriforme* Goldfuss, *Acervularia pentagona* Goldfuss, *Phillipsastraea bowerbanki* Goldfuss, *P. annas* Goldfuss, *Cystiphyllum vesiculosum* Goldfuss. *Phillipsastraea* and *Acervularia* are characteristic of the upper horizon of the Middle and the lower zone of the Upper Devonic, and their occurrence in numerous species is a feature that marks and distinguishes the three above named regions. The relationship of the Altai is also greater with the Mugadzhar Mts. than with the Urals.

In the Middle Devonic of the Altai-West-Siberia region occur *Favosites* 7, *Coenites* 1, *Striatopora* 1, *Roemeria* 1, *Alveolites* 3, *Heliolites* 1, *Aulopora* 2, *Syringopora* 1, *Amplexus* 2, *Calophyllum* 1, *Cyathopaedium* 1, *Cyathophyllum* 5, *Mesophyllum* 2, *Spongophyllum* 2, *Phillipsastraea* 3, *Cystiphyllum* 1, *Calceola sandalina.*

MIXED TYPE.

North Siberia.—The corals of this region are of species that have "immense vertical and horizontal distribution, as *Favosites cristatus, Alveolites suborbicularis, Aulopora serpens, Cyathophyllum caespitosum* and *C. hexagonum.*" Baron Toll, after a review of the entire fauna of North Siberia, concluded it to be a mixture of Ural and American types and of the Stringocephalus horizon. "It is worthy of remark that the deposits of the Stringocephalus horizon, which in West Siberia do not show any decided affinity with the same deposits of America, have, farther to the east in the region explored by Toll, forms in common to both these areas."

SUMMARY.

West European type.—The Middle and Upper Devonic coral faunas of this subprovince are marked by an abundance of *Cyathophyllum,* massive and branching *Favosites, Cystiphyllum, Acervularia, Spongophyllum, Endophyllum, Mesophyllum, Phillipsastraea, Alveolites, Campophyllum* and *Stromatopora,* a less abundance of *Pachypora, Striatopora, Heliolites, Aulopora,* and the presence of *Amplexus, Calophyllum, Coelophyllum, Metriophyllum, Hadrophyllum, Microcyclus, Pachyphyllum, Calceola, Chaetetes, Michelinia, Plagiopora, Roemeria* and *Syringopora.*

Central Russian type.—The Middle and Upper Devonic coral faunas of this subprovince are marked by a paucity of species. *Favosites* is represented by a few forms, while the genera *Coenites, Striatopora, Phillipsastraea, Acervularia* and *Endophyllum* are absent. The faunas consist essentially of *Syringoporidae* and *Cyathophyllidae.*

Ural-Altai type.—The Middle Devonic coral fauna of this subprovince in its western and eastern limits. is marked by an abundance of *Favositidae, Cyathophyllidae* and *Syringoporidae.* To these in the west (Urals) are added sparingly *Alveolites, Heliolites* and *Calceola sandalina* and an abundance of *Cystiphyllum.* In the southwest (Mugadzhar Mts.) are known only *Acervularia, Phillipsastraea, Alveolites* and *Cyathophyllum.* In the eastern region, besides the three families mentioned for the western area, there also occur in abundance *Mesophyllum, Spongophyllum, Phillipsastraea* and more rarely *Acervularia, Cystiphyllum, Calceola sandalina, Heliolites porosus,* etc. This area contrasts strongly with that of the western area. Another feature of the Ural region is the occurrence of American species, as *Favosites placenta, F. nitellus, Emmonsia hemispherica, Syringopora perelegans, S. nobilis* and *S. tabulata.*

An examination restricted to the corals of the Middle Devonic of the Ural Altai region seemingly decides that the Ural faunas have the Central Russian type. In all of these there is an abundance of *Acervularia, Phillipsastraea* and *Endophyllum,* while these genera are prominent in the other areas, as the Mugadzhar Mts., Altai Mts. and West Siberia.

The Faunal Provinces of the American Middle Devonic.*

In the Middle Devonic of North America there are probably more than 600 described species of corals,† and of these the writer has found that 106 occur in two or more widely separated places. These are listed in the appended table and their distribution noted in eleven columns.

The table shows clearly that the corals of the various Onondaga localities belong to one province. Of the 53 widely dispersed species of Mackinac, 39 are also found in New York or Port Colborne and 49 at Louisville. Together these localities have 67 species and of these but 6 occur in Iowa or Wisconsin. Five of the 6 species are cosmopolitan forms and are common to both the Onondaga and Hamilton formations.

---

* This paper is a continuation of "Seas and barriers in eastern North America." The broad problems involved in these researches the writer continues to discuss with Mr. Ulrich and desires to thank him for the great aid rendered.

† Louisville, Kentucky is said to have 438 species. Time will doubtless show that many synonyms exist.

[*Editor's Summary of Pages 144–148:* Based on the distribution of these Onondaga faunas and their general similarities, Schuchert constructs a map to show one biogeographic province (Plate XX). The succeeding Hamilton faunas indicate two biogeographic provinces, one in Missouri, Iowa, and Wisconsin and a second province in southern Illinois, Kentucky, Ohio, Ontario, New York, and Michigan. Schuchert then identifies an American or Mississippian Sea province (equivalent to Freck's North Helderberg Sea) and a Dakota sea, which has a fauna of the "Eurasiatic province" (Plate XXI; see the page following page 150). He concludes that the biological separation between these two faunas was effective.]

Plate XX

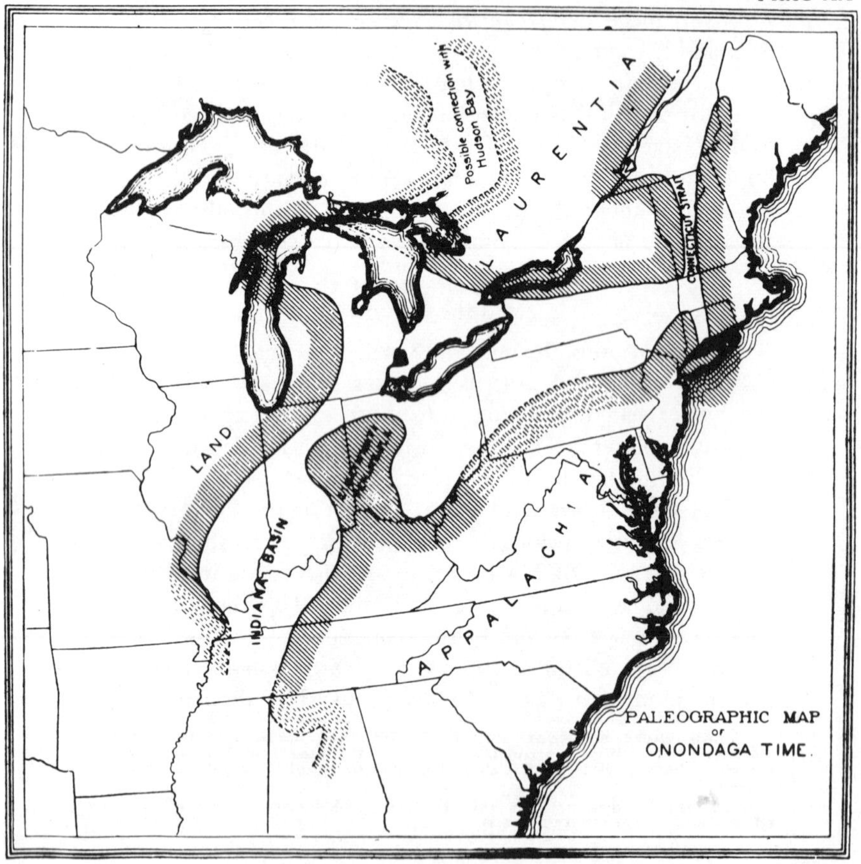

PALEOGRAPHIC MAP
or
ONONDAGA TIME.

Either Iowa with the adjoining areas was then land or the Kankakee axis was completely effective in separating the eastern and western seas.

Comparing the Mississippian Middle Devonic coral faunas with those of certain areas determined by Lebedew, it is seen that they have not the Eurasiatic facies. It is in vain that one seeks for such characteristic elements of the "West European type" as *Spongophyllum, Endophyllum, Mesophyllum, Campophyllum, Heliolites, Calophyllum, Coelophyllum, Metriophyllum, Pachyphyllum* and *Calceola.* On the other hand, a closer communication between Eastern America and Northwestern Europe is indicated by the fact that in these two regions are found in abundance massive and branching *Favosites,* some *Phillipastraea, Amplexus, Hadrophyllum, Microcyclus, Michelinia* and *Roemeria.*

The Dakota coral faunas occurring in western Michigan, Wisconsin, Iowa and Missouri and extending westward to the Pacific and northward from Arizona to the Arctic coast, are all of the Eurasiatic type. It is only in these western and northern areas of American Devonic provinces that many of the genera above noted as not found in eastern America occur, namely, *Spongophyllum, Endophyllum, Campophyllum, Heliolites* and *Pachyphyllum.*

The Devonic Eurasiatic faunas are now known in the Dakota sea from southern Minnesota south to Callaway and Moniteau counties, Missouri, north through Manitoba into the far north in the Mackenzie River basin and the Arctic coast of Alaska. In the "Cordilleran sea" (Walcott) these same faunas are found in southwestern Colorado, Bisbee and Rio Verde, Arizona (here occurs the Iowan Upper Devonic), White Pine and Eureka Districts, Nevada, and the Yellowstone Park; while in the "California sea" (Walcott) they also occur at many localities in northwestern California and southern Alaska.

The barrier that more or less effectively prevented the mingling of the faunas of the Mississippian and Dakota seas during Hamilton and Chemung times is here named the *Kankakee axis or peninsula.* To the writer it seems that during Onondaga time all of Iowa, Missouri and northern Illinois was land. After the beginning of Hamilton time, the Dakota sea invaded the area of these states and spread northward through *Traverse straits* (taking its name from Traverse formation) along the western side of the Kankakee peninsula into northern Michigan where it came into more or less unrestricted communication with the Mississippian sea. The general trend of the Kankakee axis is northeast from southern Illinois to the region of the Kankakee river where it seems to be flexed, following the general trend of that stream and again bending, strikes northerly through the western part of the Lower Peninsula of Michigan. The change of strike along the Kankakee river seems to be due to another axis having a northwest-southeast direction and named by Gorby* the Wabash axis. The name for the heretofore unnamed axis has been taken from the Kankakee valley since it is to the west, south and east of this region that the lay of the Devonic deposits against the Niagaran and Cayugan formations can be observed in outcrops or by means of well records. In the south the Ohio and Mississippi rivers have eroded through the uplift where rocks of Middle and Upper Devonic age of the Mississippian type are at the surface. Not only this, but in Hardin county, Illinois, Worthen and Englemann† describe a low arch near Elizabeth. Otherwise the greater portion of the original uplift now lies buried beneath the "Eastern Interior Coal Field" and the "Northern Interior Coal Field."

The writer has now pointed out some of the faunal features that distinguish the Dakota and Mississippian seas during the Middle Devonic. The area of these seas, the Cincinnati island, Kankakee peninsula, and the margin of the bordering lands are in a general way shown on the accompanying maps. It has also been shown that the Dakota and Mississippian seas intermingled to some extent while the intermigration of species is believed to have been most decided during the Upper Devonic. The migrants appear to have traveled from Iowa into New York along the southern shore of Laurentia. This is the generally accepted view and it has much to recommend its acceptance, but that the decidedly European facies of the Naples fauna in the Upper Devonic of New York also came along the same shore as stated by Clarke, the writer cannot believe.

* *Fifteenth Rep. Geol. Surv. Indiana*, 1886, p. 228. ORTON, *Eighteenth Ann. Rep. U. S. Geol. Surv.*, 1889, p. 580; KINDLE, *Amer. Jour. Sci.*, June, 1903, pp. 461, 463
† *Geol. Surv Ill.*, 1, 1866, p. 352.
‡ *N. Y. State Museum, Rep. State Pal.*, 1902, p. 670.

Plate XXI

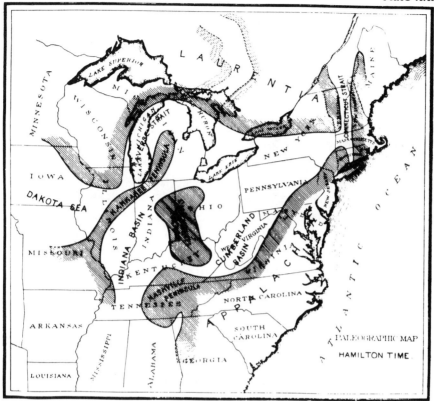

[*Editor's Summary of Pages 151–162:* Schuchert searches for the origin of his Dakota sea fauna by comparing this with other known North American faunas. He decides the relationships of his Mississippian Sea fauna with European faunas require a close connection and finds Dana's "Connecticut Valley trough" is supported by the paleontological evidence. He connects the southern Indian basin with Brazil via Alabama based on faunal similarities. He is unable to find faunal evidence to connect the Devonian of Hudson Bay with either his Dakota or Mississippian seas.]

Editor's Comments
on Papers 8 Through 11

Late Paleozoic floras have received a great deal of study and close scrutiny since the discovery in the later part of the 1800s that these floras showed marked provincialism in different parts of the world.

Paper 8 by Charles B. Read is concerned mainly with different North American floral assemblages and their similarities and contrasts, and in part with a comparison with floras of the Southern Hemisphere. A later paper by Read and Mamay (1964) gives more details for North America. An interesting additional study on the North American Carboniferous terrestrial biota by Durden (1974) examines the insect faunas from these Carboniferous forests and interprets the differences among various European, Appalachian, and Midcontinent fossil insect assemblages in terms of differences in latitude and climate.

Paper 9 by T. G. Halle is a summary of eastern and northern Asian late Paleozoic floras and includes both Carboniferous and Permian floras. Considering the literature available to Halle in the 1930s, he skillfully assembled a broad picture of late Paleozoic floral distributions that has, in general, been supported and enlarged upon by later paleobotanists (e.g., Tschudy and Scott, 1969).

Paper 10 by W. J. Jongmans looks into the questions of floral correlations and geobotanic provinces within the Carboniferous, and particularly emphasizes the difficulties of trying to establish detailed time–stratigraphic correlations between different late Paleozoic botanical provinces. His concern was mainly with plant megafossils, and at the time he wrote this paper (about 1935) the field of palynology was in its infancy. Readers interested in exploring the developments in palynology and their effect on questions of correlation and provinciality will find Tschudy and Scott (1969) a current and useful summary.

Paper 11 by Theodor Just presents a broad geographic summary of the evolution of higher plants. This paper is from a symposium on the role of the South Atlantic basin in biogeography and the evolution of plants (Mayr, 1952). Edna Plumstead (1973) presents an updated accounting of the *Glossopteris* flora without altering the general relationships used by Just.

REFERENCES

Durden, C. J. (1974). Biomerization: an ecologic theory of provincial differentiation, p. 18–53 *in* C. A. Ross, ed., Paleogeographic provinces and provinciality. Soc. Econ. Paleont. Mineral., Spec. Publ. 21, iv + 233 p.

Mayr, E., ed. (1952). The problem of land connections across the South Atlantic, with special reference to the Mesozoic. Bull. Am. Mus. Nat. Hist., 99(3), 79–258.

Plumstead, E. P. (1973). The late Paleozoic *Glossopteris* flora, p. 187–205 *in* A. Hallam, ed., Atlas of palaeobiogeography. Elsevier, Amsterdam, xii + 531 p.

Read, C. B., and S. H. Mamay (1964). Upper Paleozoic floral zones and floral provinces of the United States. U.S. Geol. Survey, Prof. Paper 454-K, 35 p.

Tschudy, R. H., and R. A. Scott, eds. (1969). Aspects of palynology. Wiley, New York, vii + 510 p.

8

Reprinted from *Jour. Geol.*, 55, 271–279 (1947)

PENNSYLVANIAN FLORAL ZONES AND FLORAL PROVINCES[1]

CHARLES B. READ

U.S. Geological Survey

ABSTRACT

Paleontologic correlation must be guided by principles, some of which are stated here, derived from a study of living organisms and their distribution. In application of these principles, nine floral zones in the Arcto-Carboniferous province of Pennsylvanian floras are considered. The floras included are all lowland, warm-climate, rain forest assemblages. In the Rocky Mountain area certain floral types appear that are thought to represent upland floras. In the Antarcto-Carboniferous province the Pennsylvanian floras differ greatly in the main and are believed to represent coal-climate, rain forest assemblages. A few representatives of these assemblages are considered, and a few Arcto-Carboniferous elements present are regarded as immigrants during an epoch of milder climate.

INTRODUCTION

The following statement, written in connection with a discussion of Devonian faunas over forty years ago, requires only slight modification to make it applicable to land floras. It contains, in fact, many of the fundamental ideas of sound stratigraphic paleontology. The import of these remarks has frequently been lost sight of but is of great consequence.

Students of geographical distribution have shown that in distant parts of the same ocean the species are widely divergent, as much difference existing between the marine faunas of the southern and northern temperate zones as between the faunas of the successive formations of a continuous geological section. It is evident from this observation that discussions of the time relations of fossils must treat not only of the genetic affinity of the forms making up a fauna, but of the geographic distribution and of the geological range of the species concerned.[2]

As a basis for the ensuing presentation of paleontologic data, it is pertinent to introduce several of the principles of the science that are of value in the study of

Paleozoic floras. Their statement may recall them to more active use, with beneficial results, and it is with this purpose that they are set down.

First is the biological principle that forms the fundamental basis for paleontologic work: The facts and accepted hypotheses of modern biology, although not always applicable, are the bases for studies of fossil organisms. The features of modern classifications that are proved to be of importance must be the elements of any classification of fossils. It is generally recognized that the paleontologist, owing to imperfect preservation of the remains with which he works, must accept, as his primary bases for determining identity and relationship, superficial characteristics that are deemed of secondary importance in the classification of most living organisms. This fact he should fully admit, and he should recognize that, in consequence, superficial homeomorphy may be wrongly interpreted and that many patterns of "evolutionary trends" may be more apparent than real. In short, the paleontologist must be fully aware, and be prepared to admit, that interpretations of generic and even specific relationships of fossil organisms are based on opinion rather

---

[1] Published by permission of the director, U.S. Geological Survey.

[2] H. S. Williams, "The Correlation of Geological Faunas," *U.S. Geol. Surv. Bull. 210* (1903), p. 15.

than on the biologic details that permit the student of living organisms to determine affinity.

Several ecological and geographical principles are of considerable importance in stratigraphic work. They are, briefly, as follows:

1. *The fundamental unit in ecologic and geographic studies is the biota or life-association.*—The biota is also fundamental in geologic work. Since the paleontologist rarely has recourse to the complete association, he should accept the largest and most representative sample available. Should he choose to make a special study of only a part of the sample, such as a family or a genus, he should admit this limitation by qualification of his results.

2. *Organisms are rarely cosmopolitan.*— There are very few known forms so adaptable that they have a general or universal distribution on continental masses.

3. *Associations living during a limited period of time in the same general region are much more alike in their constitution and the proportionate abundance of dominant species than are those of widely separated regions.*—In this principle we find a justification for the method of correlation in restricted areas through abundance of certain species in a stratal zone and an explanation of the common failure of such a zone to maintain its identity over wide areas. We likewise find an explanation of variation in fossil content in widely separated areas of strata that we believe to be contemporaneous.

4. *The association of elements of one flora with those of another suggests an intermediate environment if, on other bases, the floras are known to represent different habitats.*—If they are considered to represent similar or identical habitats, such floras are indicative of different stratigraphic positions, or they may be from some intermediate region that marks the general position of a path of intercommunication or migration.

5. *Variations in content of associations may be due to geographic distribution or geological range.*

6. *Widespread distribution of organisms or associations is to be expected in a negligible period of time only when there is adequate evidence that the areas occupied are equally and broadly continuous. If the areas occupied are recognized as discontinuous, it is then to be expected that the geologic time involved will be variable but not always negligible.*

7. *Identical genera or species may frequently be traced through several life-associations that form a geographic or a geologic succession or both. Accepting correlation by the use of associations, the oldest occurrence of a species or genus suggests the approximate locus of its origin.*

8. *Of paramount importance in geographic and geologic work is the recognition of the facts that (a) practically simultaneous changes in forms often occur over areas of great geographic spread and (b) more protracted changes in forms occur in a small area during appreciable intervals of geologic time.*—These changes may develop uniformly, or they may take place in surges. Stratigraphic variation in species, genera, or biotas may be explained by migration, changes, or hiatuses. A hiatus can be recognized, on paleontologic grounds, with certainty and properly evaluated only when the suspected local absence of a biota can be proved through its presence elsewhere in the region.

Further, I wish to present three opinions concerning associations of Pennsylvanian floras:

1. Floras of the Pennsylvanian coal swamps of the Northern Hemisphere

have many morphological characteristics that suggest that they persisted under conditions no cooler than those of present-day, warm-temperate rain forests.

2. Floras of the Pennsylvanian equivalents in the Southern Hemisphere originated under conditions no warmer than, and possibly similar to, the cool-temperate rain forests or provinces of the present day.

3. Upland floras are believed to have existed during much, if not all, of the period of existence of land plants. Their remains are not likely to be preserved in the fossil record because of distance of habitats from basins of deposition. They may be expected in the sedimentary record if local conditions were such that basins occurred adjacent to uplands.

FLORAS OF PENNSYLVANIAN AGE

Two major floral provinces have been recognized as existing during the Pennsylvanian epoch: (1) the northern or Arcto-Carboniferous province of warm-climate, rain forest coal floras and (2) the southern or Antarcto-Carboniferous province of cool-climate, rain forest floras.

ARCTO-CARBONIFEROUS FLORAS

STANDARD SEQUENCE OF PENNSYLVANIAN FLORAS

The study of Pennsylvanian fossil plants has been largely a study of the "Coal Measures" floras, groups of plants that occur in strata that are largely continental in origin and that are believed to have accumulated in extensive flood plains and in deltas under mild and humid climatic conditions. Nine successive floral zones are recognized in the eastern and Mid-Continent regions of North America. These zones provide a practical basis for general correlation of the containing rocks, "general correlation" not implying the precise establishment of boundaries of the floral zones. Obviously, in thick sequences of clastic strata the possibility of preservation everywhere of characteristic floras at the precise position of their extinction is unlikely. The stratigrapher can, therefore, only "rough in" the stratigraphic units through strict paleontologic work.

Secondary floral characteristics, such as relative abundance of certain species, may be further used for detailed local correlations, but only rarely are they valuable over large areas. Preferred for refinements of correlation within floral zones when possible are field mapping and direct tracing.

The nine floral zones have been named from characteristic genera or species. Such naming of zones does not mean, however, that knowledge of the index or marker fossils alone is sufficient for recognition. The names given are those of certain more common forms; but definite zone identification requires a study of the entire flora when possible. The nine zones and their gross distribution are given in Table 1. The following paragraphs discuss these zones in greater detail.

Zone 1.—Zone of *Neuropteris pocahontas* and *Mariopteris eremopteroides.* This zone is recognized as characteristic of the lower Lykins coals of the Pottsville formation in the Anthracite fields of Pennsylvania and of the Pocahontas formation in West Virginia. It is likewise identifiable in the southern Appalachians in areas of expansion of the Pottsville formation. Throughout much of the area the limits of the zone have not been clearly established. Correlative units probably exist in many parts of the United States but are rarely of facies suitable for preservation of fossil plants.

Zone 2.—Zone of *Mariopteris pottsvillea* and of common occurrence of *Aneimites* spp. Floras belonging to this zone occur in Lykins coal No. 4, in the lower part of the New River formation of West Virginia, and at traceable positions farther south. Such floras also occur locally at the base of the Pennsylvanian series in the Mid-Continent region. Thus the Wayside member of the Caseyville

Virginia, and is found in the upper part of the Lee formation farther south. It is likewise known from the Morrow formation and its equivalents in the Mid-Continent region.

The following list of selected plants has been made to illustrate a characteristic

TABLE 1

Zone	Name	Appalachian Region	Mid-Continent Region
9	*Danaeites*	Upper part of Monongahela formation	In Mid-Continent region Zones 8 and 9 are not separable and could together be designated the zone of *Odontopteris* spp.
8	*Lescuropteris*	Lower part of Monongahela formation and upper part of Conemaugh formation	
7	*Neuropteris flexuosa*, and *Pecopteris* spp.	Upper part of Allegheny formation	Upper part of Des Moines group
6	*N. rarinervis*	Lower part of Allegheny formation	Lower part of Des Moines group
5	*N. tenuifolia*	Major portion of Kanawha formation	Major portion of Lampasas group
4	*Cannophyllites*	Base of Kanawha formation	Base of Lampasas group
3	*Mariopteris pygmaea, Neuropteris tennesseeana, Ovopteris communis, Alloiopteris inaequilateralis,* and *Alethopteris decurrens*	Coals 2 and 3 of the type Pottsville formation, upper part of New River formation, upper part of Lee formation	Morrow formation
2	*Mariopteris pottsvillea* and *Aneimites* spp.	Lykins coal No. 4 of Pottsville formation, lower part of New River formation	Base of Pennsylvanian series in Mid-Continent region
1	*Neuropteris pocahontas* and *Mariopteris eremopteroides*	Lower Lykins coals of Pottsville formation and Pocahontas formation	

formation, southern Illinois, carries this association.

Zone 3.—Zone of *Neuropteris tennesseeana, Ovopteris communis, Alloiopteris inaequilateralis, Alethopteris decurrens,* and *Mariopteris pygmaea.* As indicated by its contained floras, this zone is widespread in the Appalachian region and in the Mid-Continent area. It occurs in the roofs of coal Nos. 2 and 3 of the type Pottsville, is characteristic of the upper part of the New River formation in West

flora of this zone. Drurey shale (Battery Rock member), NW. $\frac{1}{4}$ sec. 32, T. 10 S., R. 1 W., Carbondale Quadrangle, Illinois: *Pecopteris serrulata, Alethopteris lonchitica, A. decurrens, A. owenii, A. owenii* var. *grandifolia, A. owenii* var. *helenae, A.* sp., *Neuropteris tennesseeana, Neuropteris* spp., *Linopteris* sp., *Mariopteris pygmaea, M. speciosa, Mariopteris* spp., *Diplothmema cheathami, Diplothmema* sp., *Eremopteris inaequilateralis, Cardiocarpon* spp., *Cordaites principalis,*

Sphenophyllum sp. cf. *S. longifolium,* *S. cuneifolium, Lepidophyllum campbelleanum, Sigillariostrobus* spp.

Zone 4.—Zone of common occurrence of *Cannophyllites.* This zone may be recognized at or near the base of the Kanawha formation and its equivalents in the Appalachian region and at the base of the Lampasas group in the midcontinent region. It is frequently difficult to identify because of similarities of many of its elements to those of adjacent strata.

The following list has been compiled from several collections in western Illinois and illustrates the general characteristics of floras assigned to this zone. Tarter member, Tradewater formation in (1) SW. ¼ SW. ¼ sec. 27, T. 11 S., R. 4 E., Marian Quadrangle, Illinois; (2) NE. ¼ SW. ¼ SW. ¼ sec. 29, T. 12 N., R. 2 W., Monmouth Quadrangle, Illinois; (3) NE. ¼ NE. ¼ sec. 36, T. 9 N., R. 1 W., Avon Quadrangle, Illinois; (4) vicinity of Port Byron, Illinois: *Pecopteris serrulata* (2), *Neuropteris tenuifolia* (early form) (1), *Archaeopteris stricta* (1), *Cannophyllites marginata* (4), *Cannophyllites abbreviata* (4), *C. dawsoni* (1), *C. rectinervis* (2), *C. southwelli* (3, 4), *C. fasciculata* (4), *Alethopteris* sp., *Eremopteris grandis* (2), *Ovopteris communis* (2), *Cardiocarpon* sp. (large) (2, 3), *Trigonocarpus* sp. (2), *Sigillaria rugosa* (2), *Annularia cuspidata* (2), *Lepidodendron aculeatum* (3).

Zone 5.—Zone of *N. tenuifolia.* This zone is characteristic of the major portion of the Kanawha formation and is apparently also characteristic of rocks included in the Lampasas group in the Mid-Continent region.

The following list of species from the shale above the Cannelton coal at Tell City, Indiana, illustrates a typical flora from this zone: *Alethopteris lonchitica, A. owenii, N. scheuchzeri, N. tenuifolia,*

Mariopteris sp., *Cardiocarpon* sp., *Trigonocarpum* sp., *Lepidodendron* sp., *Lepidophyllum* sp., *Lepidostrobus* sp.

Zone 6.—Zone of *N. rarinervis.* The flora, characterized by *N. rarinervis,* occurs in the lower part of the Allegheny formation in the Appalachian region and in the lower part of the Des Moines group in the Mid-Continent region. It is the highest zone in the suite of Lower Pennsylvanian floras.

The following floras, collected from the roof of the Murphysboro coal at mines near Murphysboro, Illinois, is characteristic of this zone: *Pecopteris vestita, Alethopteris serlii, Mariopteris occidentalis, M. sillimanni, Neuropteris ovata, N. scheuchzeri, N. rarinervis, N. clarksoni, Linopteris rubella, Odontopteris* sp., *Cordaites communis, Lepidophyllum oblongifolium, Stigmaria ficoides, Calamites suckowi, Annularia stellata, Sphenophyllum emarginatum.*

Zone 7.—Zone of *N. flexuosa* and appearance of abundant *Pecopteris* spp. This zone occurs in the upper part of the Allegheny formation and is characteristic of the upper part of the Des Moines group. It is the lowest zone in the suite of Upper Pennsylvanian floras.

Zones 8 and 9.—Zones of *Lescuropteris* and *Danaeites,* respectively. These two zones are separable in the Appalachian region, where the lower, Zone 9, characterizes the upper Conemaugh and lower Monongahela formations. The upper, Zone 8, is characteristic of the upper part of the Monongahela formation. In the Mid-Continent region the two cannot everywhere be separated and perhaps should be designated as a single unit— the zone of *Odontopteris* spp.

The following list, compiled from collections from the Chanute shale, Kansas City group, illustrates a flora from the lower part of the *Odontopteris* zone:

Mariopteris pluckenetii, M. cordato-ovata,
Sphenopteris pinnatifida, Aloiopteris win-
slovii, Alethopteris virginiana, Alethop-
teris sp. cf. *A. virginiana, A. grandini,*
Odontopteris reichiana, Neuropteris ovata,
N. plicata, N. scheuchzeri, Annularia
stellata, Sphenophyllum oblongifolium,
Sigillaria camptotaenia.

FLORAL MODIFICATIONS IN THE
ROCKY MOUNTAIN REGION

With this brief sketch of the characteristic floral sequence in the typical
Pennsylvanian series, it is now appropriate to examine the floras of rock sequences some distance from Coal Measures deposition. In the southern Rocky
Mountains considerable stratigraphic
work involving the Pennsylvanian series
has been undertaken in recent years.
Physical conditions and sedimentation in
that area were unlike those of the coal
basins. A series of linear positive elements, developed in early Pennsylvanian
time, and variable quantities of continental and marine sediments were deposited in the adjacent basins. Certain of
the positive areas were sufficiently large
to provide considerable tracts of uplands.

Remains of plants that are occasionally found in the clastic portions of the
Pennsylvanian sequences afford an opportunity for comparison with floras in
the Appalachian and Mid-Continent
areas. Since these floras are still in process of investigation, I shall draw from
them only a few illustrations bearing on
an important point.

In the Weber(?) formation of the
Mosquito Range, in central Colorado, a
flora belonging to Zone 3, the zone of
Mariopteris pygmaea, has been described.[3]

[3] C. B. Read, "A Flora of Pottsville Age from
the Mosquito Range, Colo.," *U.S. Geol. Surv. Prof.*
Paper 185 (1934), pp. 79–96.

At a locality in the lower 100–200 feet
of the formation on Evans Peak, the
following species were found: *Neuropteris*
dluhoschi, N. heterophylla, N. gigantea?,
Sphenopteris hoeningshausii, Sphenop-
teris sp. cf. *S. microcarpa, Diplothmema*
cheathami, D. patentissima, Adiantites
rockymountanus, Cordaites sp., *Cordaicar-*
pon sp., *Trichopitys whitei, Dactylophyl-*
lum johnsoni, Lepidostrobus weberensis,
Stigmaria verrucosa, Calamites sp.,
Asterophyllites longifolius(?). The dominant species are those of the Coal Measures facies. Fernlike plants and representatives of the Cordaitales are abundant,
and remains of Lycopodiales are rare.
There are also present, in abundance, the
putative conifers *Trichopitys* and *Dacty-*
lophyllum.

In the Upper Pennsylvanian strata of
the McCoy formation, north-central
Colorado, occur fernlike plants associated with species of the Paleozoic conifer,
Walchia. The following florule is reported
from interval 74, McCoy formation near
Yarmony School, Eagle County, Colorado: *Odontopteris mccoyensis, Samarop-*
sis hesperius, Walchia stricta, Walchia
sp., *Walchiastrobus* sp.[4] The flora appears
to belong to Zone 7, the zone of *N.*
flexuosa, and is therefore, in terms of the
typical Pennsylvanian section, equivalent to the flora in the upper part of the
Allegheny formation.

In northern New Mexico the Sandia
formation and the lower part of the superjacent Madera limestone contain
fairly typical Coal Measures floras in
which fernlike plants and *Cordaites* sp.
are dominant. In the upper part of the
Madera limestone floras containing
Walchia spp. and other conifers occur.

[4] C. A. Arnold, "Some Paleozoic Plants from
Central Colorado and Their Stratigraphic Significance," *Univ. Mich., Contr. Mus. Paleon.,* Vol. VI
(1941), pp. 59–70.

From these notes on the Pennsylvanian floras of the Rocky Mountain region an important conclusion may be drawn. The plant associations, in the older parts of the Pennsylvanian sequences, are similar to those of the Coal Measures but, through the relative rarity of Lycopodiales, suggest a drier habitat than that indicated by the approximately contemporaneous floras in the eastern coal basins. The higher Pennsylvanian floras are striking departures from the plant associations of the same general ages, in the eastern coal basins, as inferred from index forms and independent stratigraphic data. These plants in the western area occur in suites of sediments that were deposited during a period of widespread orogeny in the southern Rocky Mountains. During this period of mountain-building many geologic data indicate a restriction of lowland and an expansion of upland habitats or areas. The floral modifications are in the direction of mesophytic associations.

For the modifications that occur in the Rocky Mountains I propose that the term "Cordilleran flora" be used.

ANTARCTO-CARBONIFEROUS FLORAS

Upper Carboniferous floras of the Southern Hemisphere contrast strikingly with those of the Arcto-Carboniferous province. An outline of the sequence and distribution of the continental Upper Carboniferous deposits is, however, not within the scope of this paper, and it is therefore sufficient to call attention only to the general features of a few representative sequences.[5]

In Argentina the Pagonzo system includes strata of Lower Carboniferous, Upper Carboniferous, Permian, and Triassic ages. Tillites occupy the basal portions of the sequence and appear to be of Lower Carboniferous and early Upper Carboniferous ages. Above are continental deposits of various types, and still higher are some marine beds.

In Brazil the Santa Catherina system is essentially identical with the Pagonzo system of Argentina. It has been divided into the Itararé, Tubarão, and Passa Dois series, which, in turn, have been divided into local units. The Itararé series is a sequence of tillite and interbedded shale and sandstone and is conformably overlain by the Tubarão series, which is the principal coal-bearing unit in the Paraná Basin.

From the lowest part of Stage 1 of the Pagonzo system, floras of probable Chester or late Mississippian age are reported. Succeeding them are associations characterized by an abundance of *Glossopteris* spp., *Gangamopteris* sp., and *Phyllotheca* sp. A similar flora is known from the Itararé tillite in southern Brazil. Itararé series, Paraná Basin, Brazil (1) near Suspiro, Rio Grande do Sul; (2) Teixeira Soares, Paraná: *Gangamopteris obovata* (1), *Glossopteris indica*, *Gl. browniana*, *Gl.* sp., *Phyllotheca* sp., *Brachyphyllum* sp. cf. *B. australe*.

In the upper part of Stage 1 of the Pagonzo system as well as from the Tubarão series in Brazil occur floras that contain, in addition to species of *Glossopteris* and *Phyllotheca*, an abundance of *Pecopteris*, *Zeilleria*, *Annularia*, *Sphenophyllum*, and *Lepidodendron*. Illustrative of the association, the following forms from a single locality are listed: Tubarão series, Cambuhy, Rio das Pedras, Paraná, Brazil;[6] *Pecopteris pedrasica* (Ar),

[5] C. B. Read, "Plantas fósseis do Neo-Paleozóico do Paraná e Santa Catarina," *Brasil, Min. Agr., Dept. Nac. Prod. Min., Div. Geol. Min., Mono. 12* (1941), pp. 1–102.

[6] In the list forms indicated by "I" are known by the writer to occur in the Itararé series. Forms indicated by "An" are considered typical members

P. cambuhyensis (Ar), *P. paranaensis* (Ar), *Zeilleria oliveirai* (Ar), *Glossopteris indica* (I, An), *Glossopteris* sp. (I, An), *Glossopteris* sp. cf. *ampla* (I, An), *Sphenophyllum oblongifolium* (Ar), *Annularia? americana* (Ar), *Phyllotheca australis* (I, An), *Phyllotheca* sp. (I, An), *Brachyphyllum* sp. cf. *B. australe* (I), *Buriadia* sp. (An), *Lepidostrobus* sp. (Ar, An), *Lepidodendron pedroanum* (Ar, An).

A preliminary conclusion is that such floras as those from the upper part of Stage 1, Pagonzo system, and from the Tubarão series, Santa Catherina system, are correlative with the upper part of the Pennsylvanian series of North America. This conclusion is based on the general similarity of the types common in the floras. It is further concluded that the tillite sequence which contains a zone of abundant *Glossopteris*, *Gangamopteris*, and *Phyllotheca* is also of Pennsylvanian age.[7]

The presence of a flora so dissimilar to those known in the Arcto-Carboniferous province, succeeded by others having many similarities to Arcto-Carboniferous floras, is perhaps explained by reference to the rock sequences. The floras of the Itararé series and its equivalents are in shale interbedded with tillite. The environment, although probably not then glacial in the area inhabited by the plant associations, was probably more rigorous than is generally characteristic of Carboniferous floras. It is apparent from a consideration of paleogeography that the Antarcto-Carboniferous area in question was separated from Arcto-Carboniferous land areas; therefore, opportunities for

development of dissimilar floras were present. Following maximum glaciation in the Pennsylvanian of South America, climatic amelioration occurred, as indicated by the presence of coal in the superjacent rocks. During, or prior to, that period of time, members of the Arcto-Carboniferous floras presumably migrated into the Southern Hemisphere and became common in the known plant associations.

SUMMARY

Two major floral provinces have been recognized as existing during the Pennsylvanian epoch: (1) the northern or Arcto-Carboniferous warm, rain forest province of coal floras and (2) the southern or Antarcto-Carboniferous cool, rain forest province. In the Arcto-Carboniferous province the dominant associations preserved are Coal Measures floras. The eastern and Mid-Continent regions appear to have been a rather continuous lowland, across which shallow seas intermittently advanced and retreated, and physical conditions must have been generally similar everywhere in the basin. Thus widespread migration of plant associations in negligible time is probable, and the floral zones may be inferred to have chronologic value.

In portions of the Rocky Mountain region there were restricted lowlands and basins of deposition adjacent to rapidly rising geanticlines. Modifications of the lowland floras are found there, these modifications being due to the inclusion of mesophytic forms. The presence of such types may be explained by the existence of upland habitats, with mesophytic floras growing upon them, adjacent to sites of deposition; it may also be interpreted in terms of regional climatic variations.

To the sequence of rain forest lowland or Coal Measures floras the term "Arcto-

---

of the Antarcto-Carboniferous flora; by "Ar" are those related closely to Arcto-Carboniferous types. Several of these are conspecific with, or closely related to, forms that occur in the Arcto-Carboniferous floras of the Pennsylvanian series.

[7] Read, ftn. 5.

Carboniferous" has been applied because of the widespread occurrence of this suite of associations in the Pennsylvanian or Upper Carboniferous rocks of the Northern Hemisphere. For the modifications that occur in the southern Rocky Mountains the term "Cordilleran flora" is recommended.

The Antarcto-Carboniferous floras were, during earlier Pennsylvanian time, dominantly cool, rain forest types that developed independently or semi-independently of the Arcto-Carboniferous floras. During later Pennsylvanian time there was sufficient migration of Arcto-Carboniferous floras southward to modify the earlier existing floras. This occurred at a time of climatic amelioration following glaciation.

More important than the discussion of these wanderings of the Pennsylvanian floras are the principles that have been suggested. Geographical distribution is a frequently neglected factor in studies of Paleozoic organisms. Furthermore, differences between floras on the two margins of a single major depositional area are sufficiently great to require the utmost care in establishing even approximate correlations. Such care consists of weighing all available evidence, organic and stratigraphic, and of study of entire associations rather than of certain selected biologic groups.

It is perhaps inadvisable to draw any generalization from these observations. Depending upon one's point of view, it might be concluded that land floras show so many variations that regional correlations based on them are of doubtful value. Equally possible is the conclusion that problems of geographical distribution and variation have not been sufficiently considered by paleontologists.

9

Reprinted from "Deuxième Congrès pour l'Avancement des Études de
Stratigraphie de Géologie du Carbonifère, Heerlen, 1935,"
Compt. Rend., **1**, 237–245 (1937)

THE RELATION BETWEEN THE LATE PALAEOZOIC FLORAS OF EASTERN AND NORTHERN ASIA.

BY T. G. HALLE.

(with one map).

The problems presented by the distribution of the late Palaeozoic floras seem to be more complicated in Asia than in any other continent. The dominant feature is the contrast between the northern A r c t o - c a r b o n i f e r o u s flora and the G l o s s o p t e r i s f l o r a of Gondwana Land. The Indian Glossopteris flora, which extends northwards at least to Kashmir, is in all respects typical and presents no regional differentiation. The northern province, on the other hand, is divided into two regions with comparatively distinct floras, not counting the Arcto-Carboniferous flora of Asia Minor.

In the north, the A n g a r a f l o r a or K u s n e z k f l o r a extended all through Siberia, from Petchora in Northern Russia to the region of Vladivostok. This flora is generally characterized as an Arcto-carboniferous flora with a strong admixture of Gondwana elements, though the dominant northern component is not quite typical of the Carboniferous or Permian floras of Europe. It has comparatively few species in common with Europe and contains many peculiar forms, even though it may be doubted whether all the new genera described by ZALESSKY are well founded. The Gondwana element in the Angara flora has in my opinion been somewhat overemphasized. *Glossopteris* is not known, and the occurrence of *Gangamopteris* is at least very doubtful.

In the south-east of Asia, the Arcto-Carboniferous Palaeozoic flora of China and Korea has been shown by EDWARDS and especially by JONGMANS and GOTHAN to extend southwards to Malacca and Sumatra. This flora is generally known as the

Gigantopteris flora. But *Gigantopteris* is only characteristic of its latest phase, and, in dealing with the entire Carboniferous and Permian plant-succession, I propose to use the
term Cathaysia flora, after the Palaeozoic land-mass
Cathaysia in GRABAU's palaeogeographic maps.

The Carboniferous division of the Cathaysia flora is on the
whole of the European type. The Westphalian is little developed: the occurrence of a Westphalian flora was indeed doubtful
until it was discovered by MATHIEU at Kaiping. The richest
Carboniferous flora and the great coal-fields belong to the Stephanian. The limit between the Stephanian and the Permian is
even more vague and difficult to fix than in Europe, chiefly because typical species of *Callipteris* are absent or rare even in beds
which, from other criteria, must be placed fairly high up in the
Permian. In agreement with KAWASAKI, I refer to the Permian
the Lower Shihhotse series in Shansi and the corresponding
Jido series in Korea. The flora of this division, too, is mainly
of the European type, but the progressing differentiation of
the Cathaysian geobotanical province is marked by the appearance of several peculiar forms, such as species of *Tingia* and
Lobatannularia. Another feature is the first appearance of elements that indicate affinity with North America, in the first
place *Gigantopteris Whitei*, which is closely allied to *G. americana*, and possibly *Emplectopteris*.

The youngest division of the Palaeozoic flora in Cathaysia
is represented by the Upper Shihhotse series in Shansi and the
Kobosan series in Korea. This is the Gigantopteris flora, in the narrow sense, characterized by *G. nicotianaefolia*,
certain species of *Lobatannularia* and *Tingia*, and several other
peculiar plants. This division contains several Mesozoic elements
and is held by KAWASAKI and other Japanese palaeobotanists
to be Triassic. I cannot accept this view but believe that the
Gigantopteris flora is not younger than Middle Permian.

ZALESSKY has recently described some few species of *Pecopteris* from West Turkestan. These species are all characteristic
of the European Stephanian, and most of them also occur in
China. The material is rather poor, and additional evidence is
needed, but it would appear as if we had here an eastern out-
post of the European Carboniferous flora. Opposite this Turkestan flora, the Cathaysia flora reaches its farthest western

point in Kansu, and it is possible that we may here have a trace of the old connexion between Europe and China which seems to be required by the close agreement between the Stephanian floras of the two regions. This belt would cut off the lines of communication that are supposed to have existed between the Indian Glossopteris flora and the Angara flora. The northward migration of the Gondwana elements would then probably not have set in until after the connexion between the Carboniferous floras of Europe and China had been broken [1]).

It is clear from the distribution of the late Palaeozoic floras, roughly shown on the sketch-map, fig. 1, that Central Asia must be a region of particular importance to the study of their mutual relation [2]). This is the region where the joint Chinese-Swedish expedition under SVEN HEDIN and Prof. Sü PING-CHANG has been working for some years. In view of the interest attaching to the fossil floras of this critical region, Dr. HEDIN kindly consented to add to his staff a young swedish geologist, Mr. GERHARD BEXELL, who was to study especially the plant-bearing deposits in the Nanshan region and collect fossil plants. What we hoped to gain was in the first place new information on the westward extension of the Cathaysia flora and possibly on its relation to the Kusnezk flora.

The difference between these two floras is well known but

[1]) The stratigraphical distribution of the Indian Gondwana elements in the Siberian Angara flora proves that the supposed northward migration of species from India must have taken place long before the end of the Permian. This matter is of considerable interest in connexion with the theory of continental drift and the Permo-Carboniferous Ice Age. WEGENER's explanation of the glaciation implies, i. a., that during the Late Palaeozoic the present Indian Peninsula was placed near the South Pole, which he believed to have been situated at that time somewhat east of South Africa. From this extreme southern position, bordering on Madagascar, India is supposed to have gradually drifted northwards. The drift is believed to have been causally connected with the great orogenetic movements in Asia, which did not close until the final uplift of the Himalayas during the Tertiary. Whether the northward drift of India continued during the whole intervening period or not, it must obviously have been very slow, and the Indian Peninsula could not have come into touch with the rest of the continent of Asia until long after the Permian Angara Flora of Siberia was extinct. The generally accepted explanation of the Gondwana elements in the Angara Flora therefore cannot be reconciled with this aspect of the drift theory. A similar difficulty, only slightly less obvious, applies to the occurrence of *Glossopteris* in the Rhaeto-Liassic of Tonkin, which demands a possibility of communication between Gondwana Land and southeastern Asia at some period before the end of the Trias.

[2]) This sketch-map differs slightly from that shown at Amsterdam and Heerlen, the northeastern boundary of the Glossopteris flora in India having been modified from information supplied by Prof. SAHNI.

Glossopteris Flora

Angara Flora

European Permo-Carb. Flora

Cathaysia Flora

STATENS REPRODUKTIONSANS

difficult to explain. We do not know to what extent it is conditioned by factors of time or of space or of both combined. In China and Korea there is an unbroken succession of floras from the upper Westphalian, or at least the Stephanian, to the Middle Permian or, if some Japanese authors are right, to the Lower Triassic.

In Angara land the conditions are less clear. ZALESSKY, in his great work of 1918, in agreement with ZEILLER, regarded the plant-bearing series of Kusnezk as falling altogether within the Permian: he thought that it is separated from the underlying Lower Carboniferous by a great disconformity. Most Russian geologists and palaeozoologists, on the other hand, have believed that it is Carboniferous. In the Kusnezk basin Miss M. NEUBURG distinguishes three series: (I) a lowermost series without *Callipteris*, which she regards as Upper Carboniferous, (II) a Permian series with *Callipteris* and (III) a purely Mesozoic, probably Lower Jurassic, series, which contains the Mesozoic elements that were formerly believed to be scattered all through the Kusnezk deposits.

The late Palaeozoic flora of China has been known for some time to extend in a westward direction as far as Kansu. Mr. BEXELL has now made extensive collections in the Nanshan region, especially in the Richthofen Mountains, and has succeeded in placing the plant-bearing horizons in a general section that extends from the Stephanian or Permo-Carboniferous to the Jurassic. Only a minor part of the collections has at present arrived at Stockholm, and the superficial examination that I have made has done little more than confirm the conclusions which Mr. BEXELL had drawn in the field regarding the nature and age of the various plant-assemblages.

Mr. BEXELL [1]) has given the following somewhat simplified table of the main divisions of strata in descending order:

8. Green or brownish sandstones with intercalations of black carbonaceous shales, about 300 m. — P l a n t-b e a r i n g   z o n e   D.

7. Greenish or reddish, coarse, barren sediments, about 500 m.

---

[1]) G. BEXELL: On the stratigraphy of the plant-bearing deposits of late Palaeozoic and Mesozoic age in the Nanshan region (Kansu). -- Geografiska Annaler. Stockholm 1935. P. 63.

6. Green sandstones and shales, about 100 m. -- P l a n t -
   b e a r i n g z o n e C.
5. Transition zone of parti-coloured (green and red) barren
   sediments, about 150 m.
4. Red, barren sediments, about 450 m.
3. Transition zone of parti-coloured (green and red) barren
   sediments, about 150 m.
2. Predominant green sediments, about 200 m. -- P l a n t -
   b e a r i n g z o n e B.
1. Marine sediments with intercalations of plant-bearing
   beds, about 90 m. -- P l a n t - b e a r i n g z o n e A.

The first division, p l a n t - b e a r i n g z o n e A, contains
marine invertebrates, which have not yet been determined, coal-
seams and the following fossil plants:
*Annularia stellata, Sphenophyllum emarginatum, Sph. oblongi-
folium, Neuropteris pseudovata, Emplectopteris triangularis,
Dicranophyllum* sp. and *Tingia Hamaguchii,* besides several
species of *Sphenopteris, Pecopteris, Linopteris, Taeniopteris* and
Cordaites.
This flora seems to correspond to that of the Yuekmenkou
series in Shansi and the Koten series in Korea, but also to the
lower parts of the succeeding Lower Shihhotse series in Shansi
and the Jido series in Korea. It therefore probably represents
the Stephanian, but perhaps also the lowermost Permian. It
should be noted that the flora is entirely of the Chinese type
and that the species, with few exceptions, also occur in Europe,
while there is no trace of any Angara elements.
The next division 2, or p l a n t - b e a r i n g z o n e B, is
entirely continental. Among the material from this level I have
identified most of the species from zone A, not, however, *Emplectopteris* and *Tingia Hamaguchii.* Among other species may
be mentioned *Sphenophyllum Thonii, Pecopteris* cf. *hirta,
Alethopteris Norinii, Taeniopteris multinervis, Lepidodendron
oculis felis, Tingia carbonica* and species of *Walchia* (?). Most
of these species are distributed all through the Lower Shihhotse
series in Shansi and the Jido series in Korea.
In the upper part of this division occur further: *Annularia*
cf. *Shirakii, Sphenopteris pseudogermanica, Odontopteris orbi-
cularis, Protoblechnum Wongii* and *Tingia elegans.* These spe-

cies are in Shansi either confined to the Upper Shihhotse series, or they are found both in the lower part of that series and in the upper part of the division below it. In Korea they occur in the Kobosan series or just below its base, at the top of the Jido series. Neither *Gigantopteris nicotianaefolia* nor typical *Lobatannularia* have been found, and the flora therefore cannot be said to correspond exactly to the *Gigantopteris* flora in the strict sense. It appears as if the true Gigantopteris flora did not extend so far to the west, and it may be that the arid conditions that terminated its existence in Shansi set in earlier in Nanshan. But it is quite clear that up to a horizon that corresponds to the lower part of the Upper Shihhotse or Kobosan series, Nanshan had a flora of the Cathaysia type.

Above the plant-bearing zone B follow red or green, arid sediments without fossils, BEXELL's divisions 3-5, comprising together 750 m. This series evidently corresponds to the arid Shihchienfeng series in Shansi, perhaps also to the upper part of the Upper Shihhotse series, since the latter may not be fully developed in the Nanshan. The geographical development has evidently run more or less parallel in Nanshan and Shansi and also in Korea, where the arid division is represented by the Green Series.

Above the arid series follows BEXELL's division 6, or the plant-bearing zone C. Most of the collections from this zone are still in China, but I have identified the following species: *Phyllotheca deliquescens, Ph.* cf. *Schtschurowskii, Callipteris* sp., *Iniopteris sibirica, Brongniartites salicifolius, Zamiopteris glossopteroides, Rhipidopteris ginkgoides, R. lobata* and *Noeggerathiopsis scalprata.*

This is a typical Angara or Kusnezk flora, corresponding to the second or Permian series of Miss NEUBURG's. Mr. BEXELL has thus shown that in Nanshan the Cathaysia flora and the Kusnezk flora met and overlapped. It follows from this that the typical Permian Kusnezk flora is, on the whole, younger than the late Palaeozoic flora of Cathaysia. It is true that the youngest phase of the Cathaysia flora—or the Gigantopteris flora in the narrow sense—may not be typically developed in Nanshan. But even if that is the case — which is not yet certain — the beds with the Gigantopteris flora in Shansi ought to correspond to the lower transition series in Nanshan, i. e., BEXELL's division 3, or possibly to the lower part of his division 4: his division 6, and with it the Angara flora with *Callipteris*, must be younger. The contrast between the Kusnezk flora with *Callipteris* and the Cathaysia flora is therefore, at least mainly, due to a difference in geological age, the Kusnezk flora being the younger. This conclusion, which I believe to be firmly established by Mr. BEXELL's work, has important bearings also on the age of the Gigantopteris flora. It is generally agreed that the typical Kusnezk flora with *Callipteris* cannot be younger than Permian. The Gigantopteris flora, consequently, cannot possibly be Triassic, as believed by the Japanese authors, even if this were probable in view of the composition of the flora, which I think it is not. The stratigraphical relations that Mr. BEXELL has found in Nanshan tend to place the Gigantopteris flora l o w e r and the Kusnezk flora h i g h e r in the stratigraphical scheme than has so far been assumed. The Gigantopteris flora cannot very well be younger than Middle Permian: it might indeed be Lower Permian, while the Kusnezk flora with *Callipteris* ought to be of fairly young Permian age. This conclusion is somewhat surprising, since the supposed Mesozoic elements in the Kusnezk flora have been shown by Miss NEUBURG to belong to a much younger Jurassic division, and the typical Kusnezk flora with *Callipteris* — Miss NEUBURG's second series — contains rather less of Mesozoic elements than the Gigantopteris flora.

The difference in age between the Gigantopteris flora and the Kusnezk flora does not, however, exclude the possibility that there existed also a phytogeographical difference between Cathaysia and Angara land. This question at present hinges on the age of the lower part of the Kusnezk series. If the Kusnezk flora is in part Upper Carboniferous or Lower Permian, then the two floras would be partly of the same age, and during a considerable length of time there would have existed a regional difference, quite striking considering the short distance between the two floras.

Finally I wish to draw attention to a curious feature in regard to the geographical relation between the Permian floras of East Asia and America. The Permian Cathaysia flora has a clear affinity to the American Permian flora, expressed in the distribution of *Gigantopteris* of the *G. americana*-type and other forms. The late Dr. DAVID WHITE, who was the first to point out this important fact, therefore assumed a migration or interchange of species between the two continents. It is generally agreed that the Pacific Ocean has existed at least since early Palaeozoic times, whatever view we may take of WEGENER's theories. Any communication that may have existed between East Asia and America was therefore probably restricted to te Behring Straits. But the Gigantopteris flora of Korea and China is cut off from this supposed land bridge by the Kusnezk flora, which reaches the coast near Vladivostok. This difficulty at first appears great; but if the Kusnezk flora is younger than the Gigantopteris flora of Cathaysia, then the latter may have been in communication with the American flora before the Kusnezk flora was established, or at any rate before it reached the Pacific coast.

DISCUSSION.

W. GOTHAN. — Prof. GOTHAN betonte besonders die Wichtigkeit der neueren Entdeckungen in Nanschan; ist die Kusnezk-Angara-flora wirklich ganz jünger als die Cathaysia-flora Ostasiens, so kann diese nur dem unteren Perm zugewiesen werden, ja kann sehr wohl schon im Stephan begonnen haben.

W. VAN WATERSCHOOT VAN DER GRACHT. — I want to emphasize that, whether or not we accept any drift of continents in the sense of WEGENER, an *enormous* distance has originally separated India from the Continent of Asia with its Cathaysia and Angara floras. Between these lies the most enormous bundle of mountain chains known anywhere on the earth, not only the Himalayas, but the entire mass of folds of the Pamirs, of Thibet, of the Tien Shan, etc. etc. I would not dare to suggest what distance would be represented by flattening out these folds and thrusts and the *enormous* geosynclinal area represented by them.

If we can find older massivs in these chains, as we know them from the Alps and other mountain systems, it may be possible to find accessible to observation parts of the substratum of the old Alpine Tethys and its fossil contents either marine or terrestrial.

10

Reprinted from *Proc. 16th Internat. Geol. Congr., Washington, D.C., 1933,*
Vol. 1, 1936, pp. 519–527

MAJOR DIVISIONS OF THE PALEOZOIC ERA:
MIDDLE PALEOZOIC

Floral correlations and geobotanic provinces
within the Carboniferous

By W. J. Jongmans

Heerlen, Netherlands

ABSTRACT

The correlation table for the Carboniferous sequence of northwestern Europe drawn up in 1927 by the international stratigraphic congress at Heerlen, Netherlands, is presented in this paper. More or less striking variations in the flora permit close subdivision and correlation of the sequence in the paralic coal-measure facies; the few but distinct marine horizons, with goniatites, are a means of assigning to these floristic divisions their place in the main pelagic sequence. For the more distant basins of central and eastern Europe correlation is more difficult, as marine beds are restricted to the lower division and belong to a different faunistic province; the flora is the only means of comparison. Much work is still needed on the floristic correlation of the eastern United States and Canada with the European sequence. Sufficiently complete and reliable American collections are lacking in Europe, and publications with good figures and descriptions are rare. For still more remote sedimentary provinces the problem becomes increasingly difficult. As a working hypothesis only, for further research, a rough floristic correlation of the Carboniferous of the United States and Canada with Europe is attempted in this paper.

Both the European and the eastern North American floras belong to the Arcto-Carboniferous province, distinct from the *Gigantopteris* flora of eastern Asia, which, however, is also represented in the western states of North America, especially Oklahoma. The two floras have many forms in common, but, notably in the upper Pennsylvanian and the Permian, special forms appear (*Gigantopteris, Taeniopteris,* etc.) that are unknown in the first province. A third assemblage, the Gondwana flora, is restricted to two opposite (possibly originally polar) provinces of the earth, and associated with glacial deposits and other indications of a colder climate. Here also mixture occurs with Arcto-Carboniferous forms.

I contend that the presence of the same specimens in widely separated regions is no proof of strictly contemporaneous deposition; certain living floral assemblages are of a type which elsewhere prevailed in the Tertiary; similar conditions may have occurred in the Paleozoic.

There is no geologic formation of which the detailed subdivision and correlation have been the subject of longer scrutiny than the Carboniferous, but at the same time agreement on this formation has been found more difficult to attain than on almost any other. The coal-bearing portions have been studied particularly, on account of their economic value, but they are in a continental or at least semicontinental deltaic (paralic) facies; other regions are entirely marine, and comparison between these two facies was difficult and was possible only where they intermingle. Many geologists have worked independently of each other with plants, fresh-water animals, or marine forms, and, naturally, uniformity of results was sorely lacking—indeed, many conclusions were manifestly contradictory.

It was the object of the Congrès pour l'avancement des études de stratigraphie carbonifère held at Heerlen, Netherlands, June 7 to 11, 1927, to seek an agreement. Here, in the center of the coal fields in the province of Limburg, representatives of botanic and zoologic paleontography, students of marine and continental forms, research workers in stratigraphy and sedimentation convened. The subjects for discussion included the general subdivision of the Carboniferous sequence, stratigraphy and structure of the principal areas, the significance of cer-

tain plants and animals for stratigraphy and correlation, the areal limits of certain subdivisions and their possible extension over larger areas, but, above all,
the principles on which correlation should be based. The necessity for a uniform
basis for subdivision of the Carboniferous was generally felt, and it was soon
agreed that only those species or distinct varieties should be accepted as guide
fossils that were short-lived, but belonged to groups that were long-lived, so that
each established and recognized subdivision should have its own characteristic
types. Finally, agreement was attained on the annexed standard subdivision and
correlation of the Carboniferous sequence of western Europe (pl. 1).

The great difficulty is that the belief that there exists an adequate variability
everywhere or for all groups is a mere assumption. As soon as we have at our dis-
posal a sufficiently complete collection of all existing forms and varieties and have
studied a conformable sequence, without important breaks either in sedimentation or in facies, we find that it is utterly impossible to draw sharply defined
limits between the various forms. For instance, when we look over the neurop-
terids of the Carboniferous in general, we can readily distinguish several types
or species, which characterize certain horizons, but as soon as we have a complete
collection of all forms contained in all strata, it becomes impossible to draw a
line between species: every form is connected by gradations with an older one.
This is true of every fully developed group of animals or plants in an unbroken
sequence of sediments of sufficiently similar facies. The influence of personal opinion and evaluation becomes very considerable, and the differences between many
so-called "species" are so minute that only a specialist can distinguish them—in
fact, it is not impossible that such "species" are valid only for a certain locality.
Notwithstanding all this, we can find forms that fulfill the principal requirements
prescribed above. In the Carboniferous of western Europe the fossil remains of
plants and of most land or fresh-water animals are found in an unbroken sequence
of sediments, and here the above-mentioned difficulties apply to their fullest extent. The fossils are very important for defining general facies and age of beds,
but only rarely do we find any that can be used for detailed subdivision and for
correlation of individual strata. Certain species of plants show a small vertical
distribution, but this is by no means normal, and it is most improbable that certain forms would be confined to strata associated with only a single coal seam.
Here, however, we can use another characteristic—namely, the rate of frequency
of certain species in a typical association of forms.

We can use nonmarine animals in the same manner and with the same restrictions as plants, and the more common animals in the coal-measure facies have a
distribution and relation very similar to those of the plants. In addition to the
common species of *Carbonicola*, *Anthracomya*, and *Najadites*, for instance, we find
genera and families that are represented almost entirely by rarer forms. The
study of these animals is by no means as far advanced as that of the plants.
Only during the last 10 years or so, particularly as a result of work by Pruvost,
Trueman, and many others, have the nonmarine animals been used for stratigraphic correlation. The results of this work indicate the importance of these fossils and they have been of much help in subdividing and correlating local sequences. If beds containing fish scales, with insects and other nonmarine forms,

Europe

Stage		Dutch coal fields	Ruhr Basin	Belgium	France (north basin)	Great Britain	Upper Silesia	Lower Silesia	Saxony	Central Bohemia	Saar Basin	Wettin Harz	Thuringia	France (central plateau)	Russia
Permian.								Rotliegendes.	Rotliegendes.	Rotliegendes.	Rotliegendes.	Rotliegendes.	Rotliegendes.	Rotliegendes.	Artinskian.
Stephanian.			Pesberg. Ubenbüren. Flammkohlen.	Flénus.	Assise de Bruay. — Mansfield —	Rödstockian (Stefordian). Upper	Chelmer Schichten.	Radowenz. Schadowitz.	Zwickau. Lugau.	Kunovaer Schichten. Nyran. Radnitzer and Wladnoer Schichten.	Ottweiler Schichten. Flammkohlen Fettkohlen	Wettiner Schichten. Grillenberg Schichten.	Ohrenkammer.	Stephanian. Lowest Gard beds.	Ouralian. C₃'
Westphalian.	C	Jabeek. Maurits.	Aegir — Petit Buisson Gasflammen-kohlen. Lingula (Domina)	Eikenberg. Asch.	Assise d'Anzin. + Gin Mine +	Yorkian.	Nikolaier Schichten.	Hangend-zug or Schatzlarer Schichten.	Flöha.						
	B	Hendrik. Wilhelmina.	Gaskohlen. Catharina ++ Fettkohlen. Lingula (Neurode) +++	Pessonnière Genck. Beyne.	Assise de Vicoigne. + Halifax Hard bed ++	Lower Yorkian.	Rudaer Schichten	Schichten.							
	A	Baar- + Foelina Io.	Upper Mager ++ Neberbank +++ Lower Kohlen. Sarnsbank-Niveau	Oupaye. Bouharmont +			Sattelflöz Schichten. Hiatus?	Wegssteiner Schichten. Hiatus?							Mjatchkovian. Samarian.
Namurian.	Epenian.	Epenian. Lower Magerkohlen. Flözleeres -		Assise d'Andenne.	Flines.	Millstone	Ostrauer Schichten.	Walden-burger Schichten.	Borna. Dobrilugk.						Moskowian.
	Gulpenian.	Gulpenian. Alaunschiefer.	(Lontzen beds) Bruille. Chokier.			grit.	Kulm Dach-schiefer.	Kulm.	Kulm.						Serpuchovian.
Dinantian.		Visean. (Kohlenkalk)	Visean.	Visean.	Visean.	Carboniferous limestone. Calciferous sandstone.									

Older Paleozoic

CORRELATION OF THE CARBONIFEROUS AND PERMIAN OF EUROPE

are studied more closely, the frequency of such forms will undoubtedly prove far greater than has yet been suspected, and it seems probable that they will prove as useful as plants. We have already found certain strata of wide horizontal distribution that contain a very typical nonmarine fauna and are very helpful in the correlation of sequences within at least a certain rather wide area. An example is the zone with very numerous *Anthrapalaemon grossarti* in the Carboniferous of the province of Limburg. An enormous number of fragments and individuals of this crustacean occur in a single stratum, everywhere in the same position above a definite coal seam, everywhere associated with an assemblage of numerous fish scales, insects, arachnids, and many other forms. This zone occurs in all the collieries in the Netherlands where this sequence is exposed, so that it has become a most important guide zone for this area; but in spite of many efforts, it has not yet been identified in either of the adjacent Belgian or German coal fields.

Most important guides for correlation in the upper Carboniferous coal measures of western Europe are the comparatively few marine strata, which, though widely spaced vertically in the sequence, have a very extensive horizontal distribution. In the lower portion of the productive coal measures their number is large, and at the very lowest horizons the beds are nearly all marine, but their spacing increases continuously toward the top of the productive sequence so that they become extremely valuable for purposes of correlation. An individual marine stratum, though only a few inches thick, may extend over hundreds of kilometers. It may be rather difficult to identify, because the marine forms are, in places, by no means abundant or conspicuous, but the specialist soon learns to distinguish a marine from a fresh-water shale by some particular appearance, very difficult to describe adequately, and diligent search may then disclose marine fossils. Inasmuch as these marine beds are so widely spaced, notably in the higher divisions, they are more likely to contain distinct forms that can be used for major subdivisions and correlations over very great distances. In the lower divisions, where these marine beds are very much closer together, their faunas are of course not so sharply differentiated, but nevertheless they are of considerable help. Nearly all of them contain goniatites, which in their evolution illustrate better than almost any other known genera the principles discussed above and therefore make excellent guide fossils.

On the other hand, in a sequence that is mainly marine but that contains few and widely spaced fresh-water or land beds, with plant or animal fossils of continental facies, these fossils will have the same importance as the marine forms have for a paralic facies, such as prevails in the coal fields of western Europe. Such a sequence is found in certain portions of the Russian Carboniferous province, in the lower part of the coal measures of western Europe, and in great areas of the Carboniferous deposits in the Midcontinent region of the United States. In these regions plants promise to be as characteristic as goniatites are where the sequence is mainly continental.

On the basis of the principles discussed above, the following general subdivision has been proposed for the Carboniferous of western Europe, the marine horizons being used as guides:

TABLE 1.—*Carboniferous of western Europe*

Permian: *Callipteris* beds.

Stephanian. (No marine forms are known in western Europe.)

Westphalian:
{
Subdivision C
Marine zones: Aegir, Petit Buisson, Mansfield.
Subdivision B
Marine zones: Catharina, Poissonnière, Gin mine.
Subdivision A
Marine zones with *Gastrioceras subcrenatum.*
}

Namurian:
{
Zone with *Reticuloceras.*
Zone with *Homoceras.*
Zone with *Eumorphoceras.*
}

Dinantian (Lower Carboniferous):
{
Visean.
Tournaisian (Étroeungt).
}

The boundary between the continental Permian and the Carboniferous is purely botanic. The presence of *Callipteris* must be regarded as the principal characteristic of the basal Permian, and in many areas the drawing of the dividing line is purely a matter of personal judgment. The distinction of the Stephanian from the Westphalian is more or less a similar matter. In certain regions the presence of conspicuous conglomerates at the base of the Stephanian is very useful, but they are the result of orogenic movements and are by their very nature of local importance only. The evolution of the flora during the Westphalian is very marked: there is practically no species of plant that occurs in both the lowest and the highest portions, and the three major subdivisions may be distinguished very definitely by their floras, but nevertheless there is also a regular progressive development from base to top, and sharp dividing lines cannot be drawn where the sequence is continuous. If we consider individual groups of species in the various genera, the evolution is very well marked: in genera like *Sigillaria, Neuropteris,* and *Mariopteris* we can recognize a regular succession all through the Westphalian. The flora of the Namurian, as known in western Europe, is rather poor. The upper portion has practically the same flora as the basal beds of the Westphalian. In the middle and lower portions plant beds are rare: the sediments are too largely marine in western Europe, and the floras present include few species. Locally conditions are somewhat better in some of the lowest strata, as in Belgium (Renier) or in the Gulpen measures in the coal fields of Dutch Limburg.

A further difficulty is that the standard section of the Heerlen Congress cannot be so readily applied to the continental coal basins outside of the paralic belt, which are in a purely limnic facies and do not contain any marine beds and therefore no goniatites—for instance, the basin of Saarbrücken. Here we have only the floral sequence to guide us, unless further research discloses characteristic land animals.

The same difficulty exists in regions, like Silesia, where marine zones, though still present in the lower divisions, are entirely absent in the higher strata. Here also correlation with western Europe may be quite feasible in the major divisions, but closer correlation becomes very difficult. The faunas of the marine beds represent a transition from the faunas of the western province to those of eastern Europe (Russia). The Heerlen Congress did not attempt to link eastern Europe to the standard section of the west, but studies by Gothan, Susta, Patteisky, Zalessky, and others have since made it possible to go further in this respect and to make a rough correlation. Difference in facies is the major difficulty: in western Europe most of the Namurian contains no coals; in eastern Europe the Namurian is a productive sequence with many and thick coal seams.

The Heerlen Congress concerned itself still less with Carboniferous provinces outside of Europe. All participants were convinced of the difficulty of tying these into the European section. Our knowledge of these outside areas is too incomplete, and it would require close cooperation and much work by local specialists of the various countries, comparative examination of the collections that various museums may have available, and very much field work to make these collections more complete and representative. The trouble with museum collections is that they are very rarely, if ever, representative of all the material present in the localities concerned but contain only selected specimens of special beauty, whereas inconspicuous or poorly preserved specimens are almost entirely lacking. They also give no information about the mode of occurrence and the relative frequency of the species. The publications are usually too incomplete in description and especially too poor in figures of the specimens, or the execution of these figures makes them unsuitable for use as a basis for comparative research: it is invariably necessary to study the collections in person at the places where they are preserved. But even then, almost without exception, the labels do not give localities or the stratigraphic position of the specimens in the sequence precisely enough.

The lack of comparative research by interchange of specialists from both sides of the Atlantic has made it impossible, so far, to correlate the sections of the Carboniferous of North America closely with those of Europe. As a working hypothesis, a tentative correlation can be given and is here attempted (pl. 2), but with every reservation as to probable future changes.[1]

The difficulties become still graver if we extend our research to more remote regions. A fairly good correlation may still be possible between the Carboniferous of the United States, Canada, Europe, and a part of western Asia, because we have reasons for assuming that in Carboniferous time these provinces were generally related and conditions of sedimentation were not too different. These assumptions do not hold for more distant regions; here we must reckon with greater climatic differences and also with different evolutionary successions in remote floral or faunistic provinces that were not at all or very slightly interrelated. We must bear in mind that the coal-measure flora was surely not uniform over the entire earth, but that floral provinces existed as they do today, though it must be admitted that these remote and more primitive floras were surprisingly similar and related the world over in comparison with the recent ones. Climatic differences dependent on latitude certainly existed. We know that living floras have

---

[1] As the result of my studies in the United States after the Geological Congress some of these changes have now been made in the following paper: Jongmans, W. J., and Gothan, W., Florenfolge und vergleichende Stratigraphie des Karbons der östlichen Staaten Nord-Amerikas—Vergleich mit West-Europa: Geologisch Bureau voor het Nederlandsche Mijngebied te Heerlen, Jaarverslag over 1933, pp. 17–44, Heerlen, 1934.

		Europe					Arcto – Carbonic province				Gigantopteris province				Gondwana province	
							United States of America									
		Dutch coal fields	Ruhr Basin	Belgium	France (north basin)	Great Britain	Appalachian region	Wichita and Ouachita region Ardmore Basin	Kansas	W. Texas Oklahoma	Chosen (Korea)	China		Southeastern Asia	South Africa	South America
												Shansi	Kaiping			
Permian.							Permian.	Per-mian.	Permian.	Permian with Gigantopteris.	Jido. Kobosan.	Upper and Lower Shihotse series. Yehmenkou series.	Coaltes de Xianxien. Kaiping.	Sumatra and Malay States.	Upper Waaihe.	Parana beds (Brazil).
Stephanian.	C	Jabeek.	Peisberg. Utbenturen. Flammkohlen.	Flenus.	Assise de Bruay.	Radstockian (Staffordian). Upper. Yorkian.	Monongohela. Conemaugh. Allegheny.	Pontotoc. Hoxbar. Francis. Deese. Mc.Alester. Hartshorne. Atoka. McCurtain.	Wabaunsee, Shawnee, Douglas, Lansing, Kansas City, Marmaton. Cherokee.			Assise de Tongshan (Taiyuanien proGigantopteris).				
	B	Maurits. Hendrik.	Gasflamm-kohlen. Gaskohlen.	Eikenberg. Asch.	Assise d'Anzin.	Upper. Yorkian.	(Mazon Creek Illinois). Upper Pottsville.	Hills.	Atoka formation.							
	A	Wilhelmina. Baar-. Io.	Fettkohlen. Upper Mager-kohlen. Lower Kohlen.	Genck. Beyne. Oupeye.	Assise de Vicoigne.	Lower. Yorkian.	Lower Pottsville. series.	Wapanucka. Jackfork. Springer. Sand-stone.	Hiatus.							
Namurian.	Epenian.	Epenian.	Lower Magerkohlen. Flözleeres.	Assise d'Andenne.	Flines.	Millstone grit.	? Parkwood. d'Alabama. Hiatus. ? Much Chunk. Pocono not present in Appalachians of Alabama.	Ardmore. Stanley. shale.	store. formation.							
	Gulpenian.	Gulpenian.	Alaunschefer.	(Lontzen beds) Chokier.	Bruille.	Carboniferous limestone. Calciferous sandstone.	Greenbrier. ? Sycamore. Hiatus. Woodford. Woodford Chattanooga	Caney. shale. Hiatus.	Chester. formation. Boone limestone.							
Dinantian.		Visean. (Hohlenkalk).	Visean.	Visean.	Visean.	grit.	Catskill.	Hiatus. Chattanooga.	Hiatus.							
Devonian.		Unknown.	Present.	Present.	Present.	Present.										

CORRELATION OF THE CARBONIFEROUS AND PERMIAN PROVINCES OF THE WORLD reached different stages of evolution in different parts of the earth. In certain regions, for example, recent assemblages present a general aspect that corresponds with that of the Tertiary of other regions. It is therefore unwarranted to conclude that fossil floras found in widely separated provinces are strictly contemporaneous merely because they are similar.

In the upper Carboniferous and the Permian (our data for the lower Carboniferous and older formations are still too incomplete) we may distinguish three geobotanic types or provinces—the Arcto-Carbonic (European-American), the *Gigantopteris*, and the Gondwana.

1. The Arcto-Carbonic flora is by far the best known. It contains the common types of coal-measure plants, abundantly described as occurring in the coal deposits of America and western Europe, including the Permian.

2. The *Gigantopteris* flora is known to occur in eastern Asia, inclusive of Sumatra and the Malay States; it also occurs in the western states of the North American continent. The lower divisions contain almost exclusively Arcto-Carbonic types, and the number of exotic forms is very small; higher in the series the number of the Arcto-Carbonic types decreases and the special forms like *Gigantopteris* and *Taeniopteris* become increasingly abundant. The typical *Gigantopteris* flora extends from the Stephanian through the Permian, perhaps into the Triassic. In the Westphalian flora of China *Gigantopteris* had not yet appeared.

3. The Gondwana flora is restricted to two antipodal, possibly at that time polar parts of the world, and must be considered the flora of a colder climate, of regions more or less close to glaciated regions, such as we know there were in that period in certain areas of the world, notably in the present Southern Hemiphere. In some localities we find *Glossopteris* elements, more or less plentiful, associated with Arcto-Carbonic types in the same strata. We must therefore conclude that the Gondwana flora existed contemporaneously with the other floras. By analogy with the Arcto-Carbonic types we may conclude that the Gondwana flora extended from Stephanian time through all of the Permian into the Mesozoic. Its occurrence in formations older than the Stephanian cannot yet be proved.

I want to emphasize again the fact that our present knowledge does not permit us to consider it proved that the floral assemblages which we would regard as Westphalian or older in Europe were strictly contemporaneous in other parts of the world: they may have lived in considerably different periods, of course within the Carboniferous but far from contemporary. Our correlations can be only relative.

As a working basis for research, not as a fact or even as a theory, an attempt at a world-wide correlation table may be made, subject to all the reservations discussed above (pl. 2).

The correlations given in plate 1 and in the European columns of plate 2 are practically identical with those arrived at in the Compte rendu of the congress at Heerlen, but the research work by Gothan, Patteisky, and Susta on the coal fields of Silesia and the adjacent region in Bohemia has made some improvement possible for that area. We may now consider it proved that the Sattelflötz division of Upper Silesia corresponds with the upper parts of the Epen division in the Netherlands, being upper Namurian. The lower part of the Weissteiner beds of Lower Silesia are at the same horizon. A considerable hiatus must be assumed to exist between the Weissteiner and the underlying Waldenburg beds, and therefore the greater portion of the upper Namurian, corresponding to the German Flötzleeres, or unproductive series, is lacking. According to Susta there seems to be a minor hiatus also between the lower portion of the Sattelflötz beds (Pochhammer or Prokop coal) and the underlying Ostrauer beds. If that is a fact, the upper Namurian of western Europe is but partly developed in Silesia.

As a great stratigraphic interval is indicated by the difference between the floral assemblages below the basal coal seam (Pochhammer or Prokop coal) and those in the roof, it is possible that this enormous coal seam represents all or a large part of this interval.

The data that would permit a comparison between the Carboniferous flora of Russia and that of other more western parts of Europe are still very incomplete. Notably those that concern the lower portions of the Namurian and the Dinantian are insufficient. I must therefore express doubt as to the correlations given by Zalessky, whose publications contain no sufficient proof that his beds C_1^1 to C_1^3 belong to the Dinantian.

Three columns in plate 2 represent an attempt to correlate the Carboniferous of North America with that of Europe. The Appalachian province is given first, because it is best known and also is nearest to the European deposits. The Mazon Creek beds of Illinois, although they belong to the interior basin, are inserted in this same column, because they are well known and their correlation is based on paleobotanic evidence. The whole of the Appalachian Pennsylvanian, except only the lowermost Pottsville, must be considered the equivalent of the European Stephanian and Westphalian. The published material is insufficient to permit trans-Atlantic correlation of smaller subdivisions. Much systematic collecting in the field and study of existing collections will be required before we can make such correlations. It seems probable that the lowest Pottsville is equivalent to the upper part of the European Namurian—namely, the lower part of the productive measures of the Ruhr Basin and part of the underlying unproductive measures (Flötzleeres) of Germany, and the equivalent upper portion of the Millstone grit of Great Britain. In the Appalachians there is, unfortunately, a break between the lower Pottsville and the Mauch Chunk. We have not sufficient information to enable us to correlate this gap with beds elsewhere in the United States. Though certain geologic considerations may lead us to suspect that some strata in Alabama, either low in the Pottsville or high in the Parkwood formation, or equivalent beds in the Ouachita region (Jackfork sandstone of Oklahoma and Arkansas, Springer formation of the Ardmore region, Oklahoma, or part of the Tesnus formation in southwestern Texas) may belong within this interval, our floral information is insufficient to substantiate this suspicion.

Our knowledge of the flora of the Mauch Chunk and Pocono formations, generally supposed to represent the lowest beds of the Carboniferous in a paralic facies, is also rather meager, and Arnold[2] even wishes to place part of the Pocono in the Devonian. At any rate, it seems that some Devonian forms occur in the

---

[2] Arnold, C. A., Fossil plants from the Pocono (Oswayo) sandstone of Pennsylvania: Michigan Acad. Sci. Papers, vol. 17, pp. 51–56, 1933.

Pocono, either because they survived into Mississippian time or because part of the Pocono is Devonian. As we have a practically unbroken succession of sediments, it is not surprising that we cannot draw a sharp dividing line. Extensive study of these formations is most desirable, because they are in deltaic facies, whereas everywhere else the equivalent sediments are purely marine. Here there is an opportunity to establish correlations between the terrestrial floras and the marine faunas of the lower Carboniferous.

Plate 2 contains two more columns in one of which are listed the deltaic Carboniferous deposits of the Ouachita foretrough and the Ardmore Basin, in the other the marine equivalents in Kansas. These correlations are not based on floral evidence but are copied from tables published by Van Waterschoot van der Gracht in his papers on the orogeny of these regions. Plant-bearing beds exist in these sequences, included in almost completely marine deposits. Further study of the fossil plants or the fresh-water animals from this transition region between the marine conditions of the West and the deltaic or continental development along the land areas created by the Permo-Carboniferous orogeny of the Ouachita and Wichita mountain systems would be most interesting. As a guide to such research I have tentatively inserted these regions in the table.

An attempt has also been made in plate 2 to correlate the *Gigantopteris* province with the Arcto-Carbonic province. The *Gigantopteris* flora is more or less known from papers by Halle, Kawasaki, Matthieu, Gothan, and me. So far as we know, the Chinese coal fields seem to contain the most complete sequence, because the upper Westphalian is represented there, as proved by both plant and animal fossils. Collections that have reached us hitherto from these lower strata do not contain representatives of the *Gigantopteris* flora but only European forms. *Gigantopteris* appears only in the higher Stephanian beds, found in Kaiping and in Sumatra. The Sumatra flora, studied by Gothan and me, contains chiefly European forms, numerous specimens of *Pecopteris*, *Sphenophyllum*, etc., which cannot be distinguished from the well-known European species; but together with these forms *Gigantopteris* and several specimens of *Taeniopteris* are present. Unfortunately both higher and lower zones are unknown in Sumatra. The lower part of the Shansian in Shansi and Kaiping (Yehmenkou series of Halle) has the same flora as we found in Sumatra. The higher zones of the Shansian contain many Permian types, and besides these several peculiar plants that are not represented in the Sumatra flora. The same condition is found in Chosen (Korea), where the lower part of the measures must also be considered Stephanian and the upper part Permian. So far we are unable to draw the dividing line between Carboniferous and Permian in China and Chosen. The species of *Gigantopteris* we find in Sumatra are not identical with those of China and Chosen; the rest of the flora is also different; even *Gigantopteris* is much rarer, apparently, in Sumatra than farther north. The Sumatra flora seems to have a more pronounced Arcto-Carbonic character than the floras of China; possibly it represents a transition between the floras of the older Tongshan beds of Kaiping, in which *Gigantopteris* has not yet been found, and the younger *Gigantopteris* floras.

In the western United States the *Gigantopteris* beds belong entirely to the Permian, as it is defined in North America. As far as we know, no trace of *Gigan-*

topteris has ever been found in beds of the American West attributed to the Pennsylvanian.

Not the slightest trace of any *Glossopteris* element has been found in the *Gigantopteris* floras. The few indications from Chosen, mentioned by Kawasaki, are certainly insufficient, as is pointed out in a paper by Gothan and me[3] presenting general considerations on the Sumatra flora.

In South Africa and in Brazil the Gondwana flora begins in the Stephanian; this is clearly indicated in localities where we find a transition from Arcto-Carbonic elements to a *Glossopteris* flora (Walton's Waukie flora in South Africa, and Lindquist's Brazilian flora). No example is known so far of any intermingling of *Glossopteris* elements with species of upper Carboniferous horizons lower than Stephanian. I consider the Gondwana flora indicative of a colder climate, having grown in the vicinity of glaciated regions, and if this view is accepted the late Paleozoic ice age must have had its beginning in upper Carboniferous (Stephanian) time. Naturally the Gondwana elements lived shortly before and after actual glaciation and followed the ice front in both its advancing and its retreating stage. Difficulty in determining the very beginning of the appearance of Gondwana elements is caused by the large break that occurs between beds with a Stephanian flora and those which contain older Carboniferous assemblages, in regions where we can expect a *Glossopteris* flora.

I desire to emphasize again that the correlations given in plate 2 for non-European regions are only very tentative and must by no means be taken as an established fact or even as an opinion; my object is exclusively to give guidance for future research—a statement of a possibility to be proved or disproved. However, I am decidedly of the opinion that a well-founded correlation between Europe and at least the Atlantic coal measures in the United States will be possible, as soon as our material is sufficient. The expanse of Carboniferous sediments of suitable facies is so wide in the United States that adequate research and collecting will naturally take very considerable time, but the closer the cooperation between workers on both sides of the Atlantic—backed by the interchange of collections obtained by the same methods in the field, and if possible by personal visits of specialists—the sooner the foundation can be laid for reliable correlations.

Discussion

Pruvost, P. E.: Je désire souligner ici l'importance des trois communications [Renier, Jongmans, and Bisat] qui sont présentées à cette séance. Sur l'initiative de M. W. J. Jongmans et celle de Messieurs A. Renier et W. Gothan un Congrès international a été tenu, il y a quelques années à Heerlen, pour les géologues spécialistes du Carbonifère européen. Ses résultats ont été d'apporter une grande précision dans les correlations des dépôts carbonifères du vieux continent.

De telles réunions spécialisées sont recommandables à la condition que leurs conclusions soient apportées devant la réunion suivante du Congrès géologique international. C'est ce que viennent de faire Messieurs Renier, Jongmans et Bisat, en y ajoutant leurs observations les plus récentes, et nous devons les en remercier.

---

[3] Jongmans, W. J., and Gothan, W., Die palaeobotanischen Ergebnisse der Djambi-Expedition, 1925: Jaarb. Mijnwezen Ned.-Indië, 1930, Verh., pp. 71–201, Batavia, 1935.

11

Reprinted from *Bull. Am. Mus. Nat. Hist.*, **99**(3), 189–203 (1952)

FOSSIL FLORAS OF THE SOUTHERN HEMISPHERE AND THEIR PHYTOGEOGRAPHICAL SIGNIFICANCE

THEODOR JUST

Chicago Natural History Museum

"IT HAS BEEN REMARKED that if the science of botany had grown up in the southern hemisphere, it would be now a very different science" (Bews, 1925). The appropriateness of this comment is constantly being demonstrated by our rapidly increasing knowledge of the living and extinct floras of the Southern Hemisphere. In recent years intensified exploration has disclosed many new and exceedingly fertile localities and the presence of new plant groups previously unknown from the Northern Hemisphere.

Despite these hopeful signs, any comprehensive treatment of the fossil floras of the Southern Hemisphere may appear daring and disproportionate, were it not for the fact that the late Paleozoic and early Mesozoic (or Mesophytic) ones have long been regarded as constituents of the *Glossopteris* flora and its descendants. Yet even a brief survey of these fossil floras, particularly those of Africa and South America, is indispensable in any discussion of their bearing on biological and geological theories (Dalloni, 1948; Darrah, 1939; Feruglio, 1951a; Frenguelli, 1949, 1950; Furon, 1949; Rodin, 1951; Seward, 1933a, 1933b; Sherlock, 1948).

Unfortunately, fewer fossil floras are known from the Southern Hemisphere and peninsular India than from the Northern. Moreover, the former do not compare favorably with their northern counterparts either in numbers of species recorded or the amount of research devoted so far to their study. For instance, Darrah (1945) published a map of South America showing the then known localities where fossil plants had been collected, and Seward (1933a) mapped some of the Mesozoic floras of the Southern Hemisphere, yet no map showing all known localities of the Southern Hemisphere is available at the moment. The limited numbers of fossil plant species reported per period from Central and South America and the Antilles are evident from Berry's (1942) calculations

(the area in question covers nearly 8,000,000 square miles): Triassic and Jurassic, one species to 160,000 square miles; Lower Cretaceous, one species to 100,000; Upper Cretaceous, one species to 300,000; Paleocene, one species to 600,000; Eocene, one species to 320,000; Oligocene, one species in 400,000; Miocene, one species in 20,000; Pliocene, one species in 40,000. Similar computations for the Paleozoic of this area or the fossil floras of Africa and other parts of the Southern Hemisphere would be even more out of proportion. Consequently conclusions based on such limited data are at best tentative rather than convincing.

To this difficulty must be added the great problem of correlating floras of roughly comparable ages but distant geographical locations. For example, after discussing "Stratigraphical problems in the coal measures of Europe and North America," Trueman (1946, pp. lxxxvii–lxxxix) concluded: "Any interpretation of the relationships between the two sides of the Atlantic is necessarily influenced by the acceptance or rejection of the hypothesis of continental drift. Certainly the explanation of the relationships is greatly facilitated by the assumption that during the Carboniferous North America was situated much closer to western Europe than at the present time, but even if the drift hypothesis is rejected, it is no less desirable to estimate the nature of the links between the two continents."

With few modifications this statement could be equally well applied to South America and Africa and indirectly to the other land masses of the Southern Hemisphere, where some problems of dating and correlating fossil floras are being resolved, but many more remain. Thus paleobotanical evidence indicates that the early vascular floras of Tasmania, supposedly of Cambro-Ordovician age, are now referable to the Silurian (Cookson, 1937), and those of Victoria are likewise referred to the Upper Silurian and Lower Devonian (Cookson, 1945, 1949). Of paramount importance are the age of the *Glossopteris* flora, formerly regarded as Permo-Carboniferous and now placed in the Permian or Permian-Triassic (Dixey, 1936; Du Toit, 1927; Gothan, 1930), and the correlation of the strata in the now widely separated areas occupied by it (Feruglio, 1951a; Sahni, 1927; Walkom, 1944, 1949).

PALEOPHYTIC FLORAS

Although the oldest plant fossils known from South Africa are of algal nature (Mac-Gregor, 1941; Vasconcelos, 1949; Young, 1949) and either Cambrian or pre-Cambrian in age, comparable records from other parts of the Southern Hemisphere seem to be quite sparse. It is therefore better to limit the following discussion to vascular floras.

Outstanding is the fact that the oldest pteridophyte (psilophyte) floras, ranging from Silurian to Lower Carboniferous, show a remarkable degree of uniformity, almost global in extent. These early vascular plants (psilophytes) ranged from the Upper Silurian to the Middle Devonian. The same degree of uniformity is seen even in the pteridophyte floras of the Upper Devonian and Lower Carboniferous, irrespective of which area is studied, whether Europe or North America, Siberia and eastern Asia, or South America, South Africa, and Australia (Hirmer, 1938).

This uniformity of early paleophytic floras and their subsequent differentiation are well illustrated by the succession of Carboniferous and Permian floras in Australia. In this succession Walkom (1937, 1944) recognizes three distinct fossil floras, namely, (1) the *Lepidodendron Veltheimianum* flora, composed of 12 species, the oldest one of this succession and very distinct from the Upper Devonian flora; (2) *Rhacopteris* flora, made up of 25 species, of Lower Carboniferous age, lacking distinctly northern Carboniferous genera, such as *Alethopteris, Neuropteris, Pecopteris*, and containing the genera *Rhacopteris, Cardiopteris, Sphenopteridium, Adiantites*, and *Noeggerathia;* and (3) *Glossopteris* flora, consisting of 48 species and varieties, abundant in Gondwanaland, and in Australia present in Lower (Creta) and Upper (Newcastle) Coal Measures. Taken as a whole, this flora is completely different from the *Rhacopteris* flora, although they have a few genera in common. But these two floras are also separated by a considerable break in time. Although the Lower Carboniferous flora of Australia can be compared with floras of the Northern Hemisphere, especially the European ones, the *Glossopteris* flora can be compared only with itself as represented in other parts of Gondwanaland. Significantly no trace of the *Rhacopteris* flora has so far been found in southern Africa, a fact that "needs some explanation" (Walkom, 1949). The absence of the *Glossopteris* flora in New Zealand also requires an explanation, if that country "occupied the position shown on Wegener's maps well into the Mesozoic Era" (Walkom, 1949, p. 12; see also Fleming, 1949; Oliver, 1950).

If the world be taken as a whole, distinct floras can first be distinguished in late Upper Carboniferous or shortly before the beginning of Permian (Darrah, 1937; Gothan, 1930; Jongmans, 1937, 1942; Read, 1947). These floras and their corresponding provinces can be traced from Stephanian to the beginning of Triassic. In accordance with Halle (1937) four major floras and provinces are now recognized, as follows (see fig. 20):

1. European Permo-Carboniferous (= Euramerican) flora, which extended throughout Europe to the Ural Mountains, Iran, Turkestan, and the eastern United States. This large area is today divided by the Atlantic Ocean.

2. Angara (=Siberian=Kusnezk) flora, which ranged from the Ural Mountains to the Pacific coast of Asia, from Vladivostok to Bering Strait, from the North Polar Sea to the Targatabai Mountains in the south and west to Turkestan and Kashmir. This is the best-defined natural area and includes the great Siberian coal basins of Kusnezk and Minussinsk.

3. Cathaysia flora, which extended from Shansi and Korea to Sumatra, New Guinea and western North America as far as Colorado, Oklahoma, and Texas, where it met the Euramerican flora; in Kansu, its western

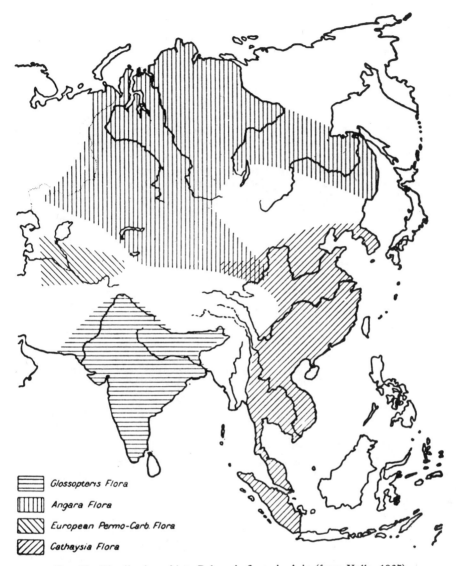

Fig. 20. Distribution of late Paleozoic floras in Asia (from Halle, 1937).

border, this flora was replaced by the invading Angara flora during Upper Permian. As is most likely, the large region occupied by this flora has always been divided by the Pacific Ocean. The Cathaysia flora was formerly called *Gigantopteris* flora, which begins in Middle Permian. The theory of continental drift does not seem to offer any better interpretation in regard to this region

than it seems to offer in connection with the problem of the origin of modern Pacific floras. This point has apparently not been stressed by other authors.

Provinces 1 to 3 are often treated collectively as the "northern or Arcto-Carboniferous province of warm-climate, rain forest coal floras" as compared with the "southern or Antarcto-Carboniferous province of cool-

119

climate, rain forest floras" (Read, 1947, p. 273). The Carboniferous and Permian floras of the large area extending from eastern Europe to eastern Asia and south to New Guinea have been carefully mapped and correlated by Jongmans (1942).

4. *Glossopteris* flora, which occupied Gondwanaland, including sections of central and southern South America, central and southern Africa, Australia, Antarctica, and India south of the Himalayan arch. During Permian the *Glossopteris* flora was in contact with the Angara flora in northern India, with the Euramerican flora in southern Rhodesia (Walton, 1929), Transvaal (Le Roux, 1949), Mozambique (Teixeira, 1947), Madagascar (Boureau, 1949), and with the Cathaysia flora in Dutch New Guinea (Jongmans, 1940). In Triassic times, elements of the *Glossopteris* flora were exchanged with Euramerican ones. One of the main arguments for continental drift is derived from the *Glossopteris* flora and its distribution.

Botanically speaking, the Angara flora is of real interest, as it contains elements of the other floras surrounding it. The mixed character of this flora is easily explained by its central position. It is noteworthy that it contains fewer elements of the *Glossopteris* flora than of the other two. Despite its synthetic composition, the Angara flora included many genera and species peculiar to it.

GLOSSOPTERIS FLORA

Although Gondwanaland was probably quite isolated, the *Glossopteris* flora had some contact, during early Permian at least, with the Angara flora in India, with the Euramerican flora in South Africa and South America, and the Cathaysia flora in Dutch New Guinea (see fig. 21). The few Euramerican elements present may be regarded as survivors of earlier cosmopolitan floras. The presence of an intermediate or transition flora in Kashmir suggests a northward migration of older Gondwana elements. The *Glossopteris* flora supposedly never invaded the Cathaysia flora despite the present proximity of areas once occupied by these floras. This fact is often construed as an argument in favor of drift (Holland, 1941, 1943; Sahni, 1936). However, a fossil flora containing Cathaysia elements has been reported from Dutch New Guinea, and *Vertebraria*, an important element of the *Glossopteris* flora, has been found in a near-by locality (Jongmans, 1940).

Comparison of the succession of Carboniferous and Permian floras in Australia with that of other parts of Gondwanaland was attempted by Walkom (1930, 1944). As pointed out above, only the Lower Carboniferous floras of Australia can be compared directly with those of the Northern Hemisphere. Permian floras of Australia, on the other hand, are directly comparable with those of western Argentina, because these Argentine floras "are almost identical with our Australian floras of the same succession." The South African succession, by comparison, is not so well represented as that of Australia. Interesting in this connection is the presence in the Lower Permian in Southern Rhodesia of "a very pronounced northern element" of Upper Carboniferous species (Walton, 1929) and of *Pecopteris* in Transvaal (Le Roux, 1949). *Sphenopteris*, *Asterotheca*, *Sphenophyllum*, and other elements of northern Permo-Carboniferous floras occur "together with the *Glossopteris* flora" in Mozambique (Teixeira, 1947, p. 28). This proves that "the gondwanean flora does not really exhibit the characters of independence which, years ago, many authors deemed it possible to ascribe to it." According to Walkom (1949, p. 12) this occurrence of northern species is "a result of slow migration southward from Europe. This migration appears to have been limited to this one line from Matthew's Holarctic region since so far this Upper Carboniferous flora is not known to have penetrated to the *Glossopteris* bearing areas of South America or Australia." Actually northern elements have been reported from Rio Grande do Sul, Brazil (Lundquist, 1919), and Dutch New Guinea (Jongmans, 1940), as indicated by arrows in figure 21. Finally, comparison with the small Carboniferous assemblage from India is barely possible, as the latter consists of only three species. How-

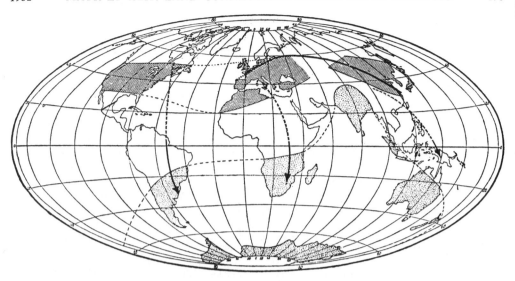

FIG. 21. Distribution of late Paleozoic floras. The continents are shown in their present position and size. The hatched area indicates "northern" floras; the stippled area shows the distribution of the "southern" or *Glossopteris* flora. The three arrows point to the localities where mixed floras of southern and northern elements have been found (modified from Zimmerman, 1930). Mixed floras have also been reported from Patagonia (Feruglio, 1951a), Mozambique (Teixeira, 1947), and Madagascar (Boureau, 1949). For further discussion, see Dalloni (1948).

ever, the Permian succession is well represented in India by the *Glossopteris·* flora, which in younger strata is associated with the *Thinnfeldia* flora (Triasso-Rhaetic). The restricted occurrence of certain species in the Indian succession may be valuable in dating isolated Australian finds.

Recent discoveries of new South American localities may eventually throw additional light on the entire succession (Dolianiti, 1946; Feruglio, 1951a; Jongmans and Gothan, 1951; Martins and Sobrinho, 1951; Mendes and Mezzalira, 1946; Mezzalira, 1949; Read, 1941). Detailed study of the plant microfossils of Gondwana deposits may greatly increase the roster of species attributed to the *Glossopteris* flora. Thus Ghosh and Sen (1948) recorded as new to the Lower Gondwanas almost all microfossils isolated by them from the Raniganj·coal field in Bengal. They described over 50 spore types, 11 wood fragments, 31 cuticles and related structures, and eight unknown plant fragments. Although the characters seen on spores may be of little use as indicators of past climates, extensive analyses of microfossils of widely

separated areas are bound to be immensely useful in correlating strata of comparable age (Virkki, 1939, 1945).

Actually our knowledge of the origin and fate of the *Glossopteris* flora is quite inadequate. Its origin can probably be connected with the cooling of the climate towards the end of the Carboniferous, resulting in large-scale extinction of pre-Gondwana cosmopolitan elements, including those of the Northern Hemisphere. Thus the *Glossopteris* flora lasted apparently from the Upper Carboniferous to the Triassic, representing a cold temperate flora in contrast to the Cathaysia flora, and becoming richer with the increasingly warmer climate of the Permian. At its maximum it was certainly not a glacial flora (Boutakoff, 1948; Du Toit, 1940; Fox, 1940; Jamotte, 1929, 1933; Sahni, 1939; Seward, 1931).

One of the greatest problems is the origin of the glossopterids. Did they originate in Argentina or South Africa, and in which direction did they migrate? Gondwanaland was well isolated from the rest of the world by the Tethys Sea occupying the geosyncline from which the Himalayas arose. However, Kashmir and Assam were promontories via which the contact with the Angara flora must have taken place (Sahni, 1936). On the whole, evolution in Gondwanaland was fairly uninterrupted, whereas major changes must have occurred in the north. A notable fact is the great impoverishment of the *Glossopteris* flora during Lowest Triassic, leaving *Dadoxylon*, two species of *Glossopteris*, and *Phyllotheca* or *Schizoneura*. During this time Brazil and South Africa probably occupied the heart of Gondwanaland, and plant migrations occurred more readily near the periphery of the continent (Du Toit, 1930). During late Triassic the flora became more numerous, reaching its peak during Upper Keuper in the Molteno stage of South Africa and corresponding stages elsewhere in the Southern Hemisphere. Pteridosperms were then largely replaced by ferns, ginkgophytes, and cycadophytes, whereas *Glossopteris* remained in South Africa and Tonkin and possibly left as its successor the genus *Linguifolium* of Chile, Tasmania, and New Zealand (Berry, 1945).

This famous Molteno flora has been dated by Du Toit (1927) as essentially of Upper Triassic age, containing some Permian survivors and many new and abundant elements, especially ferns and ginkgophytes which migrated north to form Rhaetic assemblages in the Northern Hemisphere. Or as Du Toit (1927, p. 290) puts it: "The rather poor Permo-Triassic '*Glossopteris* assemblage' of Gondwanaland has indeed now been replaced almost entirely by the more abundant and cosmopolitan Trias-Rhaetic '*Thinnfeldia* Flora.'" *Thinnfeldia* and other frond genera are now believed to represent the foli-

age of the Mesozoic pteridosperm family Corystospermaceae described from the Molteno beds by Thomas (1933).

Effective comparison of the *Glossopteris* flora as represented in various parts of Gondwanaland is greatly hindered by the profound taxonomic problems encountered. For instance, Jones (1949) and Jones and de Jersey (1947) pointed out that much of the Australian *Thinnfeldia* material might profitably be treated as a single species rather than divided up among five of the eight species customarily recognized in Australia. Generic segregation of the complex into geographical groups, as proposed by Frenguelli (1943) on external morphological grounds, is opposed by Jones and de Jersey (1947). Cuticular analysis of all, or the major part of, the genera of the *Thinnfeldia* series (T. M. Harris, 1931–1937, pt. 1) is bound to result in different generic assignments and delimitations (Lundblad, 1950). Actually, Jacob and Jacob (1950) accept the inclusion of the southern thinnfeldias under *Dicroidium* on the strength of their study of the cuticular characters of eight species of this genus. Contrary to Jones and de Jersey (1947), they recommend that the five species of *Dicroidium* combined by these authors be maintained as separate entities.

Whence did the thinnfeldias come? Did they originate in the Northern Hemisphere from Permian ancestors like *Supaia* as represented in the Hermit Shale of Arizona? "If some of the Hermit Shale plants are the ancestors of the Thinnfeldias, their distribution could be explained either on the theory of isthmian links of Willis and Schuchert or on Wegener's theory." Walkom (1949) favors the former.

MESOZOIC FLORAS

Beginning with the well-known Rhaeto-Liassic floras of the Northern Hemisphere, especially those of Greenland, it is possible to distinguish three main floral regions (T. M. Harris, 1931–1937, pt. 5; Hirmer, 1939):

1. Northern province: which ranged from eastern Greenland (latitude 70° N.), through Sweden to France, Rumania (latitudes 45°–

56° N.), Poland, Russia, Siberia within the same latitudes to western Japan (about latitude 35° N.) (see Kobayashi, 1942, for Mesozoic floras of eastern Asia).

2. Middle province: which extended from Tonkin to certain localities in China proper, Pamir, Iran, Armenia, Mexico, and British Honduras.

3. Southern province: various floras of

South Africa, Australia (Queensland), and Argentina may eventually be referred to this province after their correct age has been established (see Feruglio, 1951b, for map of Mesozoic floras of Argentina).

The northern province is marked by the presence of a uniform flora, extending 160° in longitude (from 25° W. of Greenwich to 135° E.), and 35° in latitude (between 70° and 35° N.). Characteristically this province ran obliquely across present climatic and latitudinal belts. The same phenomenon is seen with regard to the central province. Hirmer (1939) believes that this position is in agreement with the distribution of land at the end of Triassic and the beginning of Jurassic, as postulated by Köppen and Wegener (1924).

As the exact age of the floras of the southern province is still doubtful (Frenguelli, 1948; T. M. Harris, 1931–1937, pt. 5; Fossa-Mancini, 1940; Teixeira, 1943), additional information is badly needed, before this province can be satisfactorily defined. This aim

will not be materialized until critical collections from these areas have been studied in detail by methods employed by Harris on his Greenland material or by Lundblad (1950) in the description of the Rhaeto-Liassic floras of Sweden.

Not many Jurassic floras are known from the Southern Hemisphere, and none is so well represented as the famous British flora. Edwards (1934) analyzed the known Jurassic plants from New Zealand, comprising 40 to 50 species from six localities and possibly several horizons. He found that half that number (20 to 25) was also known from Australia, one-third from India, and one-fourth from Graham Land. A new Jurassic flora has been reported from the Esquel schists of Argentina (Cazoubon, 1947). Comparison of such recently discovered floras with well-known ones may provide the necessary information to resolve the long controversy concerning the reputedly uniform climate of the Jurassic which was apparently more imaginary than real (Florin, 1940b).

CRETACEOUS AND TERTIARY FLORAS

For practical as well as scientific reasons late Cretaceous and Tertiary floras are often treated together. On the whole, floras of these periods found in the Northern Hemisphere are also better known than their contemporary floras in the south. How much uncertainty surrounds many of these southern floras is seen from Gerth's attempt (1941) to appraise all Tertiary floras of southern South America. He placed the known floras in two groups, both of which have descended from a common ancestor, an Upper Cretaceous subtropical flora found throughout South America: (1) *Maytenus-Zamia* flora, which was limited to humid subtropical or tropical climates, and originally described from a locality south of Concepción, Chile; and (2) *Nothofagus-Araucaria* flora, found in humid but cool climates and first reported from Tierra del Fuego and Antarctica; *Nothofagus* and *Araucaria* are still living in the area. While this flora arose at the extreme southern tip of South America, the *Maytenus-Zamia* flora arose at about the same time but much farther north. As the climate de-

teriorated during Tertiary times the *Maytenus-Zamia* flora retreated and the *Nothofagus-Araucaria* flora advanced northward. This trend in the change of climates has apparently not been altered since the middle of Tertiary. Gerth construes this fact as an argument against the changes of the South Pole postulated in connection with continental drift.

Another interesting discovery was recently announced by Selling (1947). A collection of Cretaceous plants from Patagonia contains among conifers, various dicotyledonous leaves, etc., and flowering and fruiting spikes of a species of *Aponogeton*. This family (Aponogetonaceae) had so far not been reported from the Western Hemisphere, either living or extinct. Moreover, the species represented seems to be related to a species of *Aponogeton* endemic in tropical Africa rather than species found in Australia or Asia. This discovery as well as relationships existing among modern plants found in South America and Africa disqualifies Berry's (1942) contention that the Tertiary floras of South

America were as American as the African ones were African. But he is undoubtedly correct in stating that considerable north and south migrations occurred at least during early Tertiary both in the New as well as in the Old World. In each case, however, the Miocene seems to mark the beginning of dissimilarity, especially in the Americas; in short, the floras assume their modern aspect (Hirmer, 1939).

In all probability, our knowledge of Tertiary floras will be greatly increased as soon as more extensive studies of spores and pollen (palynology) from various parts of the world are made and supplemented by anatomical studies of fossil woods (W. F. Harris, 1950). Some examples may illustrate this. In Africa some significant data have come from Egypt where floras dated as Middle Cretaceous furnished the first fossil record of a predominantly austral angiosperm family, the Proteaceae, from the Northern Hemisphere. These floras consisted largely of forests made up of many species of this family (represented by woods and fruits). Although the distribution of the modern members of this family has often been used in support of drift, Kausik (1943) concluded that drift alone could not account for all peculiarities of distribution of this family, past or present. Consequently, he asked for more data, particularly on fossil members. Lately Cookson (1950) and Cookson and Duigan (1950) published exemplary studies on fossil members of this family as represented in Tertiary deposits of Australia, carefully based on analyses of and comparisons with living species, particularly of their cuticular characters. These authors are of the opinion that most extra-Australian records of fossil Banksieae (Proteaceae) are based on misidentifications.

Two similar instances of regressive areas are well documented. Hofmann (1948) proved convincingly that mangrove occurred in the late Cretaceous in Central Europe (Austria) by detailed comparisons of leaf fragments, pollen, and wood samples. This mangrove included species now limited either to Old or New World tropics. The second instance concerns the past and present distribution of the austral family Dipterocarpaceae. Although its present center of distri-

bution lies in southeastern Asia, fossil woods belonging to this family have been recorded from western Europe. Recently discovered fossil woods from Egypt referable to this family now link these widely separated localities (Hirmer, 1939).

These cases are so striking because they represent definitely or predominantly austral angiosperm families, which in the past occupied larger areas frequently extending into the Northern Hemisphere. By comparison, loss of area by many fern and gymnosperm groups is to be expected as a result of the general decline of these groups during the explosive expansion of the angiosperms in the Cretaceous and early Tertiary. An excellent example of loss of area was provided by Hirmer and Hörhammer (1936) who studied the past and present distribution of the Matoniaceae, a family of tropical ferns. Today this family occupies only the equatorial portion of its former large tropical area.

Antarctica has long been regarded as a major center of origin and distribution of modern plant groups (Darrah, 1936; Hill, 1929; Wade, 1941; et al.). For instance, Copeland (1947) now estimates that "practically the whole fern world of the tropics is descended from the ferns of old Antarctica"; to be specific, nine-tenths of the tropical fern flora.

Not all migrations followed this direction, as many occurred from north to south and may even be going on today. Diels (1942) has shown how far south certain pioneers of the Holarctic realm may penetrate into tropical areas such as the Philippines or Central America. These migrations were probably induced by climatic changes during late Tertiary or Pleistocene times and thus are not very old. The remarkable fact about them is that these pioneers have hardly changed in the process, and their variability has apparently not increased beyond the range known in their normal habitats and concentrated areas of distribution.

The contrast between tropical and temperate floras in the past, at least since the late Cretaceous, has remained essentially unchanged, e.g., the tropical zones and the Equator have remained in their present position, according to evidence adduced by

Kaul (1943, 1945) from the study of fossil palm stems collectively referred to the genus *Palmoxylon.* However, Kaul explains the pre-Tertiary distribution of temperate and tropical floras on the basis of drift, at least "in Asia from the Tertiary backwards to the Carboniferous period." Obviously many more data from such critical areas will be needed before such far-reaching conclusions can be regarded as safely established. Such analyses should include extensive ecological and geographical accounts of living tropical floras, notably those of the Far East (Airy Shaw, 1942; Gordon, 1949; W. F. Harris, 1950;

Richards, 1942). Gordon (1949) advocates "two lines of research; firstly the continued and intensified study of fossil angiosperm floras and contemporary faunas all over the world, with strict insistance on reliable identification or none at all, and secondly a detailed investigation of the geological structure and origin, and the botanical history, of isolated oceanic islands, especially Hawaii. When we can explain how *Astelia* and other sub-antarctic genera reached Hawaii, I don't think we will have so much difficulty in accounting for their distribution in the southern continents."

DISTRIBUTION OF GYMNOSPERMS

In addition to the tracing of the history of entire floras it is often equally instructive and profitable to examine the records of important groups. The history of the gymnosperms is of particular interest here, as their rise and subsequent decline either coincided with or, as is more likely, preceded the spectacular ascent of the angiosperms in late Mesozoic times (for summary, see Just, 1948). Moreover, some of the most important recent advances in paleobotany pertain to the Mesozoic history of the gymnosperms.

The present distribution of the Cycadaceae shows remnants of an area undoubtedly much larger in the past (Suessenguth's Schollenareale or Vester's pluricontinental area). Apparently the cycads originated from Permian pteridosperms (late Paleozoic taeniopterids). Although they were never so numerous as their contemporaries, the Cycadeoidales (Bennettitales), the cycads survived until the present, whereas the cycadeoids vanished completely in the Upper Cretaceous. This phenomenon always elicited considerable speculation regarding the causes of extinction. In preference to the old stories of insect attacks, ill effects of injurious volcanic gases, etc., Ridley (1938) offered a rather plausible explanation, namely, "The evolution of the broad-leaved dense forest is, I think, enough to account for the disappearance of the Mesozoic Cycads and gymnosperms." According to him, some living species are actually near extinction, as the Christmas Island cycads, and other species

which propagate mostly by vegetative means rather than by sexual reproduction (*Cycas Rumphiana*). The distribution of the cycads has been interpreted on the basis of drift and also used as an argument against it. As far as is known, both lines of Cycadaceae, namely, the Cycadoideae and Zamioideae, are equally old, dating from the end of the Triassic (Rhaetic) (Florin, 1933; T. M. Harris, 1941) and, paradoxically enough, are better represented by Mesozoic fossils than Tertiary records.

The case of the Ginkgoales is equally interesting and significant (Florin, 1936, 1940c, 1949). Well-established fossil representatives of this order belonging to some 15 genera ranged from the Lower Permian to the Upper Pliocene. The order originated from the same ancestors from which the Cordaitales came and not from the latter, as commonly held. *Ginkgo*, often listed as the oldest known plant genus, dates from the Middle Jurassic and comprises seven fossil species and the type species, *Ginkgo biloba.* Beginning with the Upper Triassic, the evolution of the Ginkgoales is marked by the rapid appearance of new forms represented by many individuals. The order reached its greatest development during Jurassic and Lower Cretaceous. Then all genera disappeared except *Ginkgo, Ginkgoites,* and *Torellia.* Like the cycadeoids they show a marked decline during the Middle Cretaceous. This decline, too, may well have been associated with the explosive evolution of the angiosperms during the Cretaceous.

The Tertiary conifers of the Southern Hemisphere (Florin, 1940b) have been more carefully studied than any other large systematic group. Florin (1940a) has shown that the origin of the conifers must probably be dated back to Westphalian, as the conifer floras of the Upper Carboniferous and Permian are as clearly separated as are the other elements of these periods.

During late Paleozoic the conifers of the Northern Hemisphere progressed far more than their southern contemporaries and were more widely distributed than the latter. Beginning with the Permian, and especially since the Jurassic, the conifers of the Southern Hemisphere became more and more differentiated from their northern counterparts. Apparently contact between these two groups existed across trans-tropical bridges. However, northern genera rarely over-stepped the Equator, whereas southern genera probably migrated more often to the north. Few conifer genera are represented in both hemispheres. The great uniformity of climate and world-wide distribution of Jurassic floras postulated by many paleobotanists can no longer be substantiated by recourse to the conifers.

The traditional view of the north temperate (or even Arctic) origin of conifers is apparently in need of some revision. As far as southern conifers are concerned, Florin (1940b, p. 92) assumes former land connections "or at least much closer proximity, between Antarctica and the adjacent southern ends of South America, Australia, New Zealand, and South Africa."

SUMMARY

Modified drift theories continually attract favorable comment from many paleobotanists and botanists (for summaries, see Just, 1947; Good, 1950). However, not all problems of origin and distribution can be explained on the basis of continental drift, certainly not at our present state of knowledge. In many instances, we need more fossil records of critical groups from critical areas and periods, especially of early angiosperm types. Surprises may also be in store, such as the discovery of a new gymnosperm group, the Pentoxyleae, so ably described by the late Professor Sahni from Jurassic beds of India (1948). Our rapidly growing knowledge of plant micro-fossils will be an indispensable aid in solving problems of correlation. Progress may also result from reorientation and new interpretations. The dramatic reclassification of the large cycadophyte complex by Florin and Harris is an excellent example in point (see summary in Just, 1951). Ecologists, too, may soon be able to give us more information regarding the effects of microclimate on local distribution and survival of many important species. Thus the vestiges of our kaleidoscopic picture of the fossil floras of the Southern Hemisphere will be removed, the sequence of events established, and the panorama completed.

REFERENCES

AIRY SHAW, H. K.
 1942. The biogeographic division of the Indo-Australian archipelago. 5. Some general considerations from the botanical standpoint. Proc. Linnean Soc. London, vol. 154, pp. 148–154.
BERRY, E. W.
 1942. Mesozoic and Cenozoic plants of South America, Central America, and the Antilles. Proc. 8th Amer. Sci. Congr., vol. 4, pp. 365–373.
 1945. The genus Linguifolium of Arber. Johns Hopkins Univ. Studies Geol., vol. 14, pp. 187–190.
BEWS, J. W.
 1925. Plant forms and their evolution in South Africa. London.
BOUREAU, E.
 1949. Paléophytogéographie de Madagascar. Mém. Inst. Sci. Madagascar, ser. D, vol. 1, no. 2, pp. 81–96.
BOUTAKOFF, N.
 1948. Les formations glaciaires et postglaciaires fossilifères, d'âge permo-car-

bonifère (Karroo inférieur) de la région de Walikale (Kivu, Congo belge). Mém. Inst. Géol. Univ. Louvain, vol. 9, no. 2, 124 pp., 5 pls.

CAZOUBON, A. J.
1947. Una nueva flórula jurásica en el cordón de Esquel en el Chubut meridional. Rev. Asoc. Geol. Argentina, vol. 2, no. 1, pp. 41–58, 3 figs., 2 pls.

COOKSON, I. C.
1937. The occurrence of fossil plants at Warrentinna, Tasmania. Papers Proc. Roy. Soc. Tasmania, for 1936, pp. 73–78.

1945. Records of plant remains from the Upper Silurian and Early Devonian rocks of Victoria. Proc. Roy. Soc. Victoria, new ser., vol. 56, no. 2, pp. 119–122.

1949. Yeringian (Lower Devonian) plant remains from Lilydale, Victoria, with notes on a collection from a new locality in the Siluro-Devonian sequence. Mem. Natl. Mus. Melbourne, vol. 16, pp. 117–131, 6 pls.

1950. Fossil pollen grains of proteaceous type from Tertiary deposits in Australia. Australian Jour. Sci. Res., ser. B, biol. sci., vol. 3, no. 2, pp. 166–177, 3 pls.

COOKSON, I. C., AND SUZANNE L. DUIGAN
1950. Fossil Banksieae from Yallourn, Victoria, with notes on the morphology and anatomy of living species. Australian Jour. Sci. Res., ser. B, biol. sci., vol. 3, no. 2, pp. 133–165, 9 pls.

COPELAND, E. B.
1947. Genera filicum. Ann. Cryptog. and Phytopath., Waltham, Massachusetts, no. 5, xvi+247 pp., 10 pls.

DALLONI, M.
1948. Observations sur la paléogéographie de la fin des temps primaires et la notion du "continent de Gondwana." La Rev. Sci., Paris, year 86, fasc. 9, pp. 523–526.

DARRAH, W. C.
1936. Antarctic fossil plants. Science, vol. 83, pp. 390–391.

1937. Some floral relations between the Late Paleozoic of Asia and North America. Problems Paleont., Moscow, nos. 2–3, pp. 195–200.

1939. Notas sobre la historia de la paleobotánica Sudamericana. Lilloa, vol. 6, no. 2, pp. 213–239, 3 pls.

1945. Paleobotanical work in Latin America. *In* Verdoorn, Frans, Plants and plant sciences in Latin America. Waltham, Massachusetts, pp. 181–183.

DIELS, L.
1942. Ueber die Ausstrahlungen des Holarktischen Florenreiches an seinem Südrande. Abhandl. Preussischen Akad. Wiss., Math. Nat. Kl., no. 1 pp. 1–14.

DIXEY, F.
1936. The transgression of the Upper Karroo, and its counterpart in Gondwanaland. Trans. Geol. Soc. South Africa, vol. 38, pp. 73–89.

DOLIANITI, E.
1946. Notícia sobre novas formas no "flora de Glossopteris" do Brasil meridional. Notas Prelim. e Estud., Div. Geol. Min., Rio de Janeiro, no. 34, 6 pp., 1 pl.

DU TOIT, A. L.
1927. The fossil flora of the Upper Karroo beds. Ann. South African Museum, vol. 22, no. 5, pp. 289–420.

1930. A short review of the Karroo fossil flora. Compt. Rendus 15th Internatl. Geol. Congr., for 1929, vol. 2, pp. 239–251.

1940. Climatic variations over southern Africa during later Paleozoic. Rept. 17th Internatl. Geol. Congr., USSR, vol. 6, pp. 213–221, 1 fig.

EDWARDS, W. N.
1934. Jurassic plants from New Zealand. Ann. Mag. Nat. Hist., ser. 10, vol. 13, pp. 81–109, 6 figs., 2 pls.

FERUGLIO, E.
1951a. Su alcune piante del Gondwana inferiore della Patagonia. Pubbl. Ist. Geol. Univ. Torino, fasc. 1, pp. 1–34, pls. 1–4.

1951b. Piante del Mesozoico della Patagonia. *Ibid.*, fasc. 1, pp. 35–80, pls. 1–3.

FLEMING, C. A.
1949. The geological history of New Zealand (with reference to the origin and history of the fauna and flora). Tuatara, vol. 2, no. 2, pp. 72–90.

FLORIN, R.
1933. Studien über die Cycadales des Mesozoikums, nebst Erörterungen über die Spaltöffnungsapparate der Bennettitales. K. Svenska Vetensk.-Akad. Handl., ser. 3, vol. 12, no. 5, pp. 1–134, 40 figs., 16 pls.

1936. Die fossilen Ginkgophyten von Franz-Joseph-Land nebst Erörterungen über vermeintliche Cordaitales mesozoischen Alters. ! Palaeontographica, vol. 81B, nos. 3–6, pp. 71–173, 20 figs., pls. 11–42; vol. 82B, nos. 1–4, pp. 1–72, 8 figs., pls. 1–6.

1940a. On Walkomia n. gen. a genus of Upper

Paleozoic conifers from Gondwanaland. K. Svenska Vetensk.-Akad. Handl., ser. 3, vol. 18, no. 5, 23 pp.

1940b. The Tertiary fossil conifers of south Chile and their phytogeographical significance. With a review of the fossil conifers of southern lands. *Ibid.*, ser. 3, vol. 19, no. 2, 107 pp.

1940c. On the occurrence of the genus Sphenobaiera (Ginkgoales) in the Tertiary of south Chile. Rev. Univ., Univ. Catolica de Chile, vol. 25, no. 3, pp. 147–154, 2 pls.

1949. The morphology of *Trichopitys heteromorpha* Saporta, a seed-plant of Paleozoic age, and the evolution of the female flowers in the Ginkgoinae. Acta Horti Bergiani, vol. 15, no. 5, pp. 79–109, 4 pls.

FOSSA-MANCINI, E.
1940. Los caracteres paleontológicos del Rético en la Republica Argentina y en Chile según H. Gerth. Notas Mus. La Plata, vol. 5, geol. no. 11, pp. 259–293.

1941. Los "Bosques Petrificados" de la Argentina según E. S. Riggs y G. R. Wieland. *Ibid.*, vol. 6, geol. no. 12, pp. 59–92.

FOX, C. S.
1940. The climates of Gondwanaland during the Gondwana era in the Indian region. Rept. 17th Internatl. Geol. Congr., USSR, vol. 6, pp. 187–208.

FRENGUELLI, J.
1943. Reseña crítica de los géneros atribuídos a la "Serie de Thinnfeldia." Rev. Mus. La Plata, vol. 2, sect. paleont., pp. 225–342, figs. 1–33.

1948. Estratigrafia y Edad del Llamado "Retico" en la Argentina. Gaea, vol. 8, pp. 159–309, 4 pls.

1949. Addenda a la flora del Gondwana superior en la Argentina. Physis, Buenos Aires, vol. 20, no. 57, pp. 139–158, 5 pls.

1950. Addenda a la flora del Gondwana superior en la Argentina. Rev. Asoc. Geol. Argentina, vol. 5, no. 1, pp. 15–30, 2 pls.

FURON, R.
1949. Notes sur la paléogéographie de Madagascar. Mém. Inst. Sci. Madagascar, ser. D, vol. 1, no. 2, pp. 69–80.

GERTH, H.
1941. Die Tertiärfloren des südlichen Südamerika und die angebliche Verlagerung des Südpols während dieser Periode. Geol. Rundschau, vol. 32, pp. 321–336.

GHOSH, A. K., AND J. SEN
1948. A study of the microfossils and the correlation of some productive coal seams of the Raniganj coalfield, Bengal, India. Trans. Min., Geol., Metal. Inst. India, vol. 43, no. 2, pp. 67–95, figs. A–C, pls. 3–14.

GOOD, R.
1950. Present position of the theory of continental drift. Nature, vol. 199, pp. 585–586.

GORDON, H. D.
1949. The problem of sub-Antarctic plant distribution. Rept. Australian and New Zealand Assoc. Adv. Sci., Hobart meeting, pp. 142–149.

GOTHAN, W.
1930. Die pflanzengeographischen Verhältnisse am Ende des Paläozoikums. Englers Bot. Jahrb., vol. 63, pp. 350–367.

HALLE, T. G.
1937. The relation between the Late Paleozoic floras of eastern and northern Asia. Compt. Rendu Deuxième Congr. Av. Études de Stratigraphie Carbonifère, Heerlen, 1935, vol. 1, pp. 237–245.

HARRIS, T. M.
1931–1937. The fossil flora of Scoresby Sound, East Greenland. Part 1 (1931): Meddel. om Grønland, vol. 85, no. 2, 104 pp., 37 figs., 18 pls.; pt. 2 (1932): *ibid.*, vol. 85, no. 3, 114 pp., 39 figs., 9 pls.; pt. 3 (1932): *ibid.*, vol. 85, no. 5, 133 pp., 52 figs., 19 pls.; pt. 4 (1935); *ibid.*, vol. 112, no. 1, 176 pp., 53 figs., 29 pls.; pt. 5 (1937): *ibid.*, vol. 112, no. 2, 114 pp., 5 figs., 1 pl., 3 tables.

1941. Cones of extinct cycadales from the Jurassic rocks of Yorkshire. Trans. Roy. Soc. London, ser. B, vol. 231, pp. 75–98, 5 figs., pls. 5–6.

HARRIS, W. F.
1950. Climate relations of fossil and recent floras. Tuatara, vol. 3, no. 2, pp. 53–66.

HILL, A. W.
1929. Antarctica and problems in geographical distribution. Proc. Internatl. Congr. Plant Sci., vol. 2, pp. 1477–1486.

HIRMER, M.
1938. Geographie und zeitliche Verbreitung der fossilen Pteridophyten. *In* Verdoorn, Frans, Manual of pteridology. The Hague, pp. 474–495.

1939. Paläobotanik. *In* von Wettstein, F., Fortschritte der Botanik. Berlin, vol. 8, pp. 91–130.

HIRMER, M., AND L. HÖRHAMMER
1936. Morphologie, Systematik und geographische Verbreitung der fossilen und rezenten Matoniaceen. Palaeontographica, vol. 81B, nos. 1/2, pp. 1–70, 7 figs., 10 pls.

HOFMANN, E.
1948. Das Flyschproblem im Lichte der Pollenanalyse. Phyton (Ann. Rei Bot.), vol. 1, no. 1, pp. 80–101.

HOLLAND, T. H.
1941. The evolution of continents: a possible reconciliation of conflicting evidence. Proc. Roy. Soc. Edinburgh, ser. B, vol. 61, pp. 149–166.
1943. The theory of continental drift. Proc. Linnean Soc. London, vol. 155, pp. 112–125 (discussion on pp. 119–125).

JACOB, K., AND (MRS.) CHINNA JACOB
1950. A preliminary account of the structure of the cuticles of Dicroidium (Thinnfeldia) fronds from the Mesozoic of Australia. Proc. Natl. Inst. Sci. India, vol. 16, no. 2, pp. 101–126.

JAMOTTE, A.
1929. Sur la découverte d'une flore à "Glossopteris" dans la vallée de la Lukuga, aux environs de Greinerville (Congo belge). Bull. Acad. Roy. Belgique, cl. sci., vol. 15, no. 7, pp. 635–638.
1933. Découverte de la flore à *Glossopteris* dans la cuvette charbonnière de la Luena (Katanga). *Ibid.*, cl. sci., vol. 19, no. 5, pp. 561–564.

JONES, O. A.
1949. Problems of Queensland Mesozoic paleobotany. Australian Jour. Sci., vol. 11, no. 6, pp. 192–193.

JONES, O. A., AND N. J. DE JERSEY
1947. The flora of the Ipswich coal measures—morphology and floral succession. Papers Dept. Geol. Univ. Queensland, vol. 3, nos. 3/4, pp. 1–88, 10 pls.

JONGMANS, W. J.
1937. Floral correlations and geobotanical provinces within the Carboniferous. Internatl. Geol. Congr., Washington, 1933, 16th Sess., vol. 1, pp. 519–527, pls. 1–2.
1940. Beiträge zur Kenntnis der Karbonflora von Niederländisch Neu-Guinea. Meded. Geol. Bur. Mijngeb. Heerlen, 1938 en 1939, pp. 263–274, 3 pls.
1942. Das Alter der Karbon- und Permfloren von Osteuropa bis Ost-Asien. Palaeontographica, vol. 87B, no. 1, pp. pp. 1–58, 1 map.

1951. Fossil plants of the island of Bintan with a contribution by J. W. H. Adam). Proc. K. Nederlandse Akad. Wetensch., ser. B, vol. 54, no. 2, pp. 3–10, pls. 1–3, 1 map.

JONGMANS, W. J., AND W. GOTHAN
1951. Beitrag zur Kenntnis von Alethopteris Branneri White. An. Acad. Brasileira Cien., vol. 23, no. 3, pp. 283–290, 4 pls.

JUST, T.
1947. Geology and plant distribution. Ecol. Monogr., vol. 17, no. 2, pp. 127–137.
1948. Gymnosperms and the origin of angiosperms. Bot. Gaz., vol. 110, no. 1, pp. 91–103.
1951. Mesozoic plant microfossils and their geological significance. Jour. Paleont., vol. 25, no. 6, pp. 729–735.

KAUL, K. N.
1943. Discussion. *In* Holland, T. H., 1943 (*q.v.*), p. 124.
1945. A fossil palm stem from South Africa (Palmoxylon du toitii), sp. nov. Proc. Linnean Soc. London, vol. 156, no. 3, pp. 197–198.

KAUSIK, S. B.
1943. Distribution of the Proteaceae, past and present. Jour. Indian Bot. Soc., vol. 22, pp. 105–123.

KOBAYASHI, T.
1942. On the climatic bearing of the Mesozoic floras in eastern Asia. Japanese Jour. Geol. Geogr., vol. 18, no. 4, pp. 157–196.

KÖPPEN, W., AND A. WEGENER
1924. Die Klimate der geologischen Vorzeit. Berlin, 256 pp.

LE ROUX, S. F.
1949. A fossil plant, of the genus Pecopteris, from Vereeniging. South African Sci., vol. 2, pp. 132–133.

LUNDBLAD, ANNA B.
1950. Studies in the Rhaeto-Liassic floras of Sweden. I. Pteridophyta, Pteridospermae, and Cycadophyta from the mining district of NW Scania. K. Svenska Vetensk.-Akad. Handl., ser. 4, vol. 1, no. 8, 82 pp., 13 pls.

LUNDQUIST, G.
1919. Fossile Pflanzen der Glossopteris-Flora aus Brasilien. K. Svenska Vetensk. Akad. Handl., new ser., vol. 60, no. 3, 36 pp., 2 pls.

MACGREGOR, A. M.
1941. A pre-Cambrian algal limestone in southern Rhodesia. Trans. Geol. Soc. South Africa, vol. 43, pp. 9–15.

MARTINS, E. A., AND M. S. SOBRINHO
1951. Lycopodiopsis Derby (Renault) e Glossopteris sp. no Estrada Nova (Permo-
Triassico) do Rio Grande do Sul. An.
Acad. Brasileira Cien., vol. 23, no. 3, pp.
323–326, 2 pls.

MENDES, J. S., AND S. MEZZALIRA
1946. Posição estratográfica dos novos hori-
zontes com vegetais; fósseis da forma-
ção Estrada Nova. Notas Prelim. e
Estud., Div. Geol. Min., Rio de Janeiro, vol. 30, 4 pp., 4 pls.

MEZZALIRA, S.
1949. Nova ocorrência de vegetais fósseis em
Tatuí, S. Paulo. Min. e Metal., vol. 33,
no. 77, p. 278.

OLIVER, W. R. B.
1950. The fossil flora of New Zealand. Tuatara, vol. 3, no. 1, pp. 1–11.

READ, C. B.
1941. Plantas fósseis do Neo-Paleozoico do
Paraná e Santa Catarina. Monogr.
Dept. Nac. Prod. Min., Div. Geol. Min.,
Brazil, vol. 12, pp. 1–102.
1947. Pennsylvanian floral zones and floral
provinces. Jour. Geol., vol. 55, no. 3,
pt. 2, pp. 271–279.

RICHARDS, P. W.
1942. The biogeographical division of the
Indo-Australian archipelago. 6. The
ecological segregation of the Indo-
Malayan and Australian elements in the
vegetation of Borneo. Proc. Linnean
Soc. London, vol. 154, pp. 154–156.

RIDLEY, H. N.
1938. Cause of the disappearance of the Cy-
cadeoidea in the Cretaceous period.
Ann. Bot., new ser., vol. 2, pp. 809–810.

RODIN, ROBERT J.
1951. Petrified forest in south-west Africa.
Jour. Paleont., vol. 25, no. 1, pp. 18–20,
pl. 5.

SAHNI, B.
1927. The southern floras: a study in plant
geography of the past. Proc. 13th
Indian Sci. Congr., pp. 229–254.
1936. Wegener's theory of continental drift in
the light of paleobotanical evidence.
Jour. Indian Bot. Soc., vol. 15, pp. 319–
332.
1939. The relation of the Glossopteris flora
with the Gondwana glaciation. Proc.
Indian Acad. Sci., sect. B, vol. 9, pp.
1–6.
1948. The Pentoxyleae: a new group of Jurassic gymnosperms from the Rajmahal

Hills of India. Bot. Gaz., vol. 110, no.
1, pp. 47–80.

SELLING, O. H.
1947. Aponogetonaceae in the Cretaceous of
South America. Svensk Bot. Tidskr.,
vol. 41, no. 1, p. 182.

SEWARD, A. C.
1931. Some late Paleozoic plants from the Belgian Congo. Bull. Acad. Roy. Belgique,
cl. sci., vol. 17, pp. 532–543, 2 pls.
1933a. Plant life through the ages. Cambridge.
1933b. Paleobotany. Mesozoic. Britannica
Booklet, no. 10, pp. 186–195.

SHERLOCK, R. L.
1948. The Permo-Triassic formations. A world
review. London.

TEIXEIRA, C.
1943. Notas para o estudo do "Karroo" da
região de Tete, na Africa oriental portuguesa. Bol. Soc. Geol. Portugal, vol.
2, no. 1, pp. 41–47, 1 fig., pls. 1, 2.
1947. Contribuição para o conhecimento ge-
ológico do Karroo da África Portuguesa.
An. Estudos Geol. Paleont., Ministério
das Colónias, Lisbon, vol. 2, tomo 2, pp.
5–28, 16 pls.

THOMAS, H. H.
1933. On some pteridospermous plants from
the Mesozoic rocks of South Africa.
Phil. Trans. Roy. Soc. London, ser. B,
vol. 222, pp. 193–265.

TRUEMAN, A. E.
1946. Stratigraphical problems in the coal
measures of Europe and North America.
Quart. Jour. Geol. Soc. London, vol.
102, no. 2, pp. xlix–xcii.

VASCONCELOS, P.
1949. On the occurrence of algal remains in
the ancient rocks of Angola. Amer.
Midland Nat., vol. 41, pp. 695–705.

VIRKKI, C. (MRS. C. JACOB)
1939. On the occurrence of similar spores in a
Lower Gondwana glacial tillite from
Australia and in Lower Gondwana
shales in India. Proc. Indian Acad. Sci.,
sect. B, vol. 9, pp. 7–12, 1 pl.
1945. Spores from the Lower Gondwanas of
India and Australia. Proc. Natl. Acad.
Sci. India, vol. 15, nos. 4/5, pp. 93–
176.

WADE, A.
1941. The geology of the Antarctic continent
and its relationship to neighboring land
areas. Proc. Roy. Soc. Queensland, vol.
52, no. 1, pp. 24–35.

WALKOM, A. B.
1930. A comparison of the fossil floras of

Australia with those of South Africa. Compt. Rendus 15th Internatl. Geol. Congr., for 1929, pp. 161–168.

1937. A brief review of the relationships of the Carboniferous and Permian floras of Australia. Compt. Rendu Deuxième Congr. Av. Études de Stratigraphie Carbonifère, Heerlen, 1935, vol. 3, pp. 1335–1342.

1944. The succession of Carboniferous and Permian floras in Australia. Jour. Proc. Roy. Soc. New South Wales, vol. 78, nos. 1/2, pp. 4–13.

1949. Gondwanaland: a problem of palaeogeography. Rept. Australian and New Zealand Assoc. Adv. Sci., Hobart Meeting, pp. 3–13, 8 figs.

WALTON, J.
1929. The fossil flora of the Karroo system in the Wankie district, southern Rhodesia. Bull. Southern Rhodesia Geol. Surv., vol. 15, no. 2, pp. 62–76, pls. A–C.

YOUNG, R. B.
1949. Domed algal growths in the dolomite series of South Africa, with associated fossil remains. Trans. Proc. Geol. Soc. South Africa, vol. 51, pp. 53–62.

ZIMMERMANN, W.
1930. Die Phylogenie der Pflanzen. Jena.

Editor's Comments
on Papers 12 and 13

12 **KNOWLTON**
Relations Between the Mesozoic Floras of North and South America

13 **BERRY**
Paleogeographic Significance of the Cenozoic Floras of Equatorial America and the Adjacent Regions

In 1918 the Geological Society of America sponsored a symposium on the relationships between fossils of North, Central, and South America and considerable attention was given to the terrestrial biota. Paper 12 by F. H. Knowlton treated the Mesozoic floral relations. For Mesozoic floras Knowlton favored a theory of land bridges across Antarctica to link many of the Southern Hemisphere assemblages; it is interesting to compare Knowlton's ideas of 1918 with Paper 11, Just's summary at a much later date (1952). Berry's statement (Paper 13) indicates some of the uncertainties of correlation and difficulties in establishing ancestries for some of the endemic South American floras.

12

Reprinted from *Bull. Geol. Soc. America*, **29**, 607–614 (Dec. 30, 1918)

RELATIONS BETWEEN THE MESOZOIC FLORAS OF NORTH AND SOUTH AMERICA [1]

BY F. H. KNOWLTON

(*Read before the Paleontological Society December 31, 1917*)

CONTENTS

INTRODUCTION

The natural order of sequence in considering the relations between the Mesozoic floras of North and South America must be the Triassic, the Jurassic, and the Cretaceous.

TRIASSIC

Rocks of Triassic age are known in many and widely separated parts of the world, and they are of great thickness, which implies a varied and relatively long-continued period of geologic activity. It is also evident from the thick deposits of coal known at several points, and in other ways, that vegetation must have been fairly abundant and considerably varied in character; yet the determinable forms of plant life that have thus far been recovered from the Triassic are surprisingly few in number—in fact, it is doubtful if the known flora far exceeds 300 species. This is, of course, to be largely attributed to the fact that much of the deposits were laid down under marine conditions, where one would hardly expect to find the remains of land plants preserved except near shore and more or less fortuitously; and, further, if the evidence believed by some to indi-

---

[1] Manuscript received by the Secretary of the Society August 22, 1918.

cate aridity is really valid—it is not altogether accepted—it might account to some extent for the absence of plant remains in certain very thick and generally barren deposits.

But, be these controlling factors what they may, the fact remains that the known Triassic flora is but scantily preserved to us. In addition to the scantiness of the plant remains, there is another element that must be mentioned, namely, the authenticity of the reference of certain deposits to the Triassic. Though quite generally considered as referable to this epoch, there is some lack of confirmatory data—that is, in the absence of thoroughly satisfactory information it is often difficult to decide between Upper Triassic and Lower Jurassic.

The Triassic flora of North America north of Mexico numbers 136 nominal species. Of this number about 120 species are confined to the eastern province—that is, to Massachusetts, Connecticut, Pennsylvania, Virginia, and North Carolina. The remaining 16 forms are distributed as follows: Abiquiu, New Mexico, 11 species; fossil forests of Arizona and vicinity, 4 species; and Alaska, 1 species.

Passing south into Sonora, Mexico, it is to be noted that Newberry described a small collection belonging to 9 genera and 13 species that were supposed to be in the same stratigraphic position as those at Abiquiu, New Mexico, though only one species is common to the two areas. From the so-called Mixteca Alta, or high country, on the southern edge of the Cordilleran system facing the Pacific, in the Mexican State of Oaxaca, Wieland has reported the presence of Triassic plants; but it is my opinion that these plants are younger than this, though it is not definitely settled just what their position is.

In 1888 Newberry[2] reported the presence of Rhætic plants from San Juancito, Honduras, enumerating 11 genera and 14 species. In a number of subsequent publications Carl Sapper[3] has intimated that Newberry's age determinations should be accepted only with doubt; they are possibly Jurassic.

Passing now into South America, so far as I am able to determine the first plants of supposed Triassic age were reported by Zeiller,[4] in 1875, from La Ternera, northern Chile. They were found in a rather important coal-basin and comprise only six species, belonging to the genera *Jeanpaulia, Angiopteridium, Pecopteris, Dictyophyllum, Podozamites,* and *Palissya;* they were referred to the Rhætic.

[2] J. S. Newberry: Am. Jour. Sci., 3d ser., vol. 36, 1888, pp. 342-351.
[3] Bol. Inst. geol. Mexico, no. 3, 1896, pp. 5-8.
[4] R. Zeiller: Note sur les plantes, fossiles de la Ternera (Chili). Bull. Soc. geol. de France, 3d ser., vol. 3, 1875, pp. 572-574.

In 1899 Solms-Laubach[5] reported on a much larger collection from the same locality. He was able to enumerate about a dozen genera, but did not give specific names to all. They were still considered to be of Rhætic age.

About the same time that the discovery of Triassic plants was made in Chile, plants presumed to be of the same age came to light in Argentina. They were first reported on by Geinitz,[6] in 1876. He enumerated 14 species, referred to 12 genera. Only one or two of the forms are identical with those described from Chile.

A few years later Szajnoche[7] described a small flora of 11 species, referred to 8 genera, from Cachenta, in the Province of Mendoza. Only two species appear to be identical with those described by Geinitz. Szajnoche compares this flora with beds of similar age in Queensland, New South Wales, India, Germany, and Bjuf, in Sweden. It does not appear that any of the Argentinian forms are identified with North American species.

From this hasty review it must be very evident that the data are lacking for an adequate comparison of the Triassic floras of North, Central, and South America. The localities are few, are often separated by thousands of miles, and, moreover, there is still more or less doubt in some cases as to their position.

JURASSIC

No Jurassic floras are at present known from eastern North America. From western North America—principally California, Oregon, and Alaska—about 125 species have been recognized. Of these, two localities—Kadiack Island and the Matanuska Valley, Alaska—with about a dozen species each, are referred to the Lias, and the remainder are referred to the Middle and Upper Jurassic. The latter find their close parallel with the well known Jurassic floras of eastern Asia.

Within the past year two small floras thought to be of Liassic age have been described by Lozano[8] from the States of Vera Cruz and Puebla, Mexico. Although certain of the forms described are seemingly abundant

---

[5] H. Grafen, Solms-Laubach : Zu Beschribung der Pflanzenreste v. La Ternera. Neues Jahrb., Beilage, vol. 12, 1899, pp. 593-609, pls. 13, 14.

[6] H. B. Geinitz : Ueber Rhätische Pflanzen-und Thierreste in den argentinischen Provinzen La Rioja, San Juan und Mendoza. Palæontographica, Suppl. 3, 1876, pp. 1-14, pls. 1, 2.

[7] L. Szajnoche : Über fossile Pflanzenreste aus Cachenta in der argentinischen Republik. Sitzr. d. Akad. wiss. Wien, vol. 97, 1888, pp. 1-20, pls. 1, 2.

[8] E. D. Lozano : Description de unos plantas Liasicas de Huayacocotla, Ver.. Algunas plantas de la flora Liasica de Huauchinango, Pueb. Bol. Inst. geologico de Mexico, no. 34, 1916, pp. 1-18, pls. i-ix.

and well preserved, the flora as a whole is small and poorly represented. Only eight genera are recognized, and of these several are so poorly preserved that specific identification was not attempted. A number of the forms recognized are identical with those described by Wieland[9] from the Mixteca Alta, in Oaxaca, from beds also referred to the Lias. This is a comparatively rich flora, comprising 21 genera and about 60 forms. It is especially rich in forms as well as individuals of the peculiar *Williamsonias*. This flora can hardly be accepted at its face value. It seems to me that there must have been either a mixture of horizons or a misidentification of generic types. If the genera *Noeggerathopsis*, *Trigonocarpus*, *Rhabdocarpus*, *Alethopteris*, *Sphenopteris*, and, above all, *Glossopteris*, have been correctly identified, it would certainly argue for a much older position than the Lias, and the *Williamsonias* and other types of cycads would not be out of place at a higher horizon.

The entire South American Continent is without a known locality for an undoubted or at least adequate Jurassic flora. As already pointed out, the floras from Chile and Argentina above referred to the Triassic may possibly be referable in whole or in part to the Lias instead of the Rhætic, but further data must be forthcoming before the matter can be settled. Thus, from Piedra Pintada, on the northern border of Patagonia, Kurtz has described a small collection of plants procured by Roth. They are associated in beds with marine fossils considered to be of Liassic age. Kurtz has compared the plants with the Rajmahal flora of the Upper Gondwanas of India.

The largest and by all odds the most interesting Jurassic flora is really extralimital. This is the Middle Jurassic flora described by Halle[10] from Hope Bay, Graham Land, 63° 15′ south, and just outside the Antarctic Circle. It embraces 61 forms, of which number 21 are definitely identified with previously known forms, and of these 17 are found elsewhere in strata believed to be of Middle Jurassic age, although it includes some types that are older and some that might be younger. The closest affiliation of this flora is shown to be with the well known Jurassic of Yorkshire, England, there being no less than 9 of the 21 species in common. There are no South American floras of any importance that can be considered contemporaneous with this Graham Land flora.

Possibly contemporaneous with the Graham Land deposit is a collec-

---

[9] G. R. Wieland: La flora Liasica de la Mixteca Alta. Bol. Inst. geologico de Mexico, no. 31, 1914.

[10] T. G. Halle: The Mesozoic flora of Graham Land. Wissen. Ergebnisee d. Schwedinschen Südpolar-Exped. 1901-1903, vol. 3, no. 14, 1913.

tion described by Halle[11] from Bahia Tekenika, Tierra del Fuego, which is about 60 nautical miles northwest of Cape Horn. It comprises only two generic types (*Sphenopteris* or *Coniopteris* and *Dictyozamites*), neither of which was sufficiently well preserved to admit of specific determination. It can not have very much weight in the present connection.

CRETACEOUS

Just as it has been found difficult on the basis of available data to distinguish between Triassic and Jurassic, so is it difficult to decide between Jurassic and Cretaceous. Thus, from middle Peru, Neumann[12] enumerated seven species which he held to be of Wealden age, although they include some apparently Upper Jurassic elements. The same year Lukis mentioned three poorly preserved species of plants from a coal mine near the same locality as those mentioned by Neumann, referring them to the Neocomian. Later, in 1910, Salfeld[13] procured about a dozen species of plants from the same general area, referring them in part to the extreme Upper Jurassic and in part to the lowest Cretaceous. Halle has expressed the opinion that they are probably all to be best regarded as lowest Cretaceous; also transitional between the Jurassic and Cretaceous, or possibly belonging wholly to the latter, is a small collection described by Halle[14] from Lago San Martin, central Patagonia. It embraces about twelve generic types and a slightly larger number of species. Some are older in affinity as some are somewhat younger, but on the whole Halle concludes that there is nothing to militate seriously against their extreme Lower Cretaceous age.

One of the most important discoveries bearing on the present discussion was that of an extensive dicotyledonous flora at Cerro Guido, Province of Santa Cruz, Argentina. This flora was listed in a short, unillustrated paper published by Kurtz[15] in 1902. It enumerated 31 forms, of which 21, or 75 per cent, are characteristic types of the Dakota group. Although these plants have never been figured, and it is consequently impossible to check up the identifications, they are mostly such characteristic species

[11] T. G. Halle: Kungl. Svenska Vetensk. Handl., vol. 51, 1913, no. 3, pp. 6-12.

[12] Richard Neumann: Beitrage zur Kentniss der Kreidformation in Mittle-Peru. Neues Jahrb., Beilage, vol. 24, 1907, pp. 69-132.

[13] H. Salfeld: Fossile Pflanzen aus dem obersten Jura, bzw. der untersten Kreide von Peru. Wissen. ver offenntich. d. gesell. f. Erdkunde, Leipzig, vol. 7, 1911, pp. 211-217.

[14] T. G. Halle: Kungl. Svenska Vet.-Akad. Handl., vol. 51, no. 3, 1913.

[15] F. Kurtz: Contribiciones à la palæophytologia Argentina. Sobre la existencia de una Dakota flora en la Patagonia austro-central, Revista Museo La Platà, vol. 10 (1899), 1902, pp. 43-60.

that it is hardly possible to suppose that all or even any considerable percentage have been incorrectly determined. Therefore, taking the list at its face value, it might well enough have been made of a collection from the Dakota of Kansas or Nebraska. In the 5,000 miles between Kansas and this locality in Argentina no trace of this flora has been reported. Argentine geologists regard these beds as Cenomanian; but, as Berry, Halle, and others have suggested, they are probably not older than Turonian.

Summary

With the exception of the Dakota flora from Argentina just mentioned, there is comparatively little demonstrable relationship between the Mesozoic floras of North and South America. The total known Triassic flora from South America hardly exceeds 30 species, and of these no more than two or three are specifically identical with North American forms, though it is but fair to state that when the floras of the two continents come to be revised in the light of existing knowledge more species will probably be found to be common to the two. As the matter now stands, a majority of the South American species are regarded as endemic, and many of the others are either questionably identified or referred to Old World species.

Demonstrable specific relationship between the Jurassic floras of North and South America is, if possible, less satisfactory than in regard to the Triassic floras. The Jurassic floras on the South American Continent are so fragmentary and generally unsatisfactory that it is hardly worth while to attempt comparisons. The only flora of importance is the extra-limital one found on Graham Land. This and such as we know from South America are clearly but integral parts of the great world-ranging Jurassic floras. In fact, remoteness appears to have had little influence on distribution. Witness this Graham Land flora, which finds its closest relationship with that of Yorkshire, England, and important relationship with other parts of Europe as well as India, and, it may be added, it is not greatly different from the well known Jurassic floras of California, Oregon, Alaska, and Siberia.

The relationship between the Upper Cretaceous Dakota group flora of Argentina and the Dakota flora of Texas, Kansas, and Nebraska is direct and positive. The now dominant group of dicotyledons clearly originated in the north, and in late Lower Cretaceous time had spread south over eastern North America and western Europe. By latest Lower Cretaceous and early Upper Cretaceous time they had spread west in North America, to become the Dakota flora of the central Western States. This Dakota flora undoubtedly spread southward to find its southernmost known limit in Argentina. The pathway between these areas, although now largely buried beneath later sediments, was undoubtedly open in Upper Cretaceous time. The presence in the Argentine "Dakota" flora of a very few elements of possibly later age may be sufficient to make it slightly younger than the principal Dakota floras of the north—that is, it may have taken an appreciable time for its journey, and it may not have reached there until early Turonian time.

A word may be said as to the possible, not to say probable, routes by which the Triassic and Jurassic floras reached South America. It is necessary to review briefly the plant distribution during Permo-Carboniferous time in order to get a proper perspective. In Permo-Carboniferous time the world was divided into two phytogeographic provinces—a northern and a southern—and there was extremely little commingling of plant types between them. The southern province, characterized by the so-called *Glossopteris* flora, embraced portions of India, Australasia, South Africa, Antarctica, and eastern South America. It reached the northern province at a single point in north central Russia. It has been found within 5 degrees of the South Pole. To my mind the facts all point to the origin of this flora in the south, either in Australia or on the Antarctic land-mass, and I believe there was in Permo-Carboniferous time a practically continuous land connection between the Antarctic Continent (Gondwana Land) and south Africa, Australia, India, and South America. Any attempt to derive the *Glossopteris* flora of Brazil, the Falkland Islands, and Buckley Island (85 degrees south) from the north by way of North America is without supporting data.

Some students—notably Ettingshausen—have held that the differences between the northern and southern phytogeographic provinces that are so marked in Permo-Carboniferous time continued well into the Jurassic. This appears to be true only in part, for while there are some notable differences in the floras of the two areas, the differences are by no means so sharp as in Permo-Carboniferous time. For example, the Ginkgoales, a dominant and widespread group of the north, did not reach the southern province, and *Podozamites*, abundant in the north, is but sparsely represented in the south. From available data it appears that at least the major portion of the early Mesozoic flora originated in the north, whence they spread pretty much over the globe. Their routes of travel are not always clear, however. It is possible that the Jurassic flora found on Graham Land may have reached this far southern point by way of North America, Central America, and the entire length of South America; but the fact that the obvious affinity of this flora is with the floras of India, Europe, and thence to England, rather than with western America, leads me to think it at least possible that the land bridge over Antarctica was still existent.

13

Reprinted from *Bull. Geol. Soc. America*, **29**, 631–636 (Dec. 30, 1918)

PALEOGEOGRAPHIC SIGNIFICANCE OF THE CENOZOIC FLORAS OF EQUATORIAL AMERICA AND THE ADJACENT REGIONS [1]

BY EDWARD W. BERRY

(Read before the Paleontological Society December 31, 1917)

CONTENTS

INTRODUCTION

In order not to occupy too much time or confuse my auditors with details, it has seemed best to give a brief summary of the known fossil floras bearing on the problem of intercontinental land connections between North and South America, followed by the conclusions which it has seemed might be legitimately drawn from their evidence.

This evidence is not presented in detail, since it has been given at length in publications recently printed or now in press.[2]

---

[1] Manuscript received by the Secretary of the Society August 22, 1918.

[2] E. W. Berry: The physical conditions and age indicated by the flora of the Alum Bluff formation. U. S. Geol. Survey Prof. Paper 98 E, 1916.

———: The Lower Eocene floras of southeastern North America. Idem, 91, 1916.

———: The Pliocene citronelle formation of the Gulf Coastal Plain and its flora. Idem, 98 L, 1916.

———: The Catahoula sandstone and its flora. Idem, 98 M, 1916.

———: The Upper Cretaceous floras of the eastern Gulf region. Idem (in press).

———: The Midde and Upper Eocene floras of southeastern North America. Idem (in press).

UPPER CRETACEOUS

The flora which commenced to radiate from Arctogæa in the Ceno-manian, and which during the Turonian and Emscherian covered most of North America and Europe, and presumably Asia, penetrated as far as southern South America before the close of the Upper Cretaceous, and at least 26 typical species have been recorded from Argentina.[3] This would seem to indicate that there was some land connection between North and South America throughout the greater part of the Cretaceous, during which time the prevailing direction of migration was from north to south.

MIDWAY EOCENE FLORA OF NORTH AMERICA

The small Midway Eocene flora recorded from the Gulf Coastal Plain contains five species belonging to the genera Pourouma, Cecropia?, Asimina, Dolichites, and Terminalia which I have regarded as having been derived from tropical America and as lacking direct ancestors in the Upper Cretaceous of North America.

WILCOX EOCENE FLORA OF NORTH AMERICA

IN GENERAL

This very extensive and well preserved flora comprises to date 345 described species, distributed in 136 genera. There are in common with the almost unknown Tertiary floras of South America 2 species in 2 genera. Those that I consider as having originated in Arctogæa number 179 species in 57 genera. Those whose place of origin is unknown number 65 species in 29 genera. Those that appear to have originated in the American tropics number 101 species in 50 genera.

I do not consider the relationship of the existing flora of the Antilles with that of South America to be as intimate as was the relationship of the Lower Eocene flora of southeastern North America with that of South America. This statement is not true of Central America, where the present lowland flora is a direct continuation of that of South America, while the upland flora is a mixture of survivals from the southward migration of North American types in the Miocene mixed with later immigrants from both the north and the south.

---

[2] E. W. Berry: The fossil flora of the Panama Canal Zone. U. S. Natl. Mus., Bull. 103, 1918, pp. 15-44, pls. 12-18.
————: The Tertiary flora of Peru. Proc. U. S. Natl. Mus. (in press).
[3] F. Kurtz: Revista Museo La Plata, vol. 10 (1889), 1902, pp. 43-60.

CLAIBORNE (AUVERSIAN) EOCENE FLORA OF SOUTHEASTERN NORTH AMERICA

The Claiborne flora as at present known comprises 78 species in 59 genera. Of this number 7 genera, with 8 species, are considered as new arrivals from tropical America.

JACKSON (PRIABONIAN) FLORA OF SOUTHEASTERN NORTH AMERICA

The known Jackson flora comprises 79 species in 60 genera. Of this number 6 genera, with 6 species, are regarded as new arrivals from tropical America. These are species of Phœnicites, Myristica, Burserites, Dombeyoxylon, Rhizophora, and Calocarpum. In addition to these, *Palmoxylon lacunosum* (Unger) Felix of the Jackson is probably common to the lower Oligocene (Sannoisian) of the Island of Antigua.

CATAHOULA AND VICKSBURG (OLIGOCENE) FLORAS OF NORTH AMERICA

The known flora from these formations is a small one, numbering but 24 described species in 15 genera. The only new genus derived from tropical America is the genus Embothrites. Two species of palms, however, are common to the lower Oligocene of the Island of Antigua, another is very close to an Antigua and Panama form, and a third occurs in southern Mexico. There is in addition an undescribed petrified wood in the Vicksburg that is common to Antigua.

TERTIARY FLORAS OF SOUTH AMERICA

The origin of such South American Eocene floras as are known at present may well have been from Arctogæa; but if this was the case, as seems probable, their ancestors reached South America during the Cretaceous, since they are found at a number of widely scattered localities in rocks of earlier Tertiary age, namely, at Coronel, Chile;[4] Santa Ana and Caucathale, in Colombia;[5] Tablayacu and Loja, in Ecuador,[6] and near Tumbez, in Peru[7]. Moreover, the evergreen beeches (Nothofagus) which are found in the lands bordering the present Straits of Magellan[8] appear to be of northern, probably of Asiatic, origin.

[4] H. Engelhardt: Abh. Senck. Naturf. Gesell., vol. 16, hft. 4, 1891, pp. 629-692, pls. 1-14. Abh. Sitz. Naturw. Gesell., Isis in Dresden, 1905, pp. 69-82, pl. 1.
[5] H. Engelhardt: Abh. Senck. Naturf. Gesell., vol. 19, 1895, pp. 1-47, pls. 1-9.
[6] Idem.
[7] Berry: Op. cit.
[8] C. von Ettingshausen: Sitz. k. Akad. Wiss. Wien, vol. 100, 1891, pp. 114-137, pls. 1, 2.
A. Gilkinet: Resultats voyage du S. Y. Belgica en 1897-1899, 1900.
A. Dusen: Svenska Exped. till Magellanslanderna, vol. 1, 1899, pp. 87-107, pls. 8-13.
XLVII—BULL. GEOL. SOC. AM., VOL. 29, 1917

The exact ages of these various South American Tertiary floras has never been accurately determined. De Lapparent regarded the first as Eocene, probably Sparnacian. Dusén, following Wilckens, regarded them as Oligocene. It is extremely unlikely that they all are of the same age. Those from Colombia, Ecuador, and Peru I am inclined to regard as the same age as the flora from the Isthmus of Panama,[9] which appears to represent various stages of the Oligocene plus the Aquitanian and Burdigalian, and these find their counterparts in the floras of the Catahoula, Vicksburg, and Alum Bluff formations of the United States and in the petrified woods of Antigua and others of the Antilles.

Part of the Chilean and Patagonian floras appear to be older than these, for they have a few but striking elements in common with those of the lower Eocene of the Mississippi embayment region. Later Tertiary floras from South America not previously mentioned include the Pliocene flora of Bolivia, amounting to 85 species and strictly endemic in character,[10] and a Pliocene flora from the province of Bahia, Brazil, comprising about 70 species, some of which are of North American ancestry.[11]

SUMMARY

The following somewhat categorical conclusions are indicated by a detailed study of the foregoing floras:

1. There appears to have been free intercommunication between North and South America during the Upper Cretaceous, with the invasion of the northern (Holarctic) flora into all parts of South America and probably to Antarctica.

2. Continued land connection between North and South America during the basal and lower Eocene, which, combined with the ameliorating climate of southeastern North America, resulted in the introduction of many new types in that region which were derived from the south.

3. During the middle and upper Eocene, as well as during the Oligocene, there was a continued influx of a few tropical American types into our Southern States, but these were not in sufficient force to demand a land connection between the two regions, nor can it be certain that these types came from South America and not from the intermediate region of Central America and the Antilles.

4. During the Oligocene there appears to have been a rather free interchange of plant types between Panama and the Antilles, best illustrated

---

[9] Berry : Op. cit.

[10] E. W. Berry : Proc. U. S. Natl. Mus., vol. 54, 1917, pp. 103-164, pls. 15-18.

[11] F. Krasser : Sitz. k. Akad. Wiss. Wien., vol. 112, abh. 1, 1903, pp. 852-860.
Ed. Bonnet : Bull. Mus. d'hist. Nat., année 1905, pp. 510-512.

by the extensive flora of petrified· woods found on Antigua, which has several forms common to Central America and the southeastern United States.

5. In formations correlated with the Aquitanian and Burdigalian of the European section, but generally considered as upper Oligocene by American geologists, the tropical types of plants become fewer on the North American mainland and are largely replaced by temperate forms.

6. The land emergence of the Miocene appears to have developed land connections in Central America and the Antilles reflected in the floras by a general spreading southward of the temperate flora of North America. This emergence resulted in the connection of the Windward Islands with South America, Cuba with Yucatan, and the Jamaica, Haiti, and Porto Rico axis of elevation with Honduras. To this period may be attributed the original colonization of the many North American types that still persist in the Antilles and Central America. The invasion of the latter regions, and that beyond in South America, by the oaks (Quercus), walnuts (Juglans), and many other Holarctic types probably occurred at this time, since some of them are found in the Pliocene of Brazil. The radiation of the agaves may also be more appropriately dated from the Miocene rather than from the Pleistocene, as is done by Trelease,[12] since while there was considerable elevation during the Pleistocene the major tectonic lines were established during the Pliocene by block faulting, as Vaughan has shown, and the failure of Agave to penetrate to any considerable extent into South America was due to the impenetrability of the tropical rain forest and not to geographical barriers.

7. North and South America were connected during the Pleistocene and there was considerable elevation in the Antillean region, resulting in a northward spread from South America of various elements from the rain forests of the Amazon and Orinoco basins, one or two of which occur in the late Pleistocene of southern Florida. Actual land connections among the Antilles or with Florida are not considered probable.

Note

There are a number of reasons why the arguments against a land bridge between North and South America based on the evidence of vertebrate paleontology can not be regarded as conclusive. It is opposed by the evidence derived from the study of the distribution of other animal types, as has been already pointed out by several students.

First, as regards actual changes in level. The general thesis of the

---

[12] William Trelease : Mem. Natl. Acad. Sci., vol. 11, 1913.

permanency of ocean basins, as developed by Matthew in "Climate and evolution," [13] seems unquestionably sound in the present state of our knowledge of isostatic compensation. Whether the latter was always as complete as it seems to be at present may well be doubted—certainly we know of remarkably great epeirogenic changes, and, furthermore, a rather good case can be made out for the theorem advanced by Walther and others, that the deep seas are post-Mesozoic in age. Be this as it may and while its acceptance does not directly oppose Matthew's contention, the present continental scarps can not be regarded in all cases as the metes and bounds of the continents in past times, and a very strong case can be made out for Suess's sunken block theory in a large number of areas, *and especially in the Caribbean region.*

It seems to me, as I review the various lines of evidence, that the long region of weakness extending from Graham Land to the Antilles may well indicate that the Antilles were once a part of South America, and that the latter continent was connected with Antarctica. I have recently shown that the change of elevation of the great central plateau of Bolivia and the eastern Andes in that region has amounted to a minimum of $2\frac{1}{2}$ miles since the Pliocene,[14] and the weight of this argument is not disposed of by calling these changes orogenic instead of epeirogenic. I am therefore the more inclined to believe that comparable changes of level have taken place in the Caribbean region during the Tertiary.

In all discussions of paleogeography based on the distribution of the mammals it should be constantly borne in mind that the facts, in so far as they contribute toward an understanding of the relations between North and South America, are derived almost entirely from the region of the Great Plains and Rocky Mountains in North America and from Argentina in South America—regions which were separated, at least during the Eocene and the Oligocene, as they are today, by the greatest extent of tropical rain forest on the globe. The fact that the edge of this equatorial American rain forest appears to have covered the southern shores of the United States during the first half of the Tertiary renders it obvious why the Artiodactyla and Perissodactyla of our western plains were not exchanged for the typotheres and litopterns of Patagonia.

---

[13] W. D. Matthew: Annals N. Y. Acad. Sci., vol. 24, 1915, pp. 171-318.

[14] E. W. Berry: Proc. U. S. Natl. Mus., op. cit.

Part III

ORIGIN OF PHYLETIC LINEAGES

Editor's Comments
on Papers 14 and 15

14   **MATTHEW**
Affinities and Origin of the Antillean Mammals

15   **AXELROD**
Evolution of Desert Vegetation in Western North America

Papers 14 and 15 examine the origin of faunas and floras. W. D. Matthew in Paper 14 helped lay the foundation for the study of island distributions, particularly directed toward vertebrate paleontology and zoology. Matthew was at the American Museum of Natural History for a major portion of his scientific career and was particularly interested in treating fossil vertebrates as former living entities. He, perhaps as much as any other vertebrate paleontologist of that time, drew the vertebrate neontologists and paleontologists together in philosophy and action. Now, more than fifty years later, vertebrate paleontologists have remained largely zoologically rather than biostratigraphically oriented. One of Matthew's classic papers was "Climate and Evolution" (Matthew, 1915), a 148-page article (really a book) that investigated the thoughts of that day in terms of vertebrate evolution. Paper 14 is a small sampling of the type of investigations Matthew was involved in exploring.

Several later authors have developed the theme of island populations and attached considerable general significance to island dispersals in attempting to explain the similarities and differences in biota between continents widely separated today by ocean basins. As we shall see in later papers, "island hopping" was considered by many as the logical alternative to Wegener's proposal of continental drift. Readers interested in pursuing the developments using this means of dispersal will find MacArthur and Wilson (1967), Stern (1971), and MacArthur (1972) interesting and fruitful reading.

The other side of the problem of the origin of species is examined by D. I. Axelrod in Paper 15, that is, the evolution of endemic species adapted to particular environments. Axelrod clearly shows that the evolution of our present floras in the deserts of southwestern North America has taken place as a result of ecological adaptations from an existing flora in response to long-term climatic changes toward arid conditions. Geologists tell us that the Cenozoic is a time of uplift and mountain building in western North America and that these have included as much a 3 kilometers of vertical elevation in some mountain ranges. The effects on local climates were obviously major; however, the slowness of the rate of change gave the floras the opportunities to adapt to the diversity of ecological conditions and permitted communities to expand and contract in response to geologic, short-term, climatic fluctuations, such as the Pleistocene glacial intervals. The original presentation by Axelrod was nearly 100 pages, much too long to reproduce here, and much of the detailed analysis has had to be deleted. An additional reference to this form of analysis by Axelrod (1968) examines the floral history of the Snake River Basin in Idaho.

REFERENCES

Axelrod, D. I. (1968). Tertiary floras and topographic history of the Snake River Basin, Idaho. Bull. Geol. Soc. America, 79, 713–734.

MacArthur, R. H. (1972). Geographical ecology—patterns in the distribution of species. Harper & Row, New York, xviii + 269 p.

———, and E. O. Wilson (1967). The theory of island biogeography. Princeton Univ. Press, Princeton, N.J., 203 p.

Matthew, W. D. (1915). Climate and evolution. Ann. N.Y. Acad. Sci., 24, 171–318. (Reprinted 1973, Arno Press, New York.)

Stern, W. L., ed. (1971). Adaptive aspects of insular evolution. Washington State Univ. Press, Pullman, Wash., iii + 85 p.

14

Reprinted from *Bull. Geol. Soc. America*, **29**, 657–666 (Dec. 30, 1918)

AFFINITIES AND ORIGIN OF THE ANTILLEAN MAMMALS [1]

BY W. D. MATTHEW

(Read before the Paleontological Society January 1, 1918)

CONTENTS

LIMITATIONS AND RELATIONSHIPS OF WEST INDIAN MAMMAL FAUNAS, LIVING AND EXTINCT

IN GENERAL

The indigenous land mammals of the West Indies consist of three groups: (1) Insectivora, (2) hystricomorph rodents, (3) gravigrade edentates. No perissodactyls (horses, rhinoceroses, tapirs, etcetera), no artiodactyls (peccaries, deer, antelopes, etcetera), no proboscideans (elephants, mastodons, etcetera), no true carnivores (dogs, cats, raccoons, mustelines, bears, etcetera). Nor are there any sciuromorph, lagomorph, or myomorph rodents, shrews, moles, hedgehogs, or opossums. All these large groups, most of them abundant and varied in Tertiary North America, are wholly absent. Nor do the Insectivora, rodents, or edentates include anything at all nearly allied to any North American members of

---

[1] Manuscript received by the Secretary of the Society August 22, 1918.

the order or derivable from anything known to have inhabited North America in the later Tertiary.

Most of these North American groups invaded South America in the Pliocene and are part of its later fauna. In the Miocene and early Tertiary they are not found in South America, but their place is taken by a number of other groups. In place of perissodactyls, artiodactyls, and proboscideans were a number of groups of hoofed animals peculiar to Tertiary South America—the toxodonts, typotheres, litopterns, homalodontotheres, astrapotheres, pyrotheres. In place of the true Carnivora is a variety of marsupial carnivores (Borhyænidæ) paralleling the true carnivores in structure and taking their place in the fauna.[2] All of these abundant and varied groups of ungulates and pseudo-Carnivora are lacking from the Antillean fauna, nor do the rodents represent more than two or possibly three of the numerous hystricomorph rodent stocks of Miocene South America, while the edentates represent only one group of the ground-sloths, the three or four other ground-sloth groups, as well as the several kinds of armadillos and the glyptodonts, being quite unrepresented.

THE INSECTIVORA

It appears to be reasonably certain that the Antillean rodents and edentates came from South America and from Tertiary South America. Hystricomorph rodents and edentates are unquestionably South American Tertiary types, which invaded North America when the two continents were joined, toward the end of the Tertiary. The insectivores, however, are more probably derivable from North American sources;

---

[2] A similar but distinct group of marsupial carnivores (Dasyuridæ and Thylacinidæ) developed in Australia in absence of true Carnivora and still survives there.

Note.—In this paper no account is taken of animals which may have been brought to the islands by man, whether intentionally or by accident, in post-Columbian or prehistoric time. Some of these have evolved under insular conditions into races distinct enough to be recorded as species or subspecies. The majority are identical with species of North or South America, Europe, Africa, etcetera. Some are known to have been introduced; others may be so explained by reason of associations of one kind or another. The formation of distinct races, such as are classed as species by modern mammalogists, does not necessarily take many centuries under these conditions—as witness the Porto Santo (Madeira) rabbits and other instances. Some of these supposedly introduced forms may have been brought in through natural means and be truly indigenous, although not very ancient; but it is impossible to prove such cases. It seems better here to omit all this doubtful evidence and consider only the fauna that is proved to be indigenous either by occurrence in the Pleistocene cave and spring deposits, by its sharp distinctions from any continental relatives, or by the high improbability that the animal could have been transported through human agency. Dr. G. M. Allen has compiled an annotated list of the West Indian mammals which includes both introduced and indigenous types, and Dr. Thomas Barbour has done the same for the reptilia and batrachia. The evidence therein summarized will be discussed in a memoir on Cuban fossil mammals now in preparation.

for, with a single somewhat doubtful exception, the entire order of Insectivora is absent from the Tertiary faunas of South America, while they were many and varied in North America, especially in the older Tertiary. The two Antillean insectivores are not nearly related to each other, nor to any other genera of the order; they are placed in families by themselves. It has been repeatedly stated that *Solenodon* is related to the Malagasy Centetidæ, but in fact the affinity is a very distant one. *Nesophontes* is equally peculiar, and while it has some affinities with the Soricoidea (moles and shrews), they are very distant. The nearest relatives of both—collateral ancestors, perhaps—are certain imperfectly known insectivores of very primitive type in the North American Eocene and Oligocene.

THE RODENTIA

The rodents are clearly of South American affinities. They are all hystricomorphs—a group chiefly South American since the middle Tertiary (if not before). The only North American hystricomorph is the porcupine, Pleistocene and Recent, and whose ancestors have been recognized in the South American Tertiary. No traces of the hystricomorphs have been discovered in the Tertiary of North America.[3] There are certain Old World hystricomorphs, the Hystricidæ and certain Octodontidæ, and the early Tertiary Theridomyidæ of Europe have been considered ancestral to the group, but they have no significant relations to the Antillean genera and their affinities are disputed, so that they may be passed over for the present problem.

The Antillean hystricomorphs are clearly related to the South American types, but it is equally clear that the relationship is not close. There are apparently three groups. One, including *Amblyrhiza* of Anguilla, *Elasmodontomys* and *Heptaxodon* of Porto Rico, is related to the chinchillas, but not closely related. Anthony[4] places them in separate subfamilies. Miller[5] states that they are more nearly related to the extinct *Megamys* and its allies than to the living chinchillids. These genera (*Megamys, Tetrastylus,* etcetera) are found in the Entrerian, Rio Negran, Hermosan, and Araucanian formations of Argentina accompanied by a fauna which is closely related to the Pampean, but contains a few little altered survivals from the Santa Cruz Miocene and comparatively few of the North American invading types. All should, in my

---

[3] Except Leidy's *Hystrix* (*Hystricops*) *venusta*, based on two teeth of doubtful affinities and uncertain geologic age.

[4] Anthony: New fossil rodents from Porto Rico. Bull. Am. Mus. Nat. Hist., vol. xxxvii, 1917, pp. 185-186.

[5] Miller: Bones of mammals from Indian sites in Cuba and Santo Domingo. Smithsonian Miscell. Coll., vol. lxvi, no. 12, 1916, p 3.

judgment, be referred to the Upper Pliocene; they are certainly much later than the Santa Cruz.

Although *Megamys* and *Tetrastylus* are without doubt more nearly related than any modern type to the Antillean chinchillids and are considerably older, they can not be regarded as ancestral. The common ancestral stock is to be found in the Santa Cruz chinchillids, all of which are of small or medium size. These Santa Cruz species are much more primitive, and the precise relationships will require more careful study.[6]

The remaining Antillean rodents are of South American type and broadly derivable from Santa Cruz rodents, but their more exact affinities are disputed and require more thorough and critical consideration and, if possible, more complete material. A thorough revision of the fossil rodent faunas of South America is an almost necessary groundwork for a correct estimate of their affinities. It is clear, however, that with the exception of *Capromys* they are not very closely related to the South American rodents, the common ancestral stock dating back probably to Pliocene or late Miocene, as Miller believes. *Capromys* (and *Geocapromys*) would seem to be an exception, being quite close to the Venezuelan *Procapromys*.

THE EDENTATA

The edentates include four quite distinct genera from Cuba and a fifth from Porto Rico, all referred to the family Megalonychidæ, but not closely related to any of the mainland forms. The largest, *Megalocnus*,[7] is about the size of a black bear; the smallest, *Microcnus,* about the size of a cat, and there are two of intermediate size, *Mesocnus,* with a rather long, narrow muzzle, and *Miocnus,* with a broad, square muzzle. The Porto Rican genus is related to *Miocnus,* both having heavy triangular tusks like the modern *Cholœpus;* the other three genera have large tusks, but of a peculiar dished shape, with a tendency to approach toward each other like the incisors of rodents. (They are not at all of the scalpriform gnawing type, however.) While these ground-sloths are sufficiently re-

---

[6] Miller apparently associates the Santa Cruz fauna of southern Patagonia with the very different and much later Entrerian fauna of northern Argentina, as he speaks of the two as though they were essentially one fauna, and refers to the "enormous extinct Pagatonian rodents" as the nearest relatives of the Antillean chinchillids. The largest Santa Cruzian rodents are species of *Perimys,* which is not nearly related to the Antillean genera. Most of the species are quite small. As between the Santa Cruz and the Hermosan (Pliocene) group of faunas, the tendency to rapid increase in size and specialization of numerous phyla is very marked, and is further emphasized in the Pampean (Pleistocene) group of faunas.

[7] *Megalocnus,* Leidy, 1868. Proc. Acad. Nat. Sci. Phila., 1868, p. 179.

Microcnus, etc., La Torre and Matthew, 1915. Bull. Geol. Soc. Am., vol. xxvi, p. 152 (names only; descriptions have been reserved pending the securing of more complete and better associated skeleton material).

lated to the North American genus *Megalonyx* to be placed in the same subfamily, they are quite evidently not descended from it, but contemporaneous specializations from the primitive Megalonychidæ of the South American Miocene, as represented in the Santa Cruz fauna. Among the known genera of this fauna there is only one, *Eucholœops* (including *Megalonychotherium*), which can be regarded as ancestral either to the Cuban ground-sloths or to *Megalonyx*. The others all have the caniniform teeth much reduced or vestigial and in series with the cheek teeth. To the best of my judgment, the anatomical evidence is not decisive as to whether the five Antillean genera are descended from one or from two or more nearly allied Upper Miocene or Lower Pliocene genera, but leads to the conclusion that the common ancestor or ancestors was very close to or identical with the Upper Miocene ancestor of *Megalonyx*, and was either the genus *Eucholœops* or one or more genera closely allied thereto. The Antillean genera represent, therefore, only the megalonychine division of the Megalonychidæ. The other families of ground-sloths—Mylodontidæ, Scelidotheridæ, and Megatheriidæ—are not found; nor are there any armadillos, glyptodonts, or anteaters.

BATS AND BIRDS

In addition to the terrestrial mammals, bats are numerous in the cave deposits, and a number of birds, lizards, crocodiles, and turtles have been found at the Ciego Montero locality and elsewhere.

Concerning the bats, there is very little to say. Most of them are nearly allied to or identical with species now living on the islands. The Antillean bats include a number of peculiar genera, besides others common to the continental parts of tropical America. Their relations are much the same as those of the birds, and in either case it is obvious that the intervening seas would act as a hindrance to migration, but not as an absolute barrier, and would be more of a hindrance in some groups than in others. The result would be the presence of a number of peculiar types, preserved by relative isolation and specialized in adaptation to the peculiarities of their habitat, along with other widely ranging forms closely allied to or identical with those of the mainland. The distribution of the birds has been carefully studied by Chapman and others.

REPTILES

The distribution of the lizards has been recently studied by Dr. Thomas Barbour, and his conclusions as to the paleogeography are sharply at variance with mine, owing to different methods of interpreting the data. I shall not take this part of the problem up at present.

The fossil crocodiles have been examined by Dr. Barbour, who informs me that they are all referable to *Crocodilus rhombifer,* a species still living on the island of Cuba.[8] The origin of this species might be either North or South American; but too little is known of the phylogeny and distribution of the Tertiary Crocodilia for any conclusions to be drawn as to the time or method of its arrival.

The fossil chelonians have not been carefully studied, but they include two species—one a giant tortoise, *Testudo cubensis* Leidy, which, like the giant tortoises of the Galapagos and Indian Ocean islands, has the carapace much thinned out, so that the plates are apparently more or less discontinuous. There is one North American Pliocene species, *T. pertenuis* Cope, from Texas which has a remarkably thin carapace, but apparently not discontinuous. The precise significance of this species in the paleogeographic problem must also await more careful study. The genus *Testudo* occurs sparingly in South America, and is recorded as a fossil in the Pliocene and Pleistocene formations—not earlier, so far as I know. On the other hand, species of *Testudo* are the most abundant of fossils in the Oligocene to Pliocene formations of North America; in the Pleistocene and Recent their range is restricted to the Southern States and Mexico. The indications point, therefore, preferably to North American origin for this Cuban tortoise, although not decisively.

The second fossil chelonian is one of the Emydidæ, or marsh-turtles; it appears to be a species of *Graphemys,* probably the same as the still existing Cuban species, which is said to be a close ally of *G. scripta* of the Southeastern States. Whether the two are specifically distinct has been questioned. The discovery of this species (if it be the same) fossil in the Ciego Montero locality removes any doubt as to its being indigenous to the island, as it carries it back into the Pleistocene, and probably to a time before the arrival of man. Its close relationship with *G. scripta* and limitation to the western islands, Cuba and Haiti, is a strong indication of its having come from Florida, and the time of its arrival probably would be not earlier than Pleistocene or at most late Pliocene.

Summary of Affinities and Probable Origin of the Vertebrate Groups

Summing up the indicated sources of the fossil and recent vertebrate fauna, we find it to be as follows:

---

[8] Leidy's *Crocodilus pristinus* (1868, 1. c.) was based upon a vertebra not distinguished from *C. rhombifer.* The skulls obtained by La Torre and Brown represent, in Doctor Barbour's opinion, a series of growth stages of the modern species, the largest much exceeding any modern specimens. Part of a skeleton associated with one of the largest skulls equals or exceeds Leidy's type of *pristinus* in size.

1. The Insectivora, *Solenodon* and *Nesophontes,* of very ancient arrival, probably early Tertiary, and apparently of North American origin.

2. The ground-sloths, *Megalocnus, Mesocnus, Miocnus, Microcnus* in Cuba and *Acratocnus* in Porto Rico, of moderately ancient arrival, probably late Miocene or early Pliocene, and of undoubted South American origin. The rodents, with the exception of *Capromys,* fall also into this category.

3. The peculiar groups of birds, bats, and lizards are also no doubt of comparatively ancient arrival, but their source is unknown, as we know nothing of the Tertiary distribution of related groups on the mainland. The modern distribution of such related groups can not be relied on, for we know that among terrestrial mammals various groups which are today exclusively or chiefly Neotropical were Nearctic until the end of the Tertiary and unknown in South American faunas until the late Pliocene. Presumably corresponding changes in distribution have occurred among the bats, birds, and reptiles, but we have no records as to what groups were affected. The Cuban crocodile may also be placed in this category.

4. Of the two chelonians the giant tortoise may be placed as more probably of North American than of Central or South American origin, but its time of arrival can hardly be estimated until its relations to the continental Tertiary species are known. The terrapin is almost certainly of North American origin, derived from the Southeastern States, and of comparatively late arrival, probably late Pliocene or Pleistocene.

5. *Capromys* and *Geocapromys* are undoubtedly of South American derivation, like the other rodents; but, so far as may be judged from their comparatively near affinity with the Venezuelan *Procapromys,* are of later arrival, perhaps late Pliocene or Pleistocene.

It appears, therefore, in sum, that the vertebrate fauna, fossil and recent, represents only a few selections from the continental faunas of either North or South America; that it falls into several groups of diverse origin, and judging from their degree of differentiation, of diverse times of arrival.

Is the incomplete and unbalanced Character of the Fauna Real or only Apparent?

In the absence of hoofed animals, which form the greater part of all continental faunæ, in the tendency of races normally of small size to assume relatively large size and importance, in the relative fragility, so to speak, of the fauna, leading to its early disappearance when man invades the region—in many further points of detail—it parallels the faunas of those islands which lie beyond the continental shelf, and differs from those islands which lie within the shelf. In particular, the parallelism with the Madagascar fauna is made much closer by the recent discoveries. On the other hand, the contrast with the fauna of continental islands such as Borneo or Sumatra is a marked one. On these islands the fauna, although considerably specialized by isolation in Borneo, less so in Sumatra, is a fairly representative one. It includes all or nearly all of the important mammalian groups of the mainland save those which there is reason to believe are of too recent arrival or of unsuitable habitat to be present.

It may be objected that this difference is merely apparent; that the Pleistocene fauna of the Antilles was really of continental character, but because they are islands and not continents, it has been easily exterminated by man, and that the cave and spring deposits present only two distinct and very limited facies, not including the ungulates, carnivores, etcetera, which have been present. The best reply to this objection is to test it by comparison. Sumatra or Java are islands of comparable size to Cuba; Borneo is larger; Formosa or Hainan are comparable to Porto Rico. In none of these islands has the indigenous fauna been wiped out to anything like the extent necessary to obliterate its continental character, although all have been inhabited by man for a much longer time and in much larger numbers than the Antilles. The indigenous faunas of Great Britain or Ireland are far from exterminated, in spite of the great density of population and of modern civilization.

Nor can we assume that a cave or a spring fauna is so limited in its facies as to disguise a continental fauna type. The faunas of numerous caves in Europe and North America have been examined, and wherever any considerable collection is obtained it is clearly representative of the continental type. Spring or bog faunas sometimes contain little except hoofed animals, but I never heard of one in which hoofed animals and carnivora were absent. I can not escape from the conclusion that the Pleistocene fauna of Cuba was not a normal fauna, but deficient in most of the more abundant groups and composed of a selection of a very limited number of types which had expanded to a disproportionate variety and importance, owing to the absence of the rest of the fauna.

CONCLUSIONS AS TO FORMER GEOGRAPHIC RELATIONS AND MANNER OF COLONIZATION

As to the diverse origin of the several groups and the varying time during which they have been isolated on the islands, I have stated my interpretation of the evidence. I do not feel, however, that evidence of this sort leads to positive and certain conclusions.

As to the former connection of the Antilles with each other and with the mainland, my conclusions with the proviso just stated are as follows:

1. That the Greater Antilles have probably been united with each other, as far east as the Anguilla bank, in the late Tertiary or Pleistocene. This I conclude from the near affinity of representative species of the same or closely allied genera and the general similarity of the fauna, so far as known, in the different islands.

2. That they have not at any time during the Tertiary been united with North America. If they had been we should find North American ungulates, rodents, carnivores, etcetera, differentiated in accord with the length of subsequent isolation, but of clearly recognizable affinities, and it would be a balanced or representative fauna. We might object that such a fauna had perhaps existed, but been wiped out by subsequent submergence. But the presence of *Solenodon* and *Nesophontes* negatives that, for they represent a very ancient survival, and if there had been a representative fauna it is hardly credible that submergence would have spared just two insectivores and destroyed all the rest of the fauna.

3. That they probably have not been connected with South America, either via the Lesser Antilles or via Central America, during the Tertiary; for if they had the fauna should be of continental South American type, with South American ungulate groups, marsupial carnivores, and a full representation of the rodents, edentates, etcetera.

4. The mammalian fauna appears to me to be reducible to perhaps three primary rodent stocks, one or more primary ground-sloth stocks, and two Insectivora. These I conceive to have arrived at various times during the Tertiary, the rodents and ground-sloths from South or Central America, the insectivores from North America, by accidents of transportation, of which the most probable for the mammals would perhaps be the so-called "natural rafts" or masses of vegetation dislodged from the banks of great rivers during floods and drifted out to sea. The probabilities of this method I have elsewhere discussed.[9] For birds and bats, for the smaller reptiles, amphibians, fishes, and invertebrates, the problem of oversea transportation is a much simpler one.[10] That successful colonization in this way can occur is shown by their presence on nearly all oceanic islands; for it will hardly be maintained by reasonable men that every oceanic island has been joined to the mainland and has been continuously above water since its separation. Obviously, the larger the island and the nearer to continental land, the more often such colonization will occur.

---

[9] Matthew : Climate and evolution. Annals N. Y. Acad. Sci., vol. xxiv, 1915, p. 206.

[10] Tropical storms, as Wallace pointed out years ago, probably play a principal part in transportation of very small animals or their eggs. Mammals could hardly be carried that way nor survive if they were.

5. The geology of the Caribbean region appears to me to afford no positive evidence against union of the Antilles either with South America or Central America; but neither does it afford any evidence that there ever was such union. Undoubtedly there is a line of disturbance and uplift along the Lesser Antilles, and another stretching through Haiti and Jamaica to Nicaragua; but evidence of similar and contemporaneous upheavals and similar sedimentation in two portions of this line of disturbance that are now separated by abyssal depths does not in the least prove that the intervening depths were formerly continuous land bridges. They may have been, but I do not see how any geologic evidence can prove that they were so. If we have evidence from some other source that there must have been a land bridge somewhere, then these lines of disturbance show its most probable location. That is all.

Land union with Florida appears to be distinctly against the geologic evidence, as in this region we have extensive flat-lying Tertiary marine and littoral formations which indicate that there has been very slight movement during the Tertiary, and that the present limits of the continental shelf represent probably the extreme extension of the land in the Pleistocene. Dall has shown the evidence very clearly in the case of Florida. Apparently the conditions in Yucatan are partly similar, but Vaughan has shown that its tectonic relations to the Antillean ridges are more favorable to a former union.

15

Reprinted from *Contr. Paleont., Carnegie Inst. Washington, Publ. 590,*
1950, pp. 217–219, 285–306

EVOLUTION OF DESERT VEGETATION
IN WESTERN NORTH AMERICA

Daniel I. Axelrod

INTRODUCTION

The Problem

A wide diversity of opinion exists with respect to the age and derivation of modern desert environments. According to one view, desert vegetation is "an earth-old feature." A second theory, corollary to this, is that desert floras of essentially modern character have been in existence in their present positions since angiosperms first assumed dominance during the Cretaceous period. Although both opinions have been expressed on several occasions during the past decade, no evidence supporting them has been presented. The validity of these beliefs can easily be determined from the geological record of the present desert regions. Since the continental platforms have been in their present positions since the Cretaceous, it follows that the Cretaceous and Tertiary record of the present desert regions should provide evidence of desert environments of subcontinental extent if such were in existence during those periods. Analysis of the available data shows clearly that there were no desert environments of wide extent during those times. Thus a third opinion, and the one to be elaborated here, is that the desert environments now characterizing wide subcontinental regions are a phenomenon of only the latest part of geological time.

Although certain investigators seem to feel that desert vegetation of essentially modern character and distribution has been in existence since Cretaceous time, their ideas as to derivation and origin are uncertain and obscure. At least one investigator appears to regard desert vegetation as unique, and as having arisen *de novo*. Though others are inclined to the belief that desert vegetation has always been here, they make no suggestion as to its derivation. Yet the views presented by MacDougal four decades ago (1909, p. 119), that "xerophilous types of vegetation are of comparatively recent origin, ... [and that] the movement toward xerophily may be reckoned as one of the most important

160

in evolutionary procedure," seem fully acceptable today. The thesis of this paper is that desert vegetation of modern character developed during the Tertiary period by the gradual adaptation of more mesic plants to slowly expanding dry climate. This is not a new idea, but may be inferred from the ecological studies of modern angiosperms made by Andrews in Australia (1913, 1914) and by Bews in Africa (1927). Whereas their conclusions regarding evolution in response to aridity were based primarily on inferences drawn from the relationships of modern plants in dry regions to those in adjacent, more humid areas, the present study reaches the same conclusion from an analysis of the succession of fossil floras situated in or marginal to the present desert region of western North America.

Previous Work

The history of desert environment in this region was discussed about fifteen years ago by Clements (1936). His inferences were drawn largely from the present-day distribution of plants in the desert and desert-border regions, and also from the few fossil floras and faunas then known from this area. For the earlier part of the Cenozoic, Clements postulated an essentially continuous hardwood-deciduous forest extending in the interior from Oregon into southern California. He pointed out, on the basis of paleobotanical studies by Chaney and his students, that during late Oligocene and Miocene times hardwood-deciduous forest and montane conifer forest characterized the northern Great Basin, being interrupted locally by grassland to form a general prairie region. Clements inferred that grassland dominated the southern basin and range province during the Miocene, with woodland scattered locally along stream valleys and some scrub on adjacent slopes. For the later Tertiary he reached the significant conclusion that the history of desert vegetation in the western United States is "inseparably bound up with the disappearance of grass dominants in the region of desiccation" (1936, p. 90). Clements considered that this grassland was replaced by desert vegetation which migrated northward from Mexico late in Pleistocene time, but made no suggestion as to its ultimate derivation.

The writer has already modified and supplemented this interpretation (Axelrod, 1939, pp. 90-91; 1940b; 1947; 1948, pp. 138-
140). Clements' postulate of a dominant hardwood-deciduous
forest in the region now desert during Eocene and early Oligocene times is untenable. Subtropical to warm-temperate evergreen forests characterized the lowlands of eastern Oregon
during the Eocene (Chaney, 1948, pp. 9-20), and similar forests
characterized the western Mohave area at that time (Axelrod,
1939, p. 52). The belief that grassland dominated over the present
southern desert area during the middle Tertiary is also in error.
The region was characterized then by live oak and conifer woodland, chaparral, arid subtropical scrub ("thorn forest"), sage,
and desert-border vegetation, with grassy plains locally inter-
rupting the woodland (Axelrod, 1939). By Middle Pliocene time
essentially open environments extended throughout the lowlands
of the present desert province, with trees restricted largely to
stream banks and lake borders over the lowlands in a region of
dominant grassland with semidesert shrubs on adjacent slopes
(Axelrod, 1948). Whereas Clements assumed a northern Mexican
origin for the desert vegetation which migrated northward to
displace grassland, the writer has suggested that the modern
desert floras were derived largely from species which contributed to the more mesic Tertiary Floras that formerly extended
across the present desert areas. In other words, the modern
desert species appear to be hardy plants derived from members
of Tertiary communities that earlier dominated the region. They
apparently became adapted to an increasingly drier and drier
climate during Tertiary time, a climatic trend that finally culminated in a regional desert climate at the end of the Cenozoic.

Additional information now in hand adds materially to the details of desert history. The following analysis rests upon approximately 110 fossil floras, whose geographic positions are shown
in figures 1 to 3. All the middle and late Tertiary floras from
the interior regions include species closely similar to plants
which are important members of communities now bordering the
desert, or which range cut into the desert proper to contribute
to its distinctive physiognomy; the other floras provide critical
information relative to an interpretation of those at the interior.

[*Editor's Summary of Pages 220–285:* Axelrod examines the relation of modern to fossil floras of the deserts and nearby areas. He follows Shreve (1942) in recognizing four Recent floras that differ taxonomically and ecologically: the Great Basin, Mohave, Sonoran, and Chihuahuan floras. The border areas are more diverse ecologically and include conifer forest, live oak and conifer woodland, chaparral, coastal sage, arid subtropical scrub (or "thorn forest"), and desert grasslands. The Tertiary plant fossils show that many of these border communities were much more widely distributed and more diverse in composition and their species had less restricted ecological tolerances than at present. By studying the differences in fossil floras and comparing these floras with Recent ones, he interprets the evolution of these communities. He then analyzes the Cretaceous and Tertiary, particularly the later Tertiary, floras and shows that the desert environments originated in the late Tertiary and that the flora developed slowly from existing mesic species into endemic arid species.

Upper Cretaceous floras had moist, warm-temperate to subtropical species that were widely distributed in the western United States. Early Tertiary floras indicate that tropical to subtropical conditions spread northward to encompass most of the western United States. By the Miocene two floras are present: the Arcto-Tertiary flora having temperate deciduous hardwoods and conifers in the north part of the Great Basin and Columbia Plateau, and the Madro-Tertiary flora having semiarid live oak woodlands, chaparral, thorn forest, and semidesert vegetation from the southern part of the Great Basin, Mohavian, and Sonoran areas southward into Mexico. Between these floras was a broad ecotone. During the later part of the Miocene, these floras shifted both north and south and became increasingly differentiated. The Miocene topographic relief of the region was less strong than at present, and the Sierra Nevada block was still low and had westward, through-flowing streams across it from the Great Basin.

Axelrod then examines the lower, middle, and late Pliocene floras and demonstrates the lack of widespread desert floras in lower Pliocene assemblages, an increase in desert floras in the lowlands in middle Pliocene assemblages, and more moist and cooler floras in upper Pliocene assemblages. The Pleistocene floras indicate fluctuations in temperature and moisture. For the most part the present Great Basin Desert was distinct from the Mohave–Sonoran province during the Tertiary. Dominants derived from the Arcto-Tertiary flora are common in the Great Basin Desert, and relatively few shrubs derived from the Madro-Tertiary flora became established in that area during the late Tertiary. Some species of Arcto-Tertiary derivation became established during the cooler late Pliocene and Pleistocene in the Mohave and Sonoran deserts.

Comparisons with South American floras show that, although similar in general adaptive habit, their relationships are different. The disjunct distribution of identical species. paired species, and paired genera are interpreted as having either differentiated from wider ranging ancestors or migrated long distances at times of favorable environmental conditions. These disjunct pairs are less than 1 percent of the floras studied, and Axelrod believes that they are overemphasized by plant geographers; the meridional differences greatly outweigh these slight resemblances.]

SUMMARY OF DESERT HISTORY

The preceding discussion of the temporal succession of middle and later Cenozoic vegetation of the present desert and desert-border regions of the western United States indicates that desert vegetation and desert climate, as now seen in the form of sub-continental environments, are phenomena of late Cenozoic time.

The general pattern of floral succession discussed in the foregoing pages is summarized in figure 4. It shows diagrammatically the relative development of the different types of vegetation in the present desert regions during Miocene and Pliocene times as inferred from the fossil floras now known from the region. The width of the diagram for each vegetation type indicates its probable representation over lowland areas as judged from the relative abundance of the species in each flora. Although the details will undoubtedly be modified as new fossil floras are discovered, the general pattern is considered to be reasonably well established.

All important features discussed on the preceding pages are shown clearly in figure 4. These include the following:

1. The Arcto-Tertiary Flora, characterized by conifer forest and hardwood-deciduous species, dominated the present area of the northern Great Basin Desert during Miocene time. The Madro-Tertiary Flora, with its live oak woodland, chaparral, and arid subtropical scrub vegetation, characterized the regions of the present Mohave and Sonoran Deserts some 600 to 800 miles farther south. Lying between these Floras across the central Great Basin Desert region in Nevada was an ecotone or transition region, in which species of both Floras formed a mosaic of overlapping vegetation types during Miocene time.

2. These three broad vegetation provinces persisted into the Lower Pliocene, with dominant montane forest at the north giving way gradually to woodland and scrub at the south.

3. Warm, semiarid climate during Middle Pliocene time is reflected in the widespread restriction of forest and woodland over the lowlands. Semiarid grassland and shrubland are now inferred to have dominated in the lowlands throughout the region, with forest and woodland occurring in the mountains.

4. Cooler Upper Pliocene climate is inferred to have permitted some expansion of forest and woodland, with a consequent restriction of grassland and shrubland.

5. The inferred fluctuations of lowland vegetation in response to the alternation of dry and moist climate during the Quaternary

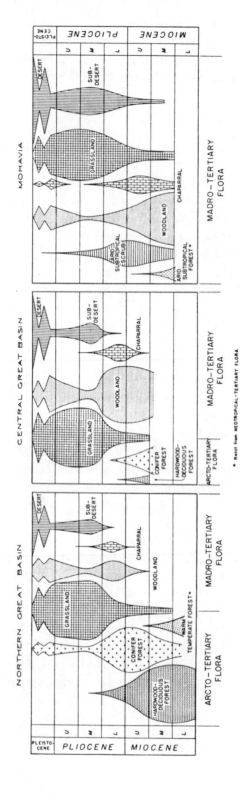

Fig. 4. Changing vegetation types in the present desert region during middle and upper Cenozoic times

166

have been shown in abbreviated form, with expanding desert
environment correlated with the dry, and grassland with the
pluvial, stages.

THE PATTERN OF DESERT EVOLUTION

The task of determining the origins of modern desert vegetation is by no means an easy one, as is amply evident from the
preceding discussion. The problem is one whose solution rests
largely with paleobotany, though it is clear that data from modern plant distribution can give important clues and that cyto-
genetic studies can aid greatly in determining centers of differentiation. The procedure followed above for interpreting the
history of the Great Basin, Mohave, and Sonoran Deserts appears
sound. Furthermore, the general pattern of desert development
elsewhere in the world seems to parallel closely that in western
North America, with respect both to age and to origin. That the
fundamental problems of desert development are probably much
alike for all the deserts is a view which finds support in the following statement by Shreve (1937, p. 214), one of the foremost
investigators of present-day desert vegetation:

> The biological problems of the seven great desert regions of the world are
> partly universal, having common relations to all of them, and partly specific,
> having reference to each considered alone. . . . The phases of structure,
> function, and behavior which have adjusted plants to the various intensities
> of desert conditions are very much the same in all the deserts. Identical or
> closely similar features in anatomy, physiology, and adjustment to environ-
> ment are found again and again in far-separated regions or in different
> continents. Some of the most fundamental desert problems arise among
> these universal features, and have to do with the ways in which evolutionary
> processes have molded the vegetative organs of plants, of whatever relation-
> ship, into successful desert patterns. The problems which are specific for
> each desert are those which concern its distinctive communities of plants,
> the relationships of the plants found there, the regions from which they have
> been derived, and the ancestral stocks from which they have descended.

Age

There appears to be no support from the geological record for
the recently stated belief that "desert floras may well have an

age and continuity comparable with floras of the wet tropics"
(Johnston, 1940). In all cases known, their present areas were
occupied by more mesic vegetation, not only in Tertiary time,
but in the Cretaceous as well. At higher latitudes, both northern
and southern, fossil floras characterized by temperate and cool-
temperate forests have been recorded from regions now desert
well into Miocene time. At lower latitudes in both hemispheres,
fossil floras dominated by species representing tropical, subtropical, and warm-temperate forests occupied regions now
desert in the Cretaceous and early Tertiary, and were succeeded
by semiarid vegetation during late Eocene and early Oligocene
times. The world-wide evidence can now be summarized briefly,
commencing with North America.

As was pointed out in the sections reviewing the Upper Cretaceous and Lower Tertiary environments in western North
America, the first indications of subhumid climate are to be
found in the Middle Eocene Green River flora of the central
Rocky Mountain region. Climate over the lowlands there was
similar to that now found in regions where arid subtropical
forest and scrub characterize the lowlands, and where temperate forest is on adjacent cooler slopes. The modern setting
near Monterrey, Mexico, where dry subtropical and temperate
plants are now mingling in a region of moderate relief in the
foothills, is strongly suggestive of Green River environment.
Earlier in the Tertiary, and during the Upper Cretaceous as well,
the present desert region of western North America largely supported hardwood-evergreen forests, grading into warm-temperate
and temperate forests farther north. The floras show a gradual
change in composition which may be correlated with a change
from warm-temperate climate at middle latitudes to subtropical
climate at lower latitudes over the areas now desert. These older
floras, which were characterized largely by species of the Neotropical-Tertiary Flora and also include warm-temperate members of the Arcto-Tertiary Flora to the northward, demonstrate
that prior to the later Eocene the present dry belt of the "horse
latitudes" was not in existence as a region of low rainfall. The
evidence suggests that the area then had a subtropical climate,
characterized by mild yearly temperature and ample rainfall

distributed largely during the summer months, with a dry winter season. The detailed evidence presented for the middle and upper Cenozoic environments shows that following Eocene time rainfall gradually decreased and extremes of yearly temperature increased, but that the region had an effective rainfall (10 to 15 inches) sufficient to support widespread grassland as late as Upper Pliocene time. The inference is that the present desert region of subcontinental extent is essentially of recent origin. This does not preclude the local occurrence of desert and subdesert arèas during the middle and later Tertiary, particularly in the lee of high ranges within the present southern desert area.

In central Asia fossil floras from Sinkiang and bordering provinces show that semiarid climate characterized the present desert regions of that area into late Tertiary time (Chaney, 1933, 1935). But earlier in the period the temperate Arcto-Tertiary Flora dominated much of central Asia, to judge from the character of the fossil floras from Han Nor in Mongolia (Florin, 1920), the Irtysh basin in Siberia (Kryshtofovich and Borsuk, 1938), and Buchtarma in the Altai Mountains (Schmalhausen, 1887). Subhumid conditions existed during Oligocene time in the present desert region of southwestern Turkestan, where there was an ecotone between the Arcto-Tertiary Flora and sclerophyllous vegetation analogous to the Madro-Tertiary Flora (Korovin, 1932). Yet earlier in the Tertiary the Paleotropical-Tertiary Flora extended through this region, ranging westward into central Europe, the Mediterranean region, and northern Africa. Although little is known of the fossil angiospermous flora of Africa, the presence of forest trees in the deserts of Libya and Egypt during the Cretaceous and early Tertiary seems well established (Berry, 1916, pp. 253-255; Krausel and Stromer, 1924; Seward, 1935; Chiarugi, 1929), and their relationships seem clearly to be with subtropical and tropical genera. Fossil trees and shrubs recorded from the present Namaqualand Desert of southwest Africa provide clear evidence of relatively humid climate there during the Tertiary (Rennie, 1931). Tertiary floras from the present desert and desert-border regions of central and south Australia (Chapman, 1935, 1937) similarly record more mesic conditions in the Tertiary than exist there today, suggesting again that the desert is not an ancient one. As for South America, the occurrence of Eocene and Miocene warm-temperate to subtropical forests representing the Neotropical-Tertiary Flora in the present desert region of Chile and Peru provides conclusive evidence for the relative recency of desert conditions in those areas (Berry, 1919a, 1938). The Neotropical-Tertiary Flora also reached southward across the present desert of Argentina (Berry, 1938). Berry has emphasized the point that the modern desert vegetation there must have developed later in the Cenozoic as the Andean chain was elevated several thousands of feet and more mesic types of vegetation were eliminated from the area in its lee (Berry, 1928, 1932). The temperate Antarcto-Tertiary Flora has been recorded from early and middle Tertiary localities in the present cold desert regions of Patagonia (Duzen, 1899; Berry, 1937), Argentina (Berry, 1928), and Seymour Island, Antarctica (Duzen, 1908). It seems clear that the present desert environment now characterizing those areas must be more recent in age.

Data from Pleistocene rocks in regions now desert elsewhere in the world agree with those reviewed above for the western United States, and suggest that the present dry regions were not entirely desert during the pluvial stages (Flint, 1947). For example, late Pleistocene leaves from the Kargha depression in Egypt include species which represent vegetation like that now in the moister savanna of the Sudan to the southward (Gardner, 1935). Flint also points out that the occurrence of fossil mammoth, rhinoceros, hippopotamus, and crocodile in the Saharan region clearly records a stage of much greater rainfall than now occurs in that area and shows that the present desert climate is essentially recent. Furthermore, the Pleistocene occurrence of fossil crocodiles in South Australia, and of fossil lungfishes in the Lake Erye region of central Australia, also indicates a climate distinctly more humid than the present one. The widespread occurrence of lakes in the present desert regions of Asia, Africa, South America, and Australia also demonstrates the existence of moister climate during the pluvial stages, and suggests that desert environments were not so prominent then as they are today.

Thus the present desert environments of subcontinental extent appear to be comparatively recent. Subhumid climate was developing over areas of certain modern desert regions by late Eocene and Oligocene times, but the present regional desert climates and floras seem to be no older than late Cenozoic anywhere in the world. Such environments are not "earth-old features." They may more appropriately be termed "climatic accidents," for they have been rare in earth history.

Origin

The present desert floras of the world appear to have had their sources chiefly in the Tertiary Floras which occupied the regions now desert. This is suggested by the time-space relations of Tertiary vegetation and is in harmony also with the relationships of the modern desert floras. Two outstanding facts shown by the succession of Tertiary vegetation in desert areas, as well as in areas outside the present desert regions, have an important bearing on an interpretation of the development of desert vegetation.

In the first place, the relation between successive Tertiary floras of any given region and the present vegetation of that region is a function of the age of the fossil flora. Older Tertiary floras rarely contain any fossil plants whose nearest representatives are found near the fossil locality today; most of these modern equivalents occur in regions hundreds or thousands of miles distant. The constituents of middle Tertiary floras often find similar species living at no great distance from the fossil locality, but many of the modern representatives are in areas far removed. The species of upper Tertiary floras have the majority of their nearest living relatives in the general region of the fossil locality. Species of Pliocene floras from the regions now desert are commonly closely related to woody vegetation now marginal to the desert. These modern communities which resemble those of the Pliocene from desert and desert-border areas regularly include species which also range out to make up an integral part of desert vegetation today. Since a number of the shrubs and small trees now typical of desert regions have close equivalents in the middle and later Tertiary floras from the regions now desert, it is believed that they were associated then with the woodland, forest, and more mesic communities which dominated the region during that period. Thus they must have become adapted to extensive regional desert environment more recently.

The second fact derived from an analysis of Tertiary floras is that the successively younger fossil floras in all regions outside the present areas of rain forest indicate that a general world-wide climatic trend toward lowered yearly rainfall, shifting seasonal distribution of rain, and increasing ranges and extremes of temperature characterized the period. The selective influence of these climatic changes acting upon the varying ranges of tolerance of each species continuously modified the composition of Tertiary vegetation. Lowered rainfall gradually restricted mesic vegetation types to humid areas more and more distant from the evolving dry areas. By mid-Pliocene time trees had largely been eliminated from the lowlands of the regions now desert, except in marginal stream-bank and lake-border habitats. Grassland and subdesert communities, which had been confined to the borders of forest and woodland earlier in the Tertiary, now attained areas apparently subcontinental in extent. Whereas the grass, herb, and semiwoody species evolved rather slowly in the more localized environments of the Miocene and early Pliocene, they seem to have differentiated rapidly as suitable environments of wide regional extent came into existence during Middle Pliocene time. Increasing topographic and climatic diversity over these areas during later Cenozoic time has resulted in the differentiation of numerous species (and varieties) well adapted to these more narrowly defined environments of the desert and adjacent regions. Thus it is not surprising to find that many desert herbs are polyploids whose nearest ancestors have equivalents in the forest, woodland, scrub, and grassland vegetations now marginal to the desert region. The desert floras must therefore have evolved since their more mesic ancestral species inhabited the lowlands of the area now desert.

With these points in mind, the following generalizations may be formulated as a working basis for investigation of the history of particular desert floras:

1. The cold deserts of the northern and southern hemispheres have largely derived their species from the Arcto-Tertiary and Antarcto-Tertiary Floras. Ancestral species seem to have developed slowly on the cooler, subhumid borders of these Floras during much of middle and later Tertiary time. They spread widely and evolved rapidly as cool-dry environments of subcontinental extent came into existence at the north and south late in the period. Though the cold northern deserts (and bordering steppes) are largely distinct as to their species, there are closely related as well as identical species in the arid regions of North America and Eurasia. They may be explained partly by latitudinal migrations along semiarid corridors linking these areas through Beringia during later Cenozoic time, and partly by derivation from wider-ranging ancestors of late Tertiary age which occupied the subhumid margins of the Arcto-Tertiary Flora.

2. The modern floras of the warm deserts find their sources chiefly in the Tropical-Tertiary Floras. The temporal succession of Tertiary vegetation indicates that subtropical savanna, arid subtropical forest, arid subtropical scrub, sage, plains grassland, and warm desert evolved successively, and more or less contemporaneously, as regional plant formations on the northern and southern margins of the tropics as drier climate expanded over low to middle latitudes during the period. Many initially tropical types of wide range thus appear to have gradually become adapted to new, drier, border-tropical environments as rain forest was confined equatorward by the radial expansion of subhumid climate over low to middle latitudes following Eocene time.

This interpretation, which is strongly supported by the fossil record, has been hinted at earlier by Andrews and Bews to explain the relationships of modern species in drier areas marginal to the wet tropics. It seemingly accounts for the facts (a) that there are today many species complexes and paired genera that extend north and south from the humid tropics into these derivative vegetations, (b) that there are many species common to several of these plant formations, and (c) that there are closely related species and paired genera in the desert, plains, sage, and arid subtropical scrub vegetations north and south of the tropics which no longer have intervening stations. This leads to the suggestion that some desert species belonging to tropical families may have evolved by the gradual adaptation of wider-ranging prototypes to essentially similar environments developing contemporaneously on the northern and southern borders of the areas of their Tertiary distribution. The selective influence of progressively expanding dry climate may in this manner have molded successful desert types from pantropic Tertiary ancestors in such genera as *Acacia, Baccharis, Cneoridium, Dodonaea, Euphorbia, Fagonia, Ficus, Lycium, Rhus, Salicornia,* and *Zygophyllum* in the widely separated warm desert regions of North America, South America, Africa, Asia, and Australia. Some of these species are closely related, especially in the warm deserts of the same hemisphere, whereas those in different hemispheres generally vary more widely. Others have apparently evolved sufficient ranges of tolerance so that they have been able to survive in some of the colder deserts at middle latitudes as well.

Such an analysis of the present meridional pattern of distribution of many plants represented by paired species and paired genera in desert and subdesert regions now separated by tropical forests seems more nearly in accord with the facts than explanations which require trans-tropic Cretaceous and Tertiary migrations along postulated land bridges or island archipelagoes in meridional directions. In the first place, there is no need to assume these earlier migrations, for there is no evidence that desert environments were in existence during those times. Furthermore, suitable routes in meridional directions apparently were not in existence then, since the mountains were lower and less continuous in north-south directions during the Cretaceous and Tertiary than they are now; actually the tropics are now more restricted and the arid corridors are greater in extent than they have been at any time since the early Cretaceous, and migration is not highly effective for the plants under consideration. It must be noted also that tropical climate was sufficiently widespread during the Cretaceous and early Tertiary so that high, continuous mountains would be required to develop arid corridors across the expanded tropical zone; there is no clear evidence that such mountains ever existed. Finally, it must also be agreed that if these meridional patterns are to be explained by normal migration along presumed mountain chains, these mountains would also have supported a humid flora on their windward slopes. We might then expect an interchange of forest elements between the northern and southern hemispheres, yet the fossil record indicates that such elements have been largely distinct throughout Cretaceous and Tertiary times.

Some of the plants now exhibiting a meridional pattern across the tropics may have attained such a distribution by long-distance migration during the later Cenozoic. This statement applies chiefly to the genera of herbs and grasses which have been evolving rapidly in the later Cenozoic, and which have produced numerous species, some of which are represented by identical or closely related species in the desert, steppe, mediterranean, and cool-temperate climates north and south of the tropics. The fact is again emphasized that these plants account for only a very small fraction of the flora of the Americas.

3. Radially expanding semiarid Tertiary Floras of low-middle latitude origin have contributed to both the warm and cold desert floras which now occupy the areas they dominated during middle and later Tertiary times.

4. Differentiation of the great desert floras into their present distinctive communities was accomplished chiefly during late Pliocene and Pleistocene times, largely in response to environmental selection as topographic and climatic diversity increased over the ever expanding dry regions into the latest state of geologic time.

5. The evolution of desert floras is continuing actively at the present time.

CONCLUSIONS

The temporal succession of angiosperm floras from areas now desert and subdesert in western North America shows that subtropical and warm-temperate forests dominated over the lowlands of the present desert regions from Upper Cretaceous into early Tertiary time; that temperate deciduous and conifer forest, arid subtropical forest, arid subtropical scrub, live oak woodland, and chaparral characterized these areas during the middle and later Tertiary; and that grassland and scrub became prominent during the Middle Pliocene. The present desert environments of subcontinental extent must therefore be of latest Cenozoic age.

The present-day desert plants apparently have been derived from species represented in the major Tertiary Floras which occupied the regions now desert. During much of Tertiary time ancestors of modern desert species were evolving rather slowly in localized subhumid climates on the borders of the Arcto-Tertiary, Madro-Tertiary, and Neotropical-Tertiary Floras. The more mesic communities of these Floras had largely been eliminated from lowland areas over the present desert region by Middle Pliocene time. As open environments of subcontinental extent now came into existence, the semiarid shrub, herb, and grass communities which were restricted earlier to the borders of the more mesic vegetation types spread widely and evolved rapidly. But even at this later date desert conditions appear to have been largely subordinate to grassland, which seems to have characterized the lowlands of the present desert region. Under the impact of fluctuating late Cenozoic climates and continued topographic differentiation, a host of new forms adapted to desert climate came into existence. Many of them find their ancestors in species which now contribute to the vegetation on the borders of the desert, communities which were well represented by equivalent species over the lowlands of the present desert area during Pliocene time.

Desert floras elsewhere in the world may have followed the same general pattern in evolution. Their age appears to be late Cenozoic, for the regions they now occupy supported more mesic types of vegetation in the Cretaceous and Tertiary than exist in those areas today. Thus there seems to be no evidence to support the belief that desert environments of essentially modern distribution have been in existence since the early Cretaceous.

Although desert environments have been çalled "earth-old features," they may more appropriately be termed "climatic accicents," for they seem to have been rare in earth history.

The desert floras of the other continents also seem to have derived their species from the major continental Tertiary Floras which occupied those areas. As gradually expanding dry climate restricted the Tertiary Floras to moister regions marginal to the areas of desiccation, subhumid communities on their borders slowly expanded and through time have produced species adapted to successively drier climate. Some related species and paired genera now occurring in desert regions widely separated in meridional and latitudinal directions may have been derived from ancestral forms represented in Tertiary Floras which earlier connected the areas. There is some evidence of interchange of essentially identical or closely related species between dry regions in meridional directions by long-distance migration since the middle Tertiary. Such migrations account for only a very small fraction of these floras, which are otherwise largely distinct.

Acknowledgments. The writer has received considerable assistance in the course of assembling this paper, which has been in preparation since 1939. The Carnegie Institution of Washington has generously provided funds which have made field work possible during many seasons, both winter and summer, over the past decade; of this assistance the writer is deeply appreciative. Much of the preliminary work on this paper was completed during tenure as National Research Fellow at the United States National Museum during the years 1939-1941. To the Council, to the staff at the Museum, and particularly to Roland W. Brown, under whom research was conducted, the writer expresses his thanks.

The substance of this report was presented first at a symposium on "The evolution of plant communities in southwestern North America" held at the Dallas meetings of the AAAS in the winter of 1941, and much discussion resulted at that time. Part of the paper was read also at a symposium on "Desert evolution" at the spring meetings of the AAAS in Berkeley in 1948. More recently the manuscript has been extensively criticized by Carl Epling, Ralph W. Chaney, Harry D. MacGinitie, and Lincoln Constance. They have added materially to its clarity, and have offered many valuable suggestions.

LITERATURE CITED

Andrews, E. C.

    1913   Development of the natural order Myrtaceae. Proc. Linn. Soc. New South Wales, vol. 38, pp. 529-568.

    1914   Development and distribution of Leguminosae. Jour. and Proc. Roy. Soc. New South Wales, vol. 48, pp. 333-407.

Axelrod, D. I.

    1937   A Pliocene flora from the Mount Eden beds, southern California. Carnegie Inst. Wash. Pub. 476, III, pp. 125-183.

    1938   The stratigraphic significance of a southern element in later Tertiary floras of western America. Jour. Wash. Acad. Sci., vol. 28, pp. 313-322.

    1939   A Miocene flora from the western border of the Mohave Desert. Carnegie Inst. Wash. Pub. 516.

    1940a   The Pliocene Esmeralda flora of west-central Nevada. Jour. Wash. Acad. Sci., vol. 30, pp. 163-174.

    1940b   Late Tertiary floras of the Great Basin and border areas. Bull. Torrey Bot. Club, vol. 67, pp. 477-487.

    1940c   The Mint Canyon flora of southern California: a preliminary statement. Amer. Jour. Sci., vol. 238, pp. 577-585.

    1944a   The Mulholland flora. Carnegie Inst. Wash. Pub. 553, V, pp. 103-146.

    1944b   The Alvord Creek flora. Carnegie Inst. Wash. Pub. 553, IX, pp. 225-262.

    1944c   The Alturas flora. Carnegie Inst. Wash. Pub. 553, X, pp. 263-284.

    1947   Pliocene environments of the Great Basin. Bull. Geol. Soc. Amer., vol. 58, p. 1246.

    1948   Climate and evolution in western North America during Middle Pliocene time. Evolution, vol. 2, pp. 127-144.

    1949   Eocene and Oligocene formations in the western Great Basin. Bull. Geol. Soc., Amer., vol. 60, p. 1935.

    1950a   Classification of the Madro-Tertiary Flora. Carnegie Inst. Wash. Pub. 590, I, pp. 1-22.

    1950b   Further studies of the Mount Eden flora, southern California. Carnegie Inst. Wash. Pub. 590, III, pp. 23-71.

    1950c   The Anaverde flora of southern California. Carnegie Inst. Wash. Pub. 590, IV, pp. 73-117.

Babcock, E. B., and G. L. Stebbins, Jr.

    1938   The American species of *Crepis*. Carnegie Inst. Wash. Pub. 504.

Ball, O. M.
  1931 A contribution to the paleobotany of the Eocene of Texas. Bull.
        Agric. and Mech. Coll. Texas, 4th ser., vol. 2, no. 5.
Berry, E. W.
  1916 The Upper Cretaceous floras of the world. *In* Upper Cretaceous,
        Maryland Geol. Surv., pp. 111-313.
  1919a Fossil plants from northwestern Peru. Proc. U. S. Nat. Mus.,
        vol. 55, pp. 279-294.
  1919b An Eocene flora from trans-Pecos Texas. U. S. Geol. Surv.
        Prof. Paper 125a, pp. 1-10.
  1928 Tertiary fossil plants from the Argentine Republic. Proc. U. S.
        Nat. Mus., vol. 73, art. 27, pp. 1-27.
  1929 A revision of the flora of the Latah formation. U. S. Geol. Surv.
        Prov. Paper 154h, pp. 225-264.
  1930 A flora of Green River age in the Wind River basin of Wyoming.
        U. S. Geol. Surv. Prof. Paper 165b, pp. 55-79.
  1931 A Miocene flora from Grand Coulee, Washington. U. S. Geol.
        Surv. Prof. Paper 170c, pp. 31-42.
  1932 Fossil plants from Chubut Territory collected by the Scarritt
        Patagonian Expedition. Amer. Mus. Novitates, no. 536, pp. 1-10.
  1935 A preliminary contribution to the floras of the Whitemud and
        Ravenscrag formations. Canada Dept. Mines, Geol. Surv. Mem.
        182.
  1937 Eogene plants from Rio Turbio in the vicinity of Santa Cruz,
        Patagonia. Johns Hopkins Univ., Studies in Geol., no. 12, pp.
        91-98.
  1938 Tertiary flora from the Rio Pichileufu, Argentina. Geol. Soc.
        Amer., Spec. Papers, no. 12.
Bews, J. W.
  1927 Studies in the ecological evolution of angiosperms. New Phytol.,
        Reprint no. 16, pp. 1-134.
Bradley, W. H.
  1948 Limnology and the Eocene lakes of the Rocky Mountain region.
        Bull. Geol. Soc. Amer., vol. 59, pp. 635-648.
Brown, R. W.
  1933 Fossil plants from the Aspen shale of southwestern Wyoming.
        Proc. U. S. Nat. Mus., vol. 82, art. 12.
  1934 The recognizable species of the Green River flora. U. S. Geol.
        Surv. Prof. Paper 185c, pp. 45-77.
  1937 Additions to some fossil floras of the western United States.
        U. S. Geol. Surv. Prof. Paper 186j, pp. 163-186.
Buwalda, J. P.
  1914 Pleistocene beds at Manix in the eastern Mohave Desert region.
        Univ. Calif. Publ., Bull. Dept. Geol. Sci., vol. 7, pp. 443-464.

Cain, S. A.
    1944   Foundations of plant geography. Harper & Bros., New York.
Campbell, D. H.
    1942   Continental drift and plant distribution. Science, vol. 95, no.
            2455, pp. 69-70.
    1944   Relations of the temperate floras of North and South America.
            Proc. Calif. Acad. Sci., 4th ser., vol. 25, pp. 139-146.
Chaney, R. W.
    1925   The Mascall flora - its distribution and climatic relation. Car-
            negie Inst. Wash. Pub. 349, II, pp. 23-48.
    1927   Geology and paleontology of the Crooked River basin, with special
            reference to the Bridge Creek flora. Carnegie Inst. Wash. Pub.
            346, IV, pp. 45-138.
    1933   A Pliocene flora from Shansi province. Bull. Geol. Soc. China,
            vol. 12, pp. 129-142.
    1935   The Kucha flora in relation to the physical conditions in central
            Asia during the late Tertiary. Svensk. Sällsk. antropol. och geogr.
            Ann., vol. 17, pp. 75-104.
    1936   The succession and distribution of Cenozoic floras around the
            northern Pacific basin. *In* Essays in geobotany in honor of
            William Albert Setchell, pp. 55-85. Univ. California Press,
            Berkeley.
    1938a   Paleoecological interpretations of Cenozoic plants in western
            North America. Bot. Rev., vol. 4, pp. 371-396.
    1938b   The Deschutes flora of eastern Oregon. Carnegie Inst. Wash.
            Pub. 476, IV, pp. 185-216.
    1940   Tertiary forests and continental history. Bull. Geol. Soc. Amer.,
            vol. 51, pp. 469-488.
    1944a   Summary and conclusions. *In* Pliocene floras of California and
            Oregon. Carnegie Inst. Wash. Pub. 553, XIII, pp. 353-373.
    1944b   A fossil cactus from the Eocene of Utah. Amer. Jour. Bot., vol.
            31, pp. 507-528.
    1948   The ancient forests of Oregon. Condon Lectures. Oregon State
            System of Higher Education, Eugene, Oregon.
Chaney, R. W., and M. K. Elias
    1936   Late Tertiary floras from the High Plains. Carnegie Inst. Wash.
            Pub. 476, I, pp. 1-72.
Chaney, R. W., and E. I. Sanborn
    1933   The Goshen flora of west central Oregon. Carnegie Inst. Wash.
            Pub. 439.
Chapman, F.
    1935   Plant remains of Lower Oligocene age from Blanche Point, Al-
            dinga, South Australia. Trans. Roy. Soc. South Australia, vol. 59,
            pp. 237-240.

Chapman, F. — *Continued*
    1937   Descriptions of Tertiary plant remains from central Australia and from other Australian localities. Trans. Roy. Soc. South Australia, vol. 61, pp. 1-16.

Chiarugi, A.
    1929   Preliminary notes on the petrified forests of Sitrica. Nuovo gior. bot. ital., vol. 38, pp. 558-566.

Clements, F. E.
    1920   Plant indicators. Carnegie Inst. Wash. Pub. 290.
    1934   The relict method in dynamic ecology. Jour. Ecol., vol. 22, pp. 39-67.
    1936   The origin of the desert climax and climate. *In* Essays in geobotany in honor of William Albert Setchell, pp. 87-140. Univ. California Press, Berkeley.

Condit, C.
    1944a   The Remington Hill flora. Carnegie Inst. Wash. Pub. 553, II, pp. 21-55.
    1944b   The Table Mountain flora. Carnegie Inst. Wash. Pub. 553, III, pp. 57-90.

Darwin, C. R.
    1859   On the origin of species by means of natural selection, or the preservation of favoured races in the struggle for life. J. Murray, London.

Dorf, E.
    1939   Fossil plants from the Upper Cretaceous Aguja formation of Texas. Amer. Mus. Novitates, no. 1015.
    1942   Upper Cretaceous floras of the Rocky Mountain region. Carnegie Inst. Wash. Pub. 508.

DuReitz, G. E.
    1940   Problems of bipolar plant distribution. Acta phytogeogr. suecica, vol. 13, pp. 215-282.

Duzen, P.
    1899   Über die tertiäre Flora der Magellansländer. Wissensch. Ergebn. Schwed. Exped. nach den Magellansländer, vol. 1, pp. 87-107.
    1908   Über die Tertiäre Flora der Seymour-Insel. Wissensch. Ergebn. Schwed. sudpolar Exped., 1901-1903, vol. 3, pp. 1-27.

Elias, M. K.
    1942   Tertiary prairie grasses and other herbs from the High Plains. Geol. Soc. Amer., Spec. Papers, no. 41.

Flint, R. F.
    1947   Glacial geology and the Pleistocene epoch. John Wiley & Sons, New York.

Florin, R.
    1920   Einige chinesiche Tertiärpflanzen. Svensk bot. Tidskr., vol. 14, pp. 239-243.

Gardner, E. W.
    1935   The Pleistocene fauna and flora of Kargha oasis, Egypt. Quart.
            Jour. Geol. Soc. London, vol. 91, pp. 479-512.
Gidley, J. W.
    1922   Preliminary report on the fossil vertebrates of the San Pedro
            Valley, Arizona. U. S. Geol. Surv. Prof. Paper 131e, pp. 119-131.
Hansen, H. P.
    1947   Post-glacial forest succession, climate and chronology in the
            Pacific Northwest. Trans. Amer. Philos. Soc., n. s., vol. 37, pt. 1.
Hay, O. P.
    1927   The Pleistocene of the western region of North America and its
            vertebrated animals. Carnegie Inst. Wash. Pub. 322B.
Hewitt, F. D., et al.
    1936   Mineral resources of the region around Boulder Dam. U. S.
            Geol. Surv. Bull. 871.
Hopper, R. H.
    1947   Geologic section from the Sierra Nevada to Death Valley, Cali-
            fornia. Bull. Geol. Soc. Amer., vol. 58, pp. 393-432.
Hudson, F. H.
    1948   Donner Pass zone of deformation, Sierra Nevada, California.
            Bull. Geol. Soc. Amer., vol. 59, pp. 795-800.
Johnston, I. M.
    1940   The floristic significance of shrubs common to North and South
            American deserts. Jour. Arnold Arb., vol. 21, pp. 356-363.
Knowlton, F. H.
    1899   Fossil flora of Yellowstone National Park. U. S. Geol. Surv.
            Monogr. 32, pt. 2.
    1900   Flora of the Montana formation. U. S. Geol. Surv. Bull. 163.
    1916   Contributions to the geology and paleontology of San Juan County,
            New Mexico. 4. Flora of the Fruitland and Kirtland formations.
            U. S. Geol. Surv. Prof. Paper 98s, pp. 327-353.
    1917   A fossil flora from the Frontier formation of southwestern Wy-
            oming. U. S. Geol. Surv. Prof. Paper 108f, pp. 73-107.
    1924   Flora of the Animas formation. U. S. Geol. Surv. Prof. Paper
            134, pp. 71-98.
    1930   Flora of the Denver and associated formations in Colorado. U. S.
            Geol. Surv. Prof. Paper 155.
Korovin, E.
    1932   Novyi tretichnyi tip semeistva Protaceae iz Srednei Azii. Bot.
            Zhur. SSSR, vol. 17 (5/6), pp. 605-622. (Summary in Biol. Ab-
            stracts.)
Krausel, R., and E. Stromer
    1924   Die fossilen floren Ägyptens. Abhandl. Bayer. Akad. Wissensch.,
            math.-naturwiss. Abt., vol. 30, no. 2.

Kryshtofovich, A. N., and M. I. Borsuk

    1938   Contribution to the Miocene flora from western Siberia. Problems Paleontol., vol. 5, pp. 375-396, 1938. (Univ. Moscow Lab. Paleontol., Publ.) (Russian with English summary.)

LaMotte, R. S.

    1936   The Upper Cedarville flora of northwestern Nevada and adjacent California. Carnegie Inst. Wash. Pub. 455, V, pp. 57-142.

Laudermilk, J. D., and P. A. Munz

    1934   Plants in the dung of *Nothrotherium* from Gypsum Cave, Nevada. Carnegie Inst. Wash. Pub. 453, IV, pp. 29-37.

Lee, W. T., and F. H. Knowlton

    1917   Geology and paleontology of the Raton Mesa and other regions in Colorado and New Mexico. U. S. Geol. Surv. Prof. Paper 101.

Lesquereux, L.

    1878   Contributions to the fossil flora of the western territories. 2. The Tertiary flora. Rept. U. S. Geol. Surv. Terr., vol. 7.

    1883   Contributions to the fossil flora of the western territories. 3. The Cretaceous and Tertiary floras. Rept. U. S. Geol. Surv. Terr., vol. 8.

Lindgren, W.

    1911   The Tertiary gravels of the Sierra Nevada of California. U. S. Geol. Surv. Prof. Paper 73.

Longwell, C. R.

    1949   Structure of the northern Muddy Mountain area, Nevada. Bull. Geol. Soc. Amer., vol. 60, pp. 923-968.

Louderback, G. D.

    1906   General geological features of the Truckee region east of the Sierra Nevada. Bull. Geol. Soc. Amer., vol. 18, pp. 662-669.

MacDougal, D. T.

    1909   The origin of desert floras. *In* V. M. Spaulding, Distribution and movements of desert plants. Carnegie Inst. Wash. Pub. 113, pp. 113-119.

MacGinitie, H. D.

    1933   The Trout Creek flora of southeastern Oregon. Carnegie Inst. Wash. Pub. 416, II, pp. 21-68.

    1941   A Middle Eocene flora from the central Sierra Nevada. Carnegie Inst. Wash. Pub. 534.

Merriam, J. C.

    1919   Tertiary mammalian faunas of the Mohave Desert. Univ. Calif. Publ., Bull. Dept. Geol. Sci., vol. 11, pp. 437-586.

Nolan, T. B.

    1943   The Basin and Range province in Utah, Nevada and California. U. S. Geol. Surv. Prof. Paper 197d, pp. 141-196.

Polynov, B. B.
  1937   The cycle of weathering. (Trans. Alexander Muir.) Murby,
         London.
Potbury, S. S.
  1935   The La Porte flora of Plumas County, California. Carnegie Inst.
         Wash. Pub. 465, II, pp. 29-81.
Reed, R. D.
  1933   Geology of California. Amer. Assoc. Petrol. Geologists, Tulsa,
         Oklahoma.
Reed, R. D., and J. S. Hollister
  1936   Structural evolution of southern California. Amer. Assoc. Petrol.
         Geologists, Tulsa, Oklahoma.
Rennie, J. V. L.
  1931   Note on fossil leaves from the Banke clay. Trans. Roy. Soc. South
         Africa, vol. 19, pt. 3, pp. 251-253.
Sanborn, E. I.
  1935   The Comstock flora of west central Oregon. Carnegie Inst. Wash.
         Pub. 465, I, pp. 1-28.
Schmalhausen, J.
  1887   Über tertiäre Pflanzen aus dem Thale des Flusses Buchtorma
         am Flusse des Altaigebirges. Paleontographica, vol. 33, pp.
         181-216.
Schultz, J. R.
  1937   A late Cenozoic vertebrate fauna from the Coso Mountains, Inyo
         County, California. Carnegie Inst. Wash. Pub. 487, III, pp. 75-109.
Seward, A. C.
  1935   Leaves of dicotyledons from the Nubian sandstone of Egypt.
         Egyptian Geol. Surv.
Shreve, F.
  1937   Desert investigations. Carnegie Inst. Wash. Year Book No. 36,
         pp. 214-215.
  1942   The desert vegetation of North America. Bot. Rev., vol. 8, pp.
         195-246.
Simpson, G. G.
  MS   Probabilities of dispersal in geologic time.
Stebbins, G. L., Jr.
  1947   Evidence on rates of evolution from the distribution of existing
         and fossil plant species. Ecol. Monogr., vol. 17, pp. 147-158.
  1948   The origin of the complex of *Bromus carinatus* and its phyto-
         geographic significance. Contr. Gray Herb., no. 165, pp. 42-55.
Webber, I. E.
  1933   Woods from the Ricardo Pliocene of Last Chance Gulch, Cali-
         fornia. Carnegie Inst. Wash. Pub. 412, II, pp. 113-134.

Went, F. W.
    1948   Ecology of desert plants. I. Observations on germination in Joshua Tree National Monument. Ecology, vol. 29, pp. 242-253.
    1949   Ecology of desert plants. II. The effect of rain and temperature on germination and growth. Ecology, vol. 30, pp. 1-13.

Wilson, R. W.
    1933   A rodent fauna from later Cenozoic beds of southwestern Idaho. Carnegie Inst. Wash. Pub. 440, VIII, p. 117-135.

Wulff, V. E.
    1943   An introduction to historical plant geography. (Trans. Elizabeth Brissenden.) Chronica Botanica, Waltham, Mass.

Zimmerman, E. C.
    1948   Insects of Hawaii. Vol. 1. Introduction. Univ. Hawaii Press, Honolulu.

Editor's Comments
on Papers 16A and 16B

16A   **AXELROD**
      Fossil Floras Suggest Stable, Not Drifting, Continents

16B   **HAMILTON**
      Discussion of Paper by D. I. Axelrod, "Fossil Floras Suggest Stable, Not Drifting, Continents", Reply (D. I. Axelrod)

By the late 1950s, systematists had developed an elaborate, but somewhat bulky and unwieldy set of models, theories, and hypotheses to explain the origin and distribution of groups of higher plants (angiosperms), mammals, and birds. Obviously, a few questions remained over the details of this or that family or genus, but there seemed to be enough models so that each distributional problem could be solved without resorting to continental drift. In the early 1960s the tectonic geophysicists and geologists proposed a mechanism, known as sea-floor spreading, that seriously challenged the concept of the fixity or permanency of continents and ocean basins. Within a very short time serious questions about the elaborate distribution models and theories of the 1950s were raised from outside the systematists's sphere. The result was a spate of articles defending the models that required, or at least assumed, fixity of continents: a valiant last-ditch stand.

Paper 16A (and its discussion, Paper 16B) illustrates one such attempt and is particularly interesting because Axelrod published his paper in the *Journal of Geophysical Research*, essentially trying to beard the lion in his den. Daniel Axelrod, after contributing for more than twenty years in helping to develop data and models on fossil and living angiosperm ecosystems, evolution, and taxonomy, rose to the challenge offered by plate tectonics. Warren Hamilton, a research geologist with the U.S. Geological Survey, working principally in structural and tectonic geology, responded from his background of paleoclimatology and paleobiogeography, which had aided him in defining past positions and motions of land masses.

Looking back on this, and other similar encounters of those years, it is now obvious that the evolution of angiosperms, mammals and birds, all of which evolved about Jurassic time, and their subsequent distribution do not give an adequate view of the late Paleozoic biota or its evolution and distribution. In fact, the general angiosperm fossil record is poor until nearly the middle of the Cretaceous, the mammal fossil record is poor until the beginning of the Cenozoic, and the bird fossil record has never been anything but poor.

After this debate between Axelrod and Hamilton, Axelrod reexamined his position and has come to accept the concepts of sea-floor spreading and plate tectonics (see Paper 26 and Axelrod, 1972).

REFERENCE

Axelrod, D. I. (1972). Ocean-floor spreading in relation to ecosystematic problems, *in* R. T. Allen and F. C. James, eds., A symposium on ecosystematics. Univ. Arkansas Mus., Fayetteville, Ark., Occ. Pap. 4, 15–76.

16A

Reprinted from *Jour. Geophys. Res.*, **68**(10), 3257–3263 (1963)

Fossil Floras Suggest Stable, Not Drifting, Continents

Daniel I. Axelrod

Department of Geology
University of California, Los Angeles

Abstract. The distribution of forests from the Carboniferous to the early Cretaceous suggests that the continents probably have been stable, not drifting as paleomagnetic evidence has indicated. (1) During this interval of geologic time the vegetation-climatic zones display a symmetrical arrangement from northern to southern hemispheres consistent with continental stability. (2) Sequences of successively younger floras at the same latitude show similarity in composition, not the changes that would be expected with continents drifting across many degrees of latitude, and hence of climate. (3) The distribution of vegetation zones on the continents does not agree with the positions which are postulated by studies of paleomagnetism: in these, temperate forests occur in the deep tropics, tropical forests in polar regions, and dry-climate scrub in the moist tropics. (4) There are gradients of change in plant structures (annual rings, leaf size) and in general composition across latitudes which, in terms of the evolutionary (adaptive) relations of plants, appear to require stable continents.

Introduction. Students of paleomagnetism believe that commencing in the later Paleozoic the arctic shore of Holarctica drifted north from low equatorial latitudes and that the southern continents (except Antarctica) drifted equatorward from high southern latitudes, all attaining essentially their present positions by the Tertiary [e.g. *Runcorn,* 1956, 1959; *Creer et al.,* 1958; *Cox and Doell,* 1960 (review article); *Blackett,* 1961; *Kropotkin,* 1962].

Fossil floras preserved in sedimentary rocks on the continents provide a wholly reliable basis for testing whether drift of the magnitude visualized has occurred because (1) the distribution of plants is controlled primarily by climate, which varies with latitude; (2) the major vegetation zones, and the climates under which they grow, show a symmetrical arrangement since successively cooler zones occur at higher latitudes on opposite sides of the tropical belt; (3) many major plant groups (families) are good indicators of climate, and hence of latitude; for example, cycads and tree ferns live chiefly in tropical to warm-temperate regions, whereas members of the pine family are confined largely to temperate climates; (4) the climate to which plants are adapted is reflected in their vegetative (leaves, wood) structures; species from moist tropical regions have larger leaves than those of dry areas. These observable and well-known relations can be used to test the degree of past continental stability because they are the result

of the close adaptation of plants to the climates in which they lived and in which they evolved.

Symmetry. If the continents have been stable, the global arrangement of vegetation (climate) zones during a particular period of geologic time will illustrate symmetry from northern to southern hemisphere. But if the continents had markedly different positions, the distribution of the zones will reflect the latitude of the land masses as reconstructed from paleomagnetic evidence. In attempting to determine their position it should be recalled that vegetation (and climate) zones were not narrowly restricted in area during pre-Tertiary, but were broad, like those reconstructed for the early Tertiary [see *Axelrod,* 1960, Figure 5; *Durham,* 1952, Figure 1]. The broad zones evidently approximate the normal climate of the past [*Brooks,* 1949; *Schwarzbach,* 1961].

1. Early Cretaceous (Neocomian) floras show that climate in the far north was temperate because they often contain conifers allied to pine, spruce, larch, and fir, and they wholly dominated the Franz Joseph Land flora at 80°N [*Nathorst,* 1899]. At higher northern latitudes cycadophytes occur only rarely, the ferns represent taxa of temperate requirements, and the genera of ginkgophytes are not diverse. Commencing near middle latitudes, cycadophytes and tropical ferns are the floral dominants. The conifers are chiefly the taxa of warm-temperate to subtropical requirements (ginkgophytes, taxo-

diads, taxads, araucarians) ; members of the pine family are rare. In the tropical belt floras regularly are dominated by cycadophytes and tropical ferns, and by coniferophytes (araucarians, taxads, cupressads) of generally tropical affinity. Symmetry is also indicated by the occurrence at middle and high southern latitudes of temperate to warm-temperate conifers of present southern alliance [*Florin*, 1940]. In addition, floras at lower middle latitudes (California, S. Australia, S. Africa, Turkestan, Chile) have species with smaller leaves than those formed by related taxa in other regions, suggesting seasonally dry climate in the southwestern parts of the continents.

2. Jurassic floras demonstrate climatic symmetry because those from high latitudes in both northern and southern hemispheres regularly contain temperate to warm-temperate conifers [*Florin*, 1940]. These are not abundant in floras at middle to lower latitudes where cycadophytes, together with tree ferns and conifers of subtropical to tropical requirements, are dominant [*Krystofovich*, 1933; *Vakhrameev*, 1958]. East-west differences across continents are apparent, and they speak for climatic zonation in terms of the poles in their present positions. In western Europe the floras are typified by numerous cycadophytes and few conifers, whereas on the east coast of Asia this representation is reversed [*Jacob and Shukla*, 1955]. This agrees with east-west differences at middle latitudes, with relatively cooler climate on the east sides of the continents much as today—a relation not possible if the Eurasian land mass was at low tropical latitudes. Ginkgoalean plants are rare to absent in tropical India but are common in the warm-temperate floras to the north [*Jacob and Shukla*, 1955; *Vakhrameev*, 1958]. Furthermore, the Indian cycadophytes largely represent the tropical alliance Nilsonniales, but to the north genera of the warm-temperate Cycadeoidales are more prominent.

3. The late Triassic vegetation zones of North America suggest a north-south symmetry. The Mixteca Alta flora of southern Mexico is moist tropical, dominated by cycadophytes and tree ferns [*Wieland*, 1914]. The Chinle flora of Arizona [*Daugherty*, 1941] represents tropic savanna vegetation in which cycadophytes, tropical ferns, and araucarians are prominent; they all have comparatively small leaves. The Newark floras, scattered from Virginia to Con-

necticut, are dominated by tropical to warm-temperate cycadophytes and ferns [*Fontaine*, 1883; *Newberry*, 1888]. The leaves are much larger than those of the Chinle flora, suggesting a considerably moister climate on the east coast. A few genera representing warm-temperate to subtropical conifers are present, but they are quite rare in the Chinle. To the north, the flora of Scoresby Sound, Greenland [*Harris*, 1926], includes about 40 per cent conifers chiefly allied to present-day northern families, and numerous ginkgophytes are present. Both groups are rare in the Newark floras to the south. Ferns and cycadophytes are present in Greenland but are not as abundantly or as luxuriantly developed as in the more tropical Newark floras.

4. In the later Carboniferous the zonation of vegetation and climate indicates symmetry. The far northern temperate region was typified generally by the development of coniferous (ginkgoid, corditalean) trees, by generally small but varied calamitean plants, by a poor development of seed ferns and ferns, and by the rarity of lepidophytes [*Krystofovich*, 1937]. A broad central tropical zone is characterized by lepidophytes, large dendroid calamites, numerous seed ferns and ferns, and few coniferophytes. The southern temperate region includes seed ferns, some ferns, a few arthrophytes and lepidophytes, and coniferophytes, but most of them differ from the plants in the north temperate region. There is ample reason to suggest that the plants typifying these major vegetation zones developed from early Carboniferous ancestors—those in the tropical zone from generally tropical taxa in the preceding older floras, the others from forerunners in the mild-temperate regions of both hemispheres. In the temperate areas to the north and south there are distinctive new taxa in the later Carboniferous that existed with the older ones, which were then relict. The southern temperate flora interfingers with the tropical in South America and Africa, the northern temperate with the tropical in central Siberia.

Local sequences. If the continents have moved as widely as paleomagnetic studies imply, successive floras in a local region should show modifications in composition attributable to changing latitude. If a continent migrates from high to low latitudes, the fossil floras may be expected to record the movement.

1. A nearly complete sequence of Triassic

to early Cretaceous floras in southeastern Australia shows that throughout this time the area was dominated by cycadophytes and ferns of tropical alliance and that the conifers were chiefly of tropical to warm-temperate affinity [see lists in *David*, 1950]. This suggests a continuously low to middle latitude occurrence for southeast Australia during the Mesozoic. Since there is no evidence of changing composition consistent with drift from cool to warm climate (Figure 1), it would appear that movement of the magnitude visualized probably has not occurred.

2. Paleomagnetic evidence suggests that during the late Carboniferous South Africa was centered over the south pole, and that by the early Permian it had drifted to 25°S, a border tropical region where it has remained (Figure 2). If South Africa migrated 60° of latitude in the Permo-Carboniferous interval, the fossil floras presumably should show a marked change in composition during this brief span. This would involve a rapid decrease in temperate plants and their replacement by taxa of warm-temperate to tropical requirements. However, the successive floras show no significant changes: plants representing the *Glossopteris* occur in the region throughout the interval, which suggests stability (see Evolutionary Relations, item 5).

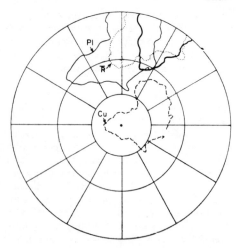

Fig. 2. Positions of Africa relative to south pole of rotation according to *Creer et al.* [1958]. Heavy outline denotes present position; C_u in late Carboniferous; P_1 in early Permian; TR in Triassic.

3. Students of paleomagnetism have placed the northern shore of Eurasia and North America near 10° to 20°N during the Permian, from which position it had drifted northward to its present area by the Tertiary (Figures 3 and 4). If such movement has occurred, the sequences

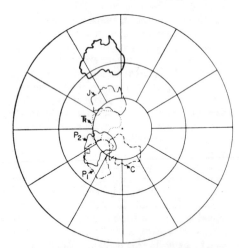

Fig. 1. Positions of Australia relative to south pole of rotation according to *Creer et al.* [1958]. Heavy outline denotes present position; C in Carboniferous; P_1 in early Permian; P_2 in late Permian; TR in Triassic; J in Jurassic.

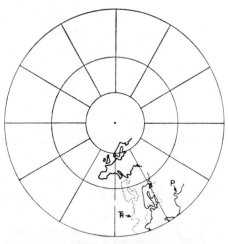

Fig. 3. Positions of Europe relative to north pole of rotation according to *Creer et al.* [1958]. Heavy outline denotes present position; P in Permian; TR in Triassic.

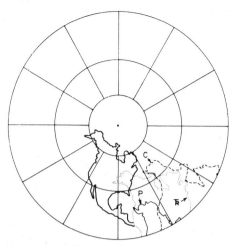

Fig. 4. Position of North America relative to north pole of rotation according to *Creer et al.* [1958]. Heavy outline denotes present position; *C* in Carboniferous; *P* in Permian; *TR* in Triassic.

of fossil floras now in the far north should record a gradual change from assemblages dominated by humid tropical plants to communities that are adapted to more temperate climates. The Permian, Triassic, and Jurassic floras of the far north contain temperate conifers (with prominent growth rings, see below), ginkgophytes, and ferns of chiefly temperate alliance; cycadophytes are not abundant. This clearly speaks for continental stability, especially since floras presently at middle latitudes are preponderantly tropical throughout the Mesozoic—dominated by tree ferns, cycadophytes, and warm-temperate to tropical conifers.

Distribution of vegetation provinces. The distribution of the continents as reconstructed from paleomagnetic evidence gives certain vegetation provinces positions that are anomalous with respect to their climatic requirements.

1. Triassic and Jurassic floras of southeast Australia (30° to 35°S) are clearly tropical. This is not consistent with the postulated position of the area at 75° to 85°S (Figure 1), especially since the Grahamland flora (67°S) is warm-temperate to temperate, with numerous conifers.

2. Late Paleozoic to early Mesozoic floras from high northern latitudes in Holarctica are composed chiefly of mild-temperate plants. This is incompatible with its postulated position at

low latitudes (Figures 3 and 4), especially since fossil floras presently at low and middle latitudes are wholly tropical.

3. Plants of the Permian Hermit flora of Arizona [*White*, 1921] reflect dry climate. Their thick, small leaves (often curled), some covered with dense hair, suggest a climate not unlike that of thorn scrub regions today. The position of the flora at 35°N is consistent with the belief that it has always occupied such a region, for this is within the area where dry climates regularly develop. But paleomagnetic evidence places the flora near 10°S (Figure 4), in a moist tropical zone where it would be out of place in terms of its existence (adaptation) and of its earlier evolution as well (see Evolutionary Relations).

4. Paleomagnetic evidence suggests that the principal belt of Carboniferous coal swamps of the northern hemisphere was near 10° to 20°N (Figures 3 and 4), which is considered consistent with their presumed moist tropical character. However, Carboniferous swamp forests very similar to those in the northern hemisphere—often including essentially the same species—also occupied South Africa, southeast Australia, and India, all of which were situated near or within the antarctic circle on the basis of paleomagnetic evidence (Figures 1 and 2). Thus while the northern swamp forests are placed under a tropical sun, essentially the same forests also have been condemned to cold, long nights in the far south.

Evolutionary relations. The fitness of living things to cope with the environment in which they live is the central theme of life: adaptation is universal. Adaptation is not accidental, but depends on the interrelationship between organism and environment; it is an evolutionary response that has developed through time.

1. Since the growing season is considerably shorter at high than at low latitudes, trees at high latitudes produce annual rings that are closely spaced as compared with those at lower latitudes which develop more widely spaced rings or none at all. Logs of *Cordaites* from the later Paleozoic rocks of Eurasia show that those from northern Siberia and the arctic islands have closely spaced annual rings, but those from central and western Europe either are widely spaced or show no development [*Krystofovich*, 1937]. This gradient of change is consistent with a short growing season at higher latitudes and a

longer one at middle latitudes. It also agrees with the distribution of the associated vegetation. Seed ferns are generally rare in the far north but are common at middle latitudes, and the large calamitean trees at middle latitudes are replaced by smaller ones in the north. Such an adaptive gradient suggests that northern Eurasia has been stable and was not in the tropical zone near 20°N in the later Carboniferous, as paleomagnetic evidence implies (Figure 3).

2. The foliar structures produced by plants show a high correlation with the environment in which they live. Plants in dry or cold regions have small leaves, species in mild-temperate regions have moderate sized leaves, and those in moist tropical areas regularly produce large leaves. Triassic ginkgophytes from localities spread southward across Holarctica show a gradual change from comparatively smaller leaves at high latitudes to larger ones at lower latitudes. This gradient is consistent with the belief that the plants at higher latitudes lived under temperate climates, those at lower latitudes in warmer ones. The gradient is accompanied by the gradual increase southward of cycadophytes, tree ferns, and conifers of warm-temperate to tropical requirements. These relations would not exist if the northern shore of Holarctica had been near 30°N during the Triassic (Figures 3 and 4).

3. As noted above, the Permian Hermit flora of Arizona [White, 1921] not only represents vegetation highly adapted to dry climate, its very existence depends on its earlier evolution in such an environment. Since evolution of vegetation of this sort takes a long time—at least a geologic period—it seems probable that it originated at low to middle latitudes, where it now occurs. It could not have evolved in the position where paleomagnetic data place it. At 5°N Pennsylvanian ancestors of these dry-climate plants would be in a moist tropical zone. Nor could they even have existed in the Permian, since for that time paleomagnetic data give the site a position near 10°S, also in the wet tropics (Figure 4).

4. In terms of numbers of taxa (species, genera, or families), the greatest diversity exists today in warm moist regions and the least in cold-temperate areas. Fossil floras from generally low to middle latitudes also display considerably greater diversity than those from very high lati-

tudes. The basal Cretaceous flora from Franz Joseph Land (80°N) includes numerous specimens, but few species are represented [Nathorst, 1899]. The presence of plants allied to larch, spruce, pine, and ginkgo (small leaves), and the absence of cycadophytes, tropical ferns, and warm-temperate conifers, indicates that it represents a cool-temperate forest. Paleomagnetic data place the region near 45°N, which gives it a position transitional between tropical and warm-temperate climate. On this basis the flora should be richer in taxa and should also contain many plants (cycadophytes, warm-temperate conifers and ferns, mild-temperate ginkgophytes) that are not represented in it. Its composition is fully consistent with a high-latitude occurrence, especially since the early Cretaceous floras at middle latitudes are wholly tropical [Vakhrameev, 1958]. The relations suggest stability for the Eurasian platform.

5. Paleomagnetic evidence implies that South Africa was centered over the south pole during the Late Carboniferous and that it drifted to 25°S by early Permian (Figure 2). No important change in the composition of the floras is recorded during this migration across 60° of latitude in about 10 million years. The persistence of the same plants in the area cannot be explained by supposing that species already adapted to living at high polar latitudes could become adjusted to a low-latitude occurrence in so brief an interval. Species that are already well adapted to one environment cannot rapidly adjust to a new and radically different one. While it may be granted that an occasional species might do so, to suppose that a whole flora—composed of plants belonging to widely disparate taxa which have had long, independent histories—would all adapt in the same direction in a short time opposes all that is known about the tempo and mode of evolution [Stebbins, 1950; Simpson, 1953] and the laws of probability. Apart from markedly different temperature-moisture relations, the change also involves a major shift in photoperiod response—from living and reproducing in a region subject to months of near darkness to one of long daylight. Since the reproductive cycle of plants is controlled largely by length of day it seems improbable that the flora could survive a shift of 60° of latitude in 10 million years.

6. If the continents are as migratory as

paleomagnetic evidence signifies, the forest belts have been carried through different climatic zones without bearing evidence of the change, which seems unlikely. India had a position near 50°S in the late Triassic according to paleomagnetic studies. During the Jurassic and Cretaceous it migrated from the southern temperate zone across the dry tropics and into the wet tropics, and it finally emerged on the north side of the tropical belt. Such changes in latitude and climate should be reflected by important changes in composition, yet none are evident [see lists in *Sitholey*, 1954]. Similarly, the successive Late Paleozoic and Mesozoic floras in the northern part of Holarctica do not record waves of changing floras, from wet tropic, to dry tropic, to mild-temperate forests, as the land mass is presumed to have drifted northward from tropical latitudes. Since the floras at higher latitudes were temperate throughout this time [*Krystofovich*, 1937; *Vakhrameev*, 1958], it appears unlikely that the northern continents have drifted as widely as has been suggested.

Conclusion. The preceding discussion, based on a small body of evidence from the extensive paleobotanical record which is available for analysis, reinforces an earlier suggestion that the distribution of ancient forests, and the climates that they indicate, appears to be consistent with continental stability since at least the Carboniferous [*Axelrod*, 1960, p. 280]. There apparently is no paleobotanical evidence that unequivocally supports paleomagnetic data which ostensibly show that the continents have drifted as much as 60° to 70° of latitude since the Carboniferous. Several lines of evidence—climatic symmetry, local sequences of floras in time, distribution of forests with latitude, evolutionary (adaptive) relations—all seemingly unite to oppose the paleomagnetic evidence for major movement. In this connection it must be emphasized that botanical data of the sort selected by *Chaloner* [1959], *Maekawa* [1960], and *Plumstead* [1961] to support the drift hypothesis are either inapplicable (discontinuous distribution patterns of angiosperm genera or families have no bearing on the problem) or inconclusive (the *Glossopteris* can be explained in other ways). To establish more firmly the relations which the fossil floras appear to indicate, it is apparent that broad regional studies of the older floras by paleobotanists competent in paleobiogeography are urgently required.

REFERENCES

Axelrod, D. I., The evolution of flowering plants, in *Evolution after Darwin*, vol. 1, pp. 227–305, University of Chicago Press, Chicago, 1960.

Blackett, P. M. S., Comparison of ancient climates with the ancient latitudes deduced from rock magnetic measurements, *Proc. Roy. Soc. London, A, 263*, 1–30, 1961.

Brooks, C. E. P., *Climate through the Ages*, 2nd ed., McGraw-Hill Book Co., New York, 1949.

Chaloner, W. G., Continental drift, in *New Biology*, pp. 7–30, Penguin Books no. 29, 1959.

Cox, A., and R. R. Doell, Review of paleomagnetism, *Bull. Geol. Soc. Am., 71*, 645–768, 1960.

Creer, K. M., E. Irving, A. E. Nairn, and S. K. Runcorn, Paleomagnetic results from different continents and their relation to the problem of continental drift, *Ann. Geophys., 14*(4), 492–501, 1958.

Daugherty, L. H., The upper Triassic flora of Arizona, *Carnegie Inst. Wash. Publ. 526*, 1–108, 1941.

David, T. W. E., *Geology of the Commonwealth of Australia*, Ed. Arnold & Co., London, 1950.

Durham, J. W., Early Tertiary marine faunas and continental drift, *Am. J. Sci., 250*, 321–343, 1952.

Florin, R., The Tertiary fossil conifers of southern Chile and their phytogeographic significance, with a review of the fossil conifers of southern lands, *Kgl. Svenska. Vetenskapsakad. Handl.,* 3rd ser., *19*(2), 1–107, 1940.

Fontaine, W. M., Contributions to the knowledge of the older Mesozoic flora of Virginia, *U. S. Geol. Surv. Monograph*, no. 6, 1883.

Harris, T. M., The Rhaetic flora of Scoresby Sound, east Greenland, *Medd. Groenland, 68*, 43–147, 1926.

Jacob, K., and B. N. Shukla, Jurassic plants from the Saighan series of northern Afghanistan and their paleoclimatological and paleogeographical significance, *Paleontol. Indica, 33*(2), 1–63, 1955.

Kropotkin, P. N., Paleomagnetism, paleoclimates and the problem of extensive horizontal movements of the earth's crust, *Intern. Geol. Rev., 4*(11), 1214–1234, 1962.

Krystofovich, A. N., The Baikal formation of the Angara group, *Trans. United Geol. & Prospect. Serv. USSR,* Fasc. 326, 1–136, 1933.

Krystofovich, A. N., Botanical-geographical zonality and stages of evolution of the Upper Paleozoic flora, *Izv. Akad. Nauk SSSR, 2*, 47–62, 1937.

Maekawa, F., The paleoequator and its relation to the recent distributional area of Coriaria, *The Quaternary Research, 1*, 212–218, 1960.

Nathorst, A. G., Fossil plants from Franz Joseph Land, *Norweg. Polar Exped. 1893–1896, 3*, 1899.

Newberry, J. S., The fauna and flora of the Triassic of New Jersey and the Connecticut Valley, *U. S. Geol. Surv. Monograph*, no. 14, 1888.

Plumstead, E. P., Ancient plants and drifting continents, *S. African J. Sci.*, *57*, 173–181, 1961.

Runcorn, S. K., Paleomagnetism, polar wandering, and continental drift, *Geol. Mijnbouw*, *18*, 253–256, 1956.

Runcorn, S. K., Rock magnetism, *Science*, *129*, 1002–1012, 1959.

Schwarzbach, M., *Das Klima der Vorzeit*, 2nd ed., Enke, Stuttgart, 1961.

Simpson, G. G., *The Major Features of Evolution*, Columbia University Press, New York, 1953.

Sitholey, R. V., The Mesozoic and Tertiary floras of India—a review, *The Paleobotanist*, *3*, 55–69, 1954.

Stebbins, G. L., Jr., *Variation and Evolution in Plants*, Columbia University Press, New York, 1950.

Vakhrameev, V. A., Stratigraphy and fossil flora of the Jurassic and Cretaceous deposits of the Viluyski depression and adjacent parts of the Verkhoyanksy depression, *Acad. Nauk USSR, Regional Stratig.* USSR, *3*, 1–136, 1958.

White, E., Flora of the Hermit shale, Grand Canyon, Arizona, *Carnegie Inst. Wash. Publ. 405*, 1921.

Wieland, G. R., La flora liasica de la Mixteca Alta, *Mexico Inst. Geol. Bull.*, *31* & atlas, 1914–1916.

(Manuscript received December 3, 1962; revised March 7, 1963.)

16B

Reprinted from *Jour. Geophys. Res.*, **69**(10), 1666–1671 (1964)

Discussion of Paper by D. I. Axelrod, 'Fossil Floras Suggest Stable, Not Drifting, Continents'[1]

Warren Hamilton

U. S. Geological Survey, Denver, Colorado

The impressive paleobotanical evidence for continental drift comes in part from such anomalies as the widespread distribution of subtropical floras in the present Arctic and of cold-climate and temperate floras in the present tropics. *Axelrod* [1963] makes no mention of these anomalies while arguing that paleofloras are arrayed in latitudinal zones and indicate continental stability; actually, the anomalies contradict those of his generalizations which are not consistent equally with drift and stability.

The remarks which follow, condensed at the editor's request from a much longer and fully documented manuscript, are based on numerous published sources of which only a few can be credited within the space allotted. Discussions with paleobotanists Sergius H. Mamay, Edna P. Plumstead, James M. Schopf, and Richard A. Scott have provided insight into the botanical problems.

Paleobotanists (among them *Andrews* [1961], *Gothan and Weyland* [1954], *Halle* [1937], *Just* [1952], *Jongmans* [1943], *Krishnan* [1954], *Plumstead* [1962], *Sahni* [1936], and *Zalessky* [1937]) have long agreed that the floras of the Upper Carboniferous and Permian fall into several great taxonomic and morphologic assemblages of obvious paleoclimatic significance. Axelrod writes in terms of these same floras without identifying them as such.

The *Glossopteris* flora occurs in the 'Gondwana' landmasses of peninsular India, central and southern Africa, central and southern South America, Australia, and Antarctica. It is Axelrod's south-temperate flora. The flora is remarkably uniform on all continents—for example, 20 of the 27 leaf species known in Antarctica [*Plumstead*, 1962; *Schopf*, 1962] are known also in India—but despite this uniform-

ity and pole-to-tropics distribution the correlative floras of the rest of the world are quite different. The *Glossopteris* flora lies above glacial tillites resting on glacially polished and striated pavements on all of the continents named, and on all with the exception of Antarctica is known to be interbedded with marine strata containing cold-water invertebrates. Annual rings are present everywhere in wood and the dominant plants were deciduous. *Glossopteris* grew at low altitudes under cold or temperate conditions.

The correlative Siberian flora [*Andreeva et al.*, 1956; *Radchenko*, 1956; summary papers noted above] occurs throughout most of Siberia and is the north-temperate flora of Axelrod. Its botanical features, annual rings, and associated sedimentary and faunal types do indeed show it to be the product of a temperate, nonglacial climate.

Between Siberian and *Glossopteris* floras is the Euramerican flora, Axelrod's tropical assemblage, known in northwestern South America, southern and middle North America, Europe, northwest Africa—and in the Canadian Arctic Archipelago, Greenland [*Halle*, 1931; *Kempter*, 1961], and Spitzbergen. Early in the Late Carboniferous the flora was present also in China, southeastern Asia (but not peninsular India), and Sumatra [*Jongmans*, 1937b], but was replaced there by the *Gigantopteris* flora of similar climatic significance before Permian time; *Gigantopteris* elements also reached western North America [*White*, 1936]. Annual rings are lacking in much of the wood, even in Spitzbergen [*Andrews*, 1961], showing uniform tropical climate; elsewhere, including other parts of the present Arctic, seasonal savannah climates and subtropical climates are indicated. Ferns and lepidophytes (scarce in Siberian and *Glossopteris* floras), giant horsetails, and other plants specified by Axelrod to infer warm climate are

---

[1] Publication authorized by the Director, U. S. Geological Survey.

abundant, even in the present-Arctic occurrences. Assemblages on opposite sides of the Atlantic are practically identical [*Darrah*, 1937; *Jongmans*, 1937a]. Correlative warm-climate sediments (thick limestones, red beds, evaporites hundreds of feet thick, etc.) and warm-water faunas (big organic reefs, fusulinids,[2] etc.) occur throughout the extent of the Euramerican and *Gigantopteris* floras, including broad regions in the Atlantic sector of the Arctic. Synthesis of evidence for continental drift assigns the present-Arctic occurrences to the northern subtropics and the warm part of the adjacent temperate zone when they formed; the slight floral gradient southward to completely tropical (near-equatorial?) assemblages in the North Atlantic region is wholly consistent with this.

The tropical-subtropical Euramerican flora extends north of any occurrence of the north-temperate Siberian flora, and yet the correlative south-temperate *Glossopteris* flora now lies along the equator. Axelrod's assertion that paleofloral distribution indicates symmetrical zoning with respect to the equator is unsupported.

The cold-water marine sediments associated with *Glossopteris* and the lack of warm-climate plants preclude the possibility that this five-continent flora grew only in tropical highlands. (In earlier papers, *Axelrod* [e.g., 1952] has used the proved rarity of old highland sediments and the assumption that there are *no* known tropical-highland Paleozoic floras as the bases for a hypothesis of angiosperm evolution.) The unique flora of modern Australia indicates the effectiveness of ocean barriers to plant migration, yet free floral interchange between the now widely separated Gondwana land masses somehow happened; speculation that vanished land bridges were responsible receives no geophysical support.

Arctic occurrences of the Euramerican flora are sometimes rationalized in terms of a warm, even climate extending to the north pole; but this is difficult to reconcile either with a spherical earth that has an inclined axis or with the widespread contemporary sea-level continental

glaciation of South America, India, southern Africa, and Australia.

Mesozoic floras also show similar anomalous distribution with respect to present continental positions.

There are many inconsistencies in the citations by Axelrod of botanical reasons for paleoclimatic assignments. The abundance of ginkgophytes in the Jurassic of Siberia and their scarcity in that of India is cited as evidence for a temperate paleoclimate in Siberia and a tropical one in India—but ginkgophytes are also lacking in the Antarctic Jurassic (which both of us consider temperate). On the other hand, *Ginkgo* is abundant in both Jurassic and Triassic of Australia, which Axelrod terms 'clearly tropical' but which appears temperate both by his own criteria and by floral similarity to Antarctic and other Gondwana assemblages. Similar discrepancies mar Axelrod's citations of giant horsetails, abundant tree lepidophytes, cycadophytes, ferns, Nilssoniales, tree rings, and coal.

Axelrod's arguments that climatic progressions of floras with geologic time in any one region negate drift are based on tenuous assumptions of worldwide climatic change in Europe, North America, and the Arctic; on use of ambiguous paleomagnetic data from Africa; and on consideration of only the record older than the specified segment of drift as normally advocated in India. The great climatic progressions demonstrable in all of these places are easily explained in terms of drift.

Paleolatitudes as deduced from the climatological evidence of marine invertebrates, land plants, and sedimentary environments, and from most paleomagnetic data, are remarkably consistent with each other and with complex continental drift. They are equally incompatible both with present geography and with crustal shift without both drift and complex internal deformation of continents by bending and strike-slip faulting. Doubting specialists attempt to discredit one aspect or another of the paleolatitude determinations, but the mutual consistency of the independent criteria is strong evidence for their general validity. Velocities required for drift are no larger than those now measured geodetically within both California and Japan, and the continents appear to be but temporary aggregates of randomly wandering and complexly deforming sialic flotsam.

---

[2] *Stehli* [1957] claimed that fusulinids, which indicate warm water, do not occur in the present Arctic; actually, they are well known in Ellesmere Island, northern Greenland, Spitzbergen, and elsewhere.

196

REFERENCES

Andreeva E. M., and others, Atlas rukovodiashchikh form iskopaemykh fauny i flory Permskikh otlozhenii Kuznetskogo basseina [Atlas of index fossil faunas and floras of Permian deposits of the Kuznetzk Basin]: *Minist. Geol. Ochrany Nedr SSSR,* 411 pp., 1956.

Andrews, H. N., *Studies in Paleobotany.* 487 pp., John Wiley & Sons, New York, 1961.

Axelrod, D. I., A theory of angiosperm evolution, *Evolution, 6,* 29–60, 1952.

Axelrod, D. I., Fossil floras suggest stable, not drifting, continents, *J. Geophys. Res., 68,* 3257–3263, 1963.

Darrah, W. C., American Carboniferous floras, *Congr. Avan. Etudes Stratigraph. Carbonifere, Compte Rendu, 2nd, Heerlen, 1935, 1,* 109–129, 1937.

Gothan, Walther, and Hermann Weyland, *Lehrbuch der Paläobotanik,* 535 pp., Akademie-Verlag, Berlin, 1954.

Halle, T. G., Younger Palaeozic plants from East Greenland collected by the Danish expeditions 1929 and 1930, *Medd. Grønland, 85(1),* 26 pp., 1931.

Halle, T. G., The relation between the late Palaeozoic floras of eastern and northern Asia, *Congr. Avan. Etudes Stratigraph. Carbonifere, Compte Rendu, 2nd, Heerlen, 1935, 1,* 237–245, 1937.

Jongmans, W. J., Contribution to a comparison between the Carboniferous floras of the United States and of western Europe, *Congr. Avan. Etudes Stratigraph. Carbonifere, Compte Rendu, 2nd, Heerlen, 1935, 1,* 1937a.

Jongmans, W. J., The flora of the Upper Carboniferous of Djambi (Sumatra, Netherl. Indies) and its possible bearing on the paleogeography of the Carboniferous, *Congr. Avan. Etudes Stratigraph. Carbonifere, Compte Rendu, 2nd, Heerlen, 1935, 1,* 363–387, 1937b.

Jongmans, W. J., Das Alter der Karbon- und Permfloren von Ost-Europa bis Ost-Asien, *Palaeontographica, 87B,* 1943.

Just, Theodor, Fossil floras of the southern hemisphere and their phytogeographical significance, *Bull. Am. Museum Nat. History, 99,* 189–203, 1952.

Kempter, Enrico, Die Jungpaläozoischen Sedimente von Süd Scoresby Land, *Medd. Grønland 164(1),* 123 pp., 1961.

Krishnan, M. S., History of the Gondwana era in relation to the distribution and development of flora, Sahni Institute of Palaeobotany, Lucknow, India, 1953 Seward Memorial Lecture, 15 pp., 1954.

Plumstead, E. P., Fossil floras of Antarctica, *Trans-Antarctic Exped. 1955–1958, Sci. Rept. 9,* 154 pp., 1962.

Radchenko, G. P., Paleobotanicheskie obosnovaniya drobnogo stratigraficheskogo raschleneniya uglenosynkh otlozheniy Kuznetskogo Basseyna i nekotorye dannye k opredeleniyu ikh vozrasta [Paleobotanical basis for the stratigraphic subdivisions of carbonaceous deposits of the Kuznetzk Basin and certain data on definition of their age], *Vopr. Geol. Kuzbassa, Minist. Ugol'noy Promyshlennosti SSSR, 1,* 119–137, 1956.

Sahni, Birbal, Wegener's theory of continental drift in the light of the palaeobotanical evidence, *J. Indian Botan. Soc., 15,* 319–332, 1936.

Schopf, J. M., A preliminary report on plant remains and coal of the sedimentary section in the central range of the Horlick Mountains, Antarctica, *Ohio State Univ. Inst. Polar Studies, Rept. 2,* 61 pp., 1962.

Stehli, F. G., Possible Permian climatic zonation and its implications, *Am. J. Sci., 255,* 607–618, 1957.

White, David, Some features of the early Permian flora of America, *Intern. Geol. Cong. Rept., 16th, 1,* 679–689, 1936.

Zalessky, M. D., Sur les zones climatiques de la terre au Carbonifere et au Permien, *Intern. Geol. Congr. Rept., 17th, Moscow, 6,* 181–185, 1937.

(Received August 29, 1963;
revised January 20, 1964.)

[Editor's Note: A reply by Axelrod follows.]

Reply

DANIEL I. AXELROD

Department of Geology, University of California, Los Angeles

Hamilton's opinion that there is 'impressive paleobotanical evidence for continental drift' is not indicated, as he asserts, by anomalous distributions of floras representing subtropical climate in the present Arctic, and temperate climate in the present tropics. They fit readily into a picture of latitudinally and altitudinally controlled climate consistent with stability. My remarks, condensed at the request of the editor from a long reply, will serve to illustrate that Hamilton's belief that fossil floras indicate drift is unfounded.

Hamilton's views stem from two misconceptions concerning paleoclimate and paleogeography. First, he asserts that particular plants are reliable indicators of climate, yet it is well known that paleoclimate must be reconstructed from an entire flora represented by an adequate sample. Floral composition is controlled by temperateness [*Bailey*, 1960], a factor which he ignores. With pronounced temperateness, as during most of geologic time, in frostless areas temperatures neither dip so low as to discourage tropical plants nor rise so high as to prevent the success of temperate ones. It largely explains the occurrence of 'subtropical' plants in the arctic Atlantic, also temperate ones at moderate altitudes at lower latitudes. Second, the plant record must be integrated with the elements of paleogeography (east-west coasts, altitude, marine currents, etc.) which Hamilton scarcely notes. This applies particularly to ancient land areas—sialic and simatic—across which forests migrated in the past, and which he merely dismisses as nonexistent in spite of ample evidence to the contrary (see below).

Hamilton asserts that evidence for drift is shown by the subtropical Euramerican flora in the Atlantic arctic during the Pennsylvanian, at latitudes north of the temperate Siberian (Angaran) flora. Their impoverished nature shows that the floras of the Atlantic arctic are not subtropical. And the ringless wood of *Cor-*

daites, and large arthrophytes and lepidophytes need only indicate mild climate and pronounced temperateness, not subtropical climate and low latitude. The presence of mild climate in the Atlantic arctic during the Carboniferous is not unique, as Hamilton intimates, but also occurred during the Triassic, Jurassic, Cretaceous, and early Cenozoic. Floras of those ages indicate that mild climate occurred north of forests of cool temperate aspect in northern Eurasia during those times. The northern extension of 'subtropical' plants into latitudes where they mingled with temperate ones may be explained by the influence of broad marine embayments, an effective 'gulf stream,' and low continents, all of which resulted in mild climate of pronounced temperateness (M 75 to 85).[1]

Stability for the Atlantic arctic is shown by Triassic floras which display a gradual southward gradient from mild temperate in Greenland, to wet tropical in Virginia, to dry tropic savanna in Texas and Arizona [*Axelrod*, 1963]. Hamilton considers the gradient consistent with low latitudes because, he states, the Greenland flora is subtropical; he totally ignores the temperate plants there and to the north where they increase. He does not understand the relation between floral composition and temperateness. That tropical and temperate plants regularly intermingle at such times can be seen today (Mexico, SE China), and it is shown also by late Cretaceous–early Tertiary floras scattered from New Mexico to the Arctic shore. Triassic, Jurassic, Cretaceous, and early Cenozoic floras of the Atlantic arctic are mild temperate. They

---

[1] A constant temperature of 14°C has a temperateness index of M 100. If climate is warmer or cooler than 14°C, or if seasonal temperature fluctuations exist, then departures are expressed by numbers that decrease from 100 as climate becomes less temperate. Temperateness at Bogota is M 90+; at San Francisco M 75; at Verkhoyansk M 11.

indicate stability because they do not reveal change from tropical to temperate climate, as demanded by drift. If we accept Hamilton's reasoning, paleobotanic data mean that North America and the Atlantic arctic only drifted north following Eocene time, when climate became cooler.

As for the *Glossopteris* flora, Hamilton asserts that since it has essentially the same composition from polar to tropical latitudes symmetrical zoning is unsupported by plant evidence. It is to be emphasized that symmetry occurred earlier in the Carboniferous, and was re-established by the Triassic [*Axelrod*, 1963] as the unique conditions which account for glaciation disappeared. Thus the *Glossopteris* problem is a special one involving the late Carboniferous-early Permian, and is not of general significance in terms of drift or symmetry. The flora evidently occupied upland areas of temperate climate at mid to low latitudes to judge from geologic evidence [*Martin*, 1961; *Plumstead*, 1961]. The relation may be invoked elsewhere since epeirogenic movements had largely eliminated the seas from Gondwanaland by late Carboniferous. Botanical evidence for an upland occurrence is provided by the numerous mixed floras composed of both tropical (Euramerican, *Gigantopteris* floras) and south temperate (*Glossopteris*) plants which represent the transition between them at different latitudes. The altitude required to support the Glossopteris flora was minimal because with pronounced temperateness altitudinal climatic zones are compressed [*Bailey*, 1960, Figure 8]. At lower latitudes, altitudes near 3000 to 4000 feet would provide temperate climate if temperateness was M 75 (like New Zealand), and if it was M 80 (like Tanganyika) altitude was lower. Hamilton's remark that the *Glossopteris* flora was not an upland community because it is interbedded with cold-water marine sediments is triply misleading. The occurrences are restricted chiefly to continental margins, whereas the flora was widespread on the continents; in most of the areas (SE Australia, S. Africa, Argentina) the sites are marginal to the ancient tropics anyway, and would thus be expected in temperate lowlands; and part of the problem is one of synchoraenity, for at the height of glaciation they were forced temporarily into lowlands.

Hamilton contends that the unique modern flora of Australia demonstrates the effective role of ocean barriers to plant migration. By analogy, he contends that the distribution of *Glossopteris* on the southern continents demands their union or close proximity in Permo-Carboniferous time, drift subsequently isolating them. Actually, the northeast coast of Australia has an Indomalayan flora, and the southeast part supports a forest like that now in New Zealand and southern Chile. Using Hamilton's argument that ocean barriers effectively prevent forest migration, then it must follow that Chile and Malaya drifted away from the east coast of Australia following Eocene time, because the floras of these now widely separated regions were much *more similar* in Eocene than at present. Actually, there is ample geologic and biologic evidence—still ignored by many geologists and most geophysicists—to indicate that the present southern continents are but fragments of wider Paleozoic and Mesozoic lands over which forests migrated in the past [see *Axelrod*, 1960, pp. 247–257, 281–284, and appendix; *Hallum*, 1963; *Martin*, 1961; *van Steenis*, 1962; *Rothe*, 1954; also *MacDonald*, 1963, for geophysical evidence against drift].

Hamilton alludes to inconsistencies in botanical reasons which mar paleoclimatic assignments. Lack of ginkgophytes in Antarctica may be ascribed to several factors, but their absence there has no climatic significance. Abundance of ginkgo in Australia during Triassic and Jurassic is understandable since most sites are at lat. 35° to 45°S, and warm temperate climate would occur there, paralleling the occurrence of abundant ginkgophytes in Oregon (lat. 42°). As for misinterpretation of arthrophytes, lepidophytes, annual rings, etc., in the Atlantic arctic, we have noted that similar conditions prevailed during the succeeding Mesozoic and early Cenozoic, times when other 'subtropical' plants lived there under mild temperate climate of high temperateness.

If there has been drift of the magnitude implied, then sequences of floras in local areas should show trends toward cooling (in Atlantic arctic) or warming (in Australia, India), yet they do not [*Axelrod*, 1963]. Hamilton incorrectly states that I did not consider India when paleomagnetic evidence specifies drift. The data place the Jurassic Rajmahal flora near 46°S, yet it is wholly wet tropical, dominated by large-leafed cycadophytes and ferns. Contemporaneous floras 500 miles (Rewah, Jabalpur) and 1000 miles (Kutch) west reveal a gradual transition to tropic savanna climate. The relation implies a general circulation system like that of today at low-middle latitudes, with drier conditions in the west, and thus suggests stability. Triassic floras of Australia (Esk, Leigh's Creek) and Arizona (Chinle) reveal dry tropic climate at low-middle latitudes. Such a relation is scarcely explicable if continents drifted from high southern (Australia) and low tropical (Arizona) latitudes. On the contrary, dry tropic Triassic climates on opposite sides of the equator would seem to nail down the continents in the present positions by that time at least.

1. Forest distribution appears to establish an Atlantic Ocean bordered by North America and Europe in their present positions by the Carboniferous: it suggests a tempering influence in the Atlantic arctic from the later Paleozoic into the early Cenozoic, with moderating effects diminishing eastward over Eurasia throughout that time. The distribution is thus consistent with climatic symmetry and continental stability.

2. Geologic and botanic data suggest that at lower latitudes the *Glossopteris* flora ascended to higher altitudes where temperate climate enabled it to live, and where it interfingered with tropical plants at lower levels. Contrasts between the dissimilar yet juxtaposed tropical (*Gigantopteris*) and temperate (*Glossopteris*) floras may thus have resulted from only moderate altitudinal (3500 to 4000 feet) differences, not from continental migration across 90° of latitude. Furthermore, similarity of the *Glossopteris* flora reflects easy migration over lands—simatic and sialic—which have since foundered, and for which there is ample evidence.

3. Paleoclimates inferred from sedimentary environments and marine invertebrates also can be interpreted as consistent with continental stability since the Carboniferous.

4. Others suggest [*MacDonald*, 1963; *Stehli and Helsley*, 1963] that geomagnetic poles may move independently of rotational poles, and hence may not indicate drift.

REFERENCES

Axelrod, D. I., The evolution of flowering plants, in *Evolution after Darwin, 1*, 227–305, University of Chicago Press, 1960.

Axelrod, D. I., Fossil floras suggest stable, not drifting, continents, *J. Geophys. Res., 68*, 3257–3263, 1963.

Bailey, H. P., A method of determining the warmth and temperateness of climate, *Geograf. Ann., 42*, 1–16, 1960.

Hallum, A., Major epeirogenic and eustatic changes since the Cretaceous, and their possible relationship to crustal structure, *Am. J. Sci., 261*, 397–423, 1963.

MacDonald, G. J. F., The deep structure of continents, *Rev. Geophys., 1*(4), 587–665, 1963.

Martin, H., The hypothesis of continental drift in the light of recent advances of geological knowledge in Brazil and in southwest Africa, *Trans. Proc. Geol. Soc. S. Africa*, Alex L. du Toit Memorial Lecture no. 7, annex to vol. 64, 1–47, 1961.

Plumstead, E. P., The Permo-Carboniferous Coal Measures of the Transvaal South Africa—an example of the contrasting stratigraphy in the southern and northern hemispheres, *Congr. Advan. Etudes Stratigraph. Geol. Carbonifere, 4th, Heerlen, 2*, 545–550, 1961.

Rothe, J. P., La zone seismique mediane Indo-Atlantique, *Proc. Roy. Soc. London, A, 222*, 387–397, 1954.

Stehli, F. G., and C. E. Helsley, Paleontologic technique for defining ancient pole positions, *Science, 142*, 1057–1059, 1963.

Van Steenis, C. G. G. J., The land bridge theory in botany, *Blumea, 11*(2) 235–372, 1962.

(Received December 5, 1963; revised January 31, 1964.)

Editor's Comments
on Papers 17A and 17B

17A COLBERT
The Mesozoic Tetrapods of South America

17B ROMER
Discussion of "The Mesozoic Tetrapods of South America"

Edwin H. Colbert's long and successful research career in the study of fossil vertebrates, particularly mammals and reptiles, was carried out in conjunction with curatorial duties at the American Museum of Natural History and the Museum of Northern Arizona and while teaching at Columbia University. He first studied Cenozoic mammals, and has done extensive research and publication of fossil Cenozoic mammals from Asia. In 1942 he shifted the emphasis of his studies to fossil reptiles and amphibians, particularly those from the Permian and Triassic and later to the rest of the Mesozoic.

At the time Paper 17A was written, Colbert was still not convinced that the vertebrate faunas of South America were so similar to those of Africa and other Gondwana continents that continental drift was needed to explain their distributions. Thus Paper 17A is typical of many papers at that time in the use of higher taxonomic categories and statistical comparisons in evaluating the two faunas, one Triassic and the other late Cretaceous. In the discussion that followed Colbert's paper, Romer (Paper 17B) raised some unsettling questions about the analysis of the Triassic fauna from South America and concluded that it did have a close resemblance to the South African Triassic vertebrates.

Alfred S. Romer's discussion of Colbert's findings came at the time when Romer had recently completed the 1945 revision of his textbook *Vertebrate Paleontology* and had recently arrived at Harvard University from the University of Chicago. Romer's early work concentrated on Permian and Triassic amphibians and reptiles, and he was a careful and cheerful debater. As seen from his discussion here, he was actively questioning the fixity of continents. More recently, as the geophysical and geologic evidence for plate tectonics grew, Romer became a more active supporter of the concepts (Romer, 1968).

In recent years Colbert also recognized the impressive vertebrate evidence that was accumulating for a late Paleozoic–early Mesozoic supercontinent of Pangaea and participated during 1969 and 1970 in the discovery of Gondwana vertebrates in Antarctica within 600 kilometers of the South Pole.

REFERENCE

Romer, A. S. (1968). Fossils and Gondwanaland. Proc. Am. Philos. Soc., 112(5), 335–343.

Reprinted from *Bull. Am. Mus. Nat. Hist.*, **99**(3), 237–249 (1952)

THE MESOZOIC TETRAPODS OF SOUTH AMERICA

EDWIN H. COLBERT
The American Museum of Natural History

INTRODUCTION

As COMPARED WITH WHAT we know about the Mesozoic land-living vertebrates in the rest of the world, our knowledge of these animals in South America is not very extensive. This is a fact that must be kept in mind by the student who attempts to make an evaluation of the Mesozoic amphibian and reptilian faunas of South America as they are related to contemporaneous faunas in other continental areas. On the other continents, with the exception of Australia, the Mesozoic faunas are known from fairly representative sequences of amphibian and reptilian assemblages that indicate various levels in the succession of Triassic, Jurassic, and Cretaceous sediments. Gaps do exist, of course, and some of the gaps are indeed large, but on the whole the Mesozoic faunas of North America, Europe, Asia, and Africa are much more complete and varied than are these faunas in South America.

In South America there are essentially only two horizons in which Mesozoic land-living tetrapods are adequately known from fossil materials. One of these is of Middle or Upper Triassic age, the other of Upper Cretaceous age. It should be said here that recently there has been reported a sauropod dinosaur of Jurassic age from southern Argentina (Cabrera, 1947). This discovery, while indicative of the possibilities for future explorations in search of Jurassic land vertebrates in South America, does not constitute a record comparable to the Triassic and Cretaceous faunas which are the subject of this paper. Therefore it can be given passing consideration but hardly more than this in the present contribution. In addition there are isolated occurrences of Mesozoic marine reptiles, such as the plesiosaurs and the ichthyosaurs, at scattered localities around the margins of the continent, but marine vertebrates are outside the scope of this particular study. Of course there is every reason to believe that as time goes on additional fossil-bearing horizons will be discovered in the continental Mesozoic sediments of South America, but at the present time our knowledge is limited, and it is upon the basis of this limited knowledge that the discussion must rest.

Triassic amphibians and reptiles have been found in Rio Grande do Sul and far along the upper reaches of the Amazon River of Brazil, and in the States of Mendoza and La Rioja in the western part of Argentina. The fossils from southern Brazil comprise a considerable fauna, which has been collected at several localities, from the Santa Maria formation of the Rio do Rasto group. (In a recent paper by MacKenzie Gordon a division of the total sequence of these sediments is made with the name Rio do Rasto limited to beds of upper Permian age, while the vertebrate-bearing Triassic sediments are called the Santa Maria formation.)

Cretaceous reptiles are known from the southern part of Argentina, especially from Nequen and from thence south through Patagonia. In this region a considerable fauna has been found, dominated by dinosaurs but containing also crocodilians, lizards, and turtles. In recent years a few Cretaceous reptiles have been found in the Baurú formation of São Paulo, Brazil, and in Uruguay.

What light do these limited Mesozoic amphibian and reptilian faunas of South America throw upon the large subject of tetrapod distributions during the age of dinosaurs? Are the Mesozoic tetrapods of South America more nearly related to the tetrapods of North America or to those of the so-called Gondwanaland continents, Africa, India, and Australia? How did the land-living tetrapods migrate back and forth between South America and other continents during Mesozoic times? Do the Mesozoic amphibians and reptiles of this region afford any information on the theory of continental drift as opposed to that of the permanence of continental blocks and oceanic basins? If the continents occupied essentially the same positions in Mesozoic times that they do today, what were the connections between them? Was there a Gondwanaland, uniting South America with Africa, India, and Australia? Do the recent amphibians and reptiles of the Neotropical region indicate anything as to the distribution of tetra-pods during the Mesozoic era?

These are some of the questions that arise when one approaches the general problem of the Mesozoic tetrapods of South America. It is doubtful whether our knowledge is complete enough at the present time to give satisfactory answers to many of these questions, but at least the evidence can be scrutinized and evaluated.

THE TRIASSIC FAUNA

The Triassic tetrapods of South America, as known from southern Brazil, and from the States of La Rioja and Mendoza in Argentina, can be listed by genera as follows:

Amphibia
  Stereospondyli, Triassic labyrinthodonts
    Pelorocephalus, a brachyopid or short-headed stereospondyl
Reptilia
  Cotylosauria, primitive reptiles
    Candelaria, a procolophonid
  Therapsida, mammal-like reptiles
    Dicynodon }dicynodonts
    Stahleckeria }
    Belesodon ⎫
    Chiniquodon ⎪
    Gomphodontosuchus ⎬cynodonts
    Traversodon ⎪
    Exaeretodon ⎪
    Theropsis ⎭
  Rhynchocephalia, rhynchocephalians, persisting to recent times
    Cephalonia ⎫ rhynchosaurs
    Scaphonyx ⎭
  Thecodontia, ancestral archosaurs
    Hoplitosuchus ⎫
    Prestosuchus ⎪
    Rauisuchus ⎬pseudosuchians
    Rhadinosuchus ⎪
    Procerosuchus ⎪
    Cerritosaurus ⎭
  Saurischia, dinosaurs
    Spondylosoma, a prosauropod dinosaur

What is the significance of this fauna in relation to Triassic faunas in other parts of the world? Perhaps the answer to this question is best found in an analysis of the assemblage not by genera but rather by higher categories, on the level of families or suborders. The several Triassic faunas of the world are so varied that extensions of a single genus from one continental area to another are not at all common. On the other hand, closely related genera are found rather frequently in different continental areas, and these genera, even though not identical, are diagnostic and valid in establishing faunal relationships between widely separated regions. Indeed, considering the world-wide distribution of the faunas under examination and the time differences that inevitably must exist between these faunas, even if they are not always apparent, it would be exceptional if many generic identities could be established among the tetrapod assemblages in different parts of the world. A tetrapod, in spreading from one continental region to another through an appreciable lapse of time, would under ordinary circumstances undergo various mutations in response to new environmental conditions, so that in the end there would be developed related but not identical genera in the two regions. Where an identity of genera in separate continental areas can be found, one must suppose that the intercontinental passage was relatively rapid, or that the genus in question was an unusually stable one. Therefore, the presence of closely related but not identical genera in different continents should be regarded as the normal condition, from which perfectly valid conclusions can be drawn.

When we turn to an examination of the Triassic land animals of South America, listed above, it is seen that to date there is known a single labyrinthodont amphibian, described by Cabrera under the generic name of *Pelorocephalus*. Although this fossil is not so well preserved as might be wished, it is sufficiently complete to show that it is a characteristic brachyopid, or short-faced stereospondyl. The brachyopids were Triassic labyrinthodont amphibians characterized by very short and broad skulls. In addition to the South American genus, the brachyopids are known from the lower or middle Triassic Moenkopi formation of the southwestern United States, from South Africa, from India, from Australia, and especially from central Europe, where a number of genera are represented. Consequently it is apparent that this South American amphibian is closely related to genera that were rather widely distributed in other continental areas.

A rather similar picture of distribution to that outlined above is seen when we examine the procolophonid reptiles. This group is represented in Brazil by the single genus *Candelaria*, recently described by L. I. Price. The Procolophonidae were the last of the cotylosaurs, the most primitive of the reptiles, and characteristically of Permian age. Although the cotylosaurs enjoyed the culmination of their evolutionary development in Permian times, there was this one family, the Procolophonidae, that originated in late Permian times and continued on through the extent of the Triassic period. The first known Permian procolophonids are found in two widely separated regions—South Africa and Russia. From these primitive forms the Triassic procolophonids evolved and spread to various parts of the world, where they have been discovered in South Africa, central Europe, Scotland, North America, and of course South America.

The dicynodonts were highly specialized therapsid reptiles characterized by the rather bizarre skull in which all of the teeth were suppressed except for two large tusks in the upper jaws of the male animals. Dicynodonts are found in incredibly large numbers in the Permian and Triassic of South Africa, and because they have been known for such a long time from that particular region they are generally thought of as characteristic Karroo reptiles. The presence therefore of two genera of dicynodonts in the Triassic of Brazil, *Dicynodon* itself and a giant form, *Stahleckeria*, might be considered as indicative of especially close relationships between South America and South Africa during Tri-

assic times. Yet it should be noted that dicynodonts are well known from the Triassic sediments of such widely separated areas as northern Russia, western China, and the southwestern United States. Therefore it is apparent that the distribution of the dicynodonts in Triassic times was similar to the distributions already reviewed for the brachyopid labyrinthodonts and for the procolophonids. In other words, these reptiles were rather cosmopolitan in their distribution.

There is a series of cynodont reptiles from the Triassic of Brazil and Argentina. These reptiles, best known from South Africa, were quite advanced towards the mammals in their structure. It is interesting to see that the only known Triassic cynodonts outside of Africa are those found in Brazil.

The rhynchocephalian reptiles are represented at the present time by a single genus, *Sphenodon*, living on a few islands off the coast of New Zealand, but during much of the Mesozoic era these reptiles were widely distributed, as is illustrated by the evidence of the Triassic genera. Triassic rhynchocephalians belonging to a distinct family, the Rhynchosauridae, are found in Brazil, where they are represented by two genera, in South Africa, in India, and in Scotland.

As in the case of the cynodonts, there are several genera of thecodont reptiles known from the Triassic sediments of Brazil. The thecodonts were characteristic Triassic reptiles, appearing at the beginning of this geologic period and becoming extinct at its conclusion. There was a considerable amount of adaptive radiation within the Thecodontia, while in addition this order of reptiles served as the stem from which several of the dominant orders of Mesozoic reptiles arose—reptiles such as the crocodiles, the two orders of dinosaurs, and the pterosaurs or flying reptiles. The Brazilian thecodonts belong for the most part to the family Staganolepidae, a group of small- to medium-sized reptiles in which the body was protected by a heavy armor plate of overlapping bony scutes. In Triassic times the staganolepids were living, on the basis of the known evidence, in South America as well as in central Europe and England, in eastern North America, in the southwestern part of the United States, and

TABLE 2

DISTRIBUTION OF TRIASSIC LAND-LIVING TETRAPODS IN SOUTH AMERICA
AND IN OTHER CONTINENTAL REGIONS

	North America	South America	Africa	Europe	Asia	Australia
Amphibians						
Stereospondyls						
Brachyopids	x	x	x	x	x	x
Reptiles						
Cotylosaurs						
Procolophonids	x	x	x	x		
Therapsids						
Dicynodonts	x	x	x	x	x	
Cynodonts		x	x			
Rhynchocephalians						
Rhynchosaurs		x	x	x	x	
Thecodonts						
Pseudosuchians	x	x	x	x		
Dinosaurs						
Thecodontosaurs	x	x	x	x	x	

in Africa. So here again we see repeated the pattern of widely distributed reptiles, living in most of the large continental areas during Triassic times.

The Triassic period was the beginning of the age of dinosaurs, but at that time the dinosaurs were for the most part rather small and primitive. It was not until later phases of Mesozoic times, the Jurassic and Cretaceous periods, that the dinosaurs attained the dominance and in many cases the gigantic size that were so characteristic of these reptiles as a group. One might say that the Upper Triassic dinosaurs show the beginnings of the evolutionary trends that were to lead to such spectacular results in the following two periods of geologic history.

A single dinosaurian genus, *Spondylosoma*, has been described by von Huene from the Triassic of Brazil on the basis of rather fragmentary evidence. This dinosaur (if the identification of the material is correct) was a prosauropod belonging to the family Thecodontosauridae. Thecodontosaurs are well known from the Upper Triassic of Europe and from eastern North America as well, while in addition they are found in the Triassic fauna of South Africa. Evidently these early dinosaurs were widely distributed in Triassic times.

Such is a summary of the Triassic land animals of South America and their relationships to closely related forms in other parts of the world. Perhaps this summary can best be presented in graphic form (table 2).

This gives us a picture of the Triassic land animals of South America as being members of a fauna of essentially world-wide distribution. In only one instance, that of the cynodont reptiles, are the distributions limited to only two continents, South America and South Africa. And it is quite possible that even in this case the limitations are caused in part by the imperfection of our knowledge.

It may be helpful to make a comparison between the several Triassic faunas here discussed, by means of the formula suggested by Simpson in 1947 whereby the common elements in two regions at a given taxonomic level are set against the total faunal units at that level in the smaller of the two faunas. The formula is

$$\frac{100C}{N_1}$$

in which C represents the number of faunal units at the given taxonomic level and N_1 is the total number of units at this level in the smaller of the two faunas.

FIG. 27. The distribution of certain amphibians and reptiles during the Triassic period. Amphibians: 1, stereospondyls, brachyopids. Reptiles: 2, cotylosaurs, procolophonids; 3, therapsids, dicynodonts; 4, therapsids, cynodonts; 5, rhynchocephalians, rhynchosaurs; 6, archosaurs, staganolepids; 7, archosaurs, thecodontosaurs.

In a comparison of the Triassic faunas of the different continents it is rather pointless to attempt the use of genera, since there are hardly any genera in common between the various continental areas. Consequently it is necessary to utilize the larger taxonomic categories, the families and suborders.

The suborders and families contained within the Triassic fauna of South America are as follows:

SUBORDERS	FAMILIES
Stereospondyli	Brachyopidae
Diadectomorpha	Procolophonidae
Dicynodontia	Dicynodontidae
Theriodontia	Chiniquidontidae
Theriodontia	Traversodontidae
Rhynchocephalia	Rhynchosauridae
Pseudosuchia	Staganolepidae
Theropoda	Thecodontosauridae

The faunal resemblances between South America and other continental areas during Triassic times on the basis of the categories listed above are as follows:

	SUB-ORDERS	FAMI-LIES
South America—Africa	100%	75%
South America—Europe	86	75
South America—North America	71	63
South America—Asia	43	37
South America—Australia	14	12

These figures must be used with some caution. They indicate that on the whole the relationships between South America and South Africa were possibly closer than those between South America and other continental areas. Yet at the family level there is no difference in the above analysis between Africa and Europe in this respect. Moreover, on the basis of the tabulation above the differences at the family level between Africa and Europe on the one hand and North America on the other are hardly significant, and such differences as do exist are largely the result of the presence of cynodont reptiles in South America.

The distribution of the Mesozoic tetrapods, as we know them at the present time, might have been brought about in part by the existence of a land connection between Africa and South America. This would have allowed a faunal flow between the continents. But the South American fauna might have been the result of migrations through Eurasia into North America and from thence southward into South America. Of course there are no cynodont reptiles and rhynchosaurs so far known in North America, so one must suppose on the basis of present evidence that these animals have been lost in the fossil record in North America if the movement was through this region.

In short, a broad analysis of the evidence of the South American land vertebrates during Triassic times seemingly is not conclusive in either direction. On this basis a South Atlantic land bridge is not absolutely essential to explain the relationships of the Triassic vertebrates of South America to those of the Old World, but the evidence would seem to be slightly in favor of such a connection. Of course, if one maintains that the two Americas were completely separated from each other throughout the extent of Triassic times, then a South Atlantic bridge is necessary to account for the presence in South America of various reptilian genera closely related to genera in Africa and in other parts of the Old World.

Did the Tethys Sea extend across the present isthmian region during the Triassic period to isolate South America completely from North America? On this point there is no general agreement. Some paleogeographers and paleontologists do indicate such a separation, in which case a South Atlantic connection between Africa and South America would be necessary (Wurm *in* Salomon, 1926; Boule and Piveteau, 1935; Swinton, 1934; Moore, 1949). With other authorities the relationships between the two continents during the Triassic are questioned (Schuchert and Dunbar, 1933). Still others would regard or consider the presence of an isthmian link between the two continents during Triassic times as probable (Grabau, 1921).

If it be supposed that North America was completely separated from South America in Triassic times by an extension of the Tethys Sea across the isthmian region, it does not necessarily follow that there was a broad Gondwana land in Triassic times, stretching between all the southern continents. But there very well may have been an isthmian or island bridge between South America and Africa, across which intercontinental movements of land-living tetrapods could have taken place. A more detailed consideration of Triassic faunas in the several continental regions would seem to indicate, as Romer points out below, that the evidence strongly favors a South American-African land bridge in Triassic times.

THE JURASSIC SAUROPOD

As mentioned above, Cabrera has described a single sauropod, *Amygdalodon*, from the Jurassic of Chubut, southern Argentina. This single genus indicates that in Jurassic times, as was the case in the preceding Triassic period, there were in South America animals of world-wide distribution. We now know that the sauropod dinosaurs were living during the Jurassic period in South America, in North America, in Europe, in Africa and Madagascar, and in Australia; consequently the evidence of *Amygdalodon* corresponds with the evidence afforded by the Triassic fauna that is discussed above and with that of the Cretaceous fauna next to be considered.

THE CRETACEOUS FAUNA

Let us now turn to a consideration of the Cretaceous tetrapods of South America, to see if they offer additional evidence on this perplexing problem. These animals are known for the most part from Argentina, especially from Patagonia and from Nequen. Four genera have been recorded from Brazil, and there is good reason to think that as more work is carried on in Brazil a considerable Cretaceous fauna will be made known from this country. The Cretaceous land-living tetrapods from South America can therefore be listed by genera as follows:

Reptilia
  Chelonia, turtles
    Niolamia, a primitive turtle
    Podocnemis   ⎫
    Naiadochelys (probably | pleurodires, or side-
        Podocnemis) | neck turtles
    Taphrosphys[a, b]   ⎬
    Osteopygis | cryptodires, or ver-
    Gyremys (very doubtful) ⎭   tical-neck turtles
  Squamata, lizards and snakes
    Dinilysia, a snake
  Crocodilia, crocodiles
    Baurusuchus,[a] an aberrant crocodilian, pecul-
                         iar to South America
    Notosuchus   ⎫
    Cynodontosuchus ⎬ mesosuchians, or archaic
    Goniopholis[a]   ⎭   crocodiles
    Leidyosuchus, a "modern" crocodile
  Saurischia, dinosaurs
    Clasmodosaurus ⎫ small, frugivorous, theropod
    Loncosaurus   ⎭   dinosaurs
    Genyodectes, a large, carnivorous, theropod
                         dinosaur
    Titanosaurus[a]
    Laplatasaurus
    Antarctosaurus, gigantic, marsh-dwelling, her-
                         bivorous sauropod din-
                         osaurs
    Argyrosaurus
    Campylodon
  Ornithischia, dinosaurs
    Loricosaurus, an armored dinosaur
    Notoceratops, a horned dinosaur (based on
                         fragmentary and ques-
                         tionable evidence)

[a] Brazil.
[b] Hitherto not known below the Eocene.

From the above list it can be seen that there was a considerable dinosaurian fauna in South America during the Cretaceous period. This fauna, known at the present time mainly from southern Argentina but supplemented by a few discoveries from Brazil, is interesting in that it contains an over-

whelming proportion of saurischian dinosaurs. In most Cretaceous dinosaurian faunas, especially those from the upper Cretaceous, the ornithischian dinosaurs are preponderant, the saurischians being represented usually by the carnivorous theropods. In the South American Cretaceous fauna, however, there is quite an array of sauropod dinosaurs, and these together with the theropods make up the bulk of the dinosaurian representation. The ornithischians are present as a single genus of armored dinosaur and a very questionable ceratopsian or horned dinosaur.

Two genera of ornithomimids are present among the theropod dinosaurs of South America. These were theropods of medium size, comparable in this respect to the modern ostriches. They had a long flexible neck and a comparatively small, bird-like skull, and it would seem that these dinosaurs probably filled an ecological role similar to that occupied by the modern "ratite" birds; they were swift-running animals that probably fed upon fruits, insects, and other such small fare. The presence of these dinosaurs in the Cretaceous fauna of South America constitutes a strong link to the Cretaceous faunas of the Northern Hemisphere, where the ornithomimids are well represented in North America and Eurasia. They have also been found in Australia.

Genyodectes is a large, carnivorous, theropod dinosaur. The ecological role played by these reptiles in Cretaceous times was a dominant one, for they were giant, meat-eating dinosaurs that preyed upon other large dinosaurs with which they were contemporaneous. They seemingly were upland forms, as were the ornithomimids, and it is not likely that they went very far out into the water or into swampy regions. Since they were upland predators they were probably capable of extensive wanderings, not only as individuals but also as species. Consequently it is not surprising to find them occupying most of the land areas of the world during Upper Cretaceous times, and the world-wide distribution of these reptiles shows that there were adequate avenues open for the dry-land dispersal of large reptiles between all of the large continents. They are found in Upper Cretaceous sediments in South America, North America, Europe, Asia, and Africa.

The sauropod dinosaurs, so characteristic and dominant in the Cretaceous fauna of South America, can be considered as a relict group in a rather isolated part of the world. By Cretaceous times the sauropods were virtually extinct in North America, although the evidence indicates that they still did persist here and there as greatly reduced remnants of what had once been a widely spread and numerous group of animals. In Eurasia and Africa the sauropods were rather better off during Cretaceous times than they were in North America, so that they formed appreciable segments of the faunas of those regions.

The giant sauropods were swamp-dwelling and swimming dinosaurs, which sought not only their food but also protection from the giant predatory theropods of the time in shallow waters. Consequently it would seem likely that the opportunities for the dispersal of these giant dinosaurs from one region to another were more extensive than for the theropods, which were confined to relatively dry ground. It is certainly true that except for North America the sauropods spread to many parts of the world during Cretaceous times, and in some cases the broad distribution of these reptiles extended to the categories of genera. Thus the genus *Antarctosaurus* is found in South America and India, the genus *Titanosaurus* is found in South America, Europe, and India, while the genus *Laplatasaurus*, so characteristic of the Cretaceous of southern Argentina, is found as well in East Africa, Madagascar, and the Indian peninsula.

These genera have been used by various authors to indicate the close relationships of the southern continents in Cretaceous times. Yet it must be remembered that they are but part of a broad distribution of Cretaceous sauropods. *Titamosaurus* is found in Europe as well as in the southern continents, while the family Cetiosauridae, to which these genera belong, is very widely distributed in Cretaceous deposits.

The only South American ornithischian dinosaur that is well enough known to permit speculation is a nodosaur, a heavily armored dinosaur that probably lived in an upland environment. The nodosaurs were cosmopolitan in their distribution during Cretaceous times, being present, on the basis of available evidence, in Europe, Asia, and North America as well as in South America. Thus the range of these dinosaurs parallels that of the other dinosaurs in that it shows essentially a world-wide distribution of the animals in question.

There was a variety of crocodilians in South America in Cretaceous times, including representatives not only of the mesosuchians, so typical of the Mesozoic, but also of the eusuchians, the modern crocodilians, which went through the initial phase of their long evolutionary history during the Cretaceous period. In addition there was one genus of the two known genera of sebecosuchians, representative of an aberrant group of crocodilians confined to South America. Since the crocodilians have been for the most part aquatic reptiles it must be supposed that facilities for the dispersal of these reptiles may have been comparatively broad during Cretaceous times. True, most modern crocodilians are confined to rivers and lakes, so that they may be fairly considered as continental vertebrates, but certain forms, such as the salt-water crocodile, cross open stretches of the sea. Therefore it is reasonable to conclude that some of the Mesozoic crocodilians followed not only drainage patterns during the course of their distributional expansion from one region to another, but also sea coasts. This may account in part for the world-wide distribution of the mesosuchians during the Cretaceous period. The eusuchians, which occur in South America, are also found in North America and Europe, and it is probable that their absence from the fossil record of Asia and Africa is the result of our imperfect knowledge—a condition that may be corrected by future collecting.

The Sebecosuchia constitute a different problem. These crocodiles were highly specialized along lines quite apart from all of the other crocodilians. Whereas the common trend among the crocodiles throughout their

TABLE 3

DISTRIBUTION OF CRETACEOUS LAND-LIVING REPTILES IN SOUTH AMERICA
AND IN OTHER CONTINENTAL REGIONS

	North America	South America	Africa	Europe	Asia	Australia
Reptiles						
Turtles						
Meiolanids		x				
Pelomedusids	x	x		x		
Thalassemydids	x	x	x	x		
Snakes						
Anilids		x				
Crocodilians						
Sebecosuchians		x				
Mesosuchians	x	x	x	x	x	
Eusuchians	x	x		x		
Dinosaurs						
Ornithomimids	x	x		x	x	x
Deinodonts	x	x			x	
Cetiosaurs	x	x	x	x	x	
Nodosaurs	x	x		x	x	
Ceratopsians	x	x?			x	

FIG. 28. The distribution of certain reptiles during the Cretaceous period. Turtles: 1, meiolanids; 2, pelomedusids; 3, thalassemydids. Snakes: 4, anilids. Crocodilians: 5, sebecosuchians; 6, mesosuchians; 7, eusuchians. Dinosaurs: 8, ornithomimids; 9, deinodonts; 10, cetiosaurs; 11, nodosaurs; 12, ceratopsians.

history was towards a flattening of the skull and the development of long jaws with many teeth, the sebecosuchian trend was in the direction of a high, narrow skull with compressed teeth, sometimes limited in number and highly specialized in form. This would point to the possibility that these crocodilians may have been more completely adapted to land life than were their contemporary relatives. If such were the case, then these crocodilians would have been more limited as to the possibility of wide distribution than were the other crocodilians living at that time. Whether or not such an explanation is valid, it seems probable that the sebecosuchians constituted a rather limited assemblage of reptiles, peculiarly adapted to certain restricted environments.

Like the crocodilians, the turtles living during Cretaceous times in South America were varied. There was one genus, *Niolamia*, representative of the *Meiolaniidae*, a peculiarly distributed group of primitive turtles, while in addition there were side-neck turtles, or pleurodires, and vertical-neck turtles, or cryptodires.

The pleurodiran *Podocnemis* was present in the Cretaceous fauna of South America, and it persists in that continent at the present time. It belongs to the Pelomedusidae, a family of turtles that was present not only in South America but also in North America and Europe in the Cretaceous period.

The thalassemydids, of which *Osteopygis* occurs in the Cretaceous fauna of South America, were likewise widely spread in late Mesozoic times. These were rather primitive cryptodirans. The description of *Gyremys*, an emydid cryptodire from the Cretaceous sediments of Argentina, is based on such doubtful evidence as to have little value in this evaluation of the Cretaceous fauna of South America.

A summary of the distribution of Cretaceous land-living tetrapods is given in table 3.

Once again, as in the case of the analysis of the Triassic tetrapods of South America, we see the Cretaceous fauna from this continent as part of a large fauna of world-wide distribution. Generally speaking, the Cretaceous land-living reptiles of South America had close relatives in most of the other conti-

nental areas, so that their presence in South America reflects the broad distribution of many reptiles during the upper part of Mesozoic times.

It may be useful at this point to attempt a comparison of South America with other continental areas in a manner similar to that utilized in the discussion of the Triassic vertebrates. Again, this comparison is made on the basis of suborders and families, which are contained within the Cretaceous fauna of South America as follows:

SUBORDERS	FAMILIES
Theropoda	Ornithomimidae
Theropoda	Deinodontidae
Sauropoda	Cetiosauridae
Ankylosauria	Nodosauridae
Ceratopsia?	Ceratopsidae?
Sebecosuchia	Sebecidae
Mesosuchia	Notosuchidae
Mesosuchia	Goniopholidae
Eusuchia	Crocodylidae
Serpentes	Anilidae
Amphichelydia	Meiolanidae
Pleurodira	Pelomedusidae
Cryptodira	Thalassemydidae
Cryptodira	Emydidae?

On the basis of these categories, the faunal resemblances between South America and other continental regions might be as follows:

	SUB-ORDERS	FAMILIES
South America—Africa	80%	33%
South America—Europe	90	58
South America—North America	90	66
South America—Asia	70	42
South America—Australia	10	8

It is seemingly apparent that in Cretaceous times the relationships between South America and the northern continental land masses were closer than those between South America and the other southern continents. Here we see a definite foreshadowing of modern faunal relationships that were inaugurated at the opening of Cenozoic times and that have developed since then. It would appear that in the Cretaceous period, as in later times, the interchange of land vertebrates between South America and other continental regions took place by way of an isthmian bridge to

North America and from thence into the Old World. Perhaps different continental relationships had been established by this geologic period as compared with the Triassic period. If there was a South Atlantic connection between South America and Africa in Triassic times, it probably had foundered by the Cretaceous period, for the evidence here is all in favor of a northern connection.

The Triassic fauna of South America has something of an African cast to it, especially by reason of the presence there of various therapsid reptiles. But the Cretaceous fauna of South America is essentially similar to the northern faunas, especially those of North America and of Europe. It is interesting to note in the distribution of the Cretaceous reptiles, as was the case in that of the Triassic forms, how little relationship there is between South America and Australia.

THE RECENT FAUNA

At this point it may be interesting to consider briefly the bearing of the Mesozoic tetrapod faunas of South America on the recent tetrapods of that continent. Of course most of the faunal categories that were present in South America in Mesozoic times have long since become extinct. Between the Triassic fauna of South America and the recent fauna there is a complete break, so that there are no animals now living in the Neotropical region which are related to the amphibians and reptiles that inhabited this area 175 million years ago. Of the Cretaceous fauna, however, there are still some representatives to be found in South America, namely, the turtles, snakes, and crocodilians.

Of the turtles, only *Podocnemis* is represented in the modern fauna. Here we see the survival not only of a family but also of a genus in modern times that was present in the same general region in Cretaceous times. The longevity of the genus *Podocnemis* is in itself a criterion of the evolutionary success of this particular form through a considerable segment of geologic history, and it is possible to say quite definitely that the presence of *Podocnemis* in the modern Neotropical fauna represents an archaic holdover that has maintained its ecologic position in spite of the influx of other faunal elements from other regions.

Niolamia is distantly related to *Meiolania* of the Pleistocene of Australia, but obviously has nothing to do with the recent fauna of South America. The supposed emydid *Gyremys* in the Cretaceous of South America is too doubtful to permit valid speculation with relation to the recent fauna.

The single snake, *Dinilysia*, described from the Cretaceous of South America, belongs to the Anilidae, a family which is still to be found in the Neotropical region as well as in Asia. Therefore it is valid to assume that the modern Anilidae in South America represent a group persisting from Cretaceous times.

Among the crocodilians, most of the Cretaceous genera belong to families and even suborders that have since become extinct. In the latter category are the Sebecosuchia and the Mesosuchia. The eusuchians, the modern crocodilians, were present in the Cretaceous of South America as the genus *Leidyosuchus*, a crocodylid. *Crocodylus*, and in addition various eusuchians belonging to the family Alligatoridae, are found at the present time in the Neotropical region. The presence of crocodylids in the Cretaceous of South America represents in effect a part of a world-wide distribution of these successful and ubiquitous reptiles.

CONCLUSIONS

Such is the evidence afforded by the Mesozoic land-living tetrapods of South America as to the relationships of this continental area to other continental regions during Mesozoic times. All in all, it would appear that during the Mesozoic era the land-living vertebrates of South America were essentially parts of widely distributed faunas of world-wide extent. In a majority of the cases, the Mesozoic genera found in South America are closely related to genera living in most of the other great continental areas. Thus it

213

would appear that there were adequate avenues for the distribution of tetrapod faunas between the several continents, and that South America shared to a very considerable degree in the movements of the terrestrial vertebrates from one region to another.

There may have been a connection of sorts between South America and Africa during the Triassic period. If other evidence indicates that South America was completely cut off from North America by the Tethys Sea throughout the extent of the Triassic period, then such a southern land connection

would be necessary to account for the South American Triassic fauna. But if the separation of the two Americas was not complete, no such southern connection is essential, though the evidence would seem to indicate that it probably existed.

By Cretaceous times the relationship of South America to North America seems to have been definitely established, and the interchange of faunal elements between these two regions foreshadowed the periodic interchanges that were to take place throughout later geologic history.

BIBLIOGRAPHY

Ameghino, Florentino
1906. Les formations sédimentaires du Crétacé Supérieur et du Tertiaire de Patagonie. Ann. Mus. Nac. Buenos Aires, vol. 15, pp. 1–568.
Boule, Marcellin, and Jean Piveateau
1935. Les fossiles. Paris, Masson et Cie.
Cabrera, Ángel
1943. El primer hallazgo de Terápsidos en la Argentina. Notas Mus. La Plata, vol. 8, paleont. no. 55, pp. 317–331.
1944. Sobre un Estegocéfalo de la Provincia de Mendoza. Ibid., vol. 9, paleont. no. 69, pp. 421–429.
1947. Un saurópodo nuevo del Jurásico de Patagonia. Ibid., vol. 12, paleont. no. 95, pp. 1–17.
Darlington, P. J., Jr.
1948. The geographical distribution of cold-blooded vertebrates. Quart. Rev. Biol., vol. 23, no. 1, pp. 1–26; vol. 23, no. 2, pp. 105–123.
Grabau, Amadeus
1921. A textbook of geology. Boston, New York, and Chicago, D. C. Heath and Co.
Huene, Friedrich von
1927. Contribución a la paleogeografía de Sud América. Bol. Acad. Nac. Cien. Córdoba, vol. 30, pp. 231–294.
1929a. Versuch einer Skizze der paläogeographischen Beziehungen Südamerikas. Geol. Rundschau, vol. 20, no. 2, pp. 81–96.
1929b. Die Besonderheit der Titanosaurier. Centralbl. Min. Geol. Paläont., Abt. B, no. 10, pp. 493–499.
1929c. Los Saurisquios y Ornitisquios del

Cretáceo Argentino. Ann. Mus. La Plata, vol. 3, ser. 2ª, pp. 1–196.
1931. Die fossilen Fährten im Rhät von Ischigualasto in Nordwest-Argentinien. Palaeobiologica, vol. 4, pp. 99–112.
1933a. Kurzer Überblick über die terrestrischen Wirbeltierfaunen der jüngeren Gondwanazeit. Centralbl. Min. Geol. Paläont., Abt. B, no. 6, pp. 345–353.
1933b. Zur Stratigraphie Brasiliens. Ibid., Abt. B, no. 7, pp. 418–423.
1933c. Auffindung und Behandlung eines fossilen Saurierskeletts. Naturwiss. Monatsschr., 46 year, no. 5, pp. 129–135.
1934. Neue Saurier-Zähne aus der Kreide von Uruguay. Centralbl. Min. Geol. Paläont., Abt. B, no. 4, pp. 182–189.
1938. Die fossilen Reptilien des südamerikanischen Gondwanalandes. Ergebnisse der Sauriergrabungen in Südbrasilien, 1928–29. Neues Jahrb. Min. Geol. Paläont., Referate 3, pp. 142–151.
1940. Die Saurier der Karroo-, Gondwana- und verwandten Ablagerungen in faunistischer, biologischer und phylogenetischer Hinsicht. Ibid., suppl. vol. 83, Abt. B, pp. 246–347.
1944. Ein Anomodontier-Fund am oberen Amazonas. Ibid., Monatshefte, Abt. B, no. 10, pp. 260–265.
1949. Gleiche Cynodontier in der Obertrias Nordargentiniens und Südbrasiliens. Ibid., for 1948, Monatshefte, Abt. B, nos. 9–12, pp. 378–382.
Moore, R. C.
1949. Introduction to historical geology. New York, McGraw-Hill.

PAULA COUTO, CARLOS DE
  1943. Vertebrados fósseis do Rio Grande do Sul. Porto Alegre, Of. Graf. da Tip. Thurmann, pp. 9–49.
PRICE, LLEWELLYN IVOR
  1945. A new reptil [*sic*] from the Cretaceous of Brazil. Notas Prelim. e Estud., Div. Geol. Min., Rio de Janeiro, no. 25, pp. 1–8.
  1946. Sôbre um novo P[s]eudosuquio do Triássico superior do Rio Grande do Sul. Bol. Div. Geol. Min., Brazil, no. 120, pp. 7–38.
  1947. Um procolofonideo do Tríassico do Rio Grande do Sul. *Ibid.*, no. 122, pp. 7–26.
SCHUCHERT, CHARLES
  1932. Gondwana land bridges. Bull. Geol. Soc. Amer., vol. 43, pp. 875–916, 2 figs., pl. 24.
SCHUCHERT, CHARLES, AND CARL O. DUNBAR
  1933. A textbook of geology. Part II—Historical geology. Third edition. New York, John Wiley and Sons, Inc.

SIMPSON, GEORGE GAYLORD
  1939. Antarctica as a faunal migration route. Proc. 6th Pacific Sci. Congr., pp. 755–768.
  1943a. Mammals and the nature of continents. Amer. Jour. Sci., vol. 241, pp. 1–31.
  1943b. Turtles and the origin of the fauna of Latin America. *Ibid.*, vol. 241, pp. 413–429.
SWINTON, W. E.
  1934. The dinosaurs. London, Thomas Murby and Co.
WEEKS, L. G.
  1947. Paleogeography of South America. Bull. Amer. Assoc. Petrol. Geol., vol. 31, no. 7, pp. 1194–1241.
WILLIS, BAILEY
  1932. Isthmian links. Bull. Geol. Soc. Amer., vol. 43, pp. 917–952, 2 figs., pl. 24.
WURM, A.
  1926. Trias. *In* Salomon, W., Grundzüge der Geologie. Band 2, Erdgeschichte. Stuttgart, E. Schweizerbart, pp. 309–340.

17B

Reprinted from *Bull. Am. Mus. Nat. Hist.*, **99**(3), 250–254 (1952)

DISCUSSION OF "THE MESOZOIC TETRAPODS
OF SOUTH AMERICA"

ALFRED S. ROMER

Museum of Comparative Zoölogy, Cambridge, Massachusetts

COLBERT GIVES ABOVE an admirable summary of the Mesozoic faunas of South America. I find myself in hearty agreement with his conclusions with regard to the Cretaceous; there is nothing in the fauna to indicate closer relationships with one continent than another, and no reason why the non-autochthonous elements present could not have arrived via a "normal" North American connection, if such existed. There is, as Colbert notes, almost no evidence regarding the Jurassic.

The Triassic is a more difficult problem. Colbert lists and describes the tetrapod groups then present in South America, tabulates their distribution in the Triassic of other continental areas, and summarizes these occurrences on a percentage basis. From this he reaches the "Scotch verdict" that the question of a possible South Atlantic connection is an open one, and that while the evidence is somewhat in favor of such a connection, arguments can be brought forward to show that the various elements of the South American fauna might have reached that continent over "orthodox" migration routes.

With this conclusion I mildly disagree. I have been reared in the tradition of continental stability and was early impressed by Matthew's demonstration (ably reënforced by Simpson in recent years) that most if not all Tertiary land bridges are not merely delusions but snares to the unwary. However, I find myself here, after consideration of the evidence, rather strongly inclined towards belief in the existence of a southern intercontinental connection between South America and South Africa in the Triassic. To my embarrassment; for in such a "leftish" position I am disturbed (like many a liberal in political circles) by the company (of bridge-builders, radical continent shifters, and Gondwanaland collectivists) which this may entail.

As Colbert tabulates the evidence, his conclusion seems reasonable enough; in a comparison of the South American fauna with that of other continents for the Triassic as a whole, it appears that the evidence for African connection is very slight. But such tabulations are valid only if the entities compared are truly comparable. This does not appear to be the case here. There is no single "Triassic fauna" with which that of South America can be compared. During the Triassic there occurred a major revolution in vertebrate life. Reptile faunas of the early Triassic are essentially late Paleozoic in type, with therapsids as the dominant forms. Those of the late Triassic are, in strong contrast, typically Mesozoic in nature, with dinosaurs as dominant elements. If such faunas are "lumped" the resultant is a meaningless composite. We must compare the South American forms, if possible, with faunas of a similar stage of the Triassic. If we were to assess the relationships of the fauna of the North American Miocene by comparing it with, say, the Pliocene fauna of Eurasia on the one hand and the combined Oligocene and Pleistocene faunas of South America on the other hand, any numerical data obtained would obviously be of little significance, if not positively misleading. It would appear that we are in danger of finding ourselves in a similar situation here.

At what positions within the Triassic lie the faunas of other continents which Colbert has compared with that of South America? For Africa, there are included two very different assemblages—a therapsid fauna of early Triassic age, from the Upper Beaufort beds,[1] and a second late Triassic fauna from the upper Stormberg with dinosaurs as the principal element (an East African fauna of intermediate age is included as well). For Europe the story is similar; there are here lumped the rather sparse fauna of the early Triassic (the Bunter) and the dinosaur fauna of the Upper Triassic Keuper which follows after the gap of the marine Muschelkalk. Remains from Asia are few but also include early and late elements. From North America the rec-

---

[1] Von Huene, in contrast to other workers, considers the Cynognathus zone at the top of the Beaufort as Middle Triassic, with a consequent effect of scaling upward other faunas advanced beyond this level.

ord is composed of late Triassic faunas, plus a very few elements of earlier age.

With which of these two major contrasting age groups, early or late, is the fauna of the South American Triassic properly compared? With neither, it would appear.

Of the Triassic tetrapods of that continent, a few are from localities of somewhat uncertain stratigraphic position in Argentina. They are not considered here; their absence does not affect the argument, since recognizable forms include only remains of cynodonts comparable to those of Brazil and the skull of a brachyopid, a labyrinthodont type found in various areas and stages of the Triassic and not of importance in a time determination. The major fauna, to be discussed here, is that from the Santa Maria beds of Brazil. Two main collecting areas are known. There is considerable difference in the faunal content of the two, but it seems to be generally agreed that there is no major difference in age between them. Von Huene (1935–1942) has described in monographic form a large collection of material from these beds; a considerable collection, mostly undescribed, was made by Price and White for Harvard in 1936; subsequently, Price has made large collections for the Brazilian geological survey but has as yet published only a small part of the data on them.

Von Huene originally claimed the Santa Maria beds to be Upper Triassic in age, basing this opinion, it would appear, on the rather advanced character of the rhynchosaurs, the first described elements of the fauna. As he has himself pointed out, however, the degree of specialization of a rhynchosaur type gives little indication of its stratigraphic position. Von Huene has since modified this opinion to some extent. He recognizes (1950) that this fauna is only a little later than that of the Cynognathus zone of South Africa, generally considered as early Triassic. But since, as noted, he considers the Cynognathus zone to be Middle, rather than Lower, Triassic, he compromises by bringing the Brazilian fauna down merely to the earliest Upper Triassic.

Analysis of the fauna clearly indicates a Middle Triassic age, somewhat above the Cynognathus zone of South Africa but far earlier than the dinosaur assemblages typical of the Upper Triassic.

Two types of reptiles make up the greater part of the material from these beds: rhynchocephalians of the rhynchosaur group, and dicynodonts. The rhynchosaurs *Scaphonyx* and *Cephalonia* (if the latter be distinct) are large and apparently rather advanced members of their family which indicates Middle or Upper, rather than Lower, Triassic age. Few of the dicynodonts have as yet been described. As far as I am familiar with them they appear as a whole to be rather primitive in nature, although some have grown to large size and would not seem inappropriate in a Lower Triassic (or even Upper Permian) fauna; none appears to be so specialized as the one genus (*Placerias*) known to have survived into the late Triassic.

Two other groups are moderately abundant: pseudosuchians and cynodont therapsids. Pseudosuchians appeared in the Lower Triassic and had become abundant and diversified in late Triassic times. The forms present are more advanced than would be expected in the early Triassic, and a middle or late Triassic age is indicated. Cynodonts are unknown beyond the Lower Triassic in any regions other than Brazil and Africa; by the Upper Triassic they had been succeeded by a miscellaneous assemblage of other therapsid forms grouped currently as the Ictidosauria, which are structurally far more advanced towards a mammalian condition. The Brazilian cynodonts appear to be a little, but only a little, more advanced than the Lower Triassic members of the group from the Cynognathus zone of the South African Beaufort series. They cannot be conceived of as later than Middle Triassic.

There remain two rarities: a dinosaur and a procolophonid reptile. The dinosaur is of a primitive saurischian type. Dinosaurs are unknown in the Lower Triassic, abundant and rather diversified in the Upper Triassic. Presumably they began their history in the Middle Triassic; the presence here of a single (and seemingly rare) form agrees well with assignment of the beds to this level in the period. The procolophonid, *Candelaria*, is a member of a cotylosaurian group that ranges widely in space and in time from the Permian to the Upper Triassic; its describer (Price, 1947) notes, however, that it appears to be closer to the Lower Triassic genera than to those of the Middle and Upper Triassic.

It is thus with Middle Triassic faunas that this Brazilian assemblage should be compared, most particularly with faunas of this age in Africa and North America. Not too many years ago we would have been blocked here through the fact that such comparable faunas were quite unknown. Today, while there are still many gaps in our knowledge, a fair body of data is at hand which bears on this subject.

In South Africa the Middle Triassic deposits are the Molteno beds, barren of vertebrates. In East Africa, however, the Manda beds of Tanganyika are considered to be equivalent to the Molteno. They have yielded in recent years a considerable fauna. Much of the material is as yet undescribed, but its general nature is clear from various accounts by Haughton, Parrington, and von Huene (summarized by von Huene, 1950). As the last writer, who is familiar with both areas, notes, the fauna is very similar to that of Brazil. As there, the most abundant forms are rhynchosaurs and dicynodonts. The rhynchosaur (a different genus) is somewhat more primitive than the rhynchosaurs of Brazil, but there is little correlation between age and degree of specialization in rhynchosaurs. The dicynodonts are varied and generally of medium to large size, as in Brazil, and von Huene states that one of them closely resembles the Brazilian *Stahleckeria*. Also present in some variety, as in Brazil, are thecodonts and cynodonts. Von Huene notes that the thecodonts are rather comparable to those of Brazil, and one genus is very similar to *Rauisuchus* of Brazil. As in the latter region the cynodonts are but little advanced over the forms of the Cynognathus zone, and one is closely related to a Brazilian form. Also as in Brazil, there is a solitary primitive dinosaur; the two are closely related and members of the same primitive family. There is thus a remarkable similarity between South American and African reptilian faunas in the Middle Triassic; the differences are no greater than might be expected between two portions of a single continental area.

The problem of comparison with North America is more difficult. Until recently, no Triassic vertebrate fauna was known from this continent other than the faunas of the Upper Triassic, with dinosaurs and their footprints, phytosaurs, and metoposaurid amphibians as the prominent components. But persistent work by University of California parties has uncovered vertebrate remains and numerous footprints in the Moenkopi formation of the Southwest, described by Welles (1947) and Peabody (1948). The Moenkopi has been regarded as a Lower Triassic formation. The amphibian material suggests, however, that the Upper Moenkopi is rather later than had been thought, and, if not so late as Middle Triassic, is certainly close to it in time. The small amount of reptilian skeletal material so far described is not revealing. The abundant footprints, thoroughly studied by Peabody, are important for our study when the time and region of their occurrence are considered. It seems that they should literally show us the footsteps of emigrants to South America if the Brazilian Triassic fauna entered from the north. Of the footprint types most are those of chirotheres and related forms. Chirothere footprints are well known from various stages of the European Triassic; it is currently agreed that most if not all of these were made by thecodonts. Thus there is evidence of the presence in North America of forms which might have invaded South America to become the pseudosuchians of the Brazilian fauna. But what of the rhynchosaurs, dicynodonts, and cynodonts which comprise the other major elements in Brazil? Had they entered South America circuitously via the northern continents, they should be in evidence in North America at the time of deposition of the Moenkopi. The material shows no trace of any of these groups. Apart from the pseudosuchian prints, there is found a great number of footprints of lizard-like appearance, but there is not a print of the sort expected of a rhynchosaur, and there is no print that could be attributed in any way to any therapsid—dicynodont or cynodont. This is negative evidence, but rather good negative evidence. If we look to the succeeding North American fauna of the Upper Triassic, cynodonts and rhynchosaurs are again absent; one large dicynodont was present in this region. The absence of cynodonts is to be expected, since they nowhere survived that late. But we may note that no cynodont is known from any Triassic stage in any northern continent. If they migrated from Africa to Brazil via the north, they must have covered an immense amount of territory without leaving a trace behind. The absence of Upper Triassic rhynchosaurs appears significant, since they were present elsewhere in late Triassic times and might reasonably be expected to have survived had they been present earlier. The one dicynodont in the American Upper Triassic would seem to have been a stray invader from some other region—Asia, or possibly from South America itself.

To sum up, it is difficult to furnish Brazil with a Triassic reptile fauna from North America, since there is no evidence that most of the groups needed to stock that fauna were present in the north at the appropriate time. On the other hand, we have in Africa a contemporary fauna almost identical in nature with that of South America. To apply to the actually comparable Middle Triassic reptile faunas the type of treatment used by Colbert, the matter of identities stands as follows:

	SUB-ORDERS	FAMILIES
South America—Africa	83%	72%
South America—North America	33	?14

The evidence as it now stands strongly supports the hypothesis that some type of connection existed between Africa and South America in Middle Triassic times; the burden of proof rests upon those who wish to deny its existence. Evidence to support or refute this hypothesis should be found from the study of sediments, of paleobotany, and of invertebrate paleontology. If, as I rather suspect, objective consideration of this evidence should tend on the whole to be favorable, we may regard the hypothesis as one reasonably well established.

While the mammalian evidence seems to show clearly and decisively that in the Tertiary as now the Atlantic Ocean formed a practically impassible barrier to the passage of terrestrial organisms, the situation becomes less clear as we pass backward in time. The vertebrate evidence from the Paleozoic, as from the Triassic, suggests that in early times this barrier was, at the most, one that was readily crossed and may then have been non-existent. I have (1945) commented on the similarity between European and North American vertebrate faunas at the general level of the Pennsylvanian-Permian boundary. American faunas of the Pennsylvanian are quite similar to those of Europe (Nopcsa, 1934; Westoll, 1944). Late Devonian fish faunas of these continents are quite similar. Possibly these similarities may be due to a nearly uniform circumpolar distribution via Asia, but this is somewhat difficult to believe.

The strongest single piece of evidence for South American transatlantic connections is, of course, the case of *Mesosaurus.* This was a small, fresh-water, fish-eating reptile, whose remains are found in rocks which were close to the Pennsylvanian-Permian boundary in age. It is known in but two parts of the world: in South Africa in the Dwyka, and in Brazil in an equivalent formation. *Mesosaurus* was adapted in a moderate degree to an aquatic life, but no one would contend that it was capable of breasting the waves of the South Atlantic from one continent to the other. If we assume that its migration route between these two areas (absolute geographical extremes under orthodox theory) passed through the northern continents, it is difficult to understand why absolutely no trace of this very distinctive form has ever been found elsewhere. It has been suggested that *Mesosaurus* may have lived under environmental conditions not represented in other deposits. But we have today a great number of faunas and collecting localities of appropriate age in both North America and Europe which appear to represent a considerable variety of environments and yet with no record of the presence of *Mesosaurus.* Maintenance of orthodoxy is difficult. In the Paleozoic, or even the early Mesozoic, the present Atlantic basin may have been passable for continent dwellers without the intervention of a Mosaic miracle.

REFERENCES

HUENE, F. VON
1935–1942. Die fossilen Reptilien des Südamerikanischen Gondwanalandes. Munich, 332 pp.
1950. Die Theriodontier des ostafrikanischen Ruhuhu-Gebietes in der Tübinger Sammlung. Neues Jahrb. Geol. Paläont., Abhandl. Jahrg. 1950, vol. 92, pp. 47–136.

NOPCSA, F.
1934. The influence of geological and climatological factors on the distribution of non-marine fossil reptiles and Stegocephalia. Quart. Jour. Geol. Soc. London, vol. 90, pp. 76–140.

PEABODY, F. E.
1948. Reptile and amphibian trackways from the Lower Triassic Moenkopi formation of Arizona and Utah. Univ. California Publ., Bull. Dept. Geol. Sci., vol. 27, no. 8, pp. 295–468.

PRICE, L. I.
1947. Um procolofonideo do Tríassico do Rio Grande do Sul. Bol. Div. Geol. Min., Brazil, no. 122, pp. 1–26.

ROMER, A. S.
1945. The late Carboniferous vertebrate fauna of Kounova (Bohemia) compared with that of the Texas redbeds. Amer. Jour. Sci., vol. 243, pp. 417–442.

WELLES, S. P.
1947. Vertebrates from the Upper Moenkopi formation of northern Arizona. Univ. California Publ., Bull. Dept. Geol. Sci., vol. 27, no. 7, pp. 241–294.

WESTOLL, T. S.
1944. The Haplolepidae, a new family of late Carboniferous bony fishes. Bull. Amer. Mus. Nat. Hist., vol. 83, art. 1, pp. 1–122.

Part IV

FIXITY OF
CONTINENTS AND OCEANS

Editor's Comments
on Papers 18 Through 21

As discussed in the introduction to this volume, until the 1960s most North American geologists and paleontologists found it difficult to accept the idea that continents were not firmly fixed in their present position since the formation of the earth's first crust.

As Schuchert, in Paper 18, and others before him such as Blanford (Paper 22) and Suess (Paper 1), had little knowledge of the composition or structure of ocean basins, he had no hesitation in assuming that the connections between continents that were needed to relate and to connect similar fossil assemblages were simply large portions of crust that had foundered in later geologic periods. Therefore, Paper 18 is particularly significant in that Schuchert had recognized the major faunal and floral associations of the Early Permian and connected different continents by broad "land bridges" (see also Paper 23). After discussing the principles of correlation and paleogeographic map construction, most of which is not duplicated here because Schuchert covered that topic more completely in 1910 (Paper 5) and 1929 (Paper 2), he examines the paleogeography of western North America during the Mesozoic. After assembling a large arsenal of data from sedimentation, paleontology, paleogeography, and structural geology, he successfully demonstrates the existence of the major seaways and clastic source areas that can be traced through Mesozoic time. In these and later maps, Schuchert established the practice of paleogeographic reconstructions drawn on maps that depict the Recent outline of continents and, in effect, firmly support the idea of fixity of continents.

In Paper 19, Ralph W. Chaney identifies five centers of floral evolution during the early Tertiary and discusses the changes that took place during the Tertiary. Considerable emphasis was placed by Chaney on climatic changes (which are further elaborated on by Dorf in Paper 28) and, of course, a possible northward movement of the whole continent was not considered. Chaney's discussion is a broad view of the evolution of angiosperms and a helpful summary.

H. D. Gordon's interest in the problem of sub-Antarctic plant distribution (Paper 20) is in part a reflection of his location in the Southern Hemisphere. Both Tasmania and New Zealand, where he has considerable firsthand background on the floras, have extensive forests of the southern beech, *Nothofagus*, which is widely distributed in the cool, wet, temperate lands of the Southern Hemisphere. Gordon presents the two conflicting schools of thought that existed in the late 1940s, transoceanic dispersal versus greater continuity of land, in an informative and logical analysis that obviously leaves him not entirely satisfied with either in their strictest form.

In Paper 21, G. G. Simpson examines Antarctica as a faunal migration route from the viewpoint of a vertebrate paleontologist in the late 1930s. He carefully presents and critically evaluates the biological and paleontological evidence available to him at that time, and reaches the conclusion that the distribution is consistent with occasional, almost random, invasions by Northern Hemisphere faunas, which is probably still a reasonably valid thesis for Recent mammals. Several of G. G. Simpson's later articles pursue the topic of dispersals, and he demonstrated convincingly the concepts of dispersals via "corridors," "filters," and "sweepstakes" for different areas at different times. In *The Geography of Evolution* (Simpson, 1965) he gathered together a number of his articles of the 1950s and early 1960s that have biogeographic emphasis. Other publications that reflect his earlier thoughts include "Mammals and the Nature of Continents" (Simpson, 1943), in which he refutes the hypothesis of drifting continents in masterful terms, and "Probabilities of Dispersal in Geologic Time" (Simpson, 1952), in which he applies probability theory and geologic time to rates of dispersal across barriers. Paper 21 and the cited articles are well worth perusing, for they help set the tone of the thinking of paleontologists and zoologists from the late 1930s to the early 1950s. Development of the theory of ocean spreading and plate tectonics in the 1960s has clarified many distributional problems and has convinced Simpson and most other paleontologists that continents were not fixed in their geographic position.

REFERENCES

Simpson, G. G. (1943). Mammals and the nature of continents. Am. Jour. Sci., 241, 1–31.

———. (1952). Probabilities and dispersal in geologic time, *in* E. Mayr, ed., The problem of land connections across the South Atlantic, with special reference to the Mesozoic. Bull. Am. Mus. Nat. Hist., 99(3), 163–176.

———. (1965). The geography of evolution. Chilton, Philadelphia, xi + 249 p.

18

Reprinted from *Bull. Geol. Soc. America*, **27**, 491, 493–497, 504–514 (Sept. 1, 1916)

CORRELATION AND CHRONOLOGY IN GEOLOGY ON THE BASIS OF PALEOGEOGRAPHY [1]

BY CHARLES SCHUCHERT

(Read before the Paleontological Society August 3, 1915)

[*Editor's Summary of Pages 491–493:* Schuchert outlines the history of geologic chronology, mentioning the laws of superposition and faunal succession, evolutionary theory, worldwide cycles of deposition, and worldwide sea-level fluctuations.]

PERMANENCY OF CONTINENTS AND OCEANS

Most of the older geologists held that the continental and oceanic areas had repeatedly changed places, and Lyell taught that all parts of the ocean bottoms had been land. James D. Dana was the first to assail this conclusion, and shortly after his trip around the world with the Wilkes Exploring Expedition announced, in 1846, that the continents and ocean basins have been practically permanent. That the continents are on the whole permanent is proved by the fact that their marine deposits are almost entirely of shallow seas. Where deep-sea formations occur they

---

[1] This is the third of four papers read at the summer meeting of the Paleontological Society at the University of California, August 3, 1915, in the symposium entitled "General consideration of paleontologic criteria used in determining time relations."

Manuscript received by the Secretary of the Geological Society April 6, 1916.

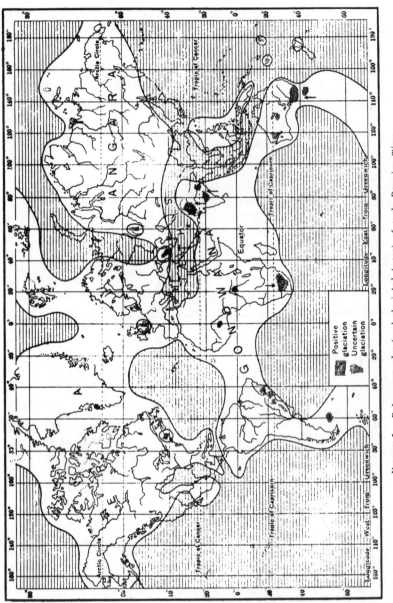

FIGURE 1.—*Paleogeography (and glaciated Areas) of early Permian Time*

Note the transverse shape and the connected condition of the continents. Arrows indicate the direction of glacier flow. From Pirsson and Schuchert's "Text-book of Geology."

are found only on the margin of the continents or on continental islands, and all of them do not total more than 1 per cent of the earth's surface. Even though the theory of the permanency of continents and oceanic basins is now of wide acceptance, it does not follow that the continents and oceans have practically retained their present outlines since the beginning of the Cambrian. On the contrary, it is held by many geologists—and more so in Europe than in America—that the continents have changed much in form and in area. My studies have convinced me that during the Paleozoic the continents were not only larger in area, but more especially that they were not then, as they are now, drawn out longitudinally. Originally there were two immense transverse or latitudinal continents (see figure 1). In the north lay Eria, the great holarctic land, at times uniting the Euro-Asiatic mass broadly with Iceland, Greenland, and North America; in the equatorial region lay Gondwana, extending from western South America across the Atlantic Ocean to unite with Africa, and broadly across the Indian Ocean to embrace peninsular India. Between these transverse lands lay the greater mediterranean, known as Tethys, uniting in the west sparingly with Poseidon (now the North Atlantic) and in the east broadly with the Pacific, the Father of Oceans. Gondwana was broken through in late Mesozoic time by Poseidon and Nereis, which together made the Atlantic, while Eria began to be fractured in the Cretaceous, though Europe, Greenland, and North America appear not to have been completely sundered until Miocene time. We therefore have here the phenomenon of oceanic realms enlarging at the expense of the continents. From this and the further evidence of volcanic activity throughout the geologic ages, it follows that the amount of water on the surface of the earth is greater now than it ever was, because enlarging basins hold increased volumes of water. The water of the enlarging hydrosphere is constantly supplied by the volcanoes and thermal springs, but what the percentage of increase has been since the Cambrian is unknown, though it has been placed as high as 25 per cent.

DISTURBANCES AND REVOLUTIONS

The continents periodically undergo elevation and mountain folding, and these times of crustal unrest all occur when the lands are the least flooded. This periodic readjustment in the earth-shell of North America is recorded by at least fourteen times of mountain-making. Eight of these are of lesser import, and may be spoken of as "disturbances" to distinguish them from the major movements that have long been referred to as "revolutions," of which there are six now named. The latter are also the "critical periods" in the history of the earth when mountains are

227

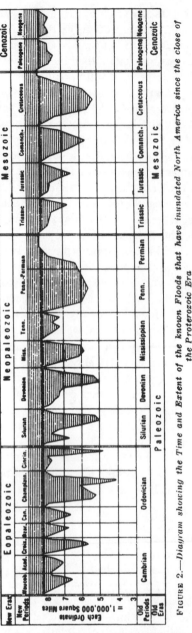

FIGURE 2.—*Diagram showing the Time and Extent of the known Floods that have inundated North America since the close of the Proterozoic Era*

made in most or all of the continents. The only established periods that in North America are not yet known to have been closed by marked crustal unrest are the Cambrian, Pennsylvanian, and Paleogene. If, however, the newer geologic chronology, which recognizes eighteen post-Proterozoic periods, is to prevail, then there are five additional times when North America is not yet known to have made mountains (Acadian, Croixian, Ozarkian, Champlainian, and Waverlian). I have no doubt that these additional disturbances will be found, but it is not essential to the newer classification that the deformation should have occurred in North America.

It is also now fairly well established that the hydrosphere has moved over the North American continent at least twelve times and in extent up to one-half of its area (4,000,000 square miles). There are, however, thirteen established periods, and the only Euro-Asiatic flood failing of record in our continent is that of Permian time (see figure 2). On the other hand, the new geologic chronology recognizes five additional times of flooding, and only one of these is as yet known to be closed by a disturbance (Lower Cambrian or Waucobian). In most cases, therefore, the periods of either the old or the new chronology are separated from one another by disturbances that, as a rule, are marked by conformable contacts. It is because of this want of marked unconformity between the

FIGURE 1.—DEVONIAN-SILURIAN DISCONFORMITY, BUFFALO, NEW YORK

Middle Devonian (Onondaga) limestone rests on the Silurian (Manlius) ; below is shown another break (Manlius-Bertie)

FIGURE 2.—PROJECTING LEDGES OF EOCENE (MIDWAY) LIMESTONE DISCONFORMABLE ON CRETACEOUS SHALE (ESCONDIDO), WHITE BLUFF, RIO GRANDE, TEXAS

Photograph by L. W. Stephenson, U. S. Geological Survey

ILLUSTRATIONS OF DISCONFORMITIES

new periods of Ulrich and Schuchert that the average stratigrapher is chary in accepting them (see figure 2).

BREAKS

The breaks in the geologic record are known to be many, and yet few stratigraphers appreciate their great number. The easily seen, marked unconformities, as, for instance, the one between the Cambrian and the Archeozoic in the Grand Canyon of Arizona, are of course accepted at full face value; but the many more apparently conformable and yet broken contacts, the disconformities, are generally overlooked, or when seen are generally undervalued (see plate 19, figures 1 and 2). It is probable that as much time is represented by the breaks as by the entire sedimentary record. This statement may appear to many as overdrawn, and yet the age of the earth estimated from geologic data is greatly at variance with the results attained by physicists on the basis of radium emanations. The differences are about as one is to eight, or even ten, and as the physicists have more reliable data on which to base their calculations, it follows that stratigraphers must either considerably elongate the geologic time-table or show that the rate of sedimentation varies greatly during the opening and closing epochs of the period when compared with the peneplained middle epoch. To emphasize the importance of breaks and their very unequal duration, it is sufficient to recall to mind the vastly long record that is absent on the Canadian Shield, where the Pleistocene drift generally reposes on the Laurentian granites of Archeozoic or Proterozoic time, or the case of the horizontal Pleistocene loess that rests conformably on Silurian limestone at Grafton, Illinois. The latter is a disconformable contact, and yet the record that is absent is equal to about one-half of the entire stratigraphic chronology.

In regard to the breaks, the statement can be made that there are at least ten disconformities for every known angular unconformity, and in the Ohio and Mississippi valleys, where the Paleozoic strata are nearly everywhere practically horizontal, there may be a hundred disconformities, and yet hardly anywhere is there to be seen a marked unconformity.

[*Editor's Summary of Pages 497–503:* Schuchert examines how sedimentation, paleontology, paleogeography, and diastrophism influence the process of stratigraphic correlation. He is aware of lithologic and faunal facies and encourages mapping these to form paleogeographic maps. He also believes changes in strand line relate to large periodic movements of the crust. (Much of this discussion is restated in Paper 2.)]

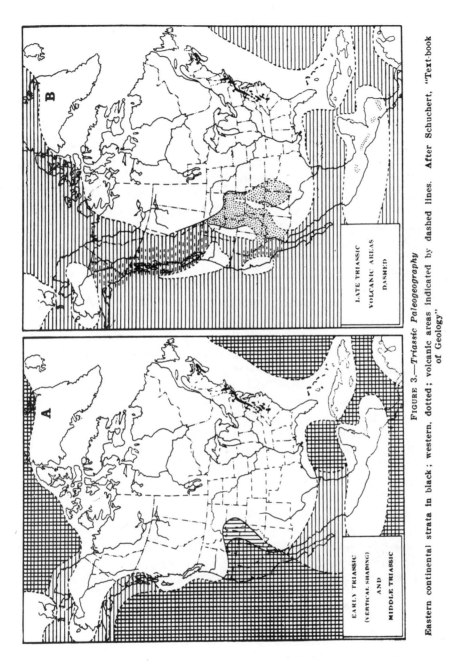

FIGURE 3.—*Triassic Paleogeography*

Eastern continental strata in black; western, dotted; volcanic areas indicated by dashed lines.   After Schuchert, "Text-book of Geology"

231

PALEOGEOGRAPHY OF WESTERN NORTH AMERICA DURING THE MESOZOIC

INTRODUCTORY

Let us now apply the methods just stated to the paleogeography of the west coast of North America during Mesozoic time.

TRIASSIC TIME

(See Figure 3)

Our knowledge of the west coast Triassic is good only for the States of California, Nevada, and Oregon, and is due mainly to the work of Prof. J. Perrin Smith. He states that the development of the Triassic of these States is unusually complete, and in thickness compares favorably with that of any other region of marine sedimentation. The deposits are usually calcareous and fairly thick (about 4,000 feet), and increase in volume with the progress of time, facts which are in keeping with the view that, as no mountain ranges were developed along the Pacific border in Permian time, there was no high land present to furnish the adjacent seas with much sand and mud.

Along the Pacific border of British Columbia, from Vancouver north to the Queen Charlotte Islands, the Triassic, and chiefly the early Upper Triassic, is of great thickness, attaining, according to Dawson, to 13,000 feet, of which more than nine-tenths is of submarine volcanic origin. With these materials are interbedded zones of marine sediments, argillites, and quartzites that are thin or even absent to the east. The volcanoes of Middle and Upper Triassic time extended from southern California into Alaska, and near Mount Saint Elias there are about 4,000 feet of basalts, followed by the same thickness of Upper Triassic limestone (Chitistone) and 2,500 feet of dark shales.

The Pacific overlap had its widest distribution in the early Upper Triassic, for its deposits occur at various points along the western border of North America and widely over Alaska and Arctic America. In the United States throughout the Rocky Mountain area the Upper Triassic is developed as a continental series of sandy red or variegated shales and cross-bedded sandstones.

In Alaska along the Pacific border, from at least Mount Saint Elias to the middle of the Aleutian peninsula, a distance of fully 800 miles, the Triassic and older formations were thrown into a folded series of mountains and injected by igneous rocks at the close of the Triassic. It seems very probable that the uplift extended across the peninsula of Alaska far southward into British Columbia. Finally, a marked break in sedimentation separates the Jurassic of California from the Triassic—a condition

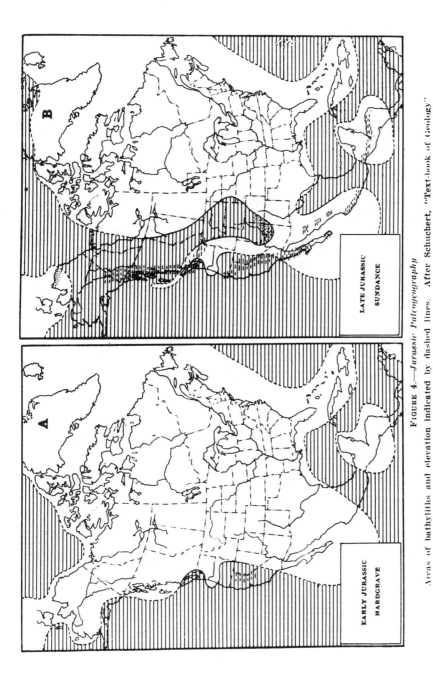

FIGURE 4.—*Jurassic Paleogeography*.
Areas of bathyliths and elevation indicated by dashed lines.   After Schuchert, "Text-book of Geology"

also common to the entire Pacific Coast region, for nowhere are there, according to Smith, any marine Rhætic strata. To emphasize this marked diastrophism, the period of orogenic mountain-making has recently been named the *Chitistone Disturbance*, after the thick limestone of the same name, so well developed in southeastern Alaska.

In this connection it is well to direct attention to the fact that the Triassic of eastern North America was also deformed at the close of this period. The Palisade Mountains of Dana, a series of faulted or block mountains, were in existence then from Nova Scotia to South Carolina, a distance of 1,000 miles. This disturbance is known as the *Palisade Disturbance*.

JURASSIC TIME

(See Figure 4)

It was stated that the Triassic period closed with crustal warping and deformation all along the entire Pacific border of North America, and that mountain-making on a considerable scale took place at least throughout Alaska. In consequence the sea appears to have been removed everywhere from the continent.

The Pacific Ocean again began to invade North America early in Jurassic time, sparingly in the Aleutian peninsula, the Cook Inlet country of Alaska, and across Vancouver Island. Of Middle Jurassic events little is as yet well known, other than that the Lower Jurassic of Alaska, with a thickness of 1,000 to 4,000 feet, continues, according to Stanton and Martin, unbroken unto the Middle (1,500 to 2,000 feet) and Upper Jurassic (5,000 feet). The total thickness of the marine Jurassic in Alaska exceeds 10,000 feet, and consists essentially of coarse deposits, such as tuffs, conglomerates, sandstones, and shales, with andesitic lava flows near the top of the series. This is largely the material from the Chitistone Mountains, formed at the close of the Triassic.

In the Californian Sea, an independent faunal province of Oregon, California, and Nevada, sedimentation appears to have been continuous throughout Jurassic time, but the detail of the formations is well known only locally. The strata of the Gold Belt series—the Mariposa and Auriferous formations—of northern California and Oregon are essentially sandstones and shales, with very little of limestone and about 500 feet of tuffaceous conglomerates. In places the thickness is 2,000 feet, rising to over 6,000 feet elsewhere in California, and if the Lower Knoxville strata of 10,000 feet thickness, with their Jurassic flora, belong here, the maximum thickness will rise considerably above the last-mentioned figure. In the Humboldt Range of Nevada there are from 1,500 to 2,000 feet of basal Jurassic limestone, followed above by 4,000 feet of slates.   Evidently the Upper Jurassic material was derived from a high land, and in places these formations are seen to rest unconformably on the Triassic.

Toward the close of the Middle Jurassic the northern Pacific, with a cool-water fauna, began to spread widely over Alaska and British Columbia, and, as the Logan Sea, continued into the States of Montana, Idaho. Wyoming, Colorado, and Utah.   In the Great Plains region the deposits of the Logan Sea have an average thickness varying between 200 and 400 feet, but increasing to the west to upward of 1,000 feet and in southwestern Wyoming to 3,500 feet.   The cross-bedded sandstones, the changeable sediments, and the general prevalence of oysters indicate that the sea was a shallow one, and, further, that it flowed over a warped land eroded to a low relief.

Volcanic activity began again locally along the Pacific border of North America early in the Jurassic and continued throughout the period, becoming more marked toward its close than at any time during the Triassic.   The eruptions were in part submarine.

Toward the close of the Jurassic the Sierra Nevadas, the Coast Range of California, and the Humboldt Range of Nevada were elevated; also the Cascade and Klamath Mountains farther north.   The making of the Sierra Nevada Mountains at this time was pointed out by Whitney in 1864 and further described by Dana.   The marked significance of this deformation has been emphasized more recently by Lawson, who regards it as having the importance of a revolution, and Smith has given it the name Cordilleran Revolution.   Last year Blackwelder called it the Nevadian movement, but it seems better to retain the older implied term of Sierra Nevada, just as we speak of the Appalachian and Laramide revolutions.   That the Sierra Nevada movement was of wide extent and that it was of greater importance than the average disturbances closing the periods is admitted, for it is probable that mountains were made extending from Mexico into southern Alaska, and yet it had not the importance of a revolution, when mountains were made in nearly all of the continents. For these reasons I prefer to call it the *Sierra Nevada Disturbance.*

The Sierra Nevada deformation also shut out the Arctic-Pacific intercommunication and prevented further wide overlaps of the Pacific Ocean over Canada and the United States.   With the rising of these mountains also began the formation of two new troughs or geosynclines.   The smaller one, which was clearly developed in latest Jurassic time, Le Conte has named the Shastan Sea, and of this the present Great Valley of California is the structural remnant   The other, of far greater extent, I have recently named the *Coloradoan geosyncline,* but it was not in full development until Cretaceous time.

235

While the Pacific border of North America was being folded in late Jurassic time, the earth-shell was also invaded by deep-seated igneous rocks (granodiorite) on a large scale. Magmas in great volume were intruded, forming the great chain of bathyliths now exposed by erosion from Lower California to the Alaskan peninsula. In comparison with this intrusion, Lindgren states that all post-Proterozoic igneous phenomena fade into insignificance. The bathylith of the Sierra Nevada is 400 miles long, with a maximum width of 80 miles. On the International Boundary there are twelve bathyliths with a width of 350 miles. Farther north appears the Coast Range bathylith, according to Le Roy probably the greatest single intrusive mass known, which extends unbroken for 1,100 miles into the southern Yukon country, with a width of from 30 to 120 miles.

SHASTAN TIME

(See Figure 5)

Into the newly made and subsiding trough of the Californian Sea the Pacific Ocean spread, while in British Columbia and Alaska the same waters gradually encroached more and more widely either as a shelf sea or, more probably, another trough—the Columbian trough. The sediments poured into these seas were coarse-grained and were delivered to them by the rivers flowing out of the highlands apparently in the main to the eastward.

The deposits are essentially sandy shales with thin bands of sandstone, local conglomerates, and rarely thin limestones. The thickness in northern California appears to be between 9,000 and 10,000 feet, of which about one-third is of Knoxville time, while the remainder is of Horsetown time.

The Shastan series of Gabb and Whitney (1869) is also known in northern Washington and along the Canadian and Alaskan coasts. The deposits are dominantly sandstones with sandy shales, and in most places include from a few hundred to 3,350 feet of lavas, tuffs, and ash beds. In the Queen Charlotte Islands, where these strata have coal beds, the depth is estimated at 9,500 feet, and elsewhere, although somewhat less, the thicknesses are rarely as low as 2,000 feet.

The sands and muds of the Shastan series in most places overlap unconformably the older and often metamorphosed formations. This unconformity is sometimes marked, as in the Klamath Mountains and the Coast Range, or is of the erosional type. However, there are also disconformable contacts. The faunas, as pointed out by Stanton, are of the Indo-Pacific realm and are remarkably distinct throughout from those

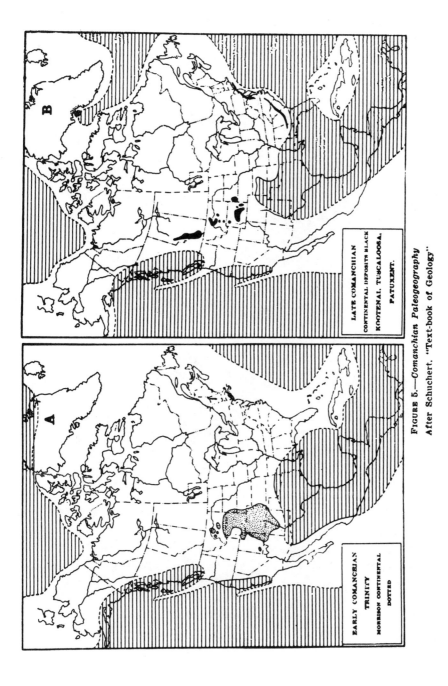

LATE COMANCHIAN
CONTINENTAL DEPOSIT BLACK
KOOTENAI, TUSCALOOSA,
PATUXENT.

EARLY COMANCHIAN
TRINITY
MORRISON CONTINENTAL
DOTTED

FIGURE 5.—*Comanchian Paleogeography*
After Schuchert, "Text-book of Geology"

of the Comanchian seas, which are of the Atlantic-Tethyian realm, a condition indicating that the two provinces were more or less, though not completely, separated from one another by a land barrier, the Mexican peninsula.

Along the Pacific coast from San Luis Obispo County, California, northward far into Oregon there is evidence of crustal movement during Shastan time. Anderson states that the Knoxville is everywhere in this

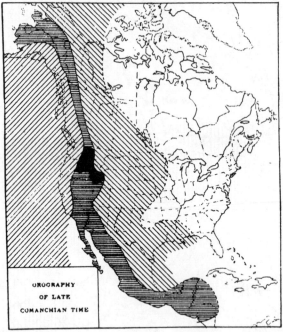

FIGURE 6.—*Map showing Regions of Elevation (horizontal Shading and solid Black), the Formation of the Coloradoan Geosyncline (right-hand oblique Lines), and the Pacific Overlap.*

Generalized from Ransome, "Problems of American Geology." The black area is that of the present Columbia River Plateau; the region of elevation to the north is known as the Northern Interior Plateaus, while that on the south is the Nevadan-Sonoran region.

area penetrated and disturbed by dikes and masses of serpentine and peridotites. Moreover, in the Coast Range of southern California, where these intruded rocks occur, the Horsetown strata are also absent, while the Chico of Cretaceous time unconformably overlies the older formations.

In western Utah, eastern Nevada, and throughout Idaho uplift was making itself felt as early as Middle Triassic time, forming the arch of the Columbia River Plateau, which persisted into late Jurassic time (the

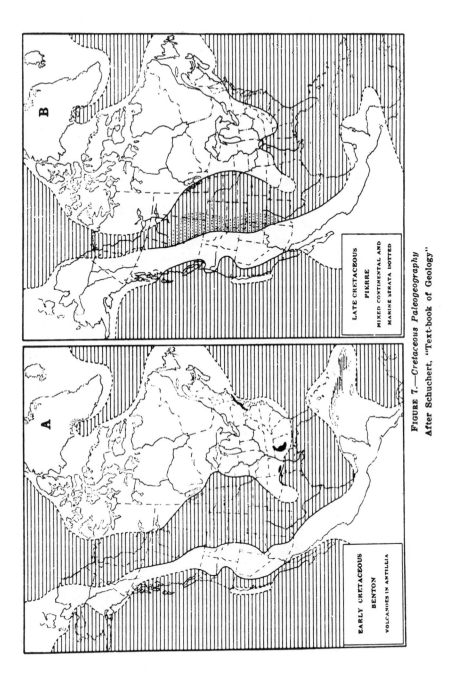

FIGURE 7.—*Cretaceous Paleogeography*
After Schuchert, "Text-book of Geology"

black area in figure 6). At the close of the Jurassic, however, the Sierra Nevada folds were thrown up to the southeast of this arch, and it was then incorporated into the area of the Sierra Nevada Disturbance. We have seen that this late Jurassic movement also forever shut out the former wide extension of the Pacific, not only from the area of the United States, but from most of western North America as well.

When we study the paleogeography of Cretaceous time, it soon becomes apparent that the conditions of oceanic spreading had been further altered toward the close of Shastan time, for subsequent to this period the Pacific overlaps were narrow (oblique lines to left in figure 6), and over what is now the site of the Rocky Mountains, and far to the east in both Canada and the United States, a new inland sea appeared—the Coloradoan Sea of Cretaceous time—extending from the Gulf of Mexico into the Arctic Ocean (oblique lines to right in figure 6). The barrier that kept these waters apart was the newly bowed-up land to the west of the present Rocky Mountains, the Cordilleran Intermontane Belt of Ransome, and this barrier continued to rise throughout all of Cretaceous and much of Tertiary time (horizontal lines in figure 6). We see here, therefore, the beginning of the process which made the Cordilleran Intermontane Belt of elevated plateaus, extending from Arctic Alaska all the way into Central America, and it is for this reason that the movement may be called the *Cordilleran Intermontane Disturbance.* Blackwelder has recently named it the Oregonian deformation, but this term is of altogether too local significance.

CRETACEOUS OR CHICO TIME

(See Figure 7)

The Chico series of sandstones and shales, with local conglomerates and coal beds, usually overlies the Shastan formations unconformably. These coarse deposits and thick formations are found all the way from the Lower Yukon, the Alaskan peninsula (1,000 feet thick), Queen Charlotte (11,000 feet) and Vancouver (5,000 feet) islands, middle and southern Oregon (4,000 feet), the Sacramento Valley (9,500 feet), and the Coast Range of California, to San Diego and the peninsula of Lower California. Stanton states that the Chico series begins somewhat earlier in time and continues longer than the Colorado series of the great Interior Sea, but does not embrace strata of youngest Cretaceous time. The Chico faunas are of the Indo-Pacific province and are markedly different from those of the Coloradoan and Mexican seas. The two provinces were separated from one another by the rising Rocky Mountain barrier.

The Cretaceous period and the Mesozoic era were closed by the Laramide Revolution. The area of folding and thrusting embraced the mountains of western North America, the Antillian region, and the Andes of South America—two grand mountain chains that extended from Cape Horn to Panama and from southern Mexico far into Alaska. Volcanic activity began late in Cretaceous time and continued into the Eocene, and the volcanoes extended from Mexico City and Arizona north far into Canada.

In conclusion, let me add that I have pointed out only the broader events and the general paleogeography. The more interesting detail and the actual ancient geography remain as the work of the rising generation of west coast geologists and paleontologists.

19

Reprinted from *Ecol. Monogr.*, **17**(2), 141–148 (1947), with permission of the publisher, Duke University Press, Durham, N.C.

TERTIARY CENTERS AND MIGRATION ROUTES

R. W. Chaney

INTRODUCTION

The rocks of the Tertiary system provide the most reliable data now available for initiating a discussion of early centers of distribution of land plants of modern types. The stratigraphic sequence of its continental deposits is more complete than in older rocks, and its fossils more readily assignable to modern families and genera than those of older floras. The ultimate origin of angiosperms is not clearly indicated by the Mesozoic record; much collecting by paleobotanists, followed by broad investigations both by paleobotanists and botanists, must precede our understanding of the nature and relationships of the earliest angiosperms, and our designation of the area or areas from which they have spread out to colonize the earth. By contrast, the composition and distribution of Cenozoic floras are already well known, particularly in the northern hemisphere.

In this paper I shall review the distribution of Tertiary vegetation in North America and elsewhere. Particular emphasis will be placed on the larger and longer-lived units of vegetation (Floras), as distinct from the more local and transient units (floras). It is these larger units, such as the Arcto-Tertiary Flora, which have migrated for thousands of miles, over a span of time to be measured in tens of millions of years. During such movements in space and time they have been altered in many details, but they have retained their essential elements and components to a marked degree. This conservatism makes possible our recognition of the Arcto-Tertiary composition of forests growing at middle latitudes in the northern hemisphere today; it provides a basis for projecting back for some 60 million years, northward for more than a thousand miles, the temperate climates now found in the United States; for the Arcto-Tertiary Flora of the Eocene epoch lived in Alaska and Greenland during a milder climatic regime than characterizes the high latitudes in our day. Similar projection backward may be made for the Neotropical-Tertiary Flora of middle latitudes; the modern forests of Mexico and Central America are its direct lineal descendents, and enable us to reconstruct an early Tertiary environment characterized by higher temperature and precipitation than is found today in Oregon and California.

Interpretation of the significance of shifting forest distribution during later geologic time must depend not upon a single flora from a limited area, nor upon the evidence of a few plants, no matter how suggestive their present habits may be. The factors which brought such profound changes must have been world-wide in operation, and must have affected whole plant populations. I shall therefore discuss the composition, changes in distribution, and inter-relations of the better known major units of vegetation of the Tertiary period. These units are the Neotropical Tertiary Flora of middle and low latitudes in the Americas, the Paleotropical-Tertiary Flora from the same latitudes in the Old World, the Arcto-Tertiary Flora of high northern latitudes, the Antarcto-Tertiary Flora of high southern latitudes, and the Madro-Tertiary Flora of the Southwest.

NEOTROPICAL-TERTIARY FLORA

PRELIMINARY REMARKS

Best known from older Tertiary deposits of the United States, this Flora is largely represented by leaf impressions of angiosperms. At the type locality of the Clarno flora in Oregon and in the Brandon lignite of Vermont, there are abundant fruits some of which show internal structure; a few stems have been studied, but relatively little is known regarding organs other than leaves.

The leaf characters of these Eocene and early Oligocene fossils suggest relationships to plants now living in the tropics. Leaves are of relatively large size, with thick texture suggesting an evergreen habit; a majority have entire margins, and species with elongate tips are numerous. Table 1 gives the percentages of these characters shown by the leaves of several Tertiary floras, with corresponding data from modern floras.

TABLE 1. Leaf characters of west American dicotyledons, showing percentages.

	Length		Margin		Dripping Point		Texture	
	Over 10 cm.	Under 10 cm.	Entire	Non-entire	Present	Absent	Thick	Thin
Panama, Recent Tropical (41 species)	56	44	88	12	76	24	98	2
Comstock, Upper Eocene (25 species)	56	44	76	24	60	40	92	8
Goshen, E-Oligocene (49 species)	53	47	61	39	47	53	98	2
Weaverville, Lower Oligocene (36 species)	60	40	47	53	49	51	57	43
Bridge Creek, Upper Oligocene (20 species)	30	70	15	85	10	90	54	46
California, Recent Temperate (22 species)	27	73	23	77	9	91	64	36

Closer resemblance of older Tertiary plants to those now living in the tropics, rather than to those of temperate America, seems readily apparent from these and other available data on characters of fossil leaves.

A second basis for establishing the nature of past floras and their environments is to determine the systematic relationships of plant fossils. Comparisons with leaves, fruits and stems of living plants have led to assignment of most older Tertiary plants from middle latitudes to genera and families now characteristic of forests living nearer the equator. Some of these are now wholly confined to low latitudes; examples are the Burseraceae (Canarium), Combretaceae (Terminalia), Dilleniaceae (Actinidia, Davilla, Dillenia, Tetracera), Meliaceae (Carapa, Cedrela), Monimiaceae (Siparuna) and Sabiaceae (Meliosma). Perhaps the most characteristic Eocene family, the Lauraceae (Acrodiclidium, Cinnamomum, Cryptocarya, Laurus, Lindera, Machilus, Nectandra, Neolitsea, Ocotea, Persea) is represented in the United States today by only a few genera, mostly shrubs or small trees which are largely confined to southern Florida; the only typically temperate members of the family are Lindera and Sassafras of the eastern United States, and Umbellularia in the West; with the exception of the last two, no arborescent members of the Lauraceae may be said to be common or widely distributed in temperate North America at the present time.

A considerable number of vines and shrubs in living forests at temperate latitudes may be interpreted as survivors of families whose Tertiary distribution extended farther north. The Araliaceae is represented today by Aralia and Echinopanax, both shrubs; Anonaceae by Asimina, a shrub; Bignoniaceae by Bignonia, a vine; Celastraceae by Celastrus and Euonymus, shrubs; Sterculiaceae by Fremontia, a shrub. All of these families were represented by additional genera during older Tertiary time, and the fossil record indicates an arborescent habit for the majority of their members. Other families such as the Boraginaceae, Euphorbiaceae and Leguminosae, whose habit was tree-like during the Eocene, are for the most part herbaceous in middle latitudes at the present time; their Eocene members, now largely tropical in distribution, were Cordia, Ehretia (Boraginaceae); Acalypha, Alchornea, Aporosa, Drypetes, Mallotus, Sapium (Euphorbiaceae); Acacia, Caesalpinia, Cassia, Canavalia, Dalbergia, Desmodium, Inga, Lonchocarpus, Pithecolobium, Pongamia, Sophora, Strongylodon, Vouapa (Leguminosae). While many Neotropical-Tertiary floras include typically temperate genera such as Diospyros, Liquidambar, Platanus, Quercus and Viburnum, all of these commonly range into the tropics today, occurring largely on mountain slopes. The composition of these older floras of middle latitudes is strongly indicative of a climate now found some twenty degrees of latitude to the south.

To botanists who are skeptical regarding the soundness of generic determinations based solely on fossil leaves—and their skepticism may reflect inadequate knowledge of leaf characters of modern plants—we may point out that even though the above mentioned generic and family assignments are incorrect, there is an independent basis for our conclusion that older Tertiary climate was subtropical in regions of North America now temperate. For the size, margin, shape and texture of the fossil leaves indicate absence of frost and high precipitation at middle latitudes, as above suggested.

The more important examples of the Neotropical-Tertiary Flora in North America are:

Wilcox flora, from the Lower Eocene of the Mississippi embayment in the southern United States (Berry 1916a, 1930, 1937, Ball 1931).—180 genera have been recognized, most of which appear to be valid; over-multiplication of species by Berry and others has brought the total to between five and six hundred, of which probably not more than half represent distinguishable and recognizable units of vegetation. "The flora is largely coastal and indicates a warm temperate climate and an abundant rainfall, more tropical in its facies than that of the late Upper Cretaceous which preceded it in this same region" (Berry 1937, p. 35). Evidence has been presented (Berry 1930, pp. 25-29, Berry 1937, p. 36) which indicates that the vegetation of the eastern side of the embayment was derived largely from the Antillean area, while that on the western side shows greater affinities with the living floras of Mexico and Central America.

Claiborne flora, from the Middle Eocene of the Mississippi embayment, and the *Jackson flora*, from the Upper Eocene of the same region (Berry 1924, 1937).—These floras, smaller than the Wilcox, indicate progressively warmer climate in a coastal habitat, and a similar origin to the south.

Brandon flora, from the Eocene of Vermont.—Berry (1919) has presented evidence that this assemblage, represented by fruits, seeds and wood, lived on the borders of an inland basin at about the same time as the Jackson, and under similar, though perhaps more temperate, physical conditions.

Catahoula flora, from the Oligocene of the southern states (Berry 1916d).—24 species assigned to 15 genera make up this relatively small assemblage, which Berry believes to indicate a continuation of subtropical strand conditions.

Raton, Denver, and Fort Union floras from the Lower Eocene (Paleocene) of the Rocky Mountain area (Knowlton 1917, 1930; Newberry 1898).—These interior floras, of approximately the same age, require restudy employing present-day methods before they can be adequately interpreted.[1] Preliminary survey by the writer (Clements & Chaney 1936, pp. 8-10) has indicated a marked decrease in subtropical representation from the Raton in New Mexico to the Fort Union in Montana. Presence of numerous temperate genera in the Fort Union indicate that it bears an ecotonal relationship between the Arcto-Tertiary Flora of western Canada and Alaska, and the more typical Neotropical-Tertiary Flora to the south.

Green River flora from the Middle Eocene of the Rocky Mountain area.—Brown's revision (1934) of this flora, together with the pollen studies of Wodehouse (Wodehouse 1933), indicates that the forest which lived near an inland basin of deposition, at an elevation of 3,000 feet or less, was of a warm-temperate type with local evidences of aridity. Several

---

[1] Study of these floras by R. W. Brown is nearing completion.

floras of possible Upper Eocene age have been collected from adjacent regions, but as yet there has been no mature consideration of this material.

Chalk Bluffs flora from the Middle Eocene of California.—MacGinitie's recent study (MacGinitie 1941) indicates that most of this flora represents a lowland-valley forest of subtropical type. Topographic diversity is suggested by the occurrence in the fossil record of upland-temperate genera. Small florules from beds of the same age to the west are interpreted as representing strand vegetation.

Clarno flora of Upper Eocene age from the John Day Basin of Oregon (Knowlton 1902), together with the *Comstock* (Sanborn 1935) and other less well known floras from western Oregon.—These floras show the occurrence of an essentially uniform forest across the state, indicating that marine control of climate was not limited during Upper Eocene time, as now, by the Cascade Range. The vegetation was subtropical in aspect, with several florules of more temperate aspect suggesting upland habitats.

Steel's Crossing and other floras commonly referred to as the *Puget*, from the Upper Eocene of western Washington (Newberry 1898).—No comprehensive study of these important units has ever been made, but large collections at hand indicate that the Steel's Crossing flora was of a subtropical character. Somewhat smaller average leaf size suggests less favorable living conditions than in Oregon and California. Material from the Bellingham Bay region appears to be older, and to represent more typically subtropical vegetation. Several floras from eastern Washington, such as the Roslyn and Swauk, seem also to be of Eocene age; like those of eastern Oregon, they indicate the absence of any climatic barrier such as the Cascades now afford.

Goshen flora from west central Oregon, occurring in beds believed to be Lower Oligocene on the basis of invertebrates, but closely related to Eocene of adjacent areas (Chaney & Sanborn 1933).—This subtropical flora appears to have lived in a valley near sea level, and to have derived most of its members from Central America and adjacent areas. The *LaPorte flora* (Potbury 1935) from northern California represents a similar forest, perhaps living at a higher elevation.

Weaverville flora from northwestern California (MacGinitie 1941).—Somewhat younger than the Goshen, this forest appears to have lived at low elevations along swampy streams, in a humid, warm temperate climate. Mingling of temperate with tropical genera suggests an ecotonal relationship with the southward-moving Arcto-Tertiary flora.

FACTORS AND EVENTS

Gradual displacement of the temperate to warm temperate Cretaceous floras of the United States by the Neotropical-Tertiary Flora during Eocene time has been considered in detail by Berry for the eastern United States (Berry 1930, pp. 15-16). Such temperate genera as Acer, Betula, Cornus, Fagus, Liriodendron, Pinus, Salix, Sassafras and Sequoia were present in Cretaceous time in the eastern United States, and most of them ranged northward into Greenland and Alaska; all of them had disappeared from the area occupied by the Wilcox flora by Eocene time when the climate became warmer, and were largely restricted to high northern latitudes[2] as members of the Arcto-Tertiary Flora. In the interior, at the southern end of the Rocky Mountain area, a similar but less pronounced change in climate may be noted, with the incoming of many new subtropical species and an accompanying northward migration of temperate types. Such Upper Cretaceous genera of the Vermejo flora as Liriodendron, Salix and Sequoia were eliminated by Eocene time, and the Raton flora which succeeded the Vermejo contains many more species of the Lauraceae, Leguminosae and Moraceae, represented by larger leaves of more tropical aspect. There is a less noticeable change from Upper Cretaceous to Eocene floras northward along the Rockies. Dorf's recent studies have shown (Dorf 1942, pp. 100-103) that the Laramie, Medicine Bow and Lance floras, in sequence from south to north, show a progressive change from subtropical to warm temperate; the Denver and Fort Union floras, which lived in the same region during Eocene time, are represented by similar climatic types. There is however another factor involved which seems indicative of rising temperatures in this region; whereas the Upper Cretaceous floras appear, from their association with marine invertebrates, to have occupied a lowland, coastal-plain environment, there was no adjacent ocean during Eocene time to moderate temperatures in the area now occupied by the Rocky Mountains. The persistence of warm temperate to subtropical vegetation into the Eocene, in spite of the less favorable geographic and topographic relations, is therefore suggestive of climatic changes like those in the Mississippi embayment, though less extreme. No floras from the Upper Cretaceous of the Pacific Coast states are available for comparison. From the Upper Cretaceous of Vancouver Island are recorded such temperate genera as Betula, Cornus, Juglans, Liriodendron, Populus, Salix, Sassafras and Sequoia, all of which ranged north into Alaska. None of them are known to have survived into the Eocene in the Puget Sound region, but all are common members of the Kenai floras of Eocene age to the north. There is indicated here, as in the case of the eastern floras, and to a lesser degree those of the interior, an early Tertiary movement northward of warm temperate to subtropical vegetation, and a concentration of temperate vegetation in high latitudes.

The causes of such wide-spread changes in climate cannot here be discussed beyond suggesting that, following the Cretaceous uplifts in the Caribbean region and along the Cordillera, there were submergences in central and northern South America, Central America, Mexico, the Mississippi embayment and the Antilles, with encroachments of the Pacific Ocean

[2] Or to higher altitudes in the eastern United States, though there is little possibility of plants being preserved as fossils in such situations.

eastward from its present shores. In Eurasia there were broad submergences in western Europe and southeastward along the Tethys trough, with an embayment extending northward along the Urals axis. Since increased circulation of ocean currents and of the atmosphere may be expected to have been general throughout the world, and to have extended mild and humid climate northward, it is appropriate to review briefly the evidence for corresponding forest migrations on other continents.

THE NEOTROPICAL-TERTIARY FLORA OF THE SOUTHERN HEMISPHERE

Available floras from low American latitudes indicate that tropical climates have prevailed there from Eocene time down to the present. An Eocene flora from Venezuela shows so close a resemblance to the Eocene of the Mississippi embayment as to suggest that the Wilcox and other floras of the southern United States had their origin in South America during the Cretaceous. Migrations in South America from low to middle latitudes are indicated by Eocene floras of Chile and Patagonia (Rio Pichileufu, Concepcion-Arauco), at latitudes ranging from 37 to 45 degrees. Like the Neotropical-Tertiary Flora of the northern hemisphere, this vegetation was made up largely of angiosperms now found nearer the equator; in addition to abundant Lauraceae, certain families now more characteristic of the southern hemisphere, the Apocynaceae, Myrtaceae and Rubiaceae, were well represented. Such conifers as Araucaria and Podocarpus give these floras an aspect as typically southern as the modern vegetation. There is no reliable evidence of the occurrence of Sequoia south of the equator.

PALEOTROPICAL-TERTIARY FLORA

In the submerged basin of western Europe, many subtropical floras are known from older Tertiary deposits. One of the best-known of these is the London Clay flora described by Reid and Chandler (1933) from well-preserved fruits and seeds occurring at 50 degrees north latitude. Among numerous angiosperms whose modern equivalents live in the tropical rain-forest of Indo-Malaysia, the record of *Nipa* is one of the most significant. Now confined to tidal flats at low latitudes, this palm has been recorded southeastward from England across Eurasia, along the shore of the Tethys Sea; it was probably along this shore that the London Clay and other floras migrated northward and westerly across India and the Near East to western Europe.[3] The Tethys Sea appears to have interposed a barrier to migration between the northern and southern hemispheres, resulting in Eocene and later differentiation of their vegetation. The Proteaceae and Myrtaceae are two typically southern families recorded largely south of the Tethys trough from Eocene down to modern times; as in South America, the conifers are typical of the southern hemisphere today, including such genera as Agathis, Dacrydium, Phyllocladus and Podocarpus,

[3] This is the Poltavian flora of Kryshtofovich (1929, p. 309).

as pointed out by Florin (1940). The Paleotropical-Tertiary Flora of the southern hemisphere is too inadequately known to provide clear evidence of southward migration routes in the Old World migrations in the fossil records of Africa, Australia and New Zealand.

SUMMARY OF TROPICAL-TERTIARY FLORAS

The older Tertiary record of these Floras indicates that they spread poleward from the tropics in the Americas and the Old World, reaching a latitude of approximately 50 degrees on the western sides of the northern continents; in the southern hemisphere their migrations are less adequately known, and appear to have been more limited. Gradual emergence, probably combined with changes in the solar constant and other factors not adequately understood, brought a reversal in the trend toward warmer and more humid climate during the Oligocene epoch. During the remainder of the Tertiary period and down to the present, these subtropical forests have been giving way to temperate members of the Arcto-Tertiary Flora in North America and Eurasia, and to those of the Antarcto-Tertiary Flora in the southern continents.

ARCTO-TERTIARY FLORA

This temperate assemblage has been so fully discussed in print during recent years that only a brief summary will here be included. Wide-spread over high northern latitudes during Eocene time, many of its genera are recorded also from the Cretaceous of Alaska, Greenland and Siberia. In fact the initial appearance of angiosperms in western Greenland, in beds referred to the Lower Cretaceous, has provided the principal evidence supporting the theory of Holarctic origin for flowering plants. The diversity of these earliest-known angiosperms suggests that they may have come into existence in pre-Cretaceous time; if their date of origin goes back to the Jurassic, they could have migrated north from lower latitudes, together with cycads, ferns and conifers which characterized the vegetation widely distributed from the Arctic to the Antarctic during this period.

Analysis of the Arcto-Tertiary Flora of the Eocene indicates that it may be subdivided into typically temperate and boreal units, the latter occurring in northern Siberia and on arctic islands to within eight degrees of the North Pole (Chaney 1940, p. 482-483). The Eocene existence of Sequoia, Taxodium, Pinus, Picea, Carex, Salix, Betula and Corylus in regions where the polar night lasts for several months has led to wide speculation regarding continental drift and the migration of poles, as well as to suggestions that the plant remains were carried northward by ocean currents. There appears to be no good reason why hardy ecotypes of these genera, many of which now range far north of the arctic circle, should not have lived in Grinnell Land at a time when winter temperatures were moderate; the existence of polar ice-caps during Eocene time is doubtful, and a wholly different regime of air and water circulation may be expected to have characterized high latitudes. Recent discovery by Wan-Chun Cheng of a living relative of Sequoia in central China, a tree which sheds its leafy twigs during the winter, provides a possible basis for explaining the occurrence of this now warm temperate plant in Grinnell Land. For without leaves, Sequoia and Taxodium might both be expected to survive a dark winter which was not cold. Occurrence of fossil remains resembling this living Chinese tree, assigned to the genus Metasequoia in a paper by Cheng and Hu not yet printed, has been reported from Eocene deposits of Greenland, Spitzbergen and Alaska, and probably in the Cretaceous of Greenland (Lesquereux 1878).

The typically temperate part of the Arcto-Tertiary Flora includes a long list of angiosperms and gymnosperms of which the more common and widely distributed are: Acer, Alnus, Betula, Carpinus, Castanea, Cercidiphyllum, Cornus, Crataegus, Fagus, Fraxinus, Ginkgo, Juglans, Pinus, Platanus, Populus, Quercus, Salix, Sequoia, Taxodium, Tilia, Ulmus and Zelkova (Hollick 1936). All of these genera except Carpinus, Castanea, Cornus, Crataegus, Fagus, Fraxinus, Salix, Zelkova had become established at high latitudes during later Cretaceous time, suggesting that the Arcto-Tertiary Flora was largely derived from an Arcto-Cretaceous Flora. Similarity in composition of the Arcto-Tertiary Flora from Alaska to Greenland, Spitzbergen and Siberia confirms geologic evidence of more extensive and continuous land masses around the North Pole.

On the southern margin of the area occupied by this temperate forest, which was largely made up of deciduous trees, there appear other genera whose modern members live under warmer conditions. A Kenai florule from Kupreanof Island, at latitude 57 degrees north in southeastern Alaska, includes such temperate genera as Castanea, Juglans, Populus, Quercus, Sequoia and Ulmus, together with several Neotropical-Tertiary genera, Ceratozamia, Dillenia, Dioon, Laurus,[4] Magnolia, Malapoenna, Pterospermites. Older Tertiary floras from southern British Columbia, Alberta and Saskatchewan, at latitude 51 degrees north, contain warm temperate to subtropical genera such as Aralia, Catalpa, Cinnamomum, Ficus, Laurophyllum, Magnolia and Pterospermites, although their dominant genera are typically temperate members of the Arcto-Tertiary Flora. Such an ecotonal relationship with the Neotropical-Tertiary Flora to the south has already been pointed out in the case of the Fort Union flora, which includes numerous temperate genera of the Arcto-Tertiary Flora. Similar ecotones may be noted in older Tertiary floras of central Europe at latitudes 48-50 degrees north, and of Manchuria at latitude 42 degrees north.

With the trend toward a cooler and less humid climate, which became pronounced at the end of the Oligocene epoch, the Arcto-Tertiary Flora migrated southward from Alaska. Reference may be made to the possibility, already mentioned, that some members of the temperate forest of the Cretaceous low-

[4] Probably Persea.

lands may have shifted up-slope and survived at higher altitudes in the United States during the Eocene. In that event they would have been in a position to have migrated down-slope during middle Tertiary time with the return of temperate climate to the lowlands. Floras of later Tertiary age in Oregon and elsewhere at middle latitudes disclose the elimination of various species and even genera during this journey, which occupied ten million years or more; other genera were added, changing the composition of the flora as known from the Eocene of Alaska. But in its general aspect, the Bridge Creek flora from the John Day Basin of Oregon (Chaney 1925), and the Miocene floras which came after it in the western United States, show essentially the same type of vegetation as the Alaska Eocene, requiring a summer-wet climate, a moderate annual range in temperature, and a valley environment. In some of the larger valleys extending in from the coast, such floras as the Latah (Knowlton 1926) and Mascall (Chaney 1925) maintained a minor relic element of the Neotropical-Tertiary Flora, including such genera as Cedrela, Gordonia, Laurus, Machilus, Magnolia, Oreopanax, and Tetracera. Along the coastal plain at the mouths of these valleys, Sabal, Catalpa, Ocotea and Persea appear to have been common, with temperate trees such as Alnus, Platanus and Sequoia largely restricted to adjacent hills. Largely as a result of the removal of continental deposits during Pleistocene glaciation, the later Tertiary floras of eastern North America are incompletely known. Small floras from New Jersey to Maryland (Berry 1916b) indicate the Miocene occurrence of such Arcto-Tertiary genera as Carpinus, Platanus, Pinus, Quercus, Salix and Ulmus, with Ficus and other Neotropical-Tertiary leaves in a marked minority. The Miocene Shanwang flora of northern China (Hu & Chaney 1938-1940), and many of the later Tertiary floras of central and western Europe, show mixtures of their dominant Arcto-Tertiary constituents with residual genera from the retreating Neotropical Flora. This sort of mingling continued even into the Pliocene in Europe, where broad submergence was maintained nearly to the end of the Tertiary period.

To some extent in Europe, and to a marked degree in most other parts of the northern hemisphere, the cool, dry climate of the Pliocene brought further shifting of the Arcto-Tertiary Flora coastward and southward. A regime of dry summer climate in western North America resulted in the disappearance of many broad-leafed, deciduous genera such as Carpinus, Fagus, Tilia, and later, Ulmus; all of these have survived in the eastern United States where rain falls during the summer. The redwood was restricted to western Oregon by the rising Cascade Range, which in the Pliocene, as now, cast its rain shadow over the John Day Basin and other regions in eastern Oregon (Chaney 1944). Along the coast of central California, the Upper Pliocene Sonoma flora is largely made up of Arcto-Tertiary species (Axelrod 1944), but a short distance into the interior these give way to the Madro-Tertiary Flora as discussed below. From causes not readily defined, but probably having to do with limited summer rainfall on the Pacific Coast, Cercidiphyllum, Ginkgo, Trapa, Zelkova and other genera were eliminated from North America altogether by the close of the epoch, although they have survived at corresponding latitudes on the western side of the Pacific. A small flora from the Citronelle formation of the Gulf states (Berry 1916e) closely resembles the existing forest of the coastal plain, with only limited representation of Asiatic and tropical forms. By the end of the Tertiary period the generalized forest which migrated south from Alaska had taken on modern regional characters in western and eastern North America; the Neotropical-Tertiary Flora had been restricted to latitudes some 20 degrees south of its Eocene occurrence. Similar southward migrations are indicated by the later Tertiary distribution of Arcto- and Neotropical-Tertiary Floras in Eurasia (Hu & Chaney 1938-40, pp. 103-117).

THE ANTARCTO-TERTIARY FLORA

Corresponding equatorward movements of the vegetation of the southern hemisphere may be noted briefly. Uplift of the Andes to their present height during the Pliocene and Pleistocene epochs, together with the general changes in climate noted to the north, appear to have been responsible for the shifting northward of the Neotropical-Tertiary Flora from Chile and Patagonia to Brazil, and for migration of the Antarcto-Tertiary Flora from high to middle southern latitudes. In his excellent paper on fossil conifers of south Chile (Florin 1940), Florin has summarized the records of temperate plants, angiosperms as well as conifers, at high southern latitudes during the Eocene; his opinion that the vegetation of the southern and northern hemispheres developed independently throughout the Tertiary period seems wholly confirmed by the evidence he presents. Such conifers as Acmopyle, Araucaria and Podocarpus, angiosperms as Knightia, Laurelia and Nothofagus, appear to have been largely or wholly southern in their Tertiary distribution as they are today. Their migration northward from Antarctica and the southern tip of South America, where they are recorded in Eocene rocks, to their present area of occurrence in south temperate latitudes, and in the case of the conifers into the tropics, appears to correspond closely to the southward movement of the Arcto-Tertiary Flora which occurred at the same time.

THE MADRO-TERTIARY FLORA

In our discussion of Pliocene floras, reference has been made to the wide extent of plants whose affinities are with those now living in dry regions. Axelrod has suggested that many of these had their center of origin in the Sierra Madre of northern Mexico, and that during the Tertiary they migrated northward along arid ranges, spreading out to the east and west in response to the trend toward reduced precipitation in the continental interior. It is possible that

vegetation at middle latitudes in the United States may have received increments of this Madro-Tertiary Flora as early as Eocene time, for the Green River flora contains several plants, including the oldest known cactus, Eopuntia, which suggest a dry southern source. During the Oligocene epoch a northward movement from northern Mexico is recorded by the presence of ten or more Sierra Madrean species in the Florissant flora of Colorado, a flora otherwise Arcto-Tertiary in composition. By Middle Miocene time, the Madro-Tertiary Flora was well established in the interior of southern California; here the Tehachapi (Axelrod 1939) and Mint Canyon (Axelrod 1940) floras show a strong development of the Southwest American Element, with small-leafed oaks and numerous other genera which make up an important part of the existing chaparral formation. Axelrod's maps of later Tertiary distribution of the Madro-Tertiary Flora (Axelrod 1939, p. 59) indicate Miocene extensions of a few of these southern forms into Washington and Idaho before the close of the epoch. At the end of the Miocene, when the East American and East Asian Elements were losing many of their members in the interior of western North America, apparently as a result of reduced summer rainfall, these hardy, drought-resistant plants from northern Mexico were moving northeastward into Texas and Oklahoma (Chaney & Elias 1936), and were becoming established on the western slopes of the central Sierra Nevada (Condit 1944). By Middle Pliocene time the Madro-Tertiary was the dominant Flora of the lowlands nearly to the coast. The Mulholland flora is an example of the oak savanna vegetation in the Coast Ranges east of San Francisco Bay (Axelrod 1944); only a few Arcto-Tertiary species had survived in it. As already mentioned, the somewhat younger Sonoma flora from the adjacent coast is largely made up of Arcto-Tertiary genera, although it also includes a few species of Madro-Tertiary origin.

SUMMARY

The fossil record during the Tertiary period, representing approximately sixty million years of earth history preceding the period in which we live, indicates progressively more temperate types of vegetation at middle and high latitudes both in the northern and the southern hemisphere; there is no evidence of marked change at low latitudes. The following major units of vegetation in North America have been considered:

The Neotropical-Tertiary Flora ranged northward to latitude 49 degrees on the west side of the continent during Eocene time, and is recorded to progressively more southerly latitudes eastward across the continent, reaching only as far north as latitude 37 degrees in the Mississippi embayment of the southern United States. During the Oligocene and Miocene epochs it was restricted southward in western North America to central and southern California, leaving relicts to the north along the Oregon coast; by Pliocene time only a few remained north of the Mexican border. In the eastern United States, most of the warm temperate and subtropical species had been eliminated by Miocene time. The Neotropical-Tertiary Flora has survived in the Antilles, northern South America, Central America and Mexico; numerous relicts occur in southern Florida and sparsely to the north, where their habit is largely as shrubs and vines. Several subtropical families are similarly represented in the living vegetation of western North America. In the southern hemisphere there was a northward shift of the Neotropical-Tertiary Flora from middle to low latitudes during the Tertiary period, and corresponding equatorward movements are indicated by the Paleotropical-Tertiary Flora.

The Arcto-Tertiary Flora ranged northward to latitude 67 degrees in Alaska during Eocene time, with a boreal unit extending to within 8 degrees of the North Pole; its occurrence at high latitudes east of Great Bear River is unknown, for Tertiary deposits have been completely removed from most of Canada by Pleistocene glaciation. During the Oligocene epoch it is recorded from Washington and Oregon, and had reached central California by the end of the Miocene; in the Rocky Mountain area, this temperate Flora extended into Colorado in Oligocene time, and in the eastern United States it was well established as far south as Maryland during the Miocene. Occurring as a relatively uniform forest throughout much of the northern hemisphere as late as Miocene time, this Flora shows evidence of regional diversity during the Pliocene epoch. Various genera disappeared from western and eastern North America, largely as a result it may be supposed, of a changing regime of summer rainfall and of seasonal temperatures; other genera were eliminated from the whole of this continent, and have survived only in Eurasia. The Arcto-Tertiary Flora has survived in North America at middle latitudes in two main provinces, an eastern characterized by broadleafed, deciduous trees, and a western characterized by conifers, broad-leafed evergreens, and broad-leafed deciduous trees and shrubs. Many details concerning the history, distribution and composition of the living forests of the eastern province have been presented by Braun, while similar material for western vegetation has been assembled by Mason and Axelrod.

The Madro-Tertiary Flora ranged northward from northern Mexico during the Tertiary period whenever favorable climatic, and wherever favorable topographic, conditions existed. It reached its widest distribution in the western United States during the Pliocene epoch, and while it is now largely confined as a dominant to the southwestern interior and adjacent Mexico, it is still represented over a much larger area in the West.

The existence of broad areas of later Tertiary grasslands has long been predicated by vertebrate paleontologists on the basis of the structures of fossil mammals. The plant record has been singularly lacking in the remains of prairie grasses until the discovery of fossil grass hulls by Elias in 1931 (1942). Since that time numerous well preserved remains, largely of the tribes Stipeae and the Paniceae, have been collected by Elias in deposits of Miocene and Pliocene age on the High Plains. The expansion of grasslands during later Tertiary time was a natural consequence of the trend toward arid climate with widely ranging temperature, a trend indicated by changes in forest composition as above described. There is little direct evidence as to the place of origin of these prairie grasses; doubtless some of them developed at the north, migrating southward on the borders of the Arcto-Tertiary Flora; others may have come up from northern Mexico with the Madro-Tertiary Flora.

While the three major units of Tertiary vegetation have developed ecotones, extending both along latitudinal and altitudinal boundaries, from the Eocene down to the present in North America, there are evidences of progressive differentiation between them and the corresponding floristic units of other continents. Until the Pleistocene or late Tertiary break in the Bering land bridge, at least the northern part of the Arcto-Tertiary Flora was co-extensive from Eurasia to North America, and the forests of northeastern Asia still show many resemblances to those of our continent. For the more southerly members of the Flora, a climatic barrier has existed since Miocene time, and numerous minor differences in composition have developed between the two continents; one of the most striking of these is the absence from the Tertiary and modern floras of Eurasia of the black oaks (sub-genus Erythrobalanus) which first appeared abundantly in the Miocene record of the United States; by this time the climate in Alaska appears to have been too cold to permit migration of the black oaks across the Bering land bridge. The Arcto-Tertiary Flora shows almost nothing in common with the Antarcto-Tertiary, from which it has been separated at least since Cretaceous time. The Neotropical-Tertiary Flora of the northern hemisphere shows little resemblance to the Paleotropical-Tertiary Flora of Eurasia, from which it must have been separated by a climatic, as well as topographic, barrier throughout Cenozoic time (Chaney 1940, pp. 481-486). The Neotropical-Tertiary Flora of North America also differs to a marked degree from that of South America, as does the Paleotropical-Tertiary Flora on either side of the equator in the Old World; the topographic barriers of the Caribbean and Tethys seaways appear to be responsible for these major differences in composition from the northern to the southern hemispheres during Tertiary time.

The distribution and composition of Tertiary floras in North America and elsewhere in the world make possible the following conclusions:

(1) Continental platforms and ocean basins have occupied essentially their present relative positions at least since the wide-spread emergence which brought the Cretaceous period to a close.

(2) During the Tertiary period and down to the present, in all parts of the world outside the tropics,

major modifications in floral composition at given latitudes have been controlled by climatic changes. Since Eocene time, a trend toward lower, more variable temperatures, and toward reduced and more seasonal precipitation, has provided the major incentive to forest migrations.

LITERATURE CITED

Axelrod, D. I. 1939. A Miocene Flora from the Western Border of the Mohave Desert. Carnegie Inst. Wash. Pub. 516: 1-129, pl. 1-12.
1940. Mint Canyon Flora of Southern California. Am. Jour. Sci. 238: 577-585.
1944. Pliocene Floras of California. Carnegie Inst. Wash. Pub. 553: 5, Mulholland Flora, 103-145, pl. 24-31.
1944. Pliocene Floras of California and Oregon. Carnegie Inst. Wash. Pub. 553: 7, Sonoma Flora: 167-206, pl. 34-39.

Ball, O. M. 1931. Contr. to Eocene of Texas. Tex. Agr. and Mech. Col. Bul., 4th ser., 2, 5: 1-137, pl. 1-48.

Berry, E. W. 1916. Lower Eocene Floras of Southeastern North America. U. S. Geol. Survey Prof. Paper 91: 21-353, pl. 9-117.
1916. Physical Conditions Indicated by the Flora of the Calvert Formation. U. S. Geol. Survey Prof. Paper 98-F: 61-70, pl. 11, 12.
1916. Flora of the Citronelle Formation. U. S. Geol. Survey Prof. Paper 98-L:193-204, pl. 44-47.
1916. Flora of the Catahoula Sandstone. U. S. Geol. Survey Prof. Paper 98-M: 227-243, pl. 55-60.
1919. Age of Brandon Lignite and Flora. Amer. Jour. Sci., 4th ser., 47: 211-216.
1924. Middle and Upper Eocene Floras of Southeastern North America. U. S. Geol. Survey Prof. Paper 92: 1-199, pl. 38-65.
1930. Revision of Lower Eocene Wilcox Flora. U. S. Geol. Survey Prof. Paper 156: 1-196, pl. 1-50.
1937. Tertiary Floras of Eastern North America. Bot. Rev. 3: 31-46.

Brown, R. W. 1934. Recognizable Species of the Green River Formation. U. S. Geol. Survey Prof. Paper 185: 45-77, pl. 8-15.

Chaney, R. W. 1925. Comparative Study of the Bridge Creek Flora and the Modern Redwood Forest. Carnegie Inst. Wash. Pub. 349: 1-22, pl. 1-7.
1925. Mascall Flora—Its Distribution and Climatic Relation. Carnegie Inst. Wash. Pub. 349: 23-48.
1940. Tertiary Forests and Continental History. Geol. Soc. Amer. Bul. 51: 469-488.
1944. Pliocene Floras of California and Oregon. Carnegie Inst. Wash. Pub. 553: 11, Dalles Flora: 285-321, pl. 49-53; 12, Troutdale Flora: 323-351, pl. 54-64.

Chaney, R. W. & Elias, M. K. 1936. Late Tertiary Floras from the High Plains. Carnegie Inst. Wash. Pub. 476: 1: 1-72, pl. 1-7.

Chaney, R. W. & Sanborn, E. I. 1933. Goshen Flora of West Central Oregon. Carnegie Inst. Wash. Pub. 439: 1-103, pl. 1-40.

Clements, F. E. & Chaney, R. W. 1936. Environment and Life in the Great Plains. Carnegie Inst. Wash. Sup. Pub. 24.

Condit, C. 1944. Pliocene Floras of California. Carnegie Inst. Wash. Pub. 553: 2, Remington Hill Flora, 21-55, pl. 1-12; 3, Table Mountain Flora: 57-90, pl. 13-21.

Dorf, E. 1942. Upper Cretaceous Floras of the Rocky Mt. Region. Carnegie Inst. Wash. Pub. 508: 1-168, pl. 1-19, 1-17.

Elias, M. K. 1942. Tertiary Prairie Grasses and other Herbs from the High Plains. Geol. Soc. Amer. Spec. Paper 41: 1-176, pl. 1-16.

Florin, R. 1940. Tertiary Fossil Conifers of South Chile and their Phytogeographical Significance. Kungl. Svenska Vetensk. Handl., Tredje Ser. Band 19, 2: 1-106, pl. 1-6.

Hollick, A. 1936. Tertiary Floras of Alaska. U. S. Geol. Survey Prof. Paper 182: 1-185, pl. 1-122.

Hu, H. H. & Chaney, R. W. 1938-40. Miocene Flora from Shantung. Carnegie Inst. Wash. Pub. 507: 1-147, pl. 1-57.

Knowlton, F. H. 1902. Fossil Flora of the John Day Basin. U. S. Geol. Survey Bul. 204. [Includes discussion and systematic treatment of the Clarno flora. See other papers of more recent date for references to Clarno plants.]
1917. Fossil Floras of the Vermejo and Raton formations of Colorado and New Mexico. U. S. Geol. Survey Prof. Paper 101: 223-349, pl. 54-113.
1926. Flora of the Latah Formation. U. S. Geol. Survey Prof. Paper 140: 17-55, pl. 8-31.
1930. Flora of the Denver and associated formations of Colorado. U. S. Geol. Survey Prof. Paper 155: 1-142, pl. 1-59.

Kryshtofovich, A. N. 1929. Evolution of the Tertiary Flora in Asia. New Phytol. 28: 303-312.

Lesquereux, L. 1878. Tertiary Flora. U. S. Geol. Survey Terr. Rept. 6: pl. 7, fig. 13.

MacGinitie, H. D. 1941. Flora of Weaverville Beds of Trinity County, California. Carnegie Inst. Wash. Pub. 465, Pt. 3: 83-151, pl. 1-15.
1941. A Middle Eocene Flora from the Central Sierra Nevada. Carnegie Inst. Wash. Pub. 534: 1-167, pl. 1-47.

Newberry, J. S. 1898. Later Extinct Floras of North America. U. S. Geol. Survey Mon. 35. [Includes systematic treatment of Fort Union and Puget plants.]

Potbury, S. S. 1935. LaPorte Flora of Plumas County, California. Carnegie Inst. Wash. Pub. 465, Pt. 2: 29-81, pl. 1-19.

Reid, E. M. & Chandler, M. E. J. 1933. London Clay Flora. Brit. Mus. Nat. Hist.: 1-561, pl. 1-33.

Sanborn, E. I. 1935. Comstock Flora of West Central Oregon. Carnegie Inst. Wash. Pub. 465, Pt. 1: 1-28, pl. 1-10.

Wodehouse, R. P. 1933. Tertiary Pollen, II, Oil Shales of the Green River Formation. Bul. Torrey Bot. Club 60: 479-521.

20

Reprinted from *Rept. Hobart Meetings, Sect. M: Botany*, Aust. N.Z. Assoc.
Adv. Sci., 1949, pp. 142–149

THE PROBLEM OF SUB-ANTARCTIC PLANT DISTRIBUTION

PRESIDENTIAL ADDRESS

By

PROFESSOR H. D. GORDON B.SC., PH.D., VICTORIA UNIVERSITY COLLEGE, WELLINGTON, NEW ZEALAND

I have chosen to speak to you on the problem of sub-antarctic plant distribution partly because we are meeting in Tasmania, which is one of the areas concerned, partly, no doubt, because I have recently undertaken airborne and seaborne migrations between this state and New Zealand, and partly because I think the problem is an interesting one; but certainly not because I wish to pose as an authority on it or offer any complete solution. I will rather consider how far existing theories take us, and what further information is most needed.

The fact that a substantial number of genera and even species of plants are common to such widely separated areas as South America, New Zealand and south-east Australia including Tasmania, or to two of these three areas, is well known, and it will suffice to name a few examples, such as the genera *Nothofagus, Astelia, Abrotanella, Colobanthus, Gaultheria, Pernettya, Ourisia, Azorella, Acaena, Uncinia, Gunnera,* and the species *Oreomyrrhis andicola, Apium prostratum, Phyllachne colensoi, Oreobolus pumilio.* Some of these occur also on the widely scattered islands of the Southern Ocean—the Falklands, South Georgia, Marion Island, the Crozets, Kerguelen, Macquarie Island, the Auckland and the Campbell Islands. There is then a sub-antarctic element in the floras of the southern lands showing close relationship, in other words common origin, as between the different areas, and the problem of apparent migration between countries so widely separated by sea has exercised the ingenuity of botanists from Hooker onwards.

There have been two conflicting schools of thought, one believing in the possibility of trans-oceanic seed dispersal, and including Darwin, Wallace, Guppy, Oliver, Gibbs, Ridley and Setchell, the other claiming that only a greater continuity of land in the past could have permitted the required migration. In this camp we find Hooker, Skottsberg, Campbell and a majority of recent authors including Wulff, Cain and Good. Some have accepted both methods in complementary roles, as Oliver (22) who accepts the former northern land-connection of New Zealand to the Malayan region, and a considerable extension to the south, but opposes the idea of any complete land bridge to the Antarctic continent. There remains however, a complete cleavage between those who believe in the effectiveness of long-distance seed dispersal and those who do not.

The affirmative case has been built up with truly Darwinian thoroughness by Guppy (14, 15) and Ridley (23) and the monumental compilation by the latter author must strongly impress all who read it. Most convincing perhaps are the examples of tropical strand-plants dispersed by floating seeds or fruits in ocean currents such as *Ipomoea biloba, Hibiscus tiliaceus* and various mangroves (in which the floating propagules are seedlings). Ridley also deals at length with dispersal by wind, birds and other animals, floating logs, drifting pumice, and leaves one with the impression that it would be very rash to deny the possibility of at least occasional long distance transport for the great majority of species. But the fact remains that dispersal does not appear to occur by such means, that species unaided by man extend their ranges slowly by marginal spread or not at all, and that observation supports the opinions expressed by Good in his recent " Geography of the Flowering Plants ", when he maintains that specialised dispersal mechanisms are significant in scattering individuals and reducing competition between themselves and with the parent, but not in extending the range, and that even the smallest degree of dispersal is sufficient from the geographical point of view. The range of a species does not seem to be correlated at all with its powers of dispersal but with the range of the appropriate environment, and discontinuity of the range of a species is not correlated with any special powers of jumping the gap, but with discontinuity of the environment.

Perhaps the only exceptions to this rule of marginal spread are of three kinds:—(1) plants which are dispersed deliberately or accidentally by man and his works, (2) the strand-plants I mentioned a moment ago, though even those usually spread gradually, and (3) some plants of fresh water like *Lemna* and the plants of the pond margin, which are carried externally by swimming and wading birds. For the rest, short-distance migration is the rule, and so trans-oceanic migration would seem to be excluded, and continuous or nearly continuous land a necessity.

I think we can be fairly certain that this is true for the genus *Nothofagus*. Nature has laid down an interesting experiment in New Zealand, which I must admit I have not yet seen but am describing from a reliable source. In the centre of the North Island is a volcanic district in which successive showers of ash and volcanic debris, on a scale no longer matched there, have produced a light and porous soil, which bears tussock grassland in an area with a beech-forest climate; the rainfall is adequate for forest, but the sharp drainage prevents its development; the grassland looks like an edaphic climax. However it is really seral for succession to beech forest is proceeding, though very slowly. The beech litter modifies the soil along the forest margin, and in this zone new seedlings become established, but seeds which fall in the tussock beyond that zone are lost; they fail to establish in the grassland soil. The forest, then, makes its soil as it goes, and its migration is strictly marginal. In the South Island there are small outliers of beech forest in the tussock grassland far to the east of the main forest zone, but these are not the result of wind-borne seeds. Here the rainfall is lower, and grassland is the true climax at the present day. The evidence of the soil here is the converse of the other case and shows that the area was formerly beech forest which has changed to grassland—presumably as a result of decreasing rainfall. The forest outliers are remnants, not pioneers.

Beech forests then advance marginally, and so appear to be precluded from crossing wide seas. I am speaking advisedly of the migration of the forest and not simply of the genus, because wherever *Nothofagus* species occur they are forest dominants, except where stunted at high altitude. A few years ago, *Nothofagus* forests were discovered on the mountains of New Guinea by the Archbold Expedition (1), whose photographs bear a strong resemblance to wet mossy beech forests in Tasmania or New Zealand, and that the evergreen beech forests of South America are also very similar is evident from the graphic descriptions of Darwin, Skottsberg and others.

Again, it is well known that certain species of *Nothofagus* in South America, Australia and New Zealand are parasitised by species of the fungus genus *Cyttaria*, and that this genus has no other host than *Nothofagus* species. Its migration has therefore been parallel to that of its host genus, which would be quite easy in the case of slow overland migration but very difficult to imagine if long-distance dispersal were involved. A further piece of evidence was provided by Dr. J. W. Evans (9) formerly Government Entomologist here in Tasmania. This concerns the Peloridiidae, a family of primitive sucking bugs which inhabit, and feed on, permanently wet moss in *Nothofagus* forests in South America, New Zealand, and from Tasmania to Queensland. They are not absolutely restricted to *Nothofagus* forests—other wet forests will do—but over most of this area it is only the beech forests which fill this role. These insects are closely tied to their permanently wet moss, and cannot survive in dry air; moreover they are primitively flightless. It therefore seems quite impossible for them to have travelled throughout their range by air or sea, or in fact by any other means than in continuous moist forest—most probably *Nothofagus* forest.

Nothofagus, on its own, seems to have little chance of trans-oceanic dispersal; the beech nuts seem quite unfitted for any such purpose. How much less can we contemplate the simultaneous mass dispersal overseas of a complex made up of the tree, its parasitic fungus, its epiphytic or associated moss saturated with fresh water, and its dependent type of insect! The species in each case are different in the different regions, but that does not remove the problem as the corresponding species are related in descent, and must have diverged after geographical separation. We may claim with confidence that the migration route of the southern beech forests has been continuous land with a cool moist climate, though these conditions need not have applied to the whole area simultaneously.

Where then was this land? The most popular suggestion is that the antarctic continent was formerly more extensive, was continuous with similar extensions of the present continents, and was the home of those plants which we now call the sub-antarctic flora, or of their ancestors. These plants are supposed to have migrated northwards to all the attached continents before they became detached, and before the Pleistocene glaciation exterminated the flora of Antarctica itself. This theory accords very well with the facts of plant and animal distribution, and was greatly strengthened by the discovery of Cretaceous and Tertiary fossil floras in Graham Land and Seymour Island (16, 8)—northern parts of Antarctica—containing *Nothofagus*, *Drimys*, *Laurelia*, *Lomatia*, and other genera which today inhabit the southern continents and form part of the circumpolar sub-antarctic flora. This proved that before the Pleistocene Ice Age Antarctica enjoyed a temperate climate, and harboured some at least—quite probably all—of the plants whose distribution we are seeking to explain. But the chain of evidence is not complete. We know that *Nothofagus* and other plants entered Antarctica, and therefore presume that there was at least one land connection, but we do not know that the plants migrated out of Antarctica by another route. Of the three or more suggested radiating land-bridges, that between Cape Horn and Graham Land presents least difficulty, though even here there is deep water. In the other positions it is true that there are submarine platforms, where the sea is shallower than the abyssal depths on either side, but it is still so deep and wide that if these areas were ever land-bridges an immense subsidence has occurred. If they were land-bridges, our task becomes so much the easier, but it is necessary to face this difficulty and admit that the fossil floras of Antarctica do not prove so much; Antarctica may have been continuous only with South America, in which case it could not serve as the migration route of the *Nothofagus* forests. We claim only to have proved continuous land; could it have been anywhere else?

Oliver (22) suggested that *Nothofagus* and certain gymnosperms "appear to have arisen in North America and migrated along the western shore of the Pacific; hence their presence in Australia and New Zealand but absence from Africa." As to the last point I think they would be absent from Africa in any case because it offers no microthermal rain-forest climate, just as they are absent from West Australia although *Nothofagus* was once spread over what are now arid parts of the Australian continent. I think the route proposed by Oliver is unlikely, because it involves migration across the tropics, whereas *Nothofagus* grows only in cool or temperate climates and has probably been so limited ever since its independent differentiation (cf. 11, 12). Nevertheless Oliver's suggestion is a valuable one in that it draws attention to a possibility which may well apply to some of the more mesothermal elements common to the Valdivian forests of South America, the east coast tropical rain forest of Australia and the northern or sub-tropical rain forest of New Zealand.

Even for the microthermal elements of the circumpolar flora we cannot, I think, entirely deny this possibility that the elusive latitudinal migrations may have occurred in the north rather than in the south, followed by independent southern migration in the different continents and a delimitation of present range by climate. This theory has had its adherents (29), and its main attraction is that latitudinal continuity of land in the northern hemisphere is well attested and involves none of the difficulties presented by these widely separated southern continents. Indeed I think all plant geographers would accept it as self-evident that many genera and families of predominantly north-temperate distribution have first spread throughout the northern hemisphere and then extended their range to the southern continents along the two great trans-tropical mountain chains. In the case of genera whose present distribution is entirely southern we

would have to make the further postulate of extinction in the northern hemisphere, following the southward spread. Well, of course, there was a convenient glaciation which did undoubtedly decimate and even exterminate the floras of substantial parts of the northern hemisphere, especially in Europe where southward escape was checked by the Mediterranean; but the glaciation was not equally destructive all round the northern hemisphere and the most characteristic present-day distribution which we have learnt to attribute to this cause is a discontinuity between North America and Eastern Asia, as shown by the well known cases of *Liriodendron* and other Magnoliaceae, which are absent from present-day Europe though well represented in its pre-glacial fossil floras. Complete northern extinction of a genus after it had spread to the southern hemisphere is quite possible, but we could not, I am afraid, apply that explanation to the whole sub-antarctic flora without experiencing that guilty awareness of a strained hypothesis which is perhaps not unknown to many of us in this room. Sir Arthur Hill, in reviewing this problem at the International Congress of Plant Sciences in 1929 (18), pointed out that such migrations from northern to southern centres must have occurred at some time—one might recall the affinity of *Nothofagus* with *Fagus* and the other Fagaceae of the northern hemisphere—but he concluded, like most other students of the problem, that the facts of distribution point strongly to a radiation of many genera from southern centres, most probably in Antarctica itself.

Yet there is one good feature of the northern origin hypothesis which I would like to stress further, and that is the essential selective role of climate in determining contemporary plant distribution. You may feel that this is familiar enough, and to be sure some authors have given it due weight, but I have been struck by the number of papers on phytogeographical problems which deal with possible geographical relationships between points A, B, and C, or with the distribution of genera and species, without mentioning climate at all. For a particularly clear exposition of the close correlation of both vegetational and floristic distribution with climate, especially in the parts of the world with which we are concerned, I would refer you to Dr. Herbert's Presidential Address to this Section at Melbourne in 1935 on "The Climatic Sifting of Australian Vegetation" (17). Dr. Herbert's main theme was that the vegetation of each area is a mixture of types from such other areas *of similar climate* as have been able to contribute elements by migration; he was not discussing the means of migration. A synthesis of viewpoints leads us to appreciate the extraordinary success of plant migration. In general, plants do seem to have found their way into the environments (climatic and edaphic) that suits them, and obvious barriers seem to have been less effective than we might have expected. In considering this problem we must preserve our sense of geological time. The very slowest marginal dispersal will suffice to spread a species, genus or other group throughout the world, provided the climate of the world is sufficiently uniform to fall within the tolerance of the group (which it is not), and provided the land surface of the world is continuous (which, thank goodness, is far from being the case). Yet the migration of plant groups to discontinuous suitable environments has been so successful, and approximates so nearly to the hypothetical ideal where barriers do not count, except in the more recent phase of local differentiation and the evolution of endemism, that we had better examine further the two apparently extravagant provisos to see whether they might have been less absurd in the past than they are at present.

As to a relatively uniform world climate, there is no difficulty at all. I think there is complete agreement that throughout at least the major part of the Cretaceous and Tertiary there was no steep temperature gradient from the equator to the poles, such as prevails today. And early angiosperm floras were notably cosmopolitan. We have no reason to believe their powers of dispersal were greater than those of living species, so we must envisage their cosmopolitan distribution as having been attained slowly by marginal dispersal. As to continuity of land, the requirements would be met by the admitted linking of the continents in the northern hemisphere.

With these facts in mind I feel that some previous discussions of migration have not sufficiently stressed the fact that the circumpolar plants occupy areas of similar climate, that they grow where they do because of the climate, and that their present distribution may be a product of segregation from a more general distribution, which was attained slowly under a more uniform world climate, without the need to cross large tracts of the Southern Ocean.

I have dwelt at some length on this theory of migration from the north because it emphasises the importance of climate and because we should consider all possibilities; but I cannot say that it satisfies me. The reason I have already explained. There are too many southern genera whose distribution suggests spread from southern centres, and which probably evolved in the southern hemisphere and have never left it. In spite of that, some of them have become circumpolar.

We have considered two possible land routes between the southern continents—a northern, and a southern by way of Antarctica. The former I have just set aside; the latter I discussed earlier and concluded that it fits the facts of distribution, but involves land-bridges across comparatively deep ocean. Before we tell the geologists we have proved the existence of these land-bridges we had better be sure we have not overlooked any other possibility. Could the continuous land we seek have been anywhere else? Not in the north, not in the tropics, because our plants are characteristic of cold to temperate climates. If not in the Antarctic then there is only one zone left; we shall have to find it right here in the

colder temperate latitudes of the southern hemisphere, the same latitudes in which the sub-antarctic plants occur today. We are brought right back to the problem that confronted Hooker a century ago—the direct separation of these southern lands by vast stretches of ocean. And we are looking for continuous land! Can we find it? Was it ever here? Believe it or not, our most recent authors say " Yes " (2, 12, 32). They do not propose a super land-bridge, circling the globe in the roaring forties, but have adopted the theory of continental drift and see in that the answer to many problems of discontinuous distribution. This theory, as most of you well know, claims that the continents float in a denser layer of the earth's crust, that they were originally one single mass—or two, a northern and a southern, according to some authorities—and that this original continent has broken up to form the separate continents which have gradually drifted apart and are still so drifting. The recent botanical application of this theory maintains that the sub-antarctic flora is that which occupied the southern part of the parent continent, including the lands in which they grow today as well as that part which is now Antarctica.

Nothing could more effectively dispose of these troublesome sea barriers, but you will be aware that the theory of continental drift has been a most controversial one among geologists ever since it was presented by Wegener. Some botanists dash in where geologists fear to tread, but I have a certain north-temperate origin which makes for caution. I will not presume to express an opinion on the truth or falsity of the theory of continental drift; my opinion could have no value. The verdict must be based primarily on geological evidence which I am not competent to judge. But while waiting for the competent judges to reach agreement I may still offer one or two comments on the relation of this theory to plant distribution.

Oceans divide the world longitudinally; temperature divides it latitudinally. Both oceans and temperate belts are potential barriers to plant distribution, especially to plants of the temperature and colder latitudes. These plants have a discontinuous potential area consisting of two temperate zones, one in the northern and one in the southern hemisphere, each of which is further divided by seas, especially, of course, in the south. Now there are many genera and even whole families of flowering plants in temperate latitudes which fail to cross the tropics; they are confined to the north or the south. But there are not so many families which are confined to one continent and fail to span the oceans from east to west within their potential area. Last night Dr. Walkom (30) described in very similar terms the latitudinal distribution of the old *Glossopteris* flora and its contemporaries, which first gave rise to the Gondwanaland hypothesis. It looks as though the cause must have been a very persistent one, for the *Glossopteris* flora has its counterpart in the floras of the present day. Latitudinal spread is more general than longitudinal.

This is due to the latitudinal zonation of temperature—the primary factor controlling plant distribution—but its realisation implies that the tropics are a more effective barrier to plant migration than the Atlantic and Indian Oceans. This fact would be readily understandable if the majority of the families of flowering plants were established and distributed at a time when the Atlantic and Indian Oceans did not exist. The Pacific is regarded as the original "outside" ocean, on which the drifting continents are encroaching.

Mention of time brings us to a major difficulty. I understand from Professor Carey, who is an ardent exponent of continental drift, that he believes the disintegration and separation to have begun not later than the Jurassic, and I believe that is the period usually suggested, whereas the flowering plants did not come into being, or at least become common, until the Cretaceous. Good (12) on the other hand, argues strongly for continental drift as the only satisfactory explanation of Angiosperm distribution, and does not hesitate to claim that the disintegration must have occurred in the later Tertiary and may even have been related to the onset of the Pleistocene glaciation. This view seems inconsistent with the absence of mammals from New Zealand. That country is generally held to have been unattached since before the origin of mammals, i.e. since the Cretaceous. Perhaps New Zealand broke away first from the margin of the continent, in the late Cretaceous, when the Angiosperms were already well distributed, to be followed by Australia after the origin of the Monotremes and Marsupials, but before the appearance of placental mammals. I do not know what the concensus of geological opinion is, or if there is one, but if the continents drifted apart before the origin of the Angiosperms it was most inconsiderate of them, and we shall have to solve our problems without any help from this attractive hypothesis.

There is another, and perhaps less important, complication. Some geologists believe, as I mentioned, that there were two original continents, a northern and a southern, though their parts are now in contact. If these two were separate during the Cretaceous, how did the Angiosperms achieve their early cosmopolitan distribution? Perhaps the continents made contact early and split longitudinally much later, or perhaps there was only one after all. We need not go further into this question at present.

The validity of continental drift and its placing in geological time we must leave to those who are qualified to judge. If its help is denied us, we will be driven back on the other theory which fits our facts—that of an enlarged Antarctica continuous with the southern extremities of South America. Tasmania and New Zealand. But we shall have to submit the required land-bridges to the same judges. If the verdict is again unfavourable we should remember the possibility of the northern route, but that would seem to

require a cooler climate in the tropics whereas our data confirm the reality of the tropics as a barrier to just such plants as we are considering.

So far, I have not considered plant migration to islands, but I must, because some members of the sub-antarctic flora are found not only on the isolated islands of the Southern Ocean but even on the Pacific Islands as far as Hawaii, and moreover the colonisation of islands re-opens the whole question of overseas dispersal. There is an extensive literature on the floras and faunas of islands, including such classics as "Island Life" (31) in which Wallace distinguished clearly between continental and oceanic islands. Continental islands like Tasmania or the British Isles are borne on a continental shelf, which means that they are really part of the continent, though the adjacent land is flooded for the time being by a shallow epicontinental sea. These seas, such as Bass Strait, come and go in geological time, and are not permanent barriers to plant migration. So continental islands need not detain us, but the problems which centre round oceanic islands are, I believe, the most difficult and probably the most fundamental in the study of plant distribution. Oceanic islands may be built up of coral, or may be volcanic. They arise in the ocean at any distance from land; they are not borne on a continental shelf and have apparently never been connected to continental land. It therefore seems a truism that their plants and animals must have reached them by overseas dispersal and not by marginal spread. Indeed that has commonly been assumed as axiomatic, but a few authors, especially Skottsberg (25), maintain that the island floras show features that refute the obvious conclusion.

Skottsberg has been the foremost student of the Pacific island floras for many years, and his case against trans-oceanic migration is a strong one. In part it rests on the negative evidence of migrations that do not take place; for example the flora of Juan Fernandez has affinities with those of Polynesia and with the extreme south of South America (in fact with the sub-antarctic flora) but does not seem to exchange any appreciable number of species with the nearest part of Chile, from which it is 360 miles distant.

But the main argument relates to endemic species and genera. It has long been realised that isolated islands have a very high percentage of endemics—i.e. species peculiar to the island, and that this is related to their isolation. The principle is that once a species becomes established on such an island its evolution proceeds independently of the individuals in other areas, because interbreeding is prevented by isolation. But if the original colonisation was trans-oceanic why does migration not continue to bring mainland individuals to interbreed with the island population and prevent the divergence? It looks as though isolation has increased since colonisation. Skottsberg is not drawing far-reaching conclusions from isolated examples: he has been most thorough in his analysis of

Pacific island floras, and the facts relating to endemism are very consistent. In Juan Fernandez 69 per cent of the flowering plants are endemic; in Hawaii the figure is about 80 per cent and there is a marked development of species peculiar to a single island of the archipelago. This last feature is also general and important. Darwin (5) was greatly impressed, and puzzled, by the conspicuous local endemism in the Galapagos Islands; he described how certain types of plants and animals were represented by distinct species on each island—all very similar and as he realised later (6), obviously related in descent, but having developed characters of their own. Later observers found the same thing in other archipelagos—in fact it is now recognised as characteristic of oceanic archipelagos. Endemism is due to isolation, and this local endemism indicates that the populations of the various islands are isolated from one another; their dispersal is not effective even across the relatively narrow seas that separate the islands. How then could the progenitors of these same plants, with the same powers of dispersal, have populated Hawaii by trans-oceanic migration from the nearest continent, which is America over 2000 miles away?

Actually Skottsberg (25) has shown that the main affinity of the Hawaiian flora is not with America at all but with Polynesia and even New Zealand—in fact partly with the sub-antarctic region, but the jump to Hawaii from the nearest islands of any consequence in this direction is still 700 miles and that would only be the last of a tremendous series of improbable hops. Yet the flora of Hawaii is no depauperate collection of waifs and strays, adapted for long-distance dispersal. It is rich, luxuriant and varied, with a high rate of endemism. Allowing for difference of area it is comparable in these respects with the flora of New Zealand, which is not a group of oceanic islands but a small continent subdivided by epicontinental seas. Skottsberg (25) and Campbell (3) have no hesitation in drawing their bold conclusion—that Hawaii, Juan Fernandez, the Polynesian and Pacific islands generally are not true oceanic islands as originally conceived, but have a continental origin. The immediately obvious objections are that Hawaii, Juan Fernandez and similar islands consist entirely of volcanic rock, and that sunken continents in deep ocean have been pretty thoroughly discredited, especially in the Pacific, which has been regarded by Marshall (20, 21) and other geologists as one of the most permanent features of the earth's surface. I note in passing that some other geologists, including Gregory (13), have disputed the permanence of the Pacific, and believe that substantial changes in level have occurred there in relatively recent geological times, but I make no attempt to adjudicate between these conflicting opinions.

Skottsberg (25) points out that Juan Fernandez, which is relatively close to the Pacific margin, is seated on a submarine ridge, separated from the nearest part of Chile by deep water, but linking to the coast further south, towards the Magellanic area with which the flora shows affinity. He suggests that this ridge is a part of the South American continent, on which the volcanic rocks were built before the subsidence, and that the future islands were thus colonised along continuous land. He and Campbell agree on a similar explanation for the Hawaiian islands and believe that there may have been a large continental area including Polynesia and Hawaii, or at least a considerable extension of land in this area.

That plants have migrated somehow over this route is certain; many genera range from New Zealand to Hawaii. One of the best examples is the Liliaceous genus *Astelia* (26), represented here in Tasmania by *Astelia alpina*, the pineapple grass, a densely tufted plant forming copious, firm mats that you can walk on in the flatter swampy areas near the summit of Mt. Wellington. The genus extends also to the Australian Alps and to higher altitudes in New Guinea—like *Nothofagus* and many other sub-antarctic plants—to Fuegia and the Falkland Islands, Réunion, New Zealand, with the Auckland, Campbell and Chatham Islands, New Caledonia, Tahiti, one or two other Pacific islands and finally a group of six endemic species in Hawaii, all closely allied. *Gunnera* and *Acaena* have very similar distributions, but slightly more extensive. All three, and others, are characteristic genera of the sub-antarctic flora with an extension of their range to some of the Pacific islands including Hawaii (27).

The coral islands, many of which are probably quite recent and could not have been in communication with any exposed continental land, have a very much scantier flora than the volcanic islands, consisting largely of strand plants and some that appear to be wind-borne or adherent to birds. Their difference from the volcanic islands may be due to dependence on overseas dispersal, and so support the idea that the volcanic islands have been stocked by other means, but it may be largely due to the edaphic difference.

The views of Skottsberg and Campbell on Pacific island floras are opposed by Setchell (24), who vigorously upholds the sufficiency of trans-oceanic dispersal. He offers an explanation of the apparent weak point that worried Guppy and was attacked by Skottsberg, namely the fact that study of the floras, and especially the incidence of endemism, demonstrates isolation. If the islands were stocked by immigration over existing seas, why has effective immigration stopped? Setchell's explanation, already suggested by Darwin (6), is the admitted fact that invaders do not easily establish themselves in a closed community. He suggests that a new-formed unoccupied island receives sea-borne, wind-borne and bird-borne seeds—and seeds have also been found on floating logs or pumice—and that this process is effective until the island is stocked. Thereafter the arrival of seeds is not in any way altered, but establishment becomes increasingly difficult as the vegetation becomes denser, until the invaders are excluded altogether. It may perhaps be doubted whether such exclusion would be rigid enough to account for the incidence of endemism, but it is a suggestive idea, and is very ably presented. In the face of this complete disagreement what we need is a brand new volcanic island, arising hundreds of miles from land, taken under the wing of Unesco or by some means reserved for observation over a period of years by scientists who would see what seeds, if any, could find their way there with no land route, and trace in detail the development of plant and animal life. It is unwise to prophesy, but although I agree on most points with those who claim that plant migration is marginal, I find it hard to believe that such an island would remain indefinitely vacant. I don't think any one has seen oceanic islands, favourable for plant growth, but lacking an appropriate plant cover just because they were too far from land.

Nature did arrange an experiment along those lines for us in 1883 when the island volcano of Krakatau in the Sunda Strait between Java and Sumatra erupted so violently that it completely destroyed its luxuriant forest vegetation. Thereafter the island was visited at intervals by scientific expeditions, and a botanical report on the first 50 years, has recently been published (7). Recolonisation was pioneered by sea-borne strand plants, and coastal communities like *Pandanus* forest gradually developed. Later, wind-borne and bird-borne plants became more numerous and the slopes of the mountain are now reclothed with dense tropical jungle, complete with lianes, ferns and mosses. Some 250 species of plants have reached the island, besides the animals which include land snails, often used as evidence of land connections; in this case they were apparently ferried on drifting logs.

We should be grateful for the Krakatau experiment, but it has one severe limitation. The island is only some 25 miles from Java and not much more from Sumatra. It is in a narrow sea partly enclosed by these richly forested lands, and it does not prove that plants can migrate to oceanic islands across a thousand miles of sea. I think the evidence of this experiment has by some been dismissed too lightly as scarcely relevant. Consider wind-borne seeds. Twenty-five miles is not a simple trajectory. A seed too heavy to be maintained in the atmosphere for a fairly indefinite period, if it were lifted high in the air by an exceptional gust, would fall to the ground long before it had travelled any such distance. In a flight of 25 miles the seeds must surely have risen and fallen repeatedly in the air according to the variation of the wind, in which case there would be no limit to their flight except that imposed by changes of wind and weather (28).

But here we are passing into the realm of speculation again. Anyone who reads the publications of the two schools of thought, the believers in effective trans-oceanic dispersal and the disbelievers, must be struck, I think, by the completeness of their disagreement, and by the apparently good case which each side has nevertheless been able to make. The

three authors of recent text-books of plant geography—Wulff, Cain and Good— are all disbelievers— probably the greatest unanimity of current opinion on this question that has ever been reached. They are all swayed largely by the important fact that migration is generally if not always slow and marginal, but Darwin's and Setchell's suggestion that this would not apply to primary colonisation of unoccupied territory, and the evidence of Krakatau, give sufficient cause for hesitation.

If I have not mis-read Good's book (12), he has run into an impasse over the Pacific Islands, like Hawaii. He has rejected the land-bridge hypothesis in favour of continental drift—pointing out with justification that to concede a land-bridge wherever somebody requires one to explain a discontinuity would involve postulating these now-sunken ridges in all directions all over the floor of the oceans. But he excludes continental drift so far as the islands are concerned, for he accepts them as truly oceanic, not continental fragments, (p. 326). Yet he will not accept overseas migration. Well, I can't see what explanation remains if all these three are excluded, but the plants are there.

In the face of this disagreement in theoretical discussions, can we look anywhere for more objective facts? I suggest the palaeontologists hold one of the best keys, but the using of it is slow and difficult. Nothing could be more solidly helpful than a great increase in our knowledge of the identities and distribution in time and space throughout the world of fossil angiosperm floras, correlated and integrated in such a way that we can begin to plot the actual course of some migrations. Can we find just where *Nothofagus* has been, and when? Has it always been confined to the southern hemisphere or has it left its foot-prints by the North Atlantic? It is easy to ask, but the task is tremendous. I look forward to hearing what Dr. Selling has to say on the " intercontinental connections between the Cainophytic floras around the South Pacific Basin ".

Obviously if this palaeontological evidence is to be of any use, the identifications must be thoroughly reliable. Undoubtedly many fossil leaf impressions have been wrongly identified; either the material 'or the describer's knowledge has been inadequate, and some Australian fossils have probably been referred to *Fagus*, *Quercus* and other northern genera on little more than the superficial appearance of a scrap of a leaf. Similarly the records of fossil *Eucalyptus* leaves in Europe have given rise to quite a bit of controversy, because the present-day distribution of the genus—wholly Australian except for a small extension into the Malavan archipelago—indicates a strong probability that it originated in Australia and has never been more widely spread than it is today. A wrong conclusion in such a case might lead us completely astray as to the former range and migration route of the genus concerned. And once a mistake gets into print it is copied from list to list and acquires a spurious authority. It is essential then, that identifications should not be attempted on inadequate evidence, and that if any unconfirmed suggestions are offered, the fact should be stated without ambiguity. On Monday we are to have a paper on " the identification of Tertiary leaves " and it will be interesting to hear something of the methods available to ensure reliability.

The position today then would seem to be that there is still a wide divergence of views on the possibilities of trans-oceanic dispersal, but that the present trend of opinion is against it, and inclines to seek continuity of land by the theory of continental drift rather than by land-bridges.

I undertook to suggest what further information is most needed. I will refrain from

demanding a new Krakatau about 1000 miles from the nearest land. I will refrain from demanding the fossil flora of a foundered land-bridge. But I suggest that the most valuable evidence might come from two lines of research; firstly the continued and intensified study of fossil angiosperm floras and contemporary faunas all over the world, with strict insistence on reliable identification or none at all, and secondly a detailed investigation of the geological structure and origin, and the botanical history, of isolated oceanic islands, especially Hawaii. When we can explain how *Astelia* and other sub-antarctic genera reached Hawaii, I don't think we will have so much difficulty in accounting for their distribution in the southern continents.

REFERENCES.

1. ARCHBOLD, R., RAND, A. L., and BRASS, L. J., 1942.—Results of the Archbold Expeditions, No. 41. *Bull. Amer. Mus. Nat. Hist.*, vol. 79, art. 3, 197-288.
2. CAIN, S. A., 1944.—Foundations of Plant Geography. New York.
3. CAMPBELL, D. H., 1919.—The derivation of the flora of Hawaii. *Leland Stanford Jr. Univ. Publ.*, Stanford University Series, 1919.
4. CROCKER, R. L. and WOOD, J. G., 1947.—Some historical influences on the development of the South Australian vegetation communities and their bearing on concepts and classification in ecology. *Trans. Roy. Soc. South Australia*, 71, 1, 91-136.
5. DARWIN, C., 1845.—Journal of researches . . . during the voyage . . . of H.M.S. Beagle. London.
6. ———, 1859.—The Origin of Species. London.
7. DOCTERS VAN LEEUWEN, W. M., 1936.—Krakatau, 1883-1933. A. Botany. Leiden.
8. DUSEN, P., 1908.—Ueber die tertiäre Flora der Seymour-Insel. Wiss. Ergebn. der schwed. Südpolar Expedition, III. 3.
9. EVANS, J. W., 1941.—Concerning the Peloridiidae. *Austr. J. Sci.*, 4, 3, 95-97.
10. GIBBS, L. S., 1920.—Notes on the phytogeography and flora of the mountain summit plateaux of Tasmania. *J. Ecol.*, 8, 1-17, and 89-117.
11. GOOD, R., 1931.—A theory of plant geography. *New Phyt.*, 30, 149-171.
12. ———, 1947.—The Geography of the Flowering Plants. London.
13. GREGORY, J. W., 1930.—The geological history of the Pacific Ocean. *Quart. J. Geol. Soc.*, 86, No. 342. Proceedings, pp. lxxii-cxxxvi.
14. GUPPY, H. B., 1906.—Observations of a Naturalist in the Pacific. II. Plant Dispersal. London.
15. ———, 1917.—Plants, Seeds, and Currents in the West Indies and the Azores. London.
16. HALLE, T. G., 1913.—Mesozoic flora of Grahamland. Wiss. Ergebn. der schwed. Südpolar-Expedition, III, 14.
17. HERBERT, D. A., 1935.—The climatic sifting of Australian vegetation. *Rep. A.N.Z.A.A.S.* Melbourne, 1935, 349-370.
18. HILL, A. W., 1929.—Antarctica and problems in geographical distribution. *Proc. Int. Cong. Plant Sci.*, 2, 1477-1486.
19. HOOKER, J. D., 1860.—Introductory essay, in Botany of the Antarctic Voyage, Part III. Flora Tasmaniae, vol. I. London.
20. MARSHALL, P., 1933.—Stability of lands in the south-west Pacific. *Rep. A.N.Z.A.A.S.* Sydney, 1932, 399-411.
21. ———, 1947.—The permanent Pacific. *Rep. A.N.Z.A.A.S.* Adelaide, 1946, 1-3.
22. OLIVER, W. R. B., 1925.—Biogeographical relations of the New Zealand region. *J. Linn. Soc. (London) Botany*, 47, No 313, 99-140.
23. RIDLEY, H. N., 1930.—The Dispersal of Plants throughout the World. Ashford, Eng.
24. SETCHELL, W. A., 1935.—Pacific insular floras and Pacific paleogeography. *Amer. Naturalist*, 69, 289-310.
25. SKOTTSBERG, C., 1925.—Juan Fernandez and Hawaii. *Bernice P. Bishop Museum Bulletin*, 16. Honolulu.
26. ———, 1934.—*Astelia*, an antarctic-pacific genus of Liliaceae. *5th Pacific Sci. Congr.*, 4, 3317-3323.
27. ———, 1936.—Antarctic plants in Polynesia. Essays in Geobotany in Honour of William Albert Setchell, 291-311. Univ. Calif. Press.
28. SMALL, J., 1918.—The origin and development of the Compositae, ix. Fruit dispersal in the Compositae. *New Phyt.*, 17, 200-230.
29. THISTLETON-DYER, 1910.—Geographical distribution of plants. Darwin and Modern Science, 298-318. Cambridge.
30. WALKOM, A. B., 1949.—Gondwanaland, a problem in palaeogeography. *Rep. A.N.Z.A.A.S.* Hobart, 1949, 1-13.
31. WALLACE, A. R., 1880.—Island Life. London.
32. WULFF, E. V., 1943.—An Introduction to Historical Plant Geography. Waltham, Mass.

21

Reprinted from *Proc. 6th Pacific Sci. Congr. Pacific Sci. Assoc., Berkeley and San Francisco*, 1939, Vol. 2, University of California Press, 1940, pp. 755–768

ANTARCTICA AS A FAUNAL MIGRATION ROUTE

By GEORGE GAYLORD SIMPSON

EVERY STUDENT of faunal and floral distribution has noticed that many non-marine plants and animals are disjunctively distributed. A given form or group may find its apparently closest allies not in an adjacent region, but in some distant land, often separated by long stretches of sea or by other barriers. some of the most striking of these disjunctive distributions occur in the Southern Hemisphere. To account for the distribution of life in the southern continents it is necessary to assume that various groups of plants and animals have somehow managed to migrate between these land areas without leaving any clear or permanent sign of their passage.

As long ago as 1847 the botanist J. D. Hooker, in working up the results of the voyage of the "Erebus" and "Terror," suggested that these anomalies might be explained by the former existence of a continuous land connection between the other southern continents by way of Antarctica. This possibility has had a powerful effect on the scientific imagination. A recent partial bibliography (Wittmann, 1934) includes 350 titles, and an exhaustive list would not fall much short of 1000. The idea of an Antarctic migration route has been opposed as strongly as it has been supported, and it has become one of the major disputed problems of earth history and biogeography.

So multiple and varied are the facts and conjectures that have been published in this field that judicious selection and emphasis of them can be made to support almost any opinion not completely irrational. No one person can hope to know at first hand all the pertinent data, and a general review of the literature leaves one with the feeling that it can be taken to prove any of a dozen conflicting theories and that it therefore proves nothing.

The discussion seems to have reached a stalemate. New facts might break this stalemate at any time. For instance, the discovery of a single fossil mammal tooth in Antarctica could at once settle some of the most disputed aspects of the problem, although of course no single fact could solve it integrally. Such a decisive discovery has not been made and apparently no one is willing to await it before forming some sort of opinion. Indeed it may be futile to expect any such simple solution, and the nondiscovery of such evidence may well be caused by its nonexistence.

In the meantime some progress can be made toward reducing this chaos to order and toward evaluating the various theories. This cannot be done simply by the literary device of amassing all the facts, supposed facts, and opinions, and attempting to strike a balance. This has repeatedly been attempted, most recently by Wittmann (1934, see references), whose condensed but fairly thorough review makes repetition by this method unnecessary at present. It is true that Wittmann, like most of the other compilers, reaches a conclusion or at least expresses a preference, but his own citations show that his preference does not really follow clearly or necessarily from his data. More progress can perhaps be made in the following ways:

1. By exact definition of the problem, which includes critical selection of the instances of disjunctive distribution to be explained and statement of the various hypotheses that might explain them.

2. By consideration of the question of authority and discarding of incorrect or inadequate data and the opinions based on them in favor of the best-informed and most logical authoritative views in each field of evidence.

3. By bringing into the discussion all new and pertinent facts as they are discovered.

4. By reconsidering the means of interpreting these facts and revaluing them in the light of consistent and scientific interpretive methodology.

A brief review along these lines, without lengthy exposition of the data available in each field, must in part be critical in a destructive sense. By weeding out the material not truly pertinent and the authority and opinions considered least reliable, a constructive and positive conclusion can nevertheless be reached if the result is to leave, as I believe it will, a consensus and a set of hypotheses that may be taken as definitely more acceptable than those adversely criticized.

The Problem

The whole question of Antarctica as a migration route arises from attempts to explain examples of disjunctive distribution of groups known only, or mainly, from the Southern Hemisphere. A point of departure, at least, is well established: types of land plants and animals do occur in two or more southern regions, for instance, in Australia and South America, and appear to be more closely similar or related to each other than they are to plants and animals of other (i.e. of northern) regions. This is an apparent factual anomaly and requires explanation. The principal biological hypotheses that arise are:

1. Structurally closely related forms of life may arise spontaneously, by processes other than those familiar as convergence and parallelism, in distinct places. I do not propose to discuss this ancient hypothesis seriously, although it has recently reappeared in somewhat more subtle form among certain European students who do not appear to me to have any grasp of the facts that they attempt to explain.

2. The apparently related forms may not be such in fact, or not to such a degree as to be anomalous under the given circumstances. They may merely be convergent or parallel forms that have arisen independently in the two regions from separate stocks or from a common ancestry that existed in intervening, northern lands. Many of the supposed items of evidence for Antarctic migration are now known to be of this sort and hence not to demand or even to suggest such a route. Examples of animals that are often used to support the theory of an Antarctic route, but that probably represent convergence and parallelism from stocks that do not suggest any such route, will be mentioned later. Despite its wide emphasis, such evidence really has no value for this problem and may be discarded in our consideration of it. It is, however, true that by no means all the apparent instances of southern disjunctive distribution can be so explained.

3. The forms of life in question may really be specially related and may have migrated more or less directly between the various southern lands, but may have done so by means other than a definite land connection. It is an established fact that both land plants and land animals cross considerable stretches of water by floating, by carriage on natural rafts, or by wind (plant spores and seeds, or animals that fly). There are also animals, including many fishes and some reptiles and mammals, that normally live in fresh water or on land, but that are capable of prolonged sojourn in the sea and are therefore capable of being carried across the ocean without a land bridge. Some of the best instances of southern disjunctive distribution belong to this class, for example the fish *Galaxias*, the meiolaniid turtles, and the sea cows. Some of the adherents of the Antarctic route complain that this process, however possible, is improbable, that repeated improbabilities may amount to a prac-

tical impossibility, and that their opponents use this sort of distribution as a *deus ex machina* to solve all their difficulties. To some degree this is true; but the process is often less improbable than it seems—for many forms of life, even highly probable. Evidence of plants and animals that might be placed in this class is to be considered as pertinent to the Antarctic problem, but the probabilities of such means of distribution must be reconsidered and it must be decided whether there are any plants and animals of this group for which Antarctic migration is definitely more probable than nonterrestrial migration.

4. The common ancestors of the disjunctively distributed forms of life may have lived but may now have become extinct in regions forming a continuous connection between the areas where the forms now occur. This is, of course, the real basis of the problem of the Antarctic route, consideration of which is very much clarified by recognition of the fact that much of the supposed evidence is not really of this crucial type but belongs rather to the nonpertinent group (2) or to the dubious group (3) of this list.

Animals that migrate by land and that cannot, within reason, be distributed in any other way are very numerous, but among them clear-cut examples of southern disjunctive distribution are rare. This is in itself a suspicious circumstance: that so few of the apparent examples of disjunctive distribution absolutely require a land bridge, and that so few of the animals that absolutely require a land bridge are disjunctively distributed.

That the southern faunas as a whole do require land connections of some sort is obvious. As to the nature of these connections, there are four main hypotheses:

1. That the southern lands (at least South America and Australia and possibly also Africa and New Zealand) were once connected by land bridges to Antarctica, and hence with each other through that continent.

2. That they were once all part of the same land mass, which also included Antarctica and which has since been split up by continental drifting.

3. That they were connected with each other more directly by transoceanic bridges, not necessarily involving Antarctica.

4. That they were connected separately to the northern continents, Australia with eastern Asia, Africa with Europe and western Asia, and South America with North America, and the northern continents were united with each other.

These hypotheses are not mutually exclusive. It would be possible for two or more of them to be true successively or simultaneously.

Only the first two hypotheses are being examined directly here, since only they involve Antarctica, and of these the first is of primary importance. The second, which is that of Wegener, seems to me and to a consensus of geologists ill supported and improbable. In any case, for present purposes it suffices to say that the faunal distributional evidence for it should be of the same sort as that for the first hypothesis, but should be decidedly stronger and more obvious because such an intimate connection should produce not merely some faunal similarities, as Antarctic land bridges might, but practically identical faunas. The scattered and incomplete nature of the evidence for either view thus does not make the first impossible, but is almost conclusively opposed to the second.

The third hypothesis, of transoceanic, non-Antarctic bridges, is peculiar in that even if a conclusive case for southern faunal migration routes were established, the absence of direct evidence in Antarctica itself would make this hypothesis as adequate to explain the facts as the hypothesis of Antarctic connections. Thus, the biological evidence alone can hardly permit an absolute choice between these two, and some adherents of the southern land bridge theory (e.g., von Huene, 1929) have rejected Antarctica as part of the supposed bridge. In common with most geologists and zoölogists, however, I feel that the idea of a transpacific bridge from Asia and Australia to South America is so nearly impossible on other grounds that if a southern bridge were established as probable I would assume, until the appearance of further evidence to the contrary, that it involved Antarctica.[1]

The fourth hypothesis, of northern connections, is in part no longer hypothetical, but a well-supported theory or indeed a fact. Obviously, South America and Africa are so connected at present, and it is all but certain that they have been so connected at various times in the more remote past. The connection of Australia with Asia has been denied, but is also almost certain—perhaps not a continuous land connection, but at least insular steppingstones that permitted the entrance of some Asiatic animals and plants into Australia. The former and frequent land connection of Eurasia and North America is definitely proved by the fossil record. Since this series of connections is almost certain and is required in any case, the principle of parsimony demands that no additional hypothesis be considered necessary or acceptable unless it clearly explains facts that are difficult or impossible to explain by these connections alone.

In further delimitation of the problem, it should be stated that the question of Gondwana Land, a South American–African–Southern Asiatic connection in the Mesozoic and earlier, does not involve Antartica, is a different or additional and not an alternative hypothesis, and so does not enter into the present discussion. The question of possible very remote Antarctic connections, in the Jurassic or earlier, is also excluded from discussion here because almost all the usual evidence is from groups of plants and animals that must have migrated in the later Mesozoic or Tertiary. The probability that Antarctica was connected with South America, at least, at some period without serving as an important land-migration route is also beside the point here.

In the present state of knowledge the best form in which to state the question appears to be this: Are there any known facts of the distribution of recent and fossil plants and animals that make it necessary or preferable to suppose that Antarctica was involved in a land-migration route between the other southern lands, or can all these facts be adequately and more probably explained by other routes for which there is more direct evidence?

The Evidence and Its Interpretation

It is not possible, and in view of the summaries of Wittmann and others it is not desirable, to attempt a review of all the evidence here. Some of the supposedly crucial evidence from each class of the vertebrates will be very briefly discussed, less for its own sake than to illustrate the conflicts of authorities and of interpretive methods and to show how a more detailed review (which I have made and on which these notes are based) can point definitely to one conclusion despite these conflicts.

GALAXIAS

The eel-gudgeon *Galaxias*, a predominantly freshwater fish, occurs in South America, South Africa, Australia, and New Zealand. As early as 1905, Tate Regan pointed out that *Galaxias* freely enters salt water and can live in it indefinitely, that it is probably of marine ancestry, and that it therefore gives no evidence of land connections. This is excellent authority and this opinion has been shared by a clear consensus of ichthyologists ever since. Nevertheless,

---

[1] Moreover, von Huene's faunal evidence for his transpacific bridge seems to me to be invalidated by more recent discoveries.

Galaxias is usually cited by adherents of Antarctic bridges as evidence for their view.

So far as any reason can be given for this disregard of authority and consensus, it is to be found in the fact that there is a partially conflicting authority, that of Eigenmann (1909), who agreed that *Galaxias* might migrate by marine routes but held this to be highly improbable. This sort of judgment of probability occurs again and again in dealing with the present problem and so merits comment.

It is indeed "highly improbable" that a given fish (or pair) should cross an ocean and colonize waters on the other side at any given time. The chances of occurrence (at a single trial) are extremely small, but probability does not depend solely on chances of occurrence, but also on opportunities for occurrence. The chances of throwing five aces with five dice in one throw are negligible, but if the opportunities for occurrence are increased, for instance by throwing one hundred dice instead of five or by throwing ten thousand times instead of once, this "highly improbable" event becomes probable and may even become certain, for all practical purposes. So with difficult migrations, such as that of *Galaxias* across the ocean. The great number of individual animals involved, usually thousands or millions, and the long span of time involved, often millions of years, give so many opportunities for occurrence that the "improbable" event becomes highly probable as long as the basic chance is real and finite, as it is granted to be for *Galaxias*. There is never an absolute certainty that the migration will be accomplished, and its time of occurrence is random—peculiarities that have a definite bearing on animal history, as will be mentioned again in discussing mammals.

THE SO-CALLED LEPTODACTYLID AMPHIBIANS

Certain amphibians, toothed toads, are frequently placed in a single family, Leptodactylidae or Cystignathidae, which occurs only in the southern continents. These forms have been claimed as having no indication of northern origin or relationships and hence as proving the existence of a purely southern land-migration route, especially by Hewitt (1922) and Metcalf (1923, 1929). Noble (1925, 1926, 1930) has claimed, however, that the supposed Leptodactylidae are merely members of the Bufonidae, a world-wide or essentially northern group, that happen to have retained teeth independently in Australia and South America. Noble's papers appear to be based on more intimate and broader knowledge of the groups concerned and to give his authority, on this particular point, greater weight. Moreover, Noble has actually identified a fossil toothed bufonid, or Leptodactylid, in the Eocene of Asia, a discovery that brilliantly confirmed his earlier zoögeographic conclusions and that also showed how precarious was the argument from absence of these animals in any northern continent.[2]

The amphibia and several analogues will doubtless continue to be cited in proof of Antarctic migration, but they cannot now be taken very seriously in this connection. Granting them all possible force, it still must be clear that alternative views of their affinities are at least as probable as those demanded by this argument.

---

[2] Metcalf emphasized the occurrence of the parasite *Zelleriella* in the so-called leptodactylids and its probable absence in the northern bufonids. This argument has generally been granted great weight, perhaps because it is so esoteric from the point of view of students of the vertebrates, but I cannot see that it has any decisive value for his thesis. Aside from other possible explanations, the survival of a parasite in toothed southern survivors or relicts and its disappearance in more advanced northern members of the same stock would be a normal phenomenon.

MEIOLANIA

One of the standard examples of disjunctive distribution and one cited by practically everyone in favor of an Antarctic or Pacific bridge is that of the fossil turtle *Meiolania* (usually incorrectly called *Miolania*), which is supposed to occur in Patagonia and in the Australian region but nowhere else, and hence to indicate a more or less direct southern connection between the two.[3] This argument is mentioned here chiefly to illustrate three failings common throughout discussions of this whole problem: faulty identification from inadequately studied materials, poor selection of authority, and temerarious leaping over enormous gaps in knowledge.

As I have elsewhere shown (1938), the Australian and South American horned turtles are not really congeneric, *Meiolania* being exclusively Australian and *Niolamia* South American. This was noticed by Tate Regan long ago (1914), and yet practically every student of zoögeography continued to state the occurrence of the same genus in the two widely separated areas as if it were a basic fact. It is not even well established that the two are especially related, although they may be, and they are tentatively placed in a single family. The only known possible ancestors of these forms are found in the Northern Hemisphere, although nothing like a direct phyletic sequence can be cited for either. *Meiolania* is found with aquatic and indeed marine associations. It was not a specialized marine turtle, but there is every reason to think that it could swim and did enter the sea. *Niolamia* and its ally *Crossochelys* are found with aquatic, but freshwater, associations, and beyond much doubt were also strong swimmers with no probable physiological barrier against entering the sea on occasion. Thus there are two alternatives—separate descent from northern ancestry and migration without land connection—each more probable than the widespread insistence on a southern land connection.

Moreover, *Niolamia* is either Cretaceous or Eocene in age and *Crossochelys* is Eocene, whereas *Meiolania* is Pleistocene. Coupled with the uncertainty of degree of relationship, this enormous time gap entirely removes the supposed disjunctive spacial distribution from the sphere of sober evidence into that of pure speculation. Anderson, who is beyond question the best-informed first-hand authority on *Meiolania*, pointed out the worthlessness of its evidence in this respect in 1925, yet the great majority of students of the land-bridge problem have continued to prefer or to give equal weight to older and less informed authorities or second- and thirdhand citations that are not, on this point, authorities in any proper sense.

THE RATITE BIRDS

The so-called ratite birds, including ostriches, rheas, emus, cassowaries, *Apteryx*, and the extinct moas and *Aepyornis*, are often believed to form a natural group derived from a common flightless ancestry and explicable only on the basis of southern land connections. Even in the last century Fürbringer, a great authority, suggested that the assemblage may really be unnatural and that the various ratites may have been derived from different flying ancestors.

---

[3] As an example, one of many that could be ferreted out in this baffling literature, of accretionary errors in quoting authority, Hesse (*Tiergeographie*, p. 113) said, correctly, that fossil pleurodires occur in the Northern Hemisphere. Arldt (*Naturwissenschaften*, XII), reviewing Hesse, agreed and added, also correctly, that Meiolaniidae are not known in the north, assuming, incorrectly but with the agreement of the consensus of that time, that *Meiolania* was a pleurodire. Wittmann (1934, p. 284) apparently read and misunderstood Arldt's review and did not read Hesse's book, for he cites Hesse as saying that *Meiolania* has been found fossil in the northern continents, which is not true and which Hesse did not say. Similarly, Wittmann incorrectly cites Hesse as claiming that fossil Chelyidae are found in the north.

This has been widely accepted, and, as for Antarctic connections, about as far as most cared to go was to agree with Blaschke (1904) that if the Antarctic route were considered probable on other grounds, then its possible effect on flightless bird distribution would have to be considered. More recently, Lowe (1928) has asserted that the group is natural and that its ancestors never were capable of flight; but Gregory (1934) and Murphy (1934) have shown that Lowe's excellent observations are more logically interpreted as favoring the opposite conclusion. It is also known that every large land area has at some time developed flightless birds and that several quite distinct ancestral types have so been modified (for general review and exhaustive references see Lambrecht, 1933). The present situation is that the flightless birds cannot be accepted as a valid example of disjunctive terrestrial distribution and cannot be used as evidence for or against the Antarctic route.

MAMMALS

Most of the fishes, most of the amphibians, most of the reptiles, and most of the birds not only do not give evidence for an Antarctic connection, but definitely oppose this. The argument for such a connection is sustained not by considering the fauna as a whole, but by selecting a few apparently anomalous forms, like *Galaxias*, the "leptodactylids," the "ratites," or *Meiolania*, and considering them as isolated things. Usually when these are examined more carefully, it is found either that their peculiarity is wholly and more simply explicable on some different basis (e.g., *Galaxias*) or that they really are not exceptional but could really have had the same geographic history as most of the fauna (e.g., the "leptodactylids"). Similarly, among mammals the marsupials are usually singled out for attention and their associations with a large number of other animals that cannot have had the history claimed for marsupials alone is ignored.

As for the marsupials, space will not be taken to argue the question, and only the categorical (but very well-grounded) statement may be made that no connection between South American and Australian marsupials has been demonstrated or is at all likely beyond common descent from undifferentiated types such as are known to have occurred in Holarctica at an appropriate time.

Australia.—The native Australian mammalian fauna consists of monotremes (of entirely unknown but surely very ancient origin), marsupials (with no known allies in Asia, but related to ancient North American and, to a less degree, European forms and with parallel or convergent relatives in South America), rodents allied to the rats (clearly of Asiatic origin), bats (which require no land connections), the dingo (of Asiatic origin and probably introduced by man), and man. For able summaries of this fauna see Anderson (1925 and 1936).

The classic explanation of this fauna was that marsupials (and monotremes) entered Australia before placentals existed and formed the whole mammalian fauna there until the few placentals were incidentally introduced at a much later date. Now it appears that placentals must have been in existence when the marsupials entered Australia,[4] and a serious problem arises, which might be solved by one of these hypotheses:

---

[4] The work of Pfeffer (1927), often given as authority in discussion of this problem, is an instructive example of how the psychology of scientists can interfere with the solution of scientific problems. He supported the outmoded view that the Australian fauna was marsupial because Mesozoic, placentals first evolving at the beginning of the Tertiary. While he was working, unquestionably placentals were found in undoubted Mesozoic beds in Asia (Gregory and Simpson, 1926), absolutely disproving his already generally discarded thesis. Pfeffer was aware of this discovery, but instead of discarding or modifying the theory that the fact disproves, he discarded the fact.

1. That placental mammals did enter Australia with the marsupials, but became extinct there.

2. That a route practicable for marsupials, but not for placentals, existed.

3. That a route practicable for both, but at any one time unlikely for either, existed, and that marsupials happen to have been the ones to cross it, or to cross it first.

The first hypothesis is entirely possible but is unlikely. Numerous marsupials are known to have become extinct in competition with placentals, but no placentals are known to have become extinct in competition with marsupials. On this basis, if on no other, the mooted South American connection is most improbable. If marsupials came from South America to Australia over a practicable bridge, placentals should have come with them, and placentals that were entirely capable of surviving along with these marsupials. If marsupials went from Australia to South America, they must have done so on a strictly one-way bridge and have eliminated the earlier native carnivores of South America. Either is just barely conceivable, but neither is acceptable as a serious theory on present evidence.

The second hypothesis would clear up the difficulties just expressed, but there is no evidence whatever that such a route ever existed, and I for one am at a loss to draw up specifications for one. The Antarctic bridge, if it had any climatic or enviromental selective action on animals passing over it, would probably have the opposite effect. It is hard to conceive of its being more easily passed by marsupials than by placentals, and the reverse is much more likely.

In contrast with these two hypotheses, of which the first is improbable and the second barely short of impossible, the third appears to explain the facts in a simple and probable way, even though the evidence is incomplete. This hypothesis postulates that mammals did not enter Australia by a continuous and fully practicable land route, but by a route such that any migration of terrestrial animals over it was very difficult. A chain of scattered islands, such as actually exists between Australia and Asia, would provide such a route. This route would be passable to small animals and especially to arboreal forms, such as many or most of the primitive marsupials and placentals were. It would be highly improbable that any mammals would follow this route at any given time, but, given many opportunities for occurrence, it would be probable that some mammals would negotiate it sooner or later. The chances that two or more different groups of mammals, such as marsupials and placentals, would successfully traverse the route at the same time would be very slight. Which group happened to traverse it first would be largely the result of random chance, except that the chances would somewhat favor the more numerous group (which may well have been marsupial at the time when they did enter Australia) and the group which had more opportunities for being carried over the stretches of sea (which may also have been the primitive marsupials). It is not, however, necessary to postulate that marsupials were favored by these chances, for it could well happen that the less-favored group was in fact the first one to succeed in the journey.

According to this view, the marsupials happened to be the first group to reach Australia and it would then probably be a long time before any other mammals succeeded in reaching that continent. When they did arrive, they would do so in impoverished and precarious condition and would find a continent already well stocked with diverse and well-adapted marsupials. Under these conditions placental invaders would have the odds against them, whereas under the ordinary conditions of migration the odds favor them in competition with marsupials. Eventually, no doubt, some placental group would get a foothold, and so in fact one did: the rodents. It is generally conceded that these Australian rodents are of Asiatic origin and reached Australia without a continuous land bridge; indeed, no other inference can be seriously supported. The whole problem of the Australian fauna is beautifully simplified if it be supposed that the ancestral didelphoid-dasyuroid stock reached that continent in the same way as the rodents, but at an earlier time.

In this case the postulation of a difficult migration route, instead of being a disadvantage of the hypothesis, is really a great advantage. The difficulty of such a route makes the hypothesis more, and not less, probable, because this difficulty itself helps to explain facts that would be almost inexplicable if a continuous and easy route had existed and been used.

As to where this route was, surely the simplest and preferable hypothesis is that it lay between Asia and Australia. Such a route has long existed in that position and it has been used by the rodents, by lower vertebrates, and evidently also by invertebrates and plants.[5] The only strong argument against this is the fact that marsupials are unknown in Asia. All the indirect evidence, however, favors the belief that marsupials did exist there in the late Mesozoic and early Tertiary, and the negative direct evidence is practically valueless.

Some students have argued from the distribution of marsupials within the Australian region that they must have come from the south. Most of the data used in this argument are really not pertinent to this problem and those that are could just as well be used to support northern origin. An example of the occasional complete absence of all logic or reason in such arguments is the serious statement that the occurrence in Tasmania of the only known Australian Tertiary marsupial, a late Tertiary specimen belonging to a specialized living group, favors southern origin for the marsupials in that continent.

Africa.—The African mammals have no relationships at all with Australia, but mainly with the north, primarily with the Mediterranean region, especially its eastern part, and with Asia, especially its southern part. Some groups arose in Africa from earlier immigrants or from ancient autochthonous stocks, and others arose in the north and invaded Africa. For both, the land connections definitely suggested are northern, and, taking its fossil faunas as a whole, Africa seems historically to be rather another division of the Holarctic mass than a distinctly southern continent in the sense in which this is true of South America. At least, its mammals not only do not support, but also tend to oppose, the existence of an Antarctic route open to them. Possible connections of Africa with South America are mentioned below.

South America.—Turning to South America, there is no mammalian faunal resemblance to Australia except for the marsupials and these are more consistent with connections through Holarctica than through Antarctica. For various ecological and other reasons it is very unlikely that placentals of South American type ever lived in Australia.

The mooted resemblances between South American and African mammals are partly mythical. The South American pyrotheres were not proboscideans, and the South American primates have less African than North American affinities. The only good evidence is that of the hystricomorph rodents. Some of the South American and African forms do resemble each other very closely and do not have very close known relatives, fossil or recent, in the north. Here again, however, the question arises why a means of migration open to these rodents did not cause greater and less equivocal faunal resemblances in other

---

[5] The Wegenerians alone would strongly combat the belief that an island bridge lay between Australia and Asia in, say, late Cretaceous times, but they base their argument in part on the very fact of which such a connection is the best explanation and their theory is also dubious in many other respects.

respects. There is also the curious fact that these rodents are absent from the known early South American faunas and appear suddenly in the Oligocene, as if by adventitious entrance independent of the rest of the fauna. It certainly stretches the imagination to suppose that they came by a transoceanic, non-land route, but it is not impossible, and if any land mammals could use such a route it would be forms like these rodents. On the other hand, there is the more definite but still quite inconclusive fish evidence for a land route from Africa (see Myers, 1938). I cannot hazard more than a guess toward the solution of this mystery, which is perhaps the greatest involved in the problems of southern mammalian faunas. It is, however, evident that even this very dubious evidence of a South American–African connection is not strong, or even valid, evidence of an Antarctic connection. It is generally conceded that if such a connection existed it was probably transatlantic rather than Antarctic.

Evidence for a North American connection with respect to early South American mammals is much stronger.[6] A very archaic North American mammalian fauna could have given rise to the South American fauna. The two continents shared condylarths, and all the other South American ungulates could have come from these or from similarly archaic ungulates like those of North America. The annectant steps are missing, for the most part, but this is a reasonable view and there is no equally probable alternative. There were also archaic primates, edentates, and marsupials in North America that strongly indicate union with the source of South American mammals, even if North America was not itself that source. Despite many troubling gaps in the evidence, gaps that necessitate caution in reaching definite conclusions, they show that North America might have been that source, and no other has been shown to be anything like as probable.

This does not cast any light on the hystricomorph problem. If South American mammals came from North America, they did so before there were rodents in the latter place. Then the hystricomorphs are a real as well as an apparent exception and any theory must postulate for them a migration different from that of the majority of South American mammals and not necessarily from the same region. The other great mystery is negative: outside of South America there is no known source of placental herbivores unaccompanied by placental carnivores. But at present this is as great a mystery, whatever source be postulated for the herbivores, so that it is no stronger against North American than against any other origin. On the whole it argues against rather than for an Australian connection. South America did somehow get a wholly marsupial carnivore fauna accompanied by a placental herbivore fauna, and at present one can do no more than state the fact.

The Southern Predominance of Disjunctive Distribution

Even when all instances are examined and it is shown that an Antarctic migration route is an unnecessary hypothesis, the fact remains that the most striking instances of disjunctive distribution are southern. This observation undoubtedly has given most zoögeographers a predisposition toward belief in lost southern land connections. The cumulative effect of these cases may appear to be impelling, even though it be granted that each separate item is exceptional or is explicable by purely northern routes.

---

[6] Of course the many recent South American mammals, like the camelids, peccaries, placental carnivores, tapirs, etc., that are known to have come from North America at the end of the Tertiary do not enter into this problem.

Several possible explanations of this predominance of disjunction in the south have been proposed and much of the literature attempting to establish the reality of a southern route has been devoted to attacking alternative explanations. Of these the most discussed and on the whole the most successful is that set forth in Matthew's *Climate and Evolution* (1915). Some students seem to feel that if they could demolish this theory they would, *ipso facto*, establish the reality of the Antarctic route. Of course this is not so. Matthew's theory depends on Holarctic origin and migration of the various southern groups, but the theory of Holarctic as opposed to Antarctic migration routes by no means depends on Matthew's theory, or even on that of Holarctic origin.

Matthew's theory is that most forms of life have radiated from northern centers of evolution, that more advanced forms arise in such centers, that more ancient or primitive forms are peripheral in distribution, and that most disjunctively distributed southern animals are such peripheral ancient forms. Most criticisms of the theory have been given over to pointing out exceptions or in showing that the theory falls short of explaining all the facts of distribution. It may be granted that Matthew tended to oversimplify and also to extend the theory to some groups of animals not really distributed in accordance with it. It cannot, however, be denied that Matthew's theory does afford the best explanation so far proposed for many of the important and broad features of animal distribution. Both its critics and its adherents need to remember that the theory does not insist that the primitive stocks must survive in a southern area or that they cannot also give rise there to diverse and specialized forms different from those of the north. It does suggest that the southern animals are likely to have split off from the main line of evolution at an earlier date than the northern and that ancient and once cosmopolitan groups are likely to survive longer in the south. This is in accord with many known facts which this theory most successfully explains. To this extent it also helps to explain why some striking instances of disjunctive distribution are southern rather than northern.

Regardless of the truth or falsity of this particular theory, there is an explanation of the apparent predominance of such disjunction in the south, an explanation so obvious that it seems almost puerile to state it and yet one that some students appear to overlook. The southern land areas are disjunctive in fact and have, in all probability, been still more so during the later part of geological history. The northern land areas are nearly connected now and clearly have been fully connected at many stages in geological history. Many genera of mammals are common to Asia and North America, but the explanation of this fact is so plain that these are not, like the far fewer and weaker but somewhat analogous instances of a common occurrence in the Southern Hemisphere, anomalies demanding special explanation. Generally speaking, anomalously disjunctive distribution is uncommon in the north simply because northern animals occupy what is, historically, a single land area, and it is less uncommon in the south simply because southern animals occupy what are historically separate land areas. If it be supposed that all the continents once had the same fauna—a gross oversimplification but one that is valid in illustration of the process—and that they then were affected by random extinction, some groups becoming extinct in one area and not in another, the result would necessarily be to produce disjunctive distribution in the south but not, or to less extent, in the north. No additional postulate with respect to the relative primitiveness or the origin of the various groups is necessary to explain this inevitable result.

ANTARCTICA AS A SOURCE OF NORTHERN FAUNAS

The idea is not new, but relatively little attention has been paid to the hypothesis that Antarctica was not merely part of a migration route but also was a center of evolution, particularly for the mammals. The question is different from the principal subject of this paper, but it is pertinent here and must be considered briefly.

All three of the northern continents, North America, Europe, and Asia, show two major and mysterious mammalian invasions. One of these was at about the beginning of the Paleocene. It was also this invasion, by whatever route, that reached South America. It included relatively few of the ancestors of modern mammals. A second invasion occurred around the beginning of the Eocene, and the recent faunas are in the main the highly modified descendants of this Eocene fauna, descendants of which straggled into South America during the Oligocene and flooded in at the end of the Tertiary. The second invasion cannot have come from South America or Australia, and almost surely did not as a whole come from Africa, although certain of its elements were probably elaborated there. Africa had shared also in the first invasion, although the evidence is indirect, and like South America it was a center of evolution for some (but different) elements of that group and later sent their modified descendants northward; but this suggests quite the opposite of its being the source of the second invasion. Although the source is really quite unknown otherwise, the second invasion almost surely did not come from the south, and Antarctica can be almost categorically ruled out on that score.

The first invasion cannot be quite dismissed even in this unsatisfactory and negative way. If these mammals were in the Northern Hemisphere at the end of the Cretaceous, when their immediate and recognizable ancestry must have been somewhere, it is certainly strange that no trace of them has turned up any clearer than the few insectivore-carnivore skulls of the Mongolian Cretaceous. One would expect evidence of them among the many rich deposits of terrestrial reptiles of that time. There is room in the north for lack of discovery, because of facies or because of geographic isolation, and a northern origin is not impossible, but it is tempting to think that Antarctica, unknown paleontologically so far as land animals are concerned, might have been the mysterious source.

Again the possibility cannot be absolutely ruled out, but again what evidence there is opposes it. If the source of this mammalian invasion was in Antarctica, the fauna must have moved north by routes involving one or all of the present southern continents, for to suppose Antarctica connected with the northern and not with the southern continents is too fantastic for consideration. If Australia was on the route, I cannot conceive why or how all the many basic and potent placental stocks died out there without the marsupials also dying out. If South America was on the route, it is difficult to see why its abundant early faunas do not give evidence of this. Their evidence, as now known, is definitely one way: all the South American mammals might, in a general way, have been derived from groups with known North American representatives, but the converse is decidedly untrue. Even in the first invasion of North America there are many different groups of mammals that could not be derived in even the broadest sense from groups represented in the early Tertiary of South America. As for Africa, there is little good evidence one way or the other. As pure speculation, it might itself have been the source of the first great mammalian invasion. Even in such a speculative way, however, there is little reason to involve Antarctica, for Africa alone would be adequate as such a source, spread from it would almost surely be northward, and the Antarctic connection would again have only the status of an unsupported and superfluous hypothesis. There is no special evidence of an Antarctic connection specifically with Africa, but only, for what it is worth, of Antarctica as a means of connecting two or more other southern land masses.

CONCLUSION

The preceding revaluations of facts and arguments and collations of new and old data are only given as samples of reconsideration of the whole problem far too lengthy for presentation here. The conclusion reached is not based solely on what has been said here, but also on this more extensive background.

It seems to me that a definite answer can now be made to the question given above as summarizing the Antarctic migration problem. There is no known biotic fact that demands an Antarctic land-migration route for its explanation and there is none that is more simply explained by that hypothesis than by any other. The affinities of the southern faunas as a whole are what would be expected from the present northern connections of those continents and from similar connections known, or with considerable probability inferred, to have existed at appropriate times in the past. There are certain troublesome anomalies and exceptions in the evidence, but none of these can be adequately explained by postulating an Antarctic connection. The general weight of the evidence is against such a connection.

This does not prove than an Antarctic migration route did not exist; such negative proof is practically impossible. It cannot be denied that such a route may have existed or that its existence may some day be proved. But at present its existence is merely a complicating and unnecessary hypothesis additional to other hypotheses for which there is more and better evidence and which are adequate to explain the facts so far as these are now capable of explanation. In scientific theory the best-supported and most nearly self-sufficient hypothesis should be preferred and unnecessary additional hypotheses should be rejected or held in abeyance. On this basis the Antarctic migration route hypothesis remains simply a hypothesis with no proper place in present scientific theory.

ANDERSON, C.
1925a. "The Australian Fauna." *Jour. Proc. Roy. Soc. New South Wales*, LIX.
1925b. "Notes on the Extinct Chelonian Genus *Meiolania*, with a Record of a New Species." *Rec. Australian Mus.*, XIV223:-242.
1926. "The Origin of Australian Mammals." *Australian Mus. Mag.*, October–December, 1936, pp. 133–137.
BLASCHKE, F.
1904. "Ueber die tiergeographische Bedeutung eines antarktishen Kontinents." *Verh. zool.-bot. Ges. Wien*, LIV.
EIGENMANN, C. H.
1909. "The Freshwater Fishes of Patagonia and an Examination of the Archiplata-Archhelenis Theory." *Rept. Princton Univ. Exp. Patagonia*, III:225–374.
GREGORY, W. K.
1934. "Remarks on the Origins of the Ratites and the Penguins." *Proc. Linn. Soc. New York*.
GREGORY, W. K., and SIMPSON, G. C.
1926. "Cretaceous Mammal Skulls from Mongolia." *Am. Mus. Novitates*, No. 225, pp. 1–20.
HEWITT, J.
1922. "On the Zoölogical Evidence Relating to Ancient Land Connections between Africa and Other Portions of the Southern Hemisphere." *So. African Jour. Sci.*, XIX.
HUENE, F. VON
1929. "Los saurisquios y ornitisquios del cretáceo argentino." *Ann. Mus. La Plata*, Ser. 2, III.

LAMBRECHT, K.
1933. *Handbuch der Palaeornithologie* (Berlin: Gebrüder Borntraeger).

LOWE, P. R.
1928. "Studies and Observations Bearing on the Phylogeny of the Ostrich and Its Allies." *Proc. Zoöl. Soc. London*, I:185–247.

MATTHEW, W. D.
1915. "Climate and Evolution." *Ann. New York Acad. Sci.*, XXIV:171–318.

METCALF, M. M.
1923. "The Origin and Distribution of the Anura." *Am. Nat.*, LVII:385–411.
1929. "Parasites and the Aid They Give in Problems of Taxonomy, Geographical Distribution, and Paleogeography." *Smithsonian Misc. Coll.*, 81, No. 8.

MURPHY, R. C.
1934. [Discussion of Gregory, 1934]. *Proc. Linn. Soc. New York.*

MYERS, G. S.
1938. "Fresh-water Fishes and West Indian Zoögeography." *Smithsonian Rept. for 1937*, pp. 339–364.

NOBLE, G. K.
1925. "The Evolution and Dispersal of the Frogs." *Am. Nat.*, LIX:265–271.
1926. "An Analysis of the Remarkable Cases of Distribution among the Amphibia, with Descriptions of New Genera." *Am. Mus. Novitates*, No. 212, pp. 1–24.
1930. "The Fossil Frogs of the Intertrappean Beds of Bombay, India." *Am. Mus. Novitates*, No. 401, pp. 1–13.

PFEFFER, G.
1927. *Die Frage der Grenzbestimmung zwischen Kreide und Tertiär in zoogeographischer Betrachtung* (Jena, Gustav Fisher).

REGAN, C. TATE
1905. "A Revision of the Fishes of the Family Galaxiidae." *Proc. Zoöl Soc. London.*
1914. "Fishes." *British Antarctic ("Terra Nova") Exp. Nat. Hist. Repts.*, Zoölogy, I:1–45.

SIMPSON, G. G.
1938. "*Crossochelys*, Eocene Horned Turtle from Patagonia." *Bull. Am. Mus. Nat. Hist.*, LXXIV:221–254.

WITTMANN, O.
1934. "Die biogeographischen Beziehungen der Südkontinente." *Zoogeographica*, II: 246–304.

Part V

GEOLOGIC INTERPRETATIONS
OF PALEOBIOGEOGRAPHY

Editor's Comments
on Paper 22

As one searches through the early paleontological and paleobotanical literature for clear statements of the complex problems of
paleobiogeography, reference is commonly found to H. F. Blanford's 1875 publication on the plant fossils of the coal-bearing series
of India and their significance. Paper 22 is a compilation and synthesis of H. F. Blanford's extensive exploration and studies while
with the Geological Survey of India. His brother, W. T. Blanford,
recognized the glacial origin of the Permocarboniferous tillites
in the lower part of the Talchir Group as early as 1856. H. F. Blanford's
comprehensive coverage of the data and his discussion of the rami-
fications leading from these data are masterful. The presentation
is interwoven into a cohesive theory of a now dismembered continent. Blanford's "Indo-oceanic continent" was modified through
addition of other southern continents by Suess to form a large
Gondwana continent that was thought to have broken up into
blocks, parts of which foundered in the South Atlantic and Indian
Ocean basins.

Paper 22 is the first of several included in this concluding section that uses paleobiogeographic distributional data to reconstruct
a previous continent. It sets the ground rules for paleogeographic
reconstructions until Wegener's (1912a and b, 1915, 1924, 1966,
1967) controversial hypothesis of continental drift and Taylor's
(1910) provocative paper on tectonics. Earlier, Snider (1858) had
presented a fit of Europe and Africa against North and South America by closing the Atlantic Ocean. Warring's (1887) attempt to show
the closeness of this same fit across the North and South Atlantic

was an independent effort. Neither Snider nor Warring were able to add much supporting documentation to their geographical fit, and the hypothesis was not accepted based on either of their studies.

REFERENCES

Snider, Antonio (1858). La création et ses mystères dévoilés. Paris. (Reprinted in A. Holmes, 1944, Principles of physical geology. Nelson, London, p. 489–490.)

Taylor, F. B. (1910). Bearing of the Tertiary Mountain belt on the origin of the earth's plan. Bull. Geol. Soc. America, 21, 179–226.

Warring, C. B. (1887). The evolution of continents. Trans. Vassar Bros. Inst., 4(2), 256–274.

Wegener, A. (1912a). Die Entstehung der Kontinente. Peterm. Mitt., 185–195, 253–256, 305–309.

———. (1912b). Die Entstehung der Kontinente. Geol. Rundsch. 3, iv, 276–292.

———. (1915). Die Entstehung der Kontinente und Ozeane. Braunschweig, 94 p.; 2nd ed. (1920), 135 p.; 3rd ed. (1922), 144 p.; 4th ed. (1929), 231 p.

———. (1924). The origin of continents and oceans. (English transl. of 3rd ed.) Methuen, London, 212 p.

———. (1966). The origin of the continents and oceans. Dover, New York, 246 p.

———. (1967). The origin of continents and oceans. (English transl. of 4th ed.) Methuen, London, 248 p.

22

Reprinted from *Quart. Jour. Geol. Soc. Lond.*, **31**, 519, 525–526, 533–536, 538–540 (1875)

On the AGE *and* CORRELATIONS *of the* PLANT-BEARING SERIES *of* INDIA, *and the former* EXISTENCE *of an* INDO-OCEANIC CONTINENT. By HENRY F. BLANFORD, Esq., F.G.S. (Read December 16, 1874.)

[PLATE XXV.]

THE Peninsula of India (by which term I denote the whole of the country lying to the south of the Indo-Gangetic plain) has but little in common, in point of geological structure, with the ranges that encircle that plain on the north, or, indeed, with any of the neighbouring countries beyond, as far as these are known. The marine fossiliferous formations of Palæozoic age which are known to be largely developed in the Himalaya and the Salt range of the Punjáb, have no assignable representatives to the south of the Ganges; and the Neozoic marine formations of the same mountains, of the western ranges, and those of Eastern Bengal are represented in the Peninsula only by the Jurassic and Tertiary rocks of Cutch, the small Cretaceous formation of Trichinopoly, the somewhat older deposits of Bagh, &c. on the Lower Nerbudda, and the Nummulitic conglomerates of Broach and Surat. With these exceptions and the probably estuarine deposit of Rajamundry (of the date of the Deccan traps), and some recent coast-formations, the peninsula is formed exclusively of crystalline (chiefly metamorphic) rocks of high antiquity, of volcanic rocks, and great sedimentary formations, which are either unfossiliferous, or which contain the remains of plants and animals such as, with rare exceptions, indicate a freshwater origin and the immediate proximity of land.

The oldest of these, termed by the Geological Survey the Vindhyan and Infra-Vindhyan formations, have hitherto proved quite unfossiliferous; and their age is consequently unknown. But overlying these (very unconformably), and occupying much of Central India and the north-eastern part of the peninsular region, come the formations sometimes designated as the great Plant-bearing Series, the age of which and of their fossil contents has frequently been discussed in the pages of the Society's Journal and elsewhere. The latest general and comprehensive discussion of this subject is that given by Dr. Oldham in the second and third volumes of the 'Memoirs of the Geological Survey of India;' but later notices of parts of the series have appeared in the 'Memoirs and Records of the Geological Survey;' and several excellent papers on the geology of other countries have been published, chiefly in the Society's Journal, which help to throw light on the correlations of the Indian rocks. Having lately had occasion to investigate this question, with the help of this later evidence, I venture to submit the results to the Society, and at the same time to put forward some speculations respecting the ancient physical geography of an adjoining region, which, though not entirely original, have not, I believe, as yet been published in a definite form.

[*Editor's Summary of Pages 520–524:* Blanford reviews the lithologic and fossil similarities and differences in coal-bearing successions in India.

1. *Damuda Valley and Rajmahal Hills*:

Coarse sandstone and conglomerates	500 feet thick
Panchet group	1500
Raniganj group	5000
Ironstone shales	Damuda series	1400
Barakar group		2000
Talchir group		800

2. In Orissa, Sirguja, South Behar, and South Rewah, the Barakar and Talchir groups are overlain by coarse sandstone.

3. *Satpura field*:

Jabalpur group		500 feet thick
Bagra group		800
Denwa group	Mahadeva series	1200
Pachmari group		8000
Almod and Bijori groups		3000 to 4000
Motur group		?
Barakar group		400 to 500
Talchir group		?

4. In the Godavery Valley the Barakar and Talchir groups are overlain by the Kanithi group.

5. In Trichinopoly the Rajmahal group is overlain by the middle Cretaceous Octatoor group.

6. In Cutch, the succession is better studied and the upper part is Jurassic.]

The whole thickness of the Jurassic series is estimated at 6300 feet, of which 3000 feet belongs to the upper group.

The facts above given suffice to establish approximately the age of the Cutch representatives of the highest members of the plant-bearing series, as being probably not older than the youngest member of the Jurassic formation—the Tithonian. The evidence of the Trichinopoly plant-beds, as far as it goes, is corroborative; and it may be mentioned that the lower members of the Cretaceous series themselves, though not containing leaf-remains, abound in fragments of wood, which Dr. Thomson recognized as Cycadaceous[*]. The stratigraphical relations of the Trichinopoly and Cutch plant-beds, apart from the palæobotanical evidence of their contents, indicate that they belong to about the same geological horizon, even if not absolutely equivalent; and on this point we may expect further evidence when the associated marine fossils shall have been fully worked out. Meanwhile they may be regarded as probably Tithonian, possibly of Wealden age. This being determined, the identity of some of the characteristic plants in these and the Jabalpúr and Rájmahál rocks of Central India and Bengal justifies the inference that these latter are also approximately of similar age; and, indeed, the former cannot be much newer, since they are overlain unconformably by the Laméta group, with which Mr. W. T. Blanford identifies[†] the fossiliferous limestones of Bágh, which contain marine fossils of a Lower Cretaceous type[‡]. Some bones, apparently Reptilian, have been found in these Laméta-beds near Jabalpúr, but too imperfect for determination[§].

The age of the lower groups is a much more difficult question. The general opinion, deduced chiefly from the fossils sent home by the late Mr. Hislop (from the Kámthi and Panchét rocks), appears to be that they belong to the lower part of the Mesozoic series; but, as Dr. Oldham has remarked, this opinion appears to have been influenced very much by the supposed Oolitic affinities of the genus *Glossopteris*. This view was combated by Dr. Oldham in the 2nd volume of the 'Memoirs of the Geological Survey'[||], when he showed that the plants from the European Oolites referred to that genus have no claim to be so considered, and consequently that any inference of the Jurassic age of the Coal-bearing rocks of Bengal based on such supposed affinities is invalid. Sir Charles Bunbury mentions[¶] one *Phyllotheca* as occurring in the Oolitic rocks of Italy, on the authority of Baron de Zigno; but while inclining to the opinion that the Kámthi rocks are Mesozoic, he insists but little on the evidence of the plant-remains, and concludes that, "such as it is, it might be outweighed by the discovery of a single well-marked and characteristic fish, shell, or coral."

In considering the question of age we must be careful to bear in

---

[*] Mem. G. S. I. vol. iv. art. 1. p. 49.
[†] Mem. G. S. I. vol. vi. art. 4, p. 56, and vol. ix. art. 2, p. 36.
[‡] Duncan, Quart. Journ. Geol. Soc. vol. xxi. p. 349.
[§] Mem. G. S. I. vol. ii. p. 199.   [||] See p. 328.
[¶] Quart. Journ. Geol. Soc. vol. xvii. p. 344.

mind that we have to deal with a series of deposits of very considerable aggregate thickness. It has been shown that in the Dámúdá fields, omitting the upper sandstones, which both lithologically and by their position would seem to represent some part of the Mahádéva series, we have a thickness of 10,700 feet; and in the Sátpúra field, omitting the groups five to eight, which are probable equivalents of the former, we have an equal thickness (10,600 feet), making a total maximum of 21,300 feet of deposits, in which are several interruptions (evidenced by unconformity), though not perhaps of any great amount. There is certainly, then, no need, on stratigraphical grounds, to compress the plant-bearing series into very narrow limits of time; and we must be careful not to confuse the fossil evidence, such as it is, that is afforded by the different groups.

The Panchét and Rániganj groups with the Kámthi group, which we may assume provisionally to be of intermediate age to the two former, are those the fossil evidence of which is the least imperfect. Following Mr. Medlicott, I shall regard the Bijóri group as the local equivalent of the Rániganj-beds of Bengal, though it may perhaps with equal probability be regarded as the equivalent of the Kámthi rocks.

The Bijóri and Rániganj-beds together have furnished the Carboniferous genus *Archegosaurus* and plant-remains which, when compared with those of European formations, admittedly throw very little light on the question of age. The Kámthi group has yielded a Labyrinthodont not nearly allied to any European form, an *Estheria*, and plant-remains in part identical with those of the Rániganj-beds. The Panchét group has furnished *Hyperodapedon* and *Ceratodus*, both genera of Triassic affinities; also two new genera of Labyrinthodonts, a *Dicynodon*, and a fragment of a Thecodont saurian. With respect to these last, Professor Huxley remarks, " at present I think there is no evidence to decide whether they are older Mesozoic or newer Palæozoic;" and in another paper on *Hyperodapedon* he guardedly observes, " even now that *Hyperodapedon* is distinctly determined to be a Triassic genus, the possibility that it may hereafter be discovered in Permian, Carboniferous, and even older rocks remains an open question in my mind. Considerations of this kind should have their just weight when we attempt to form a judgment respecting the Reptiliferous strata of the Karoo formation in South Africa, and of Malédi and elsewhere in India." Perhaps the same caution, with an opposite tendency, may be observed with respect to the evidence afforded by *Archegosaurus.*

The evidence of age, then, afforded by a comparison of the fossil contents of these formations with those of European rocks leads only to somewhat vague conclusions. It indicates little more than that these formations probably range somewhere between the Carboniferous horizon and that of the Trias. The occurrence of *Archegosaurus* would tend to depress the Bijóri and with it the Rániganj group to a Carboniferous horizon, and that of *Hyperodapedon* and *Ceratodus* to raise the Panchet group to a Triassic horizon, in which case the Kamthi group might be supposed to represent the Permian formation.

[*Editor's Summary of Pages 531–533:* Blanford compares the Indian fossil plants with those of Australia and shows their close generic similarity and correlation. This is followed by a discussion of glacial till beds at the base of the Indian and South Africal coal-bearing groups and a comparison of the South African Karoo series and its fossils to those of India. He reviews the glacial deposits of the Pleistocene and discusses elevation, temperature, and precipitation as related to the topography of the Himalayas and to glaciation.]

If the probability be admitted that the reduction of temperature evidenced by the glaciation of Postpliocene times was general, we may by analogy argue that such was also probably the case in early Permian times, in which case the physical evidence of glaciation becomes of at least equal value with palæontological evidence in questions of correlation, perhaps even of greater value when it is a question of correlating formations in very distant regions; and since the palæontological and stratigraphical evidence warrants the conclusion that the Ecca conglomerates, the Talchír boulder-bed, and the Permian breccias of England are not very far distant in time, I think we may provisionally refer them with some probability to the same glacial period.

The comparatively small thickness of the Talchír boulder-bed in latitudes 17° to 23°, as compared with the great thickness of the Ecca conglomerate in latitudes 30° to 32°, would perhaps imply a shorter duration of glacial conditions in the former; but mere thickness is, of course, no safe criterion of duration. Moreover, as suggested by my brother, ground-ice in the winter time would be quite competent to produce all the effects to be accounted for in the Indian area, though in Natal Dr. Sutherland appears to consider that the evidence is in favour of a general glaciation of the country.

The assumed return of warmth during Permian times, after a general reduction of temperature, would be eminently favourable to that extension of similar or nearly allied forms of life which we have evidenced in the Australian, Indian, and South-African plant-beds, provided the distribution of the land were such as to admit of their diffusion, a subject which I shall discuss presently. The prevalence of a cold climate would also help to explain the comparative absence of animal and vegetable life from the deposits of the Talchír period—deposits which by their lithological character seem well fitted to receive and preserve such remains. This Koonap group in South Africa would seem to be equally deficient in fossil remains, these beds in lat. 30° to 32° being the supposed representatives of our Dámúdá series in lat. 17° to 23°, in which a flora abundant in individuals, but poor in species, at that time flourished. With regard to the Australian plant-beds I see no evidence to show whether they preceded or followed the glacial period. Australia may have been the home in later Carboniferous times of a flora which was afterwards driven towards the equator, and subsequently, on the return of a genial climate, spread to Africa and India; or since, for aught we know, the marine forms *Orthoceras, Eurydesma, Spirifer, Conularia,* and *Fenestella,* &c. may have been fitted to live in cold seas; the beds containing them may be in part of early Permian age: this supposition gains some support from the lithological resemblance of the rocks containing them to those of the Talchír group, a resemblance noticed by Dr. Oldham, and which implies some similarity of physical conditions. On this supposition *Glossopteris, Phyllotheca,* &c. may have spread to Australia in Permian times, and the extinction of the preceding Devonian flora (*Lepidodendron* &c.) may have occurred during the glacial period. Perhaps the extinction of the Carboniferous flora of Europe may have been due to the same cause.

On the review of all the probabilities (and partial probabilities are the only guides we have), I am inclined, then, to relegate the lowest groups of the Indian plant-bearing series to Permian times, and to correlate them with the lower groups of the Karoo formation of Africa and, possibly, the Wollongong sandstones and Newcastle-coal series of Australia. But I admit that the validity of this view depends in a very great measure on that of my speculation that there was a general decrease of temperature over the earth's surface between the Carboniferous and Permian epochs; and this must stand or fall by the evidence of future investigation.

The affinities between the fossils, both animals and plants, of the Beaufort group of Africa and those of the Indian Panchéts and Kámthis are such as to suggest the former existence of a land connexion between the two areas. But the resemblance of the African and Indian fossil faunas does not cease with Permian and Triassic times. The plant-beds of the Uitenhage group have furnished eleven forms of plants, two of which Mr. Tate has identified with Indian Rájmahál plants. The Indian Jurassic fossils have yet to be described (with a few exceptions); but it has already been stated that Dr. Stoliczka was much struck with the affinities of certain of the

Cutch fossils to African forms; and Dr. Stoliczka and Mr. Griesbach have shown that of the Cretaceous fossils of the Umtafuni river in Natal, the majority (twenty-two out of thirty-five described forms) are identical with species from Southern India. Now the plant-bearing series of India and the Karoo and part of the Uitenhage formation of Africa are in all probability of freshwater origin, both indicating the existence of a large land area around, from the waste of which these deposits are derived. Was this land continuous between the two regions? and is there any thing in the present physical geography of the Indian Ocean which would suggest its probable position? Further, what was the connexion between this land and Australia, which we must equally assume to have existed in Permian times? And, lastly, are there any peculiarities in the existing fauna and flora of India, Africa, and the intervening islands which would lend support to the idea of a former connexion more direct than that which now exists between Africa and South India and the Malay peninsula. The speculation here put forward is no new one. It has long been a subject of thought in the minds of some Indian and European naturalists, among the former of whom I may mention my brother and Dr. Stoliczka, their speculations being grounded on the relationship and partial identity of the faunas and floras of past times, not less than on that existing community of forms which has led Mr. Andrew Murray, Mr. Searles V. Wood, jun., and Professor Huxley to infer the existence of a Miocene continent occupying a part of the Indian Ocean. Indeed, all that I can pretend to aim at in this paper is to endeavour to give some additional definition and extension to the conception in its geological aspect.

With regard to the geographical evidence, a glance at the map will show that from the neighbourhood of the west coast of India to that of the Seychelles, Madagascar, and the Mauritius extends a line of coral atolls and banks, including Adas bank, the Laccadives, Maldives, the Chagos group, and the Saya de Mulha, all indicating the existence of a submerged mountain-range or ranges. The Seychelles, too, are mentioned by Mr. Darwin[*] as rising from an extensive and tolerably level bank, having a depth of between 20 and 40 fathoms; so that, although now partly encircled by fringing reefs, they may be regarded as a virtual extension of the same submerged axis. Further west the Cosmoledo and Comoro Islands consist of atolls and islands surrounded by barrier reefs; and these bring us pretty close to the present shores of Africa and Madagascar. It seems at least probable that in this chain of atolls, banks, and barrier reefs we have indicated the position of an ancient mountain-chain, which possibly formed the back-bone of a tract of later Palæozoic, Mesozoic, and early Tertiary land, being related to it much as the Alpine and Himalayan system is to the Europæo-Asiatic continent[†],

---

[*] 'Coral Reefs,' Appendix, p. 185.

[†] This idea, based of course on Mr. Darwin's discovery, has been also specifically suggested by Mr. Andrew Murray (see the 'Geographical Distribution of Mammalia,' p. 25), and by Mr. Searles V. Wood, jun. (see Phil. Mag. 1862, vol. xxiii. p. 388).

and the Rocky Mountains and Andes to the two Americas. As it is desirable to designate this Mesozoic land by a name, I would propose that of Indo-Oceania. Professor Huxley has suggested, on palæontological grounds *, that a land connexion existed in this region (or rather between Abyssinia and India) during the Miocene epoch. From what has been said above, it will be seen that I infer its existence from a far earlier date. With regard to its depression, the only present evidence relates to its northern extremity, and shows that it was, in this region, later than the great trap-flows of the Dakhan. These enormous sheets of volcanic rock are remarkably horizontal to the east of the Gháts or the Sahyádri range, but to the west of this they begin to dip seawards, so that the island of Bombay is composed of the higher parts of the formation †. This indicates only that the depression to the westward has taken place in Tertiary times; and to that extent Professor Huxley's inference, that it was after the Miocene period, is quite consistent with the geological evidence.

[*Editor's Summary of Pages 536 (part), 537, and 538 (part)*: The Recent faunas of India are compared with those in Africa and southern Asia.]

---

* Anniversary Address to the Geol. Soc. 1870, p. lvi.
† Wynne, "Geology of the Island of Bombay," Mem. G. S. I. vol. v. Art. 3.

Palæontology, physical geography, and geology, equally with the ascertained distribution of living animals and plants, offer then their concurrent testimony to the former close connexion of Africa and India, including the tropical islands of the Indian Ocean. This Indo-Oceanic land appears to have existed from at least early Permian times, probably (as Professor Huxley has pointed out) up to the close of the Miocene epoch; and South Africa and Peninsular India are the existing remnants of that ancient land. It may not have been absolutely continuous through the whole of this long period. Indeed the Cretaceous rocks of Southern India and Southern Africa, and the marine Jurassic beds of the same regions, prove that some portions of it were, for longer or shorter periods, invaded by the sea; but any break of continuity was probably not prolonged; for Mr. Wallace's investigations in the Eastern Archipelago have shown how narrow a sea may offer an insuperable barrier to the migration of land-animals. In Palæozoic times this land must have been connected with Australia, and in Tertiary times with Malayana, since the Malayan forms with African alliances are in several cases distinct from those of India. We know as yet too little of the geology of the eastern peninsula to say from what epoch dates its connexion with Indo-Oceanic land. Mr. Theobald has ascertained the existence of Triassic, Cretaceous, and Nummulitic rocks in the Arakan coast-range; and Carboniferous limestone is known to occur from Moulmein southwards, while the range east of the Irawadi is formed of younger Tertiary rocks. From this it would appear that a considerable part of the Malay peninsula must have been occupied by the sea during the greater part of the Mesozoic and Eocene periods. Plant-bearing rocks of Rániganj age have been identified as forming the outer spurs of the Sikkim Himálaya; the ancient land must therefore have extended some distance to the north of the present Gangetic delta. Coal, both of Cretaceous and Tertiary age, occurs in the Khasi Hills, and also in Upper Assam, but in both cases associated with marine beds; so that it would appear that in this region the boundaries of land and sea oscillated somewhat during Cretaceous and Eocene times. To the north-west of India the existence of great formations of Cretaceous and Nummulitic age, stretching through Balúchistán and Persia, and entering into the structure of the North-west Himálaya, prove that in the later Mesozoic and Eocene ages India had no direct connexion with Western Asia; while the Jurassic rocks of Cutch, the Salt range, and the Northern Himálaya show that in the preceding period the sea covered a large part of the present Indus basin; and the Triassic, Carboniferous, and still more ancient marine formations of the Himálaya indicate that, from very early times till the upheaval of that great chain, much of its present site was for ages covered by the sea.

Table of Indian Plant-bearing Series and Equivalents.

INDIA					S. AFRICA.	AUSTRALIA.	EUROPE.
CUTCH.	MADRAS.	GODAVERY.	BENGAL.	NERBUDDA.			
Upper Jurassic (of Wynne). Lower Jurassic (of Wynne).	Sripermatúr &c., Zamia beds.	Sandstones with Zamias &c.	Rájmahál group.	Jabalpúr group.	Uitenhage series.	Wollumbilla series.	JURASSIC.
		Kota-beds. Panchéts (Malédi).	Upper sandstones of coalfields. Panchét group.	Bágra group. Denwa group. Pachmari group. Almód group.		Wyanamatta group.	LIASIC. TRIASSIC.
		Kámthi group (Mángli).	Rániganj group.	Bijori group.			
			Ironstone shales.	Motúr group.	Karoo series.	Hawkesbury group. Newcastle coal. Wollongong sandstone (part)?	PERMIAN.
		Barákar group.	Barákar group.	Barákar group.			
		Talchír group.	Talchír group.	Talchír group.			

To sum up the several views advanced in this paper :—

1st. The plant-bearing series of India ranges from early Permian to the latest Jurassic times, indicating (except in a few cases, and locally) the uninterrupted prevalence of land and freshwater conditions. These may have prevailed from much earlier times.

2nd. In the early Permian, as in the Postpliocene age, a cold climate prevailed down to low latitudes, and, I am inclined to believe, in both hemispheres simultaneously. With the decrease of cold the flora and reptilian fauna of Permian times were diffused to Africa, India, and possibly Australia; or the flora may have existed in Australia somewhat earlier, and have been diffused thence.

3rd. India, South Africa, and Australia were connected by an Indo-oceanic continent in the Permian epoch; and the two former countries remained connected (with at the utmost only short interruptions) up to the end of the Miocene period. During the latter part of the time this land was also connected with Malayana.

4th. In common with some previous writers, I consider that the position of this land was defined by the range of coral reefs and banks that now exists between the Arabian Sea and West Africa.

5th. Up to the end of the Nummulitic epoch no direct connexion (except possibly for short periods) existed between India and Western Asia.

Quart. Journ. Geol. Soc. Vol XXXI Pl XXV

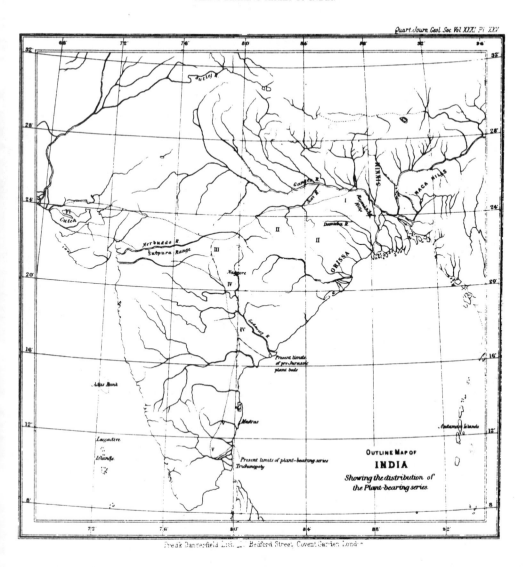

Fred.k Dangerfield Lith. 22. Bedford Street Covent Garden London

EXPLANATION OF PLATE XXV.

Outline Map of India, showing the Distribution of the Plant-bearing Series referred to in Mr. Blanford's paper.

Editor's Comments
on Papers 23 and 24

23   SCHUCHERT
The Paleogeography of Permian Time in Relation to the Geography of Earlier and Later Periods

24   HILL
Sakmarian Geography

Papers 23 and 24 examine Permian paleogeography as deduced by biostratigraphy and stratigraphy. Paper 24 was written about thirty years after Paper 23 and, because of the great increase in data, considered only the early part of the Permian Period.

Charles Schuchert, in Paper 23, shows his continued and growing interest in Carboniferous–Permian glaciations in the Gondwana continents (see also Schuchert, 1928) and the recognition of provincial fossil provinces, which he appeared to prefer to ignore as far as possible in this paper. Schuchert's analysis of world paleogeography for the early Cambrian, middle Silurian, early Permian, and middle and late Cretaceous was quite detailed considering when it was constructed.

A few years later Schuchert (1932) and Willis (1932) examined the question of "land bridges" and "isthmian links" that were once thought to have connected the Gondwana continents across what are now ocean basins. By this time, geophysical data were starting to accumulate to indicate that most of the crust was close to isostatic equilibrium (the Tethyan and circum-Pacific fold belts being the most notable exceptions), so that the idea of large blocks of sialic crust foundering to form the South Atlantic ocean basin had to be rejected. Thus Schuchert (1932) and Willis (1932) together, compromising between the fossil record and the geophysical and structural data, arrived at the conclusion that narrow land bridges were needed, which probably resembled the present Cuban–Antillean chain. A sunken Brazilian–Guinean ridge including St. Paul's Rocks was proposed, as well as other ridges across other oceans; this lead to a major problem in later literature as neontologists and paleontologists freely, without constraint of geological data, alternately constructed and severed isthmian links with abandon for the next decade.

Dorothy Hill in Paper 24 examines the paleogeography of the Sakmarian (which is used here in the sense of the early part of the Early Permian) and discusses primarily a biogeography defined by marine invertebrates. Interestingly, there is only one world map, and that indicates the location of Sakmarian glacial deposits; the rest show individual continents without attempting to connect different continents. In a later article, (Hill, 1973) a summary of Lower Carboniferous corals, she divides the world into coral provinces and subprovinces.

Dorothy Hill became interested in fossil corals because of a chance discovery she made while visiting friends near Munduberra, Queensland, Australia; this led her to Cambridge University and association with British coral workers. In England, after finishing a study of Australian fossil corals, she next unraveled the then existing confusion among Scottish Carboniferous corals. In 1937 she returned to the University of Queensland in Australia to continue further studies on Australian corals in conjunction with teaching commitments. She served with distinction in World War II in the Australian Navy. In addition to work on corals and other fossils, Dorothy Hill's interests have expanded to paleogeography and regional geologic mapping. She was founder of the Queensland Palaeontological Society and served as a university administrator. In 1971 the Queen conferred the award of C.B.E. on her for her services to geology and paleontology.

REFERENCES

Hill, Dorothy (1973). Lower Carboniferous corals, *in* A. Hallam, ed., Atlas of palaeobiogeography. Elsevier, Amsterdam, p. 133–142.

Schuchert, Charles (1928). Review of the late Paleozoic formations and faunas, with special reference to the ice-age of middle Permian time. Bull. Geol. Soc. America, 39, 769–886.

———. (1932). Gondwanaland bridges. Bull. Geol. Soc. America, 43, 875–915.

Willis, Bailey (1932). Isthmian links. Bull. Geol. Soc. America, 43, 917–952.

23

Reprinted from *Proc. Pan-Pacific Sci. Congr.*, Australia, *1923*, Vol. 2,
Pacific Sci. Assoc., Aust. Nat. Res. Council, Melbourne, 1924, pp. 1079–1091

The Paleogeography of Permian Time in Relation to the Geography of Earlier and Later Periods.

By Charles Schuchert, Yale University.

It is a great honour to have the geologists of Australia call on the writer, living 10,000 miles away from the scene of their activity, to help them solve their various Permo-Carboniferous problems. Of course, it goes without saying that I cannot at this distance do anything on the actual relationships of the field evidence, nor give any valuable opinion in regard to the floral and faunal relations to the rest of the world. It is, however, one of my dreams that I may retain the strength and ambition to visit Australia, and see the paleontologic collections in your many museums, and something of your actual field geology. Such a visit would be immensely stimulating to me, and I hope this expectation may attain realization. On the present occasion, all that I can do is to present some general impressions that I have gained from reading Australasian paleontology and stratigraphy, and to offer some world maps showing the paleogeography of Permian times and its bearing on the geography of earlier and later periods. Toward this end I am submitting five maps of Lower Cambrian, Silurian, early Permian and Middle and Upper Cretaceous times.

The Father of Australian Geology, Professor T. W. E. David, has in his time not only seen to it that the geology of your Commonwealth became known, but has developed a school of progressive and active workers, and now the clear ringing of their geologic hammers is heard near and far. Your leader is to be congratulated upon his great success.

I hope no Australian paleontologist will feel hurt when I say that until about ten years ago, most of your paleontology was of the pioneer kind. I know what the difficulties of pioneers are, since first interpretations of paleontologic ghosts usually turn out to be wide of their actual relationships and of what the "medals of creation" really have recorded in themselves. It is now nearly 30 years since a small lot of marine invertebrates collected in Alaska were handed me, and so strange were they that I soon was lost in an unknown paleontologic world. My determinations were, I thought, carefully done, and yet I was two periods away from the actual truth, as was later demonstrated to me by another. Therefore with the feeling of a fellow sinner I am glad to note that the pioneer days in Australian paleontology and stratigraphy are a matter of the past, and I hope we may soon say the same of New Zealand. Pioneer paleontology means that the generic, and especially the specific identifications are wide of the mark, since a form that has gone around the world rarely, if ever, retains the same binomial characters. Only a very few forms have done this, and these are very plastic species around circumscribed characters; such are *Atrypa reticularis, Leptœna rhomboidalis, Productus semireticulatus,* &c. None of these has any accurate meaning as to time relations or faunal affinities, and yet they have been depended upon to determine at least period values and provincial connexions. It is probably safe to say that practically every American species of brachiopod identified as occurring in Siberia, India, and Australia is not such, and that even in most cases they are not even of the same subgenus. Among species of the two hemispheres, we should expect many parallel developments, but not exact identities.

Long ago the great Huxley warned against synchronous correlations in widely separated places, and said that all we can expect in these cases are homotaxial or general relationships. In theory he was correct, but in actual practice among paleontologists dealing with marine life, it is now certain that their age determinations are everywhere to all intents and purposes of the same time. Huxley was thinking of correlations in years, or rather in some hundreds of years, but we stratigraphers know that any long-range correlation cannot be exact even to some tens of thousands of years. To paleontologists working with marine forms, the related faunas are, however, everywhere of the same age, since geology has not as yet, and probably never will have, a means for discriminating time values closer than within a few tens of thousands of years. On the other hand, we Americans know that the living *Littorina littoralis,* accidentally imported from Europe, has in about 75 years spread along the Atlantic shore more than 2,000 miles in a straight line, which means that marine species under favorable circumstances can spread many times around the whole earth in the least time measurable by stratigraphers. Therefore chronogenesis to the stratigraphers of the southern hemisphere should in practice be the same yardstick as that used by the workers north of the equator.

My hope is that the Second Pan-Pacific Congress will expunge the term Permo-Carboniferous, since all the rest of the world has rid itself of this confession of ignorance. The illustrations so far presented of Australian marine fossils give me the impression that you have some early Lower Carboniferous, possibly of two horizons (earliest and middle Lower Carboniferous), but I see no marine Upper Carboniferous anywhere. If this is true, it is a very significant fact, and all the more so because in eastern Australia, where the Devonian and Permian are so well developed, there should be no late Lower and no Upper Carboniferous. This can only mean a marked time of orogeny beginning in late Lower Carboniferous time and continuing into the Upper Carboniferous. Such a crustal movement would be in harmony with the Hercynian ones of Europe and North America.

291

To the writer, nearly all of the Australian " Permo-Carboniferous " is Permian, and younger than the earliest Permian or Artinskian. In other words, you have a grand development of what appear to be genuine Permian deposits laid down near shores adjacent to highlands unloading a·great amount of clastics into the cool to cold water Tasman Sea. The life assemblage is a marine one, peculiar to south-eastern Australia, or at least one sees no direct biotic connexion with Tethys. The marine faunas of north-western Australia I do not know, but on the basis of the paleogeography they should have decided connexions with those of the Himalayas, Timor, and the Banda Sea.

What brought on the decided Permian ice age in the southern hemisphere? Certainly never the impossible Wegenerian hitching together in close embrace of the present masses of South America, Africa, India, and Australia. In this hypothesis I see only a drunken salic supercrust hopelessly floundering upon the sober sima. Let us return to a tangible geology. We all now know that the earth underwent one of its greatest crustal disturbances during the Carboniferous, beginning late in the Lower Carboniferous (toward the close of Viséan and Chesterian time), attaining a first climax in middle Upper Carboniferous time, and a second maximum early in the Permian. The supercrust was undergoing one of its periodic revolutions, and the world was then as scenically grand as it is to-day. It is in the youthful topography, the enlarged continents, and the peculiar connexions of the continents that seemingly are to be sought the reasons for the Permian ice age. We will return to this problem when the Permian world paleogeography is shown on the screen.

In presenting five world paleogeographic maps, let me say that the making of such charts was begun ten years ago, and that I now have in preparation about 50 of them, beginning with the Cambrian and closing with the Cretaceous. Of the North American continent alone I have at least 115 for as many different times, and of South America ten more. My work on the Australasian region has been made easy through the paleogeographic studies of Dr. A. B. Walkom and Professor W. N. Benson. Therefore, with all of this information at hand, I feel safe in offering the following broader outlines of our present knowledge of the earth's changing topography.

Dana long ago said that North America is the " type continent," meaning that it is the best expression of an inherent continental mass, since it has a basin-like form. The basin includes the immensely large depressed centre that is enclosed by a rim of mountains of ancient and modern construction. On the east, the elevated rim is made by the 2,000-mile-long Appalachians, on the south by the very short Ouachitas, on the west by the very wide and 5,000-mile-long Rocky and Pacific systems, and on the Arctic side by at least the United States Mountains.

My work is also making it clear that this basin structure came into being in later Proterozoic time, and accordingly I am led to add that all continents appear to be very ancient, originating as early as the Proterozoic.

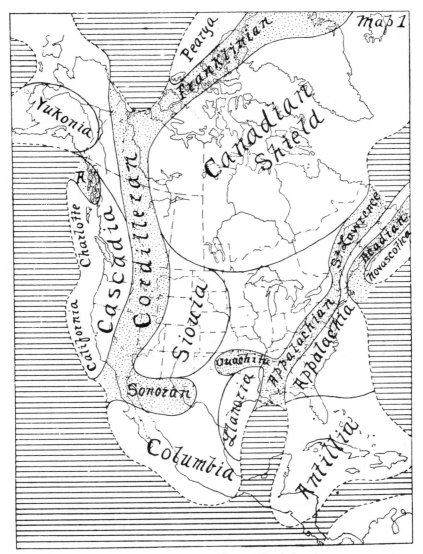

Fig. 1.—Paleozoic Geosynclines of North America.

The depressed medial area or nucleus of North America, consisting of very ancient granites and more or less highly deformed strata, came into existence as a vast peneplain before the Ordovician, and more than nine-tenths of it before the Cambrian. It is the nucleal and neutral area in regard to post-Proterozoic deformation, and until recently has lain but a few hundred feet above sea-level. On the other hand, the

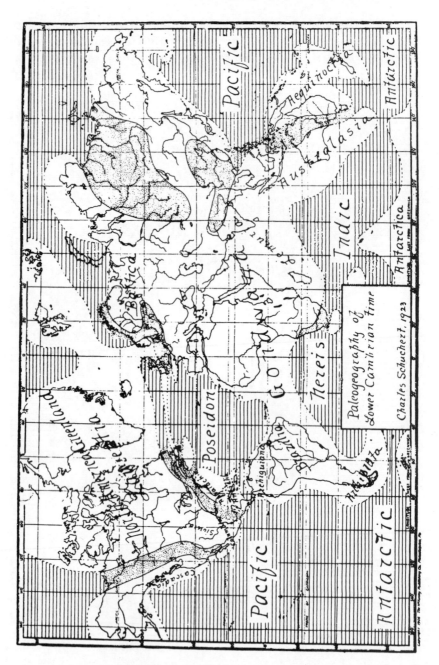

FIG. 2.—Paleogeography of Lower Cambrian time.

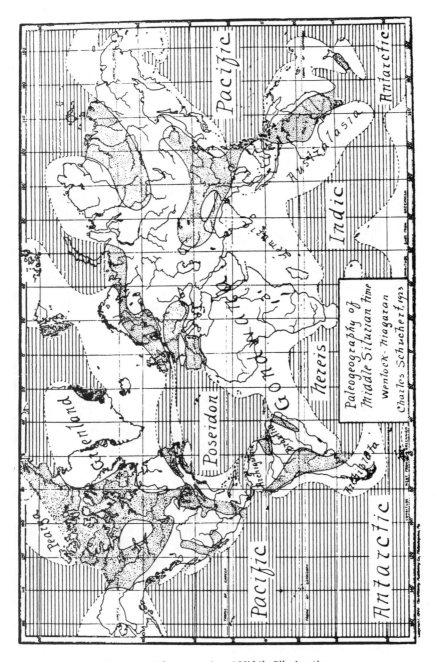

Fig. 3.—Paleogeography of Middle Silurian time.

295

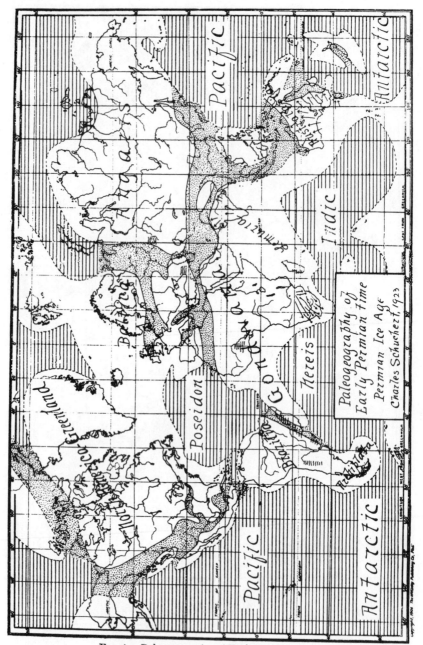

Fig. 4.—Paleogeography of Early Permian time.

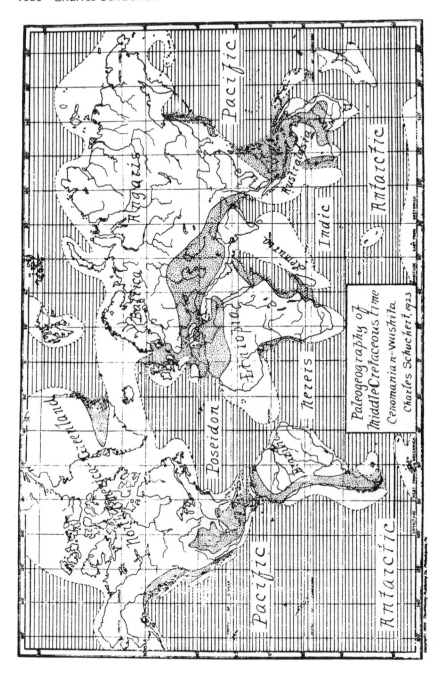

FIG. 5.—Paleogeography of Middle Cretaceous time.

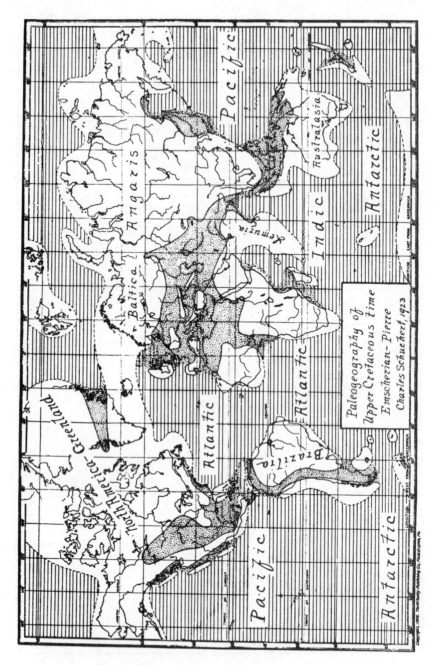

FIG. 6.—Paleogeography of Upper Cretaceous time.

elevated rims of the basin that I am calling the borderlands are the most mobile and positive parts of the continent, and are the areas from which have come nearly all the sediments for the epeiric seas that formed on their inner sides and over the neutral region. Immediately inside of the borderlands are the geosynclines that spill more or less widely over the nucleal area, while on their outer sides are the oceans with their ingression gulfs. During the marine history of the geosynclines, the borderlands are vertically elevated time and time again, but eventually they are pushed against the geosynclines, folding the deposits of the latter against and upon the nucleus or kratogen. To me it is evident that the larger portions of the mobile borderlands have during the Mesozoic and Cenozoic been foundered into the oceans.

South America is another good example of the "type continent," since its general geologic construction is very much like that of North America, with this difference, that its eastern geosyncline is a very shallow evanescent one, and hence without the long marine sequence of the Appalachian trough. Then, too, the eastern geosyncline of South America had no borderland, since the land to the east of the trough was a part of Gondwanaland, which fractured into the depths of the Atlantic very early in Cretaceous time. South America, therefore, had no eastern rim to its Brazilian nucleus, and is thus a "typical continent" with its eastern part sunk into the Atlantic.

Africa, even though still as large a land mass as either present North or South America, is a very imperfect continent geologically, since all of its western and eastern margins are broken down into the adjacent oceans. The north-western margin went down into the Atlantic during the early Cretaceous, though its south-western part appears to have sunken earlier. On the other hand, the eastern "Ethiopian Mediterranean" of Neumayr, of Mesozoic making, along with most of its eastern borderland, sank into the Indian Ocean during the later Mesozoic, early Paleozoic marine records being known only from the Sahara and Cape Colony. From these statements we see that present Africa is but part of a greater continent, since its western Gondwana extension is now the medial Atlantic Ocean, while eastern Lemuria has gone into the making of the Indian Ocean. Just when the present highly elevated condition of Africa developed is unknown to me, but seemingly during the later Cenozoic.

Euro-Asia, as the name already implies, is composed of at least two continents that are now welded into one. Europe developed around the Baltic nucleus, while Asia did the same about the Angaran nucleus of Siberia. These two continents are separated by the north-south trending Uralian geosynclinorium. On the other hand, the far north-eastern extremity of Russia, along with a part of Alaska, may be another geologic unit, in which case Euro-Asia would be composed of three continental masses separated by two geosynclines, both of which are of the type that I have elsewhere called polygeosynclines. Along the southern side of these lands lay Tethys, a vast mediterranean extending from Spain to at least Burma. Even though Euro-Asia is composed of two or three continents, we must not omit to add, as has long been recognized, that during much of Paleozoic and Mesozoic times Euro-Asia and North America were united into one vast north land, Eria.

Australia is the remainder of a vast Paleozoic land, long known as Australasia, and more recently named by Abendanon Aequinoctia. The western half of Australia appears to me to be one of the two nuclei about which the geologic history of Australasia has developed, the second one being southern Annam and the shallow sea to the south of it and east of Siam and Malacca. This vast continental mass, however, really appears to be composed of two continental units, of which the western one is Australasia proper, including most of Australia, the Dutch East Indies, Celebes and Borneo, Annam, Siam, and Burma; while the eastern, now greatly foundered continent of Aequinoctia embraces the Philippines, New Guinea, eastern Australia, New Caledonia, and New Zealand. If it ever had a nucleus, it now lies in the depths of the Tasman Sea. Australasia and Aequinoctia are separated by the medial polygeosyncline of Australia, which appears to be continued northwesterly into the present shallow Arafura Sea and the very deep waters of Banda, Celebes, Sulu, and South China seas.

Now let us look at the maps in succession to see how greater Australasia developed into its present maltreated remainder. The observer is warned at once against accepting the outer boundary lines of the continents as here shown, since their assumed shores now all lie in the great depths of the oceans where no geologist can disprove the assumptions of the paleogeographer. In our studies, however, we must have bounding limits to our land conceptions so that we may know their interrelations and their relations to the oceans as well.

The most striking fact of Lower Cambrian time is the east-west alignment of the continents, and the immense equatorial transverse land of Gondwana, with Australasia and Aequinoctia on the south-east and Brazilia to the south-west. They all appear to have come into existence during the Proterozoic, to have endured throughout the Paleozoic, and to have begun breaking up after middle Mesozoic time. The marine Paleozoic and Mesozoic faunas of western South America alone prove the existence of Gondwanaland across what is now the Atlantic, while the marine faunas of Australia show a Pacific assemblage more often than one of India and the Tethyian mediterranean.

The medial geosyncline of Australia, with marine waters, is present at least as early as the Lower Cambrian and reappears in the Middle Silurian; then there is no evidence of it until late Jurassic into Middle Cretaceous times.

The southern Tasman Sea is certainly present as early as the late Cambrian; and thence to the end of Paleozoic time this ingression gulf overlaps across eastern Australia, as is clearly evidenced by the many fossiliferous formations. After this, the eastern shore of Australia was enlarged, since it is not again overlapped until the Middle Cretaceous, and even then only in a very limited area of Queensland.

The waters of eastern Tethys do not appear in Australia definitely until Permian time, when they encroached widely across the Northern Territory and especially Western Australia. Further records are known in the Jurassic and Middle Cretaceous, though it is probable that yet other overlaps will be found.

Permian time in Australia is not only interesting for the long and peculiar marine record with its periods of coal-making swamps, but even more so because of the clear evidence of an ice age throughout the southern hemisphere. In Australia the ice masses moved to the northward, indicating high land where to-day is the Great Australian Bight. Peninsular India was also a highland, since the ice sheets here also moved to the north into the Tethyian sea. For South Africa, Du Toit has recently summed up the evidence of the Permian ice work, and here the ice sheets flowed from the highlands of the north-east towards the south and the south-west into the South Atlantic Ocean. In South America, there are tillites along a great part of eastern Brazil in the area of the Franciscan geosyncline, of wide extent in north-western Argentina, and on the Falkland Islands. Undoubtedly much of Antarctica was also ice-covered. If it can be shown that Australasia was connected in Permian times with Antarctica, then this holding in of so much of the cold waters of the Antarctic Ocean, combined with moist climates in the southern hemisphere and the general highland condition of much of the world in early Permian time, will be the explanation for the peculiar position of the continental ice masses of the southern hemisphere.

The severance of Aequinoctia from greater Australasia appears to have taken place in the Middle Devonian, and it will probably be shown to have done so even earlier. In Permian time this old land was already much broken up, and New Zealand certainly became an island mass, though its isolation may go back even to Lower Carboniferous time. The pull of the sinking Pacific and Antarctic appears to me to be the cause for this grandest of all continental destructions.

Australia became an island-continent definitely early in the Upper Cretaceous, and probably as early as Permian time. Floral and faunal migrations from Asia have been difficult probably ever since Upper Carboniferous times, and yet far more difficult intermigrations appear to have taken place between Australia and South America during the Cretaceous, in which case a land bridge is hypothecated like the one for Permian time. Some day the floras and faunas of Mesozoic and Cenozoic

Australia will prove the existence or absence of this Antarctic land bridge, and if it existed during the Cretaceous, it will be far easier to believe that the bridge was also present in Permian time.

The Permian, Jurassic, and Cretaceous maps show when, and a hint as to where, parts of greater Gondwana began to founder into the oceans, and when the Indian and Atlantic oceans, and the present continents as well, took on their familiar configurations.

In conclusion, let me add that I have written this paper in a positive manner, just as if all the things said came to pass as stated. Far be it from me to be actually dogmatic, but I hold that it is better to set up a definite target at which future work may be directed, and when all the evidence that can be gathered is at hand, a clear vision of the events of Historical Geology will be the more quickly attained.

24

Reprinted from *Geol. Rundsch.*, **47**(2), 590–629 (1958)

SAKMARIAN GEOGRAPHY

By DOROTHY HILL, *University of Queensland*

With 8 Figures

Abstract

The known palaeontological and stratigraphical evidence is used as a basis for the construction of maps of the continents showing the extent of their inundation by the sea in Sakmarian time in the Upper Palaeozoic. In the northern hemisphere apart from India the evidence is sufficiently reliable to give reasonable maps, and the great extent of the inundations suggests that the climate would be considerably modified from that of today; no undoubted Sakmarian glacials occur there. In Southern continents and India, the "Gondwana" biogeographical province has made correlation with the northern continents controversial, but reasons are given for assuming that "Gondwana" glacial deposits were at least in part Sakmarian; the resultant maps show that the "Gondwana" land surfaces were but little reduced in area, and that the main glacials (except for India) lie within a belt between 40° S and 20° S. Present lack of knowledge of Sakmarian conditions in Antarctica makes reconstructions of climatic belts too hazardous for possible use in enunciating or checking hypotheses of continental drift and polar wandering.

This paper offers provisional maps showing the land and sea surfaces deduced from the known sediments, fossils and earth movements for as short as possible a period of geological time during the upper Palaeozoic glaciation. Where knowledge warrants, the text discusses the relief of the land. The time chosen is the Sakmarian. Obviously the first essential is the reliable correlation of the strata used, from basin to basin, and from continent to continent, and the Sakmarian is particularly interesting but difficult, because of the apparent development of distinct biogeographical provinces — "Gondwanaland" and the continents of the Northern Hemisphere.

Factors affecting surface temperature in the past (as now) in different parts of the earth are numerous; but they include some about which only Historical Geology can supply evidence, such as the relation between land and sea surfaces and the relief of the land. These together affect the atmospheric and oceanic circulation; the former is the more important in weather and climate, but the latter is important in the migration of marine species. FLOHN (1952 Geol. Rundsch. p. 153) has brought to the attention of geologists modern three-dimensional studies on the atmospheric circulation in relation to land and sea surfaces, relief, and weather and climate. He indicates that glaciation under present day and Pleisto-

303

cene conditions is favoured by meridional types of circulation as against zonal types, and that in the N. Hemisphere there is a relation between the magnetic poles, the quasi-stationary high-level troughs of the meridional circulation and the centres of glaciation. He suggests that past glaciations may best be explained by fluctuations in solar radiation combined with increases in meridional type circulation due to variations in the surface relief of the earth.

If we wish to examine the upper Palaeozoic glaciation in the light of modern ideas on orography and three dimensional atmospheric circulation or vice versa, or to assess current geophysical speculations on the wanderings of the pole or continental drift, obviously a first requirement is a palaeogeographic world map that is as accurate as Historical Geology can make it. We must know what these speculations are required to explain. The maps presented here are offered primarily to historical geologists for improvement; when we are reasonably agreed upon emended maps, these can be used in the meteorological and geophysical speculations, either to check on a given hypothesis, or to help in formulating an hypothesis.

That period of the earth's history arbitrarily known as Sakmarian time takes its name from a series of sediments deposited in the environs of the Sakmara River, a tributary entering the Ural River at Chkalov (o l i m Orenburg) on the South West slopes of the Ural Mountains. The limits and equivalence of the Sakmarian, even in Russia, are still somewhat fluid, and to give some fixity for this essay I have arbitrarily taken the limits given by Ruzhentsev (1954, 1956) the Russian cephalopod specialist, which appear to be accepted also by Shimansky (1954), Rozovskaya (1949) and Maximova (1952). These are from the top of Ruzhentsev's Orenburgian stage to the base of his Artinskian stage; the Sakmarian stage thus includes Ruzhentsev's Asselian and Sakmarian substages, the latter (and upper) being divisible into the Tastuba and Sterlitamak horizons. It would thus seem to include Dunbar's (1940, 1942) usage of Sakmarian and upwards to the base of the beds with *Parafusulina lutugini;* and in these extended limits it appears to be almost exactly equivalent to the Wolfcampian of North America (Ruzhentsev, 1956). Ruzhentsev earlier had grouped the Sterlitamak horizon in the Artinskian.

Invertebrate palaeontologists have done their most precise work for this general time on goniatites, fusulines and corals, neglecting, unfortunately, the brachiopods which are to be found almost everywhere. Goniatites are common only in the Urals and North America; fusulines are common in North America, the Urals, Eurasiatic Tethys and South East Asia, and corals are common only in the Urals and in Eurasiatic Tethys. Floras show marked geographic preferences also, while vertebrate remains are not common enough for extensive use.

Using goniatites, fusulines and corals, reasonably safe correlations have been found between Europe and Asia on the one hand and North America on the other; but the correlations with and within Gondwanaland (the southern continents and India) are far from secure, and one looks for improvement to future intensive subgeneric work on the brachiopod and lamellibranch faunas of both hemispheres and particularly of south and central America, and West Australia and Indonesia, where the Gondwanan and other faunas come closest together geographically. In the sections on Europe, Asia and North America I have not discussed the correlations of strata nor the reasons for them, on which the maps offered are based, as these correlations are familiar in the relevant areas; but I have given more detail for southern continents.

It might be helpful to give lists of the genera known at present in the Sakmarian (of RUZHENTSEV) in Russia.

Goniatites (generic diagnoses as by RUZHENTSEV 1951, 1954, 1956):

1. Lasting from pre-Sakmarian times into the Asselian substage only, *Aristoceras, Eosianites, Glaphyrites, Neoaganides, Neoglyphyrites.*
2. Lasting from pre-Sakmarian times through the Asselian into the Sakmarian substage only, *Boesites, Daixites, Metapronorites, Prothalassoceras, Somoholites.*
3. Lasting from pre-Sakmarian into post-Sakmarian times, *Agathiceras, Artinskia, Kargalites (Kargalites), Marathonites (Almites), Neopronorites.*
4. Confined to Asselian substage, *Protopopanoceras, Shikhanites.*
5. Confined to Asselian and Sakmarian substages, *Akmilleria, Juresanites, Properrinites, Prostacheoceras, Tabantalites.*
6. Confined to Sakmarian substage, *Preshumardites, Propopanoceras, Synartinskia, Synuraloceras.*
7. First appearing in Asselian substage and continuing into post-Sakmarian times, *Paragastrioceras, Sakmarites, Waagenina.*
8. First appearing in Sakmarian substage and continuing into post-Sakmarian times, *Crimites, Medliocottia, Metalegoceras, Thalassoceras, Uraloceras.*

MILLER and FURNISH (1957 in [L] Mollusca 4 of the Treatise on Invertebrate Paleontology) have placed 20 of these genera in synonymy with earlier genera, but RUZHENTSEV's narrower definitions are used herein since they appear to have stratigraphic value in the standard sequences in Russia. On MILLER and FURNISH's definitions, the Russian Sakmarian would contain *Eothalassoceras, Eoasianites, Imitoceras, Boesites, Prouddenites, Neopronorites, Thalassoceras, Agathiceras, Artinskia, Peritrochia, Propopanoceras, Properrinites, Synuraloceras, Paragastrioceras, Stacheoceras, Crimites, Medliocottia, Metalegoceras* and *Pseudogastrioceras.*

Nautiloids in the Urals (generic diagnoses of SHIMANSKY, 1954, RUZHENTSEV and SHIMANSKY, 1954).

1. Known in Asselian but not later, *Belemnitomimus, Cycloceras, Kionoceras, Tabantaloceras.*
2. Known in Asselian and extending into Sakmarian substage: *Bactrites.*
3. Known in Asselian and extending into post-Sakmarian times, *Bitaunioceras, Ctenobactrites, Dolorthoceras.*

4. Known only in the Tastuba horizon of the Sakmarian substage, *Simorthoceras*.

5. Entering in the Tastuba horizon and continuing into post-Sakmarian times, *Gzheloceras, Mooreoceras, Pseudorthoceras, Rhiphaeoceras, Sholakoceras, Uralorthoceras.*

6. Entering in the Sterlitamak horizon of the Sakmarian substage and continuing into post-Sakmarian times, *Dentoceras, Hemibactrites, Mariceras, Mosquoceras, Scyphoceras.*

F u s u l i n i d s in the Urals (list not exhaustive, generic diagnoses as by Rozovskaya and other Russian authors):

1. Not known above Asselian substage of the Sakmarian, *Pseudoschwagerina, Triticites (Jigulites), Triticites (Rauserites).*

2. Extending through the Asselian and Sakmarian substages into post-Sakmarian times, *Pseudofusulina* (very numerous species), *Rugosofusulina, Schwagerina.*

3. Not appearing in the Sakmarian, but common in the Aktastian (lower) substage of the Artinskian, *Parafusulina.*

R u g o s e C o r a l s (generic diagnoses as by Fomichev, 1953) *Amplexocarinia,* "*Campophyllum*", "*Caninia*", *Caninophyllum, Timania* and the colonial "*Diphystrotion*", [*Dobrolyubovia*], *Eolithostrotionella,* [*Gorskya*], *Protowentzelella* and *Tschussovskenia.* All except *Protowentzelella* are known in the Orenburgian or earlier stages, and all except "*Campophyllum*" continue into the Artinskian. The Russian colonial Rugose genera are provincial; none have been certainly identified elsewhere. I do not know if type species have yet been named for the genera cited in square brackets.

Europe

Judging from the outcrops of strata carrying known Sakmarian fossils, the dominating region for marine sedimentation was the meridional Ural geosyncline. From this, epicontental (shelf) seas spread westerly into the Moscow Basin, the Donetz Basin and the Crimean Gulf; to the south (probably around the Caspian Sea), the Ural geosyncline presumably joined Tethys, the Mediterranean of the time; and to the north, the Polar Sea.

The Sakmarian shelf sea spreading on to the Moscow Basin reached at least as far west as the Oka uplift (Ignatiev, 1952), but may have been less extensive than earlier, Moscovian and Dinantian seas. Terrigenous material is scant in the Sakmarian deposits of the Moscow Basin, leading to the conclusion that the exposed areas of and around the Russian platform were of very low relief. The thin limestones of the Moscow Basin expand at the hinge region with the Ural geosyncline, and organic reefs grew at the edge of the shelf, from about 65° N to at least 48° N, where the Palaeozoic rocks of the Urals disappear under Mesozoic and Tertiary cover. These reefs contain corals, but cannot be called coral reefs; the dominant organisms were brachiopods, polyzoa, crinoids, forams and

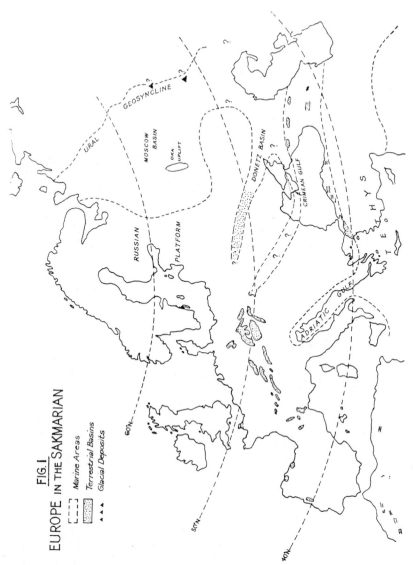

FIG. I

EUROPE IN THE SAKMARIAN

☐ Marine Areas

▦ Terrestrial Basins

▲▲▲ Glacial Deposits

algae (TOLSTIKHINA, 1937). Whether they indicate similar minimal sea surface temperatures to the coral reefs of today (16° C) is arguable.

In the Ural geosyncline itself, limestones give place eastwards to argillaceous and arenaceous sediments, derived from rising mountains fronting its eastern boundary.

The Moscow Basin seems to have been separated from the Donetz Basin by a massif of older rocks. The Donetz Basin is a relatively narrow trough; Sakmarian deposits in it are mainly gypseous dolomites, sandstones and shales, and seem to indicate less free circulation of the marine waters than formerly. A study of bore cores and river-bed rock bars has suggested that the Donetz Basin continued far to the W.N.W. almost to Pinsk, though no Sakmarian marine strata have been recognised as such in this underground extension (Sujkowski, 1946).

The Crimean Gulf is separated from the Donetz Basin by an Archaean massif though the two probably joined in the east before meeting the Ural Geosyncline in the area of the Caspian Sea. Its Sakmarian sediments are argillaceous mainly, rich in goniatites (Toumansky, 1937). It may conceivably have continued to the N.W., more or less parallel with the Donetz extension, to supply marine waters to the Carpathian region (Bükk Mountains and Dobsina, Rakusz, 1932), but there is no actual evidence for this. The facies of the Carpathian deposits is rather like that of the Adriatic Gulf.

The Crimean Gulf is separated from the more or less parallel trending Tethys by Caucasian fold mountains. Tethys lay across Asia Minor except along the north, left calcareous deposits in many places in Anatolia, in the Aegean Isles, and in southern Greece, and sent an arm up the Adriatic Sea to front on the Carnic Alps with an eastern shore in Dalmatia. The *Schwagerina* — limestone banks were formed in this sea, though much argillaceous matter was poured in also (Heritsch, 1939). These banks, which are possibly analogous to the "reefs" of the Urals, are known as far north as 47°.

How Tethys joined the Atlantic Ocean in Sakmarian times is not known, for no Sakmarian fossils are known in the western Mediterranean. Such a junction seems called for, however, to explain faunal similarities with North America.

The entire European Sakmarian is characterised by rich fusuline faunas; goniatites are common, brachiopods are often numerous, and corals occur. All the marine deposits may be considered of one major faunal province, though the Carnic Alps fauna is somewhat different from that of the Urals.

It should be noted that within Sakmarian deposits near the Sim Works in the Urals, in latitude 55° N approximately, the Dommenaya Breccia was regarded by Nalivkin (1937) as a marine glacial deposit, dropped from icebergs. Dunbar (1940) however, regards it as a landslide breccia related to submarine faulting. Similar breccias occur in the Orenburg region (52° N approximately), and Backlund (1930) drew attention to a possible late-Palaeozoic tillite in the Kara River area, N.W. Urals.

As for that part of Europe not covered by the sea, we can distinguish two main regions:— the regions where no folding later than Caledonian had occurred, and the regions affected by the Variscan orogenies. The former regions are those of the western part of the Russian Platform, Scandinavia, and most of the British Isles, and on these, continental sediments equivalent in age to the marine Sakmarian are thin and unimportant, or unknown. We may deduce that relief was slight over these areas.

Variscan orogenies, including the Asturian phase, which was probably just pre-Sakmarian, affected Spain and Portugal, France, Germany, Austria and Czechoslovakia and the eastern Balkan States. Detritus from the uplifted areas was washed down and deposited in two types of basins, one, inland, well above sea level, the limnic basins; and the other, paralic basins on coastal plains, subject to inundations by small changes of sea level. Most of these basins in existence in Sakmarian times were relics or extensions from the highly important coal basins of earlier Carboniferous times. Palaeobotanists have shown that the fossil floras of the two types of basin were distinct, and formed two different types of the swamp flora at that time characteristic of Europe and North America, particularly eastern North America (JONGMANS, 1952 a, 1952 b). In the inland basins of France, the deposits which palaeobotanists now equate with the Sakmarian are generally called Autunian, and in Germany the Lower Rotliegende. Coal seams are few, in contrast to the wealth of middle and early Upper Carboniferous times, and many of the beds (conglomerates, sandstones and shales) are red in colour, though green and violet beds also occur. In the opinion of palaeobotanists, these features indicate a period of gradually increasing aridity, though the whole of Europe had not been reduced in rainfall to that low level characteristic of later Permian times. It was a time of transition.

Thus the topography of Europe during the Sakmarian is fairly clear. A meridional sea lay in the east, a mediterranean sea in the south. From the eastern sea, shallow bodies of water spread on to low lands in the north east and central east, and along the feet of mountains in the south. The continent may have extended out beyond its present western and northern shores, possibly as far as the 2000 metre line. The northern and eastern part was of very low relief, the western and southern parts of high relief with intramontane basins.

Rainfall was probably seasonal, not of the rain belt type, for coal swamps were rare. The red colour of the terrestrial sediments could indicate either laterite soil conditions, or conditions of alternate flood and drying out, oxidation of iron occurring during drying out. If redistributed laterite soils were widely prevalent, rainbelt conditions could be assumed in the period preceding the Rotliegende, but would scarcely be arguable for the Sakmarian itself.

Several lines of attack on the problem of climatic zones have been suggested in the Sakmarian literature, but none of them have been taken up sufficiently widely yet to yield any valuable results. It has been suggested that if the Rotliegende were derived from erosion of laterite soils, the Rotliegende clays should be of types appropriate to this origin. Seasonal rings in trees have been investigated in a few isolated instances. Possibly a greater knowledge of temparate and tropical peats than we have at present will prove applicable to the study of Autunian coals (FRANCIS, 1954). So far, however, I have found nothing on these lines in the literature on which one can base sound palaeoclimatic argument.

In the marine sediments, the predominance of limestones and especially the development of organic reefs has been used as an argument for the tropical to subtropical temperature of the Ural geosyncline waters. But the limestone of the Moscow Basin may only be a reflection of the lack of terrigenous matter supplied from the Russian platform. Brachiopod — polyzoan — algal reefs may conceivably form in temperatures lower than the coral reefs of today. The fusulines, being larger forams, and developed in great numbers, are often regarded as typical warm-water forams, but no fusulines are alive today, and the argument from analogy is not very strong.

Table 1. RUZHENTSEV's subdivision of the Sakmarian stage in the Southern Urals.

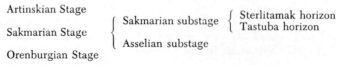

Table 2. Approximate equivalents of Sakmarian in Northern Hemisphere.

N. America	Europe marine	Europe terrestrial	Asia
Leonardian	Artinskian	U. Rotliegende	Amb Group
Wolfcampian	Sakmarian	L. Rotliegende and Autunian	Warcha Group. Talchir Group. (also Taiyuan Fm. in N. China and Maping and Chuanshan Fm. in S. China
U. Pennsylvanian	Orenburgian	Stephanian	

Asia

Sakmarian Asia appears to have been land north of 45° N except perhaps for marginal seas in Ussuriland and for a narrow (Taimyr) trough along the southern edge of the Kara Sea massif, in about 73°—75° N and 80° to 115° E (OBRUTSCHEW, 1926) rich in brachiopods (LIKHAREV, 1937; USTRISHKY, 1955). In the West this land (Angaraland) was separated from the Russian platform by the Ural geosyncline. It was probably of low or moderate relief in the north, with some fold ranges near its southern borders. Great basins of coastal plain type, not far above sea level, were present along the southern coast line, such as the Basins of Kusnetsk and Minussinsk. In some of these basins, short inundations of the sea occurred in Gshelian (Orenburgian) times, and even, perhaps, in Sakmarian time, though no satisfactory proofs of the latter exist. To the north, in the Tunguska, a great inland basin was present, unproductive of coals, rather like the Great Artesian Basin of Australia today. The flora found in the basins of Angaraland is a swamp flora of distinct province, the Angara flora, and seems to have had little or no connection with that of Europe or India or N. America. The non-marine lamellibranchs too are very different (WEIR, 1945). The sediments forming in Sakmarian (Autunian or Rotliegende) time are consequently difficult to identify, but if we follow JONGMANS (1939, 1952) they are those of the Ungian and Koltschugian and their equivalents; if we follow TSHIRKOVA and ZALESSKY (1939) they are those of the earlier Tomian and Abian Stages. These sediments in the latitudes of the forties contain productive coals. It would thus seem that precipitation was sufficiently great, or drainage sufficiently poor, to support peat swamps in latitudes 40—45° N in these times, and one gathers that Asia in these latitudes had better rainfall than Europe. Whether they were temperate peats or tropical peats is not yet clear. The only record of possible Sakmarian glacials in Angaraland is by NORIN (1930) in the Kuruk-tagh Mts. (between 88°—90° E and 40—42° N).

South of Angara lay a great mediterranean sea. This may have been connected with the Ural geosyncline, but evidence is lacking owing to Mesozoic and Tertiary cover; the connection, if any, seems to have been broken by Zechstein (Kazanian) times. Whether the Ural connection existed or not, Asiatic Tethys was continuous with the Asia Minor part of European Tethys through southern and western Iran, Baluchistan and Afghanistan (DOUGLAS, 1950; STOYANOW, 1942; THOMPSON, 1946); but whether this connection stretched northwards to the northern boundary of Iran and Afghanistan is still unclear; there may indeed have been two seas, one in the south and one in the north, but direct evidence in the form of Sakmarian fossils is absent in the north (HERITSCH, 1939).

Along meridian 70° E, this Sakmarian sea stretched from the Salt Range northwards to Ferghana, possibly with some island arcs within it parallel to its shorelines. The Salt Range deposits seem to be shelf deposits, the deeper parts of the sea lying to the north. North of Ferghana Variscan fold mountains occurred, but coal swamps lay in places between them and the sea (OBRUTSCHEW, 1926; LEUCHS, 1935).

From this wide sea in the vicinity of 70° E, there extended eastwards two short arms and one very long one, the shortest in the north. Here a sea which we might call the Dzungarian sea stretched some distance towards Urumchi in Dzungaria along the N. rim of the Tarim massif, but was only a fragment of the former Viséan and later *Neoschwagerina* seas which stretched from Dzungaria into the Mongolian Basin and thence to the Pacific (GRABAU, 1931, KAHLER, 1940, MINATO, 1953, KOBAYASHI, 1952). No certain Sakmarian marine fossils are known east of Urumchi district; and large areas of rather flat lying land were added to Angara in Sakmarian time, including the Mongolian geosyncline.

A middle or Nan Sea stretched from the Pamirs eastwards around the S. rim of the Tarim Basin, and north of the Tibetan massif, along the Nan Shan region, transgressing northwards on to the Tarim massif almost as far as Yarkand and Cherchen (OBRUTSCHEW, 1926, LEUCHS, 1935). Marine Sakmarian with fusulines is known from north of the Koko Nor and S.E. of Suchow, and it is possible that this sea was in intermittent

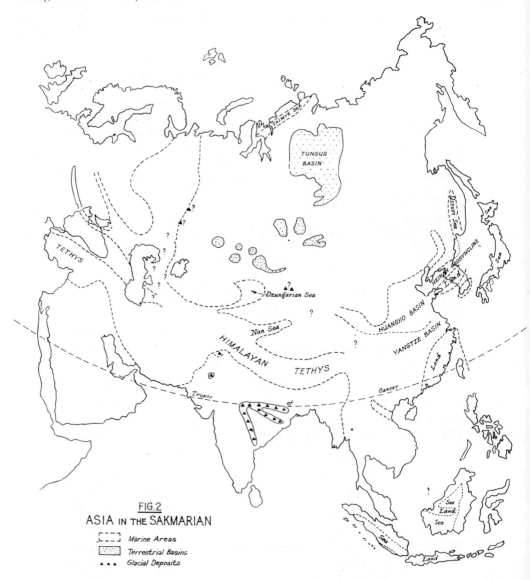

FIG.2
ASIA IN THE SAKMARIAN

Marine Areas
Terrestrial Basins
▲ ▲ ▲ Glacial Deposits

connection eastwards, through the Huangho Basin, with the Taiyuan sea of N.E. China (KOBAYASHI, 1952).

The third great sea is Himalayan Tethys, which curved eastwards from Kashmir through the Himalayan region into W. China, perhaps with shallow extensions northwards on to the Tibetian massif. In W. China (DUNBAR and MISCH, 1947) it became continuous with shallow seas over the Yangtse Basin, and with geosynclines which ran southerly through Burma and Indo-China (FROMAGET, 1952).

This sea and its deposits are of great importance to Sakmarian correlation, for it lies along the northern shore of the Indian portion of Gondwanaland, and *Eurydesma,* one of the most important and characteristic lamellibranch elements of the Gondwana marine fauna occurs in its deposits in the Salt Range and eastwards. In the lower parts of the Salt Range sequence, goniatites, fusulinids and corals that might be used to tie the sequences in with the Eurasian Sakmarian, are absent; the only help we get from these groups is in the upper part of the Amb Group in the Lower *Productus* limestone, where *Parafusulina kattaensis* (WAAGEN) is generally considered to indicate an Artinskian (Leonard) age for this limestone (DUNBAR, 1933). Beneath this lie dark sandy brachiopod limestones with interbedded clays, with the N. Hemisphere genus *Richthofenia,* which GERTH (1950) considers Artinskian. *Richthofenia* is not known in the Gondwana faunas. Below this again is the Warcha Group of brown clays and speckled sandstones, olive sandstones and clays with concretions, with the Gondwana association *Eurydesma* and *Conularia* and the plants *Glossopteris* and *Gangamopteris.* Beneath lies the Talchir Group of tillites and melt water deposits. GERTH (1950) regards both Warcha and Talchir groups as Sakmarian, and in the absence of evidence to the contrary, this seems reasonable. It is important to realise however, that the Sakmarian age for the Talchir Boulder Beds here is not strictly determined, but rests on inference, the Beds lying stratigraphically below the Lower *Productus* limestone which on the strength of one fusulinid species is regarded as Artinskian. If the Tillites are continental moraines, the *Eurydesma* fauna must have invaded the land on the waning of the ice.

In Kashmir, no sign of glacial deposits is seen, but instead vast quantities of volcanic materials appear, part of which (Agglomeratic Slate Series) is generally correlated with the *Eurydesma* Beds of the Salt Range; the Gondwana brachiopod *Taeniothaerus* occurs also, and adds to the evidence for the existence of the Gondwana marine fauna along the southern shore of Himalayan Tethys, possibly, as we have seen, in the Sakmarian. No fusulines occur on this shore. Eastwards from Kashmir and the Salt Range, no distinctive fossils are known in the Indian beds that are correlated with the Sakmarian (WADIA, 1953).

India south of the Himalayas was continental throughout Sakmarian times except for a small seaway to Umaria, in which brachiopods and possibly *Eurydesma* lived (THOMAS, 1954), its marine deposits overlying glacials. Continental glacial beds, the Talchirs, occur today in 3 main lines of outcrop (WADIA, 1952), meeting roughly in the vicinity 30° N and 20° N, the arrangement suggesting great valley deposits from huge mountain glaciers; that Indian glaciation was not of the continental ice sheet type, but of alpine type, appears to be the view of Fox (1937), and JACOB (1952), and is a view that I accept. Glaciated pavements are known in Rajputana, but the age of the glaciation is unknown; the rocks glaciated are Pre-Cambrian, so the high mountains could scarcely have been fold mountains. The *Glossopteris-Gangamopteris* flora, it seems, did not appear till the highest of the glacial beds were deposited, and many, therefore, not have entered until the succeeding amelioration of climate. This Indian flora, the typical "Gondwana" flora, has nothing in common with those of Europe and North America, so that it cannot be dated by reference to them.

In placing both the Salt Range and Indian glacials in the Sakmarian we may easily be in error. Possibly maps should be constructed showing them as Orenburgian and Moscovian as well, to see into which pattern they fit best.

The *Schwagerina* facies of Iran, Baluchistan (DOUGLAS, 1950), Chitral and the Pamirs (DUTKIEWICZ, 1937), also known in the Dzungarian and Nan Seas (THOMPSON and MILLER, 1935), is seen again in the Himalayan Sea in the N. Shan States (CHIBBER, 1934), and in this area the E. Asian Sakmarian seas meet the Himalayan Sea.

In E. Asia (except Indo-China) two main basins of Sakmarian deposition may be recognised — the Huangho, and continuous with it over the extension of the present Nan Shan, the Yangtse Basin, which merged westward with the Himalayan Sea, and opened into the Japan Sea through S. Korea. An arm of the Huangho Basin, closed eastward, extended N. E. from the head of the Gulf of Laiotung (KOBAYASHI, 1942, 1952). The Huangho Basin opened into the Japan Sea by the Heinan geosyncline across N. Korea. Japan was inundated, probably completely, during Sakmarian times, and fusulinid limestones are common (MINATO, 1955) but no goniatites are known (HAYASAKI, 1954). Sakmarian deposits follow an Orenburgian period of non-deposition or emergence in most parts of Japan.

Sakmarian deposits of the Yangtse Sea were epicontental limestones rich in fusulines and corals, with some interbedded shales — the Maping and Chuan Shan limestones. In the Huangho Basin, the Taiyuan Formation (Huangchi) is paralic, with coal measures interbedded with limestones rich in fusulinids and productids (LEE, 1939). The Huangho sea may have breeched the barrier into E. Mongolia in the vicinity of Hsin King (ENDO, 1953).

The development of coal measure swamps in the Huangho Basin shows that this was a region of high rainfall. Such deposits are not known in the Mongolian geosyncline, whose relief may not have been suitable for swamps to form, or which perhaps lay north of the rainfall belt.

In Ussuri Land (vicinity of Vladivostock and northwards along the coast) *?Taeniothaerus* of the Kashmir and Australian Sakmarian may occur

(Grabau, 1931), with other forms regarded as "southern, cold water" forms by Fredericks.

No Sakmarian deposits are known from the Philippines, which were probably land at the time (van Bemmelen, 1949); most of Indonesia and New Guinea appear to have been land also (van Bemmelen, 1949), though marine deposits rich in Sakmarian ammonites and with some fusulines are known from Timor and Sumatra (Gerth, 1950; Thompson, 1936, 1949), and some Permian smaller forams and brachiopods have been recorded from the Western Highlands of New Guinea (Rickwood, 1955; Glaessner, Llewellyn and Stanley, 1950).

North America

Most of the present land areas north of the Arctic Circle seem to have been emergent during the upper Palaeozoic, but no Upper Palaeozoic glacials are recorded therefrom, and there is no evidence bearing on the height at which the land stood. The only marine Sakmarian identified with reasonable certainty is in Ellesmereland, in a S.W.—N.E. belt through Greely Fjiord (Troelsen, 1950), where 2000 feet of impure limestones with *Schwagerina* alternate with shales and sandstones and lie discordantly on mid-Pennsylvanian marine beds; there are no coral reefs, but some brachiopod and polyzoan biostromes. This sea extended N.W. on to Axel Heiberg I., where volcanics are interbedded with its limestone deposits.

In N.E. Greenland, there is in Peary Land, Holm's Land and Amdrup's Land a thin narrow fringe of marine conglomerates, red and grey sandstones and marly shales with limestones containing fusulines followed by dolomitic and in part silicified limestones; these may be in part Sakmarian (Thompson, Wheeler and Hazzard, 1946; Frebold, 1950; Troelsen, 1950) or in part Zechstein (Dunbar, 1955). In central E. Greenland, similar rocks formerly (Frebold, 1950) regarded as Sakmarian are now all referred to the Zechstein (= Kazanian) (Dunbar, 1955).

W.S.W. from Southern Ellesmereland stretches a narrow belt of Upper Palaeozoic sediments, through Byam Martin, Melville and Bathurst Islands (Fortier, McNair and Thorsteinsson, 1954); white non-marine sandstones with coals are overlain by limestones and possibly Sakmarian deposits are represented in the sequence. The 300 ft. Sadlerochit sandstone with marine fossils in the Canning River and U. Porcupine River areas of N.E. Alaska may represent a continuation of this Parry Island Belt.

In cold temperate Canada and Alaska all but a Pacific belt some 500—600 miles wide was emergent in post-Mississippian Palaeozoic time. No glacials are recorded, and there is no evidence bearing on the height of the land. This belt seems divisible into two, the boundary lying along the Kuskokwim River in Alaska and in the Canadian Rocky Mt. Trench. In the outer belt, thick limestones interbedded with ribbon quartzites, shales and greenstones (together up to 20,000 ft.) and much intruded by Mesozoic granites (Lord, Hage and Stewart, 1947) have suggested to some (Kay, 1951) that eugeosynclinal conditions existed there, with volcanic island

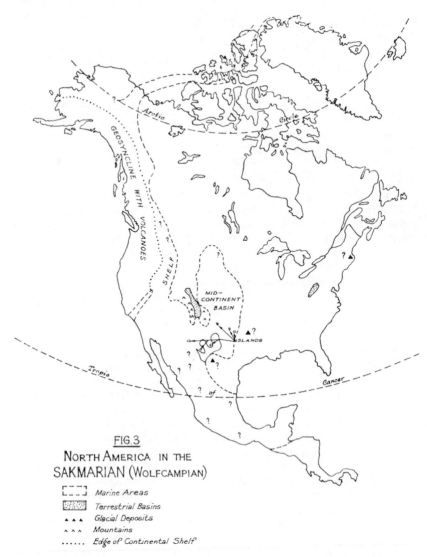

FIG. 3

NORTH AMERICA IN THE
SAKMARIAN (WOLFCAMPIAN)

└ ─ ─ ┘   Marine Areas
▨   Terrestrial Basins
▴ ▴ ▴   Glacial Deposits
^ ^ ^   Mountains
. . . . . .   Edge of Continental Shelf

arcs. The deposits (Cache Creek Group and correlatives) contain fusulines
and brachiopods somewhat rarely, some probably Sakmarian (THOMPSON,
WHEELER and HAZZARD, 1946; MOFFITT, 1954; ROOTS, 1954). The greenstones listed by WHEELER (1940) as Permian seem to be post-Sakmarian
(BLACKSTONE, 1954). Coral or other reefs are not recorded. The inner belt
contains thinner sediments interpreted as shelf sediments lacking volcanics; one of the best sections is probably that south of the Yukon River
in Alaska, where the Tahkandit Limestone, with conglomerates, sandstones

316

and shale, is abundantly fossiliferous (Smith, 1939; McLearn and Kindle, 1950).

Possibly half the present warm temperate and subtropical U.S.A. was under the sea during the Wolfcampian Sakmarian; the Pacific eugeosynclinal belt of deposits of late Palaeozoic age stretches down from Canada (Kay, 1951) to Baja California which may have been land, and into Mexico, where connection with the Atlantic may have occurred. The boundary between eugeosyncline and shelf regions ran through Idaho and central Nevada. On the shelf the sea transgressed as far the western boundary of Wyoming and the centre of Utah, then through the S.W. corner of Colorado along coastal flats at the foot of a high range the Uncomprahgee High; thence back north again round a similar high range the Front Range, through eastern Colorado, Wyoming and Montana into N. Dakota, its northmost point in the Sakmarian. The shore line then returned south through easternmost Nebraska, eastern Kansas and Oklahoma, central Texas, where high ranges (the Wichita and Marathon Ranges) fronted it, and thence swung sharply S.E. back into North Eastern Mexico. The shelf appears to have been continuous into the Sonora District of Mexico. This great epicontinental sea was thus constricted by land in Colorado, Wyoming and Montana, with high ranges and an intervening valley in Colorado.

Parts of the floor of this epicontental sea were depressed more than others, and great thicknesses of sediment were deposited for instance in the Oquirrh Basin of S.W. Utah and S.E. Nevada. Sakmarian sediments show the same range in facies throughout these seas. Near-shore, red, green and grey sandstones, shales, sometimes with anhydrite, and salt beds predominate; further from shore limestones formed. No coral reefs are known, but some small structures may be algal-bryozoan reefs. The limestones are often rich in fusulines, and these with ammonoids give the time scale for the Wolfcampian. Knowledge of the extent of the different lithological formations is far advanced in the U.S.A. and in most states is well ahead of the palaeontological knowledge necessary to give the time planes. Advances and retreats of the limestone facies towards and from the nuclear land areas seem to indicate oscillations of sea level, by which the sea advanced and retreated across wide coastal flats at the feet of the ranges [1].

The absence of coal seams from the coastal plains of the Sakmarian

[1] The following literature should be consulted for the names and descriptions, faunas and bibliographies of Wolfcampian formations in the various states. I d a h o, W. M o n t a n a and N. E. N e v a d a : Blackstone (1954), Cressman (1955), McKelvey, Swanson and Sheldon (1953), Nolan (1943) and Thompson (1954). U t a h : Baker, Huddle and Kenney (1949), Eardley (1951), Newell (1948), Ogden, 1951, B.A.A.P.G. p. 62, Thompson, Wheeler and Hazzard (1946). E. N e v a d a and S. C a l i f o r n i a : McNair (1951), Nolan (1943), T. W. and H. (1946). C o l o r a d o : Baker and Williams (1940), Brill (1952), Lovering and Goddard (1950), Maher (1954), Wengerd and Strickland (1954). A r i z o n a, N e w M e x i c o, T e x a s and O k l a h o m a : Eardley (1951), Hills (1942), King (1948), Lloyd (1949), McKee (1951), Thompson (1954), T. W. and H. (1946). K a n s a s and N e b r a s k a : Moore et al. (1951), Thompson (1954). W y o m i n g : Agatson (1954), Wanless et al. (1955). N. D a k o t a : McCabe (1954).

317

shelf seas in the western half of the U.S.A. together with the red colour of much of the sandstone and shales, the occasional beds of anhydrite, and the absence of coral reefs, seems to indicate warm temperate regions with seasonal rainfall.

The only deposits in the U.S.A. which might conceivably have been regarded as Upper Palaeozoic glacials are at present discredited as such. Thus the Sqantum Tillite near Boston is regarded as probably Devonian (EARDLEY, 1951), and the "tillites" in the Wichita and Marathon Mountains as piedmont conglomerates with slickensides induced by differential movement (DUNBAR, 1924).

Wolfcampian fusuline genera identified by THOMPSON (1954) are *Boultonia, Dunbarinella, Nankinella, Oketaella, Ozawainella?, Paraschwagerina, Pseudofusulina, Pseudofusulinella, Pseudoschwagerina, Schubertella, Schwagerina* (numerous species) and *Triticites*.

Wolfcampian nautiloid genera identified by MILLER and YOUNGQUIST (1949) are *Bitaunioceras, Coelogastrioceras, Domatoceras, Endolobus, Ephippioceras, Foordiceras, Knightoceras, Liroceras, Metacoceras, Mooreoceras, Pseudorthoceras, Solenocheilus, Stearoceras, Stenopoceras, Tainoceras* and *Temnocheilus*.

Wolfcampian ammonoid genera mentioned by MILLER and YOUNGQUIST (1949) are *Agathiceras, Artinskia, Daraelites, Marathonites* (and *Peritrochia*), *Medlicottia, Metalegoceras, Neopronorites, Paragastrioceras, Properrinites, Prothalassoceras, Pseudogastrioceras*.

Wolfcampian coral genera are *Heritschia, "Lithostrotionella", Lophamplexus, Lophophyllidium, Malonophyllum, "Palaeosmilia"* and *Sochkinophyllum*. Total ranges of the genera named are not indicated in these lists, and the lists are not exhaustive.

In the eastern United States, no marine Sakmarian seas occurred; there were probably high mountains in the Appalachian region (KING, 1951; EARDLEY, 1951); but inland areas of sedimentation were restricted compared with those of Pennsylvanian time. The only deposits reasonably to be attributed to Rotliegende time are those upper parts of the Dunkard Fm. in Pennsylvania, Ohio and West Virginia (JONGMANS, 1952). Here repeated sequences of strata occur ("cyclothems") as in the Pennsylvanian, including coal seams. The shales are mostly grey, but when calcareous may be greenish-grey and greyish-purple to purplish red (CROSS and ARKLE, 1952). The flora is European in type. This seems to indicate a humid climate for the eastern U.S.A. with persistent, rather than markedly seasonal rainfall. When contrasted with the lack of coals in the Western shelf basins, it would seem to indicate that U.S.A. was in a region of dominantly easterly winds.

In Mexico and tropical Central America Upper Palaeozoic seaways are hard to interpret, owing to our small knowledge of the few exposures. Mexico could indeed have been almost entirely submarine, an exception being that part along the border with S. Texas, where mountains must have existed to supply the coarse sediment along the margins of the basins. In S.W. Coahuila the lowest exposed upper Palaeozoic beds are greenish black greywackes with possibly some lava, and fine grained quartzite, overlain by blue grey, extremely massive crinoidal limestone, possibly in part reef, though evidently not coral reef. These are probably Pennsylvanian or possibly Wolfcampian. Unfossiliferous clastics with layers of coarse conglomerate, shale, and thin layers of sandstones and greywackes were thought to be tillites by KELLY, but this view has been discredited by KING, DUNBAR, PRESTON CLOUD and MILLER (1944). This sea was continuous with the Texas seas, and opened into the Gulf of Mexico. It may also have been continuous with the seas of Sonora, in the N.W. of Mexico.

In the extreme S.E. of Mexico (Chiapas) and running through Guatemala E—W and into British Honduras are marine clastics, followed by Wolfcampian (Grupera Fm. 400 ft) and Leonardian shales and limestones with fusulines (SAPPER, 1937; SCHUCHERT, 1935; MULLERRIED, MILLER and FURNISH, 1941; THOMPSON and MILLER, 1944). How this sea was related to that of Coahuila and that of Sonora is unknown; its fossils indicate relation with those of Texas.

In the Sonora seas, massive limestones, some reefy, graded laterally into thin-bedded, dark granular limestones with abundant crinoidal fragments and fusulines; Wolfcamp may be present, Leonard and Word certainly are; the limestones are continuous with those of S. Arizona (IMLAY, 1939; KING, 1939; DUNBAR, 1939).

South America

Wolfcampian (Sakmarian) faunas have been identified with reasonable certainty only in a narrow belt in the Andes, stretching from Cochabamba (S. of L. Titicaca) in Bolivia through Peru and doubtfully into Equador (NEWELL, CHRONIC and ROBERTS, 1953). The strata containing them are black, bituminous shales and massive limestones with some dolomite, sandstone and siltstone and are referred to the Copacabana group which lies unconformably on the Devonian near L. Titicaca and is overlain unconformably by the clastic and volcanic Permian Mitu group. The faunas consist chiefly of brachiopods, mollusca and fusulinids (KOZLOWSKI, 1914; NEWELL, CHRONIC and ROBERTS, 1953) and are like those of N. America, with *Juresania* and *Kiangsiella*, and unlike those of Australia. They possibly continue through Colombia (GERTH and KRÄUSEL, 1931) into the Maracaibo region of Venezuela (THOMPSON and MILLER, 1949), where some fusulines and cephalopods may be Wolfcampian, others Leonardian. In the Amotape Mountains and Tarma region of Peru, a mid-Pennsylvanian fauna unknown in Bolivia occurs (NEWELL, CHRONIC and ROBERTS, 1953).

This mid-Pennsylvanian fauna has been recognised in several places between 60° and 50° W. in a seaway which stretched along the Amazon in Palaeozoic times, in shale referred to the Itaituba Series (CASTER and DRESSER, 1954; MENDES, 1957), and in the R. Parnohyba basin (about 43° W) in the Piaui Series (KEGEL, 1953). Whether this seaway lasted into Wolfcampian times is unproved, but from KEGEL's work it would seem to have withdrawn at least from the Parnohyba region by then.

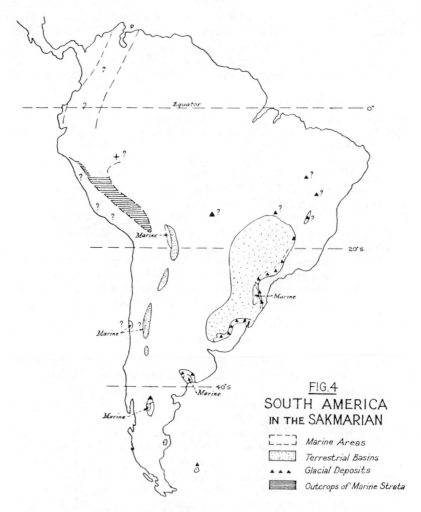

FIG.4
SOUTH AMERICA
IN THE SAKMARIAN

⌐ _ _ ⌐   Marine Areas
[∷∷∷∷]   Terrestrial Basins
▲ ▲ ▲   Glacial Deposits
▬▬▬▬   Outcrops of Marine Strata

Whether the Peruvian Sakmarian sea, which NEWELL, CHRONIC and ROBERTS (1953) regarded as transgressing from the west, extended south along the Andes into Chile and the Argentine is uncertain. Marine strata in the R. Chiapa valley in Chile have a small brachiopod and molluscan fauna generally regarded as either Gshelian (Orenburgian) or Sakmarian (FOSSA-MANCINI, 1944) as is the neighbouring fauna from the Argentinian precordillera at Barreal (REED, 1927) though CASTER (1954) advocates an equation to the Pennsylvanian Itaituba fauna of the Amazon and the Tarma fauna of Peru. Precise faunal studies are required. These faunas are very important, for in the Argentinian pre-Cordillera there are florules which FRENGUELLI (1944, 1946) interprets as indicating the gradual replacement of the cosmopolitan *Rhacopteris* flora by the *Glossopteris* flora.

The sections, like those with the marine faunas, are discontinuous in the field, and there is only indirect evidence on sequences. FRENGUELLI deduces the El Saltito flora (*Rhacopteris* and *Lepidodendron*) to be lower Westphalian; that of La Playita (similar but without *Rhacopteris* itself and with *Gondwanidium plantianum*, the earliest representatives of the *Glossopteris* flora) upper Westphalian, that of Retamito (in varves) lower Stephanian, and that of Bayo de Velis, which is almost a pure *Glossopteris* flora although the early *Gondwanidium* ist still present, upper Stephanian. He prefers to draw the base of the Permian (presumably, to him, basal Rotliegende) above the last appearance of *Gondwanidium*, though BARBOSA (1952) draws it below the base of the Bayo de Velis Fm., both positions of course, being arbitrary.

Glacial deposits such as tillites, fluvio-glacial conglomerates and varves, are found interbedded in these discontinuous pre-Cordilleran sections (HEIM, 1945) and their number and relation to floras and faunas can only be deduced indirectly. At present we can say no more than that the "*Spirifer supramosquensis*" beds of Barreal may be Sakmarian, or possibly Pennsylvanian; the plant beds of Bayo de Velis and of Totoral may be Rotliegende or Pennsylvanian. Two sets of supposed glacials at least, that below the *Rhacopteris* flora of el Tupe, and that of Retamito are pre-Sakmarian. BARBOSA (1952) considers that in the Bayo de Velis, the red Pataquia and the Totoral formations, the only glacial material is that derived from the erosion of the earlier tillites and fluvioglacials. Clearly we can as yet draw no safe palaeogeographic and palaeoclimatic picture of the Sakmarian of this region.

Further south, in Patagonia at about 43° S in the Sierra de Tepuel etc., the Tepuel system of black shales and quartz sandstones contains in its upper parts *Eoasianites* and *Anthracoceras* referred to the Mid-Pennsylvanian (MILLER & GARNER, 1953) and *Strophalosia, Mourlonia, Aviculopecten* and *Spirifer* cf. *octoplicata* (SUERO, 1952) suggesting by their names Australian Permo-Carboniferous faunas. The lower Tepuel system contains *Fenestella* and *Productus* in glacio-marines. Relation to the Barreal faunas is uncertain, as is the Sakmarian palaeogeography and climatology of the region.

East of this narrow Andean zone where upper Palaeozoic marine strata may be found there is in the sub-Andes of Bolivia (AHLFELD, 1946) and north west Argentina (BARBOSA, 1952) an unfossiliferous "Gondwana" development of continental rocks, generally called Oquita group in their lower parts, (with greyish sandstones and shales, tillites and fluvioglacials) but followed by the red Mandiyuti Formation, in which primary glacials are probably absent. Some part of this continental sequence may be Sakmarian.

In the remainder of S. America, upper Palaeozoic strata that may contain Sakmarian equivalents were deposited in N.E. Argentina and in the great inland Parana Basin, which stretches over south eastern Brazil, eastern Paraguay, into Uruguay, and possibly into the Diamantina area of Matto Grosso, S.W. Brazil.

Throughout the Parana basin, there is an important marker formation, the Irati or White Band, of black shales and white dolomites with concretions and beds of chert, and containing bones of the reptile *Mesosaurus*, seemingly deposited in the basin when it stood low enough to be invaded by the sea. Below the Irati the beds contain important glacials and are grouped as the Tubarao Formation. BEURLEN (1957) contrasts the absence of evidence of glacial and interglacial erosion in these deposits with its presence in the European Pleistocene glaciation, and concludes that the ice advanced north into a subsiding Parana Basin and south into N.E. Argentina from a high shield region in Rio Grande do Sul. MAACK (1957) considers the ice advanced South, South west and West into Brazil, Uruguay and Argentina from one centre in North east Brazil and others outside the continent. The Tubarao varies greatly in facies along and across the Basin. Thus five tillites (with interbedded coal seams) are recognised in the east in Sao Paulo, 4 in the south of Parana, 2 or 3 in Santa Catarina and only 1 in Rio Grande do Sul. In the north and west (Minas Geraes and Matto Grosso) possibly two or three tillites occur (ALMEIDA, 1952; BEURLEN, 1956). Coals are best developed in the central east and south between the tillites. PUTZER (1957) has discussed the palaeogeographic relationship of these coals and interprets them as formed in a tundra-like milieu, in interstadials or post glacial times; this would suggest the *Glossopteris* flora was a cold-temperate flora. Perhaps detailed studies on the coals would show whether their vegetable matter altered in the manner of modern cold-temperate peats. On the north and west, the beds are red. In this dominantly continental development, marine beds occur along the eastern outcrop, and principally in Santa Catarina and Parana (MENDES, 1952), the oldest at Capivari in Sao Paulo, the most widespread in Parana, Santa Catarina and Rio Grande do Sul. None of them permit any secure correlation with either Pennsylvanian or Wolfcampian faunas of America, and one, that of Taio (Santa Catarina) has a lamellibranch fauna with some genera known in Australia, though lacking *Eurydesma*. According to BARBOSA (1952), NEWELL (1949) saw correlation between the Capivari and the Sakmarian of Peru, while according to CASTER (1952) this is a depauperate representative of the Pennsylvanian of the Amazon. No *Rhacopteris* occurs in the Tubarão, but *Gondwanidium* is recorded (DOLIANTE, 1952) moderately high in the series, suggestive, perhaps, of a Bayo de Velis correlation for the Tubarão.

Above the Irati follows the Passa Dois Series, also of variable but mainly continental facies, without glacials except possibly for varves in the Theresina member; in the north and west it is developed in red facies. A marine or brackish band with the lamellibranchs *Pinzonella* previously regarded as Triassic is now referred to the Permian (MENDES, 1952).

It is thus clear that only subjective views are available or possible on the ages of the various strata of the Parana Upper Palaeozoic, and it is impossible to draw a secure Sakmarian map. If the Tubarao is Sakmarian, glaciated mountains probably existed to the east of South east Brazil; the sea also lay to the east, and spread westwards into the Basin, possibly through an opening in the ranges east of Santa Catarina. If the Irati is Sakmarian, the ranges to the east were not glaciated at that time; if the Passa Dois is Sakmarian, likewise, glaciation to the east may have been absent.

In the mountains to the south of Buenos Aires in Argentina, the Pilla-huinco Group has the Sauce Grande Formation, 900 metres of tillite and glacial conglomerates with massive quartzites at the base which many correlate with the Tubarao Formation of Brazil. Some distance higher occurs the Bonete Formation of dark green marine quartzites and mud-stones with 3 fossiliferous beds; the total fauna is very Australian in type, with *Eurydesma, ? Atomodesma* (= *Aphanaia*), *? Stutchburia, Allorisma, Schizodus, Promytilus* and *? Liopteria,* and the brachiopods *Notospirifer darwini* and *"Chonetes"* (HARRINGTON, 1955). On the known ranges of these genera, it could be Sakmarian but it could be younger. The flora of the Bonete is a pure *Glossopteris* flora, no *Gondwanidium,* or any relics of the *Rhacopteris* flora being present, and FRENGUELLI (1944) considered it younger than the U. Stephanian Bayo de Velis flora.

None of the invertebrate species of the Bonete Formation are found in any of the Tubarão faunas, and the Irati White band is unknown in the Sierra Bonaeres. Whether the glacial deposits of both areas should be correlated as contemporaneous is arguable; 5 glacial stages have been recognised in some parts of the Parana Basin, only 1 in others; the number of stages in the Sauce Grande Formation is unknown to me.

It must be concluded, I think, that evidence by which we could safely correlate any section of the Parana or Pillahuinco or Barreal faunas with the Wolfcampian of America is lacking; should it become available, then we would have a means of dating (in terms of the Northern Hemisphere sequences) the development of the *Glossopteris* flora in South America. Failing any reasonable marine correlations, we might perhaps use the floral correlations of FRENGUELLI (1944, 1946), who appears to regard the Bonete flora as Rotliegende; the Rotliegende is reasonably correlated with the Sakmarian and this would suggest a Sakmarian equivalence for the Bonete Formation. Whether the Sauce Grande glacials below it would be Wolfcampian or Pennsylvanian would still be uncertain and arguable. Whether the Tubarão floras of the Parana Basin would be equivalent to or older than the Bonete floras is arguable; since *Gondwanidium* is still present, it might be suggested that they are pre-Bonete. It is clear then, that we have no reasonable palaeontological evidence on the precise age of the Tubarão or the Sauce Grande glaciations. We might observe that the *Eurydesma* fauna entered after the Sauce Grande glaciation had dis-appeared.

In the Falkland Islands, continental glacial deposits followed by beds with *Glossopteris* are known; and in two small areas it is suggested that marine glacials are present (ADIE, 1952). There is no evidence to show whether any of these deposits could be contemporaneous with the Sak-marian marine beds of Europe. SUERO and ROQUE (1955) have shown that the Falkland Is. beds may be related to those of Bahia Laura and the

Sierra de Tepuel, so that there is no longer any need to drift them up towards Brazil to make them "fit", as DU TOIT did.

Table 3. Approximate correlation of S. African standard sequence with eastern S. American sequences.

	S. Africa and S. W. Africa	S. America Parana Basin	S. America
?Artinskian	Upper Ecca Shales		
?Sakmarian	Lower Ecca Shales White Band Boulder mudstones with *Eurydesma* Dwyka Tillite	Irati White Band Tubarão Fm. with glacials	Bonete Fm. with *Eurydesma* Sauce Grande glacials.

Africa

No marine beds with the fusulines, goniatites and brachiopods regarded as characteristic of the European Sakmarian or its North American equivalent, the Wolfcampian, have been found anywhere in Africa. Africa indeed, was almost entirely continental during the upper Palaeozoic, short and narrow marine transgressions being known only in South West Africa and in Madagascar. It is impossible to say as yet, which of the African deposits if any, are Sakmarian equivalents, so that views on the Sakmarian palaeogeography are not of much value.

In the North, in Morocco, Algeria and Tunis the only deposits found are intramontane continental deposits with some coals, the flora being that of the Autunian or Rotliegende of the Euramerican province. In Egypt, Sudan, the Sahara, Upper Guinea and French Equatorial Africa no deposits which could possibly be referred to Sakmarian times are known, and it could be that the region was a desert at the time, receiving insufficient precipitation for transport of sediment.

South of the equator, continental deposits are widespread, being formed between late Carboniferous and Jurassic times, and constituting the Karroo System. In the main part of the Karroo Basin in the Cape Province, Orange Free State and Natal, which might be regarded as standard, the sequence requiring consideration for the recognition of Sakmarian equivalents is, at the top, the Beaufort Beds, then the Ecca Shales, then the Upper Dwyka Shales with the White (*Mesosaurus*) Band near its top, and, at the base, the Dwyka Tillite.

The shales under the Tillite, formerly called the Lower Dwyka Shales, are now known to be disconformable with the tillite and have been transferred to the Witteberg Series (HAUGHTON et.al., 1953). The Dwyka Tillite attains its maximum thickness in the belt of country north of the northernmost Witteberg outcrop, where it has an average thickness of 2,000 feet. It thins again further north. The surface on which it lies was in general one of pronounced relief, with fairly deep, approximately north-south valleys running between ranges, the tillites being preserved within the valleys, but absent or very thin on the bounding heights. Much of the material for the southern facies of the tillite appears to have been derived from a (now vanished?) land mass lying to the south (Haughton, 1952). North of latitude 28° the tillite is almost completely absent, the Ecca beds

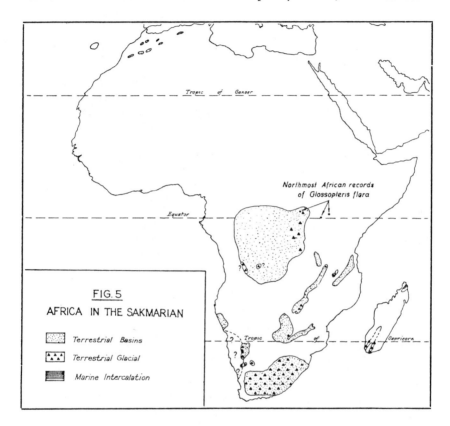

FIG. 5

AFRICA IN THE SAKMARIAN

Terrestrial Basins

Terrestrial Glacial

Marine Intercalation

Northmost African records of Glossopteris flora

resting directly on the basement, due, according to Rogers et. al. (1929) and Macgregor (1947), to its having been removed by pre-Ecca erosion.

In places, isolated developments of coarse sediments at the base of the Karroo sequence, particularly in the rift valleys, have been identified as glacials; these are in Bechuanaland (Green and Poldervaart, 1954), Southern Rhodesia (Bond, 1953) Tanganyika (Harkin, McKinley and Spence, 1954), Belgian Congo (Lower Lukugu Series, Cahen, 1954) and Angola (Mouta and Cahen, 1951). The nature of these glacial deposits in S. Rhodesia and Belgian Congo suggests to Bond and to Cahen that they were deposited by valley glaciers rather than by a continental ice sheet. But Macgregor (1947) in accordance with du Toit's view that the striated pavements of the Karroo indicated S. Rhodesia as a centre from which flowed the ice that formed the great boulder beds in the Union, suggested that the ice here may have formed a stationary conical core from which upper layers slid outwards in all directions, and that such glacial deposits as were formed were probably thin and had been mostly washed away before the Wankie coal measures were formed.

South West Africa too has a dissected pre-Dwyka surface, and the tillite is thickest in the pre-Dwyka valleys (MARTIN, 1954) at least one of which, the Kunene, has a "U" shape inherited from the Dwyka ice. MARTIN (1950) considered the glacials of S.W. Africa spread out from two high regions, lying N and S of the Damara.

It seems probable that valley glaciers and possibly mountain ice caps formed on the highlands surrounding separate basins of deposition in the area S. of the Equator, and that the continental ice, if in fact such existed, was developed as a continuous sheet only in the Karroo Basin. Du Toit's interpretation of much of the higher Dwyka Tillite of the southern outcrop of the Union as water-laid material on low land may be accepted, but flooding by the sea seems to have been confined to the western edge only, in the south of S.W. Africa only; there may have been highlands to the south, along the Palaeozoic folded belt of the Union.

Our safest line of evidence in correlating the African glacials is the marine fauna of South West Africa. However, apart from the fishes, this is a *Eurydesma*, i.e. a Gondwana fauna; the fishes give no better indication of age, according to GÜRICH (1923) than end of the Carboniferous — beginning of the Dyas.

According to MARTIN (1954) the *Eurydesma* fauna occurs in the marine-glacial boulder mudstone that follows the dark-grey or black bituminous shale with the Ganikobis fish fauna, which itself lies directly above the basal continental ground moraine. Marine glacials follow above the *Eurydesma* boulder mudstone, so that in S.W. Africa we have proof that the *Eurydesma* fauna was a cold-water, near-shore fauna, glacier ice dropping boulders into the deposits in which the *Eurydesma* shells are found.

These boulder-mudstones pass upwards into the Upper Dwyka shales, whose upper parts are white weathering and contain the White (*Mesosaurus*) Band, a remarkably constant development of the Karroo and the southern part of S.W. Africa, post glacial, and evidently deposited in a low lying coastal or estuarine region. Conditions and fauna were closely similar to those of the Irati White Band of S. America, and although the species of *Mesosaurus* are distinct, no great difference in age would seem to be apparent between the two. Unfortunately there is no reasonably safe correlation of either White Band with the European or North American succession, since *Mesosaurus* is confined to S. Africa and S. America and its relationships are obscure. Von HUENE (1940) correlates the S. African Band with the basal Rotliegende which would be Sakmarian, and the S. American with the later Rotliegende, which might be Artinskian.

The Ecca Series, entirely continental, and non-glacial, consists of bluish and greenish shales, sandstones and some coal seams, seemingly of drift origin. They transgress across the Dwyka on to older rocks, but in the Karroo Basin are thickest and coarsest in the South (HAUGHTON, 1952). *Glossopteris, Schizoneura* and *Phyllotheca* are recorded from their southern outcrops, and in the Middle Ecca Shales the *Glossopteris* flora is apparently intermingled with northern or relict forms such as *Sigillaria, Bothrodendron, Lepidodendron* and *Psygmophyllum* (ROGERS et.al., 1929).

The Beaufort beds which follow are continental sandstones and shales, rich in vertebrate fossils, chiefly reptilian. The Beaufort beds have been divided into 6 zones on the reptilian fossil assemblages; three in the lower, one in the middle and two in the upper beds. The three lower Beaufort Zones are correlated with zones II, III and IV of the Russian reptilian sequence by VON HUENE (1940), i.e. with Kazanian and Tatarian horizons, much younger than Sakmarian. HAUGHTON (1953) also equates the L. Beaufort zones with II, III and IV of Russia, but shows that similarity is great only between Zone III of S. Africa and Zone IV of Russia. EFREMOV and VJUSCHKOV (1955) equate the lowest S. African reptilian fauna of the Beaufort Beds with the upper Kazanian. If the reptile authorities are right, then the Beaufort beds are wholly younger than Kungurian.

Fresh-water mollusca (*Carbonicola, Kidodia, Palaeoanodonta*, and *Palaeomutela*), (HAUGHTON et. a., 1953) support the reptilian age determinations for the Lower Beaufort Beds, following determinations by COX (1936) of fresh water mollusca from Lower Beaufort and Ecca or L. Beaufort correlatives in Tanganyika.

North of the main Karroo Basin, older Karroo strata are non-marine and in general have considerable thickness only in rift valleys (GREEN and POLDERVAART, 1954; BRANDT, 1954). They are of smaller extent than and are overstepped by the Mesozoic upper Karroo beds, thus possibly indicating the existence of ancestral rift-valleys in pre-Karroo times. The basal beds, exposed in isolated areas, may be coarse or conglomeratic; some of these, in Bechuanaland (GREEN and POLDERVAART, 1954), Southern Rhodesia (BOND, 1953), Tanganyika (HARKIN, McKINLAY and SPENCE, 1954), Belgian Congo (CAHEN, 1954) and Angola (MOUTA and CAHEN, 1951) have been recognised as glacial and correlated with the Dwyka. But in many places, in Northern Rhodesia (BRANDT, 1954), Mozambique (BORGES, 1952), Nyasaland (DIXEY and WILLBOURN, 1951), Portuguese Nyasa (BORGES, NUNES and FREITAS, 1954), Kenya (TEMPERLEY, 1952) and Uganda (DAVIES, 1952), no glacials have been recognised and the basal beds are correlated with the base of the Wankie Series of sandstones, shales and coals seams of S. Rhodesia, which Series is itself correlated with the Ecca. The Wankie Coals are mainly drift coals (BOND, 1952). The isolated glacial deposits in these northern states would appear to be from valley glaciers rather than from any extensive ice sheets. The sources of the sediments correlated with the Ecca Series would seem to be from the highlands immediately surrounding the separate basins of deposition.

Plant fossils are practically the only direct source for an age determination of the Dwyka and Ecca correlatives in the north. Thus TEIXEIRA

(1951) records *Neuropteridium (Gondwanidium) validum* and *Noegge-rathiopsis* from the formation with glacials in Angola. CAHEN (1954) lists *Cyclodendron, Voltzia, Cyclopteris, Artesia, ? Ullmannia, Gangamopteris, Noeggerathiopsis* and *Phyllotheca* from the Lower Lukuga Series of Belgian Congo.

The Upper Wankie sandstones of S. Rhodesia contain *Glossopteris, Gangamopteris, Noeggerathiopsis* (as *Cordaites*), *Cladophlebis,* and *Sphenopteris* and the Euramerican *Sphenophyllum, Pecopteris, Chansitheca* and cf. *Cyclodendron* (BOND, 1952). This flora was considered correlatable with the Rotliegende (i.e. Sakmarian) or older by WALTON (1929). But JONGMANS (1952) considers it upper Stephanian, i.e. still within the Pennsylvanian.

The beds at Tete in Mozambique, correlated with the Upper Wankie, contain a similar assemblage and are considered "L. Permian" by TEIXEIRA (1951).

In Uganda, at Entebbe, the most northerly occurrence of the *Glossopteris* flora, Euramerican forms are present also (DAVIES, 1952). On the evidence of spores from coal seams in Natal, RILETT (1955) considers the Middle Ecca coals to be not later than Artinskian.

Thus in many of these presumably pre-Beaufort i.e. pre-Kazanian beds, *Glossopteris* and Euramerican floras are admixed. Botanists regard the Ecca-Wankie floras variously as upper Stephanian or Rotliegende.

We thus have as yet no firm basis for concluding whether Sakmarian correlatives are present in S. Africa, or if so, which they are.

BOND (1952) considers the climate in Rhodesia, during Dwyka and Ecca times to have been wet. CAHEN (1954) considers the Belgian Congo climate to have been cold and humid in Lower Lukugu times, and humid and less cold in upper Lukugu times.

In Madagascar, BESAIRIE (1952) shows that the Sakoa Group outcropping in a limited area in the south, has black shales and tillites at the base, which are quite analogous with the Dwyka of S. Africa, followed by beds with coal containing a pure *Glossopteris* flora, red beds with silicified wood, and at the top a marine transgression with *Productus, Spirifer* and lamellibranchs. These offer a prospect of future correlation with European marine sequences. The Sakamena group is transgressive and discordant on the Sakoa Group, and is essentially shaley and continental in facies, with some marine intercalations, the lowest of which contains the alga *Anthracoporella,* and the lamellibranchs *Gervillea* and *Modiolopsis* in species which are known in lower Beaufort equivalents in Tanganyika. Higher beds contain a *Glossopteris-Thinnfeldia* flora, and reptiles regarded as Beaufort equivalents. In the North of the Island (Ankitokazo) beds with *Productus* and *Spirifer* contain a fauna quite distinct from the beds with *Productus* and *Spirifer* in the south (Vohitolia) (ASTRE, 1934). The northern *Productus* fauna contains *Xenaspis carbonarius* of the Upper Permian; but whether the southern *Productus* fauna is of similar age, as BESAIRIE concludes, is questionable and deserves further study.

In the map of Africa offered, I have made the not really justifiable assumption that the Dwyka tillites and shales and Lower Ecca Shales are Sakmarian.

Antarctica

No Sakmarian strata can be safely identified in Antartica, and no marine beds are known in outcrop. The continental Beacon Sandstone, a horizontal mass at least 1,500 feet thick, developed in Victoria Land and possibly under the Ross Sea (ADIE, 1952), may however contain Sakmarian equivalents. *Glossopteris* and *Vertebraria* have been collected from calcareous sandstones, siliceous shales and coaly beds on Mt. Buckley nunatak that projects through the Beardmore Glacier, presumably high in the Beacon formation, which outcrops along the western shore of the Ross Sea and along both sides of the Beardmore Glacier (SEWARD, 1914); Upper Devonian fish scales are known from the moraine at the foot of Mt. Suess. Slabs of Beacon sandstone show ripple marks and sun cracks and the inference from the combined lithological and palaeontological evidence is that continental strata were deposited in this region continuously from the Upper Devonian to the Permian, including the Sakmarian. No tillites or varves have been reported from within the Beacon sandstone, but it does not seem reasonable to infer from this that upper Palaeozoic glaciation did not occur in Antarctica. We still know far too little about that continent.

Australia

Of the "Gondwanaland" countries, Australia appears to have the earliest Gondwana marine faunas; but since its characteristic Gondwana elements are unknown in the northern hemisphere (except in the Salt Range), and since there are few goniatites and corals and no fusulines, the recognition of Sakmarian horizons is difficult. There is however considerable agreement amongst Australian geologists on the position of the boundary between the Sakmarian and Artinskian Stages, as evidenced in the A.N.Z.A.A.S. symposium reviewed by HILL (1955).

The eastern Australian Permian fauna has fewer similarities to the N. Hemisphere fauna than the western Australian, and it is to the west that we look for our best chance of intercontinental correlation, leaving the eastern faunas to be correlated via the western.

In the Symposium mentioned above, it was agreed that in Western Australia the boundary between Sakmarian and Artinskian might fall at the base of the marine Nura Nura member with *Metalegoceras* and *Thalassoceras* in the Fitzroy Basin, at the base of the Callytharra formation in the Carnarvon Basin, and at the base of the Fossil Cliff beds in the Irwin R. Basin. *Metalegoceras* and *Uraloceras* occur in a marine intercalation in the Holmwood Shale above the Nangetty glacials; these two are not known before the Upper or Sakmarian substage of the Sakmarian stage in the Urals. If the boundary lies at the base of the Nura Nura Member, part or all of the Grant Sandstones with glacials in the Fitzroy

Basin, the Lyons (?marine) with glacials in the Carnarvon Basin, and the
Holmwood Shale and Nangetty continental glacials in the Irwin River Basin
are Sakmarian; the W.Australian Sakmarian fauna would then be very poorly
known. *Eurydesma* occurs in transitional beds between the Callytharra Fm.
and the Lyons Group, and 1,500 feet below the top of the Lyons there is
a marine fauna with productids and spiriferids that THOMAS (1954) con-

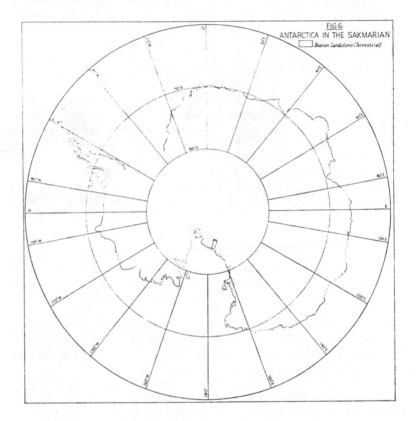

FIG 6
ANTARCTICA IN THE SAKMARIAN
☐ *Beacon Sandstone (Terrestrial)*

siders a local representative of the Umaria fauna of India, which however
is not as yet safely correlated with the Ural sequence; these productids
are not known in eastern Australia. As a result of the above, marine glacials are shown in the map of Australia in the Carnarvon, Fitzroy and
Bonaparte Gulf Basins, and terrestrial glacials (probably Alpine type) in
the Irwin River, Collie and Wilga Basins, and the Canning Desert Shelf.

Recently however RUZHENTSEV (1956. p. 143) has placed the W. Australian species *Metalegoceras clarkei* MILLER from the Nura Nura Member
in the synonymy of *M. tschernychewi* (KARPINSKY) which in Russia enters
in and characterises the upper Artinskian. If, for this reason, an upper
Artinskian horizon should be preferred for the Nura Nura Member and

equivalents, the top of the Sakmarian might be well down in, or even below the glacials.

Whether or not a continental ice sheet was present in W. Australia in the Sakmarian is yet to be established. Some authors have assumed it, but the evidence seems to me insufficient.

S. Australian upper Palaeozoic glacials are shown by Campana (1954) to be of Alpine type. Victoria also, though possessing land glacials and peri-

FIG. 7

AUSTRALIA in the SAKMARIAN

[- -]   *Marine Areas*

[::::]   *Terrestrial Basins*

▲ ▲ ▲   *Glacial Deposits*

glacials, seems to me to have had at the most a small mountain ice cap, not a continental sheet. There is no certainty that either the Victorian or S. Australian glacials are coeval with the W. Australian ones, which as seen above we assume at present to be at least in part early Sakmarian. Ludbrook (1957) has found Permian arenaceous foraminifera in claystones interbedded in the fluvio-glacial sequence in Yorke Peninsula, S. Australia.

In New South Wales the main glaciation (Upper Kuttung) was long pre-Sakmarian, probably Moscovian. But marine glacials are known in the Allandale Fm. of the Lower Marine (Dalwood) Group with a rich but incompletely described *Eurydesma-"Aviculopecten"* fauna, and these are generally correlated with the West Australian main glacials; they could indeed be Sakmarian. *Pseudogastrioceras*, in the somewhat younger Farley Formation has been compared with Artinskian species (Teichert, 1953).

Table 4. Tentative Placing of the Boundary between Sakmarian and Artinskian in Australia.

	Sakmarian	Artinskian
Carnarvon Basin W. Australia	Lyons Gp. with glacials (*Eurydesma* at top). Umarian productid fauna 1,500 ft. below top	Callytharra Fm.
Irwin R. Basin W. Australia	Holmwood Shales Nangetty glacials	Fossil Cliff Beds
Fitzroy Basin W. Australia	Grant Sandstone with glacials	Nura Nura Member
Tasmania	*Eurydesma* zone over Basal glacials	Grange mudstone and Berriedale limestone
Hunter River, New South Wales	Allandale Fm. with *Eurydesma*	Rutherford Fm.
Bowen Basin, Queensland	Cattle Creek Fm. with *Eurydesma* and marine glacials	Aldebaran Sandstone
Artesian Basin, Queensland	Joe Joe Fm. with terrestrial glacials	Colinlea Fm.

In Tasmania the Basal Glacials and the *Eurydesma* zone just above them may well be Sakmarian, but again we have no serviceable descriptions of the fauna.

In Queensland, HILL (1953) and MAXWELL (1954) have suggested that the Cracow fauna at the base of the marine sequence in the S.E. of the Bowen Basin is Sakmarian; the fauna includes *Taeniothaerus, Aulosteges, Cancrinella, Terrakea, Anidanthus, Horridonia, Strophalosia, Lissochonetes,* "*Martiniopsis*" (2 genera), *Neospirifer, Trigonotreta, Spiriferellina* and several other spiriferoid genera; *Eurydesma,* pectinoids, platychismoids, and many other undescribed forms, but no goniatites or fusulines. Terrestrial glacials with *Glossopteris* that may be coeval form the Joe Joe Formation of the Springsure district; but they could equally be Orenburgian, for all the direct evidence we have. Marine glacials occur in the Cattle Creek Formation of the S.W. exposure of the Bowen Basin, this fauna being thought to be roughly equivalent to the Cracow fauna.

Because of this rather weak evidence, the map of Australia shows marine areas in the south end of the Bowen Basin, with glaciated highlands south of the Springsure region, and marine areas in the Yarrol and Gympie Basins, and in the Northeastern part of New South Wales and in Tasmania. No Sakmarian coals or salt deposits are known. There must have been sufficient precipitation to nourish the glaciers of the West and the East, and the climate must have been somewhat cooler than the present.

In New Zealand neither Sakmarian faunas nor upper Palaeozoic glacials have been distinguished, and in view of this ignorance it has been left blank on the world map.

Palaeogeography of the Sakmarian

It will have been seen above that our Sakmarian palaeogeographical reconstructions can be reasonably accurate at present only for the continents of the northern hemisphere; difficulties of stratigraphic correlation with and within the continents of the southern hemisphere render these southern reconstructions subject to the bias of individual authors, and I have given the reasons for my preferred correlations above.

The extent of marine areas on the world map (fig. 8) shows that, unless parts of the continents that were exposed during the Sakmarina are now hidden from us under the present sea, the land masses of North America, Eurasia and South America wer considerably reduced by epicontinental and geosynclinal seas, while those of Australia, Antarctica and Africa were relatively very little affected in this way. The decrease in area of land surface in the Northern Hemisphere may have profoundly modified the circulation of the atmosphere and of the oceans, and the changes in circulation may have affected regional temperatures.

The topographic relief of the continents during the Sakmarian is still too difficult to assess, and the height and trends of the Sakmarian ranges are still too problematical for us to deduce their possible effect on the atmospheric circulation and on the belts of precipitation. Both Europe and N. America would seem, judging from the development of coal seams, to have had less precipitation in the Sakmarian than in the Pennsylvanian, and more than in the Kazanian. Heights and positions of plateaus are even more difficult to establish than those of fold mountain ranges.

Biogeographical provinces may give some lead by indicating the positions and perhaps the nature of barriers to the migration of floras and faunas.

What we might call the *Eurydesma* province is separated from what we might call the *Pseudoschwagerina-Pseudofusulina* province by only the width of Tethys to the north of India, by a land barrier in S. America and by Africa, while Australia lay wholly within the *Eurydesma* province; the most likely barrier between these marine provinces seems to be temperature; and the *Eurydesma* fauna does seem to be present where marine glacials are known, or in the marine inundation following land glaciation. Further precision is required, however, particularly in relation to the northward extent of the Sakmarian fusulines.

The *Glossopteris* flora seems present in Gondwanaland everywhere before or with the *Eurydesma* fauna, and in Sakmarian time is not known in the Northern Hemisphere except perhaps along the equator of Africa, and in northern India. The northward barrier may have been a desert in Africa, and may have been Tethys elsewhere. Temperature may also have played a part. The Angara flora may have been separated by a Ural marine barrier from the European and North American flora.

H u m i d r e g i o n s may be recognised from the coals formed in them, and from the glacials resulting from excessive precipitation in a frigid area, whether this was in high latitudes or of high relief in low latitudes.

C o a l s that appear to have been formed in peat swamps occur in the Dunkard Formation of eastern U.S.A., in the Autunian and Lower Rotliegende in W. Europe, Turkey and N.W. Africa, in the Kuznetsk and other Siberian Basins, and in the Taiyuan of China. It may fairly be deduced that all these areas had a high rainfall, but the coals have not been studied to find whether they were formed as tropical or as cold temperate peats.

In the Southern Hemisphere, coals occur in the Tubarão Series of the Parana Basin of S. America, interbedded with glacials. This region too, if we regard the Tubarão as Sakmarian, must have had high precipitation; perhaps the coals are from cold temperate peats; possibly the winds bringing the moisture were from the Atlantic to the east. Some of the Ecca coals of S. Africa may have been Sakmarian, but they seem to be drift coal.

A r i d i t y, probably intermittent, is suggested by anhydrites etc. in the shallow edges of the mid-continent seas of western U.S.A., and in some of the European continental basins.

G l a c i a l c l i m a t e s. It is still doubtful whether the Dwyka Tillite, the Talchir Boulder Beds, the Tubarão and Sauce Grand Glacials, and the Lyons and Grant glacials are Sakmarian or younger or older, or whether indeed they are all equivalent in age. On the maps offered, however, I have made the assumption that they are Sakmarian as the most reasonable deduction from the present totally inadequate evidence. On this assumption glacial climates appear to have been present in South America south of 15° S, in Africa south of 10° S, in India between 15° and 30° N and in Australia. In the Andean region Alpine glaciers or mountain ice caps may have developed; in eastern Brazil near Buenos Aires, a mountain ice cap or a continental ice sheet, possibly the former, may have been present. In Africa north of about 30° S to about 5° N, alpine glaciers or mountain ice caps may have developed, and in the Karroo to the south there may have been a continental ice sheet. In India a mountain ice cap possibly existed about and to the north of 20° N. In Australia alpine glaciers or mountain ice caps seem to have been present, but to my mind the presence of a continental ice sheet is extremely doubtful.

Belts of loess have not been distinguished associated with any of these ice masses.

Our knowledge of Antarctica and the Arctic regions is very deficient. Our inability to say what conditions were in these areas during Sakmarian times makes our speculations on the positions of climatic belts and consequently on the positions of the poles, equator and continents unrewarding. It is to be hoped that the great scientific activity in Antarctica this year will throw light on the occurrence of *Glossopteris*-bearing sediments

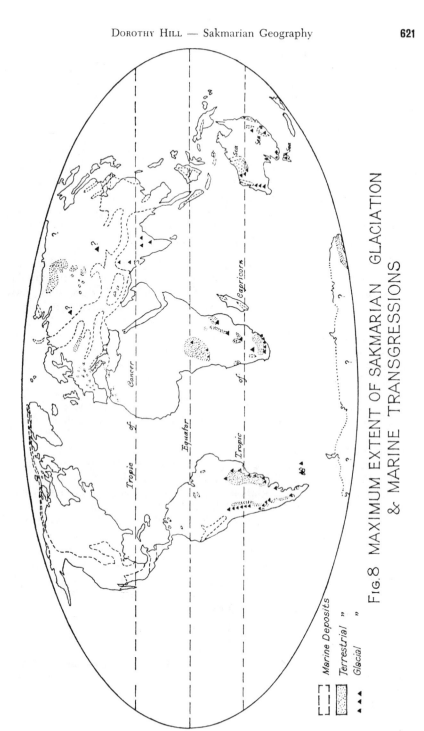

Fig. 8 MAXIMUM EXTENT OF SAKMARIAN GLACIATION & MARINE TRANSGRESSIONS

Marine Deposits
Terrestrial   "
Glacial   "

there, and show whether, as in South America, such are interbedded with glacials as yet unrecognised.

Climatic belts. If we accept the assumption of Sakmarian age for these glacial deposits, sufficient precipitation, low enough temperatures and sufficient relief for mountain ice caps or small continental ice sheets to form must have occurred between 20° and 40° S.

Asia was humid enough for coals to form between 40° and 60° N, and in W. Europe coal formed mainly in intramontane basins; the eastern part of the U.S.A. (about 40° N) was humid enough for coal formation, while western U.S.A. between 20° and 50° N was probably a rather arid temperate region of seasonal rainfall. At present we know of no Sakmarian glacials in either Antarctica or the Arctic Regions. Beyond these generalisations we cannot go as yet.

Conclusions

1. A reasonable northern hemisphere map of Sakmarian land and sea surfaces can be constructed from the known palaeontological and stratigraphical data. Sea surfaces were very much enlarged and land surfaces very much reduced. This could have had considerable influence on circulation and on climate.

2. A similar map for the southern hemisphere depends on somewhat risky assumptions being made in the correlation of faunas and strata. Land surfaces are but slightly reduced.

3. If these assumptions in correlation are made, glacial deposits were developed mainly in a belt between 40° S and 20° S; but no undoubted glacials are known in the northern hemisphere, except in India.

4. The glacials appear to be mainly from mountain ice caps and valley glaciers, though small continental ice caps may have been present in the Karroo and Parana Basins.

5. The roughly concentric development of the glacials about the present South Pole suggests that they form a climatic zone about a Sakmarian Pole approximately in todays position, but the present evidence gives no clear-cut support for either fixity or mobility of poles or continents.

6. Two problems need solution before a really reliable map of Sakmarian climatic belts can be drawn and hypotheses of polar wandering and continental drift tested. These are the Sakmarian history of Antarctica and the Arctic Regions, and the correlation of the "Gondwana" faunas and floras with those of the northern hemisphere.

References

Europe

BACKLUND, H. G., 1930. On a probable tillite of late Palaeozoic age from the Kara-River, northern-most Ural. C. R. 15th Int. geol. Congr., Vol. 2, pp. 77—82. — BANKOVITCH, V. A., and REDISYKIN, N. A., 1955. [*Schwagerina* horizon in the North East of the Donetz Basin.] C. R. Acad. Sci. U.R.S.S. 1955, 104, pp. 456—458. — DUNBAR, C. O., 1940. The type Permian: Its classification and correlation. Bull. Amer. Assoc. Petrol. Geol., 24, pp. 237—281. – 1942.

Artinskian Series (Discussion). Bull. Amer. Assoc. Petrol. Geol., **26,** pp. 402—409. — Fomichev, V. D., 1953. [Rugose corals and stratigraphy of the Middle and Upper Carboniferous and Permian deposits of the Donetz Basin.] Trudy vsesoyuz. nauch.-issled. geol. Inst., pp. 1—622, pls. 1—44. — Francis, W., 1954. Coal; its formation and composition. E. Arnold, London. — Heritsch, F., 1939. Karbon und Perm in den Südalpen und in Südosteuropa. Geol. Rdsch., **30,** pp. 529—588. — Ignatiev, V. I., 1952. Das Relief der permischen Zeit im Einzugsgebiet der unteren Oka. C. R. Acad. Sci. U.R.S.S. (2) **84,** pp. 1037—1040, 1/9. — Jongmans, W. J., 1952 a. Some problems on Carboniferous stratigraphy. C. R. 3ième Congr. Stratig. Géol. Carb. I, pp. 295—306. - 1952 b. Discussion on paper by G. Mathieu. C. R. 3ème Congr. Stratig. Géol. Carb. II, pp. 449—451. — Kopeliovitch, A. B., and Eventov, Y. S., 1956. Permian deposits of Astrakhan. C. R. Acad. Sci. U.R.S.S., **106,** No. 2, pp. 320—323. — Likharev. B. K., 1937. Permian System of the U.R.S.S., certain problems of its stratigraphy and correlations with other countries. Abstr. Pap. 17th Int. geol. Congr., p. 81. — Maximova, S. W., 1952. [On the palaeofaunistic character of the Sakmarian Stage.] Bull. Acad. Sci. U.R.S.S. (Geol.), **3,** pp. 118—127, 1 pl. — Miller. A. K., 1938. Comparison of Permian ammonoid zones of Soviet Russia with those of North America. Bull. Amer. Assoc. Petrol. Geol., **22,** pp. 1014—1019. — Nalivkin, D. V., 1937. The Sim Works. 17th Int. geol. Congr. Perm Excursion (S. Part). p. 123. - 1938. Scientific results of the Permian conference. Bull. Amer. Assoc. Petrol. Geol., **22,** pp. 771—776. — Rozovskaya, S. E., 1949. [Stratigraphic distribution of fusulinids in Upper Carboniferous and Lower Permian localities of the Southern Ural.] C. R. Acad. Sci. U.R.S.S., **69,** pp. 249—252. — Rakusz, G., 1932. Die oberkarbonischen Fossilien von Dobsina und Nagyvisnyo. Geol. hung. (b), **8,** pp. 1—224, pls. 1—9. — Ruzhentsev, V. E., 1951. Lower Permian Ammonoids of the Southern Urals. I. Ammonites of the Sakmarian stage. Trav. Paleont. Inst. Acad. Sci. U.R.S.S., **33,** pp. 1—188, pls. 1—15. - 1954. [Asselian stage of the Permian System.] C. R. Acad. Sci. U.R.S.S., **99,** pp. 1079—1082. - 1956. Lower Permian Ammonites of the Southern Urals, II. Ammonites of the Artinskian Stage. Trav. Paleont. Inst. Acad. Sci. U.R.S.S., **60,** pp. 1—275, pls. 1—39. — Ruzhentsev, V. E., and Shimansky, V. N., 1954. [Lower Permian coiled and curved nautiloids of the Southern Urals.] Trav. Paleont. Inst. Acad. Sci. U.R.S.S., **50,** pp. 1—152, pls. 1—15. — Shimansky. V. N., 1954. [Straight nautiloids and bactritids of the Sakmarian and Artinskian stages of the southern Urals.] Trav. Paleont. Inst. Acad. Sci. U.R.S.S., **44,** pp. 1—156, pls. 1—12. — Sujkowski, Z., 1946. The geological structure of East Poland and West Russia: a summary of recent discoveries. Quart. J. geol. Soc., **102,** pp. 189—201. — Sultanaev, A. A., 1954. Stratigraphic scheme for the upper Palaeozoic of the Kolvo-Vishersk region. C. R. Acad. Sci. U.R.S.S., **98,** pp. 257—258. — Tolstikhina, M. M., 1937. Upper Paleozoic reefs on the western slope of the Urals. Abstr. Pap. 17th Int. geol. Congr., p. 97. — Toumansky, O. G., 1937. Les dépôts permiens de la Crimée. Abstr. Pap. 17th Int. geol. Congr., p. 108. — Wilser, J. L., 1927. Die Steinkohlen in der Schwarzmeer-Umrandung, insbesondere bei Heraklea-Zonguldag (Nordanatolien). Geol. Rdsch., **18,** pp. 1—37.

Asia

Bemmelen, R. W. van, 1949. The Geology of Indonesia. 2 vols. The Hague. — Chibber, H. L., 1934. The Geology of Burma. Macmillan & Co., London. — Douglas, J., 1950. The Carboniferous and Permian faunas of South Iran and Iranian Baluchistan. Palaeont. Indica, **22,** No. 7, pp. 1—57, pls. i—v. — Dunbar,

C. O., 1933. Stratigraphic significance of the fusulinids of the Lower Productus Limestone of the Salt Range. Rec. geol. Surv. India, **66**, pp. 405—413, pl. 22. — DUNBAR, C. O., and MISCH, P., 1947. Fusuline-bearing Permian rocks of northwestern Yunnan. Bull. geol. Soc. China, **27**, pp. 101—110. — DUTKIEWICZ, G. A., 1937. Le paléozoique supérieur du Pamir et du Darvaz. Abs. Pap. 17th Int. geol. Congr., p. 105. — ENDO, R., 1953. A summary of the columnar section in Manchuria. Proc. 7th Pan-Pacif. Sci. Congr., Vol. 2, pp. 316—326. — FOX, C. S., 1937. The climates of Gondwanaland during the Gondwana era in the Indian region. Abs. Pap. 17th Int. geol. Congr., p. 216. — FROMAGET, J., 1952. Aperçu de nos connaissances sur la géologie de l'Indochine. Rept. 18th Int. geol. Congr., XIII, pp. 63—77. — GERTH, H., 1950. Die Ammonoideen des Perms von Timor und ihre Bedeutung für die stratigraphische Gliederung der Permformation. Abh. N. Jb. Min., Geol., Paläont. (B), **91**, pp. 233—320. — 1952. Das Klima des Permzeitalters. Geol. Rdsch., **40**, pp. 84—89. — GRABAU, A. W., 1931. The Permian of Mongolia. Amer. Mus. nat. Hist. New York. — HAYASAKA, I., 1954. Younger Palaeozoic cephalopods from the Kitakami Mountains. J. Fac. Sci. Hokkaido Univ. Ser. IV, **8**, pp. 361—374. — JACOB, K., 1952. A brief summary of the stratigraphy and palaeontology of the Gondwana system. Symposium sur les Series de Gondwana. 19th Int. geol. Congr., pp. 153—174. — JONGMANS, W. J., 1939. Die Kohlenbecken des Karbons und Perms in U.S.S.R. und Ostasien. Geol. Sticht., Jaarsverslag 1934—37, pp. 15—192. — KAHLER, F. and G., 1940. Fusuliniden aus dem Tienschan. N. Jb. Min. Geol. Paläont. Abh. (B), **83**, pp. 348—362, pls. 8—10. — KOBAYASHI, T., 1942. The Akiyoshi orogenic cycle in the Mongolian geosyncline. Proc. imp. Acad. Tokyo. **18**, pp. 306—319. – 1952. On the southwestern wing of the Akiyoshi orogenic zone in Indochina and South China and its tectonic relationship with the other wing in Japan. Jap. J. Geol. Geogr., **22**, pp. 27—37. — LEE, J. S., 1939. The Geology of China. Murby & Co., London. — LEUCHS, K., 1935. Geologie von Asien. Vol. 1, parts 1 and 2, Borntraeger, Berlin. — LICHAREV, B. K., 1937. Permian brachiopods of the U.R.S.S. Abs. Pap. 17th Int. geol. Congr., p. 97. — MA, T. Y. H., 1955. Climate and the relative positions of the continents during the Lower Carboniferous. Res. past Climate continental Drift, **7**, pp. 1—99, pls. 1—50. — MINATO, M., 1953. Palaeogeographie des Karbons in Ostasien. Proc. Jap. Acad., **29**, pp. 246—253. – 1955. Japanese Carboniferous and Permian corals. J. Fac. Sci. Hokkaido Univ., Ser. IV, **9**, No. 2, pp. 1—202. pls. 1—43. – 1956. Paleogeography of the Japanese Islands and their adjacent lands in the Upper Palaeozoic era. Reprint in Japanese, pp. 1—13. — NORIN. E., 1930. An occurrence of late Palaeozoic tillite in Kurak-tagh Mts., Central Asia. C. R. 15th Int. geol. Congr., Vol. 2, pp. 74—76. — OBRUTSCHEW, W. A., 1926. Geologie von Sibirien. Fortschr. Geol., Heft 15, 572 pp., 1 map, 10 pls., 60 text figs. — STOYANOW, A., 1942. Revision of the Permo-Triassic sequence at Djulfa, Armenia. Bull. geol. Soc. Amer., **53**, p. 1823. — TCHIRKOVA, H. T., and ZALESSKY, M. D., 1939. Carte de l'extension du continent de l'Angaride d'après les découvertes de la flore Permienne. Probl. Sov. Palaeont., **5**, pp. 311—322. — THOMPSON, M. L., 1936. Lower Permian fusulinids from Sumatra. J. Paleont., **10**, pp. 587—592, figs. 1—13. – 1946. Permian fusulinids from Afghanistan. J. Paleont., **20**, pp. 140—157, pls. 23—26, text fig. 1. – 1949. The Permian fusulinids of Timor. J. Paleont., **23**, pp. 182—192, pls. 34—36. — THOMPSON, M. L., and MILLER, A. K., 1935. *Schwagerina* from the western edge of the Red Basin, China. J. Paleont., **9**, pp. 647—652. — USTRISHKY, V. I., 1955. Brachiopod complex of the Permian deposits of East Taimyr. C. R. Acad. Sci. U.R.S.S., **105**, pp. 805—807. — WADIA, D. N., 1937. Note on the Palaegeography and climate of Kashmir during the Permo-Carboniferous. Abs. Pap. 17th Int. geol. Congr., pp. 218—219. - 1953. The Geology of India. 3rd Edit. Macmillan & Co., London. — Weir, J., 1945. A review of recent work on the Permian non-marine lamellibranchs and its bearing on the affinities of certain non-marine genera of the Upper Palaeozoic. Trans. geol. Soc. Glasg., **20**, pp. 291—340.

North America

Agatson, R. S., 1954. Pennsylvanian and Lower Permian of northern and eastern Wyoming. Bull. Amer. Assoc. Petrol. Geol., **38**, pp. 508—583. — Baker, A. A., Huddle, J. W., and Kinney, D. M., 1949. Paleozoic geology of north and west sides of Uinta Basin, Utah. Bull. Amer. Assoc. Petrol. Geol., **33**, pp. 1161—1197. — Baker, A. A., and Williams, J. S., 1940. Permian in parts of Rocky Mountain and Colorado Plateau regions. Bull. Amer. Assoc. Petrol. Geol., **24**, pp. 617—635. — Blackstone, D. L., 1954. Permian rocks in Lemhi Range, Idaho. Bull. Amer. Assoc. Petrol. Geol., **38**, pp. 923—925. — Brill, K. G., 1952. Stratigraphy in the Permo-Pennsylvanian zeugogeosyncline of Colorado and northern New Mexico. Bull. geol. Soc. Amer., **63**, pp. 809—880, 3 pls. — Cressman, E. R., 1955. Physical stratigraphy of the Phosphoria Formation in part of southwestern Montana. Bull. U.S. geol. Surv., 1027-A, pp. 1—31, pls. 1—5. — Cross, A. T., and Arkle, T., 1952. The stratigraphy, sedimentation and nomenclature of the Upper Pennsylvanian-Lower Permian (Dunkard) strata of the Appalachian area. C. R. 3ieme Congr. Stratig. Carbon., I, pp. 101—111. — Dott, R. H., 1955. Pennsylvanian stratigraphy of Elko and northern Diamond Ranges, N.E. Nevada. Bull. Amer. Assoc. Petrol. Geol., **39**, pp. 2211—2305. — Dunbar, C. O., 1924. Was there Pennsylvanian-Permian glaciation in the Arbuckle and Wichita Mountains of Oklahoma? Amer. J. Sci., **8**, pp. 241—248. - 1939. Permian fusulines from Sonora. Bull. geol. Soc. Amer., **50**, pp. 1745—1760, 4 pls. - 1955. Permian brachiopod faunas of central east Greenland. Medd. Grønland, **110**, No. 3, pp. 1—170, pls. 1—32. — Eardley, A. J., 1951. Structural geology of North America. 624 pp. Harper, New York. — Fortier, Y. O., Mcnair, A. H., and Thorsteinsson, R., 1954. Geology and Petroleum Possibilities in Canadian Arctic Islands. Bull. Amer. Assoc. Petrol. Geol., **38**, pp. 2075—2109. — Frebold, H., 1950. Stratigraphie und Brachiopodenfauna des marinen Jungpaläozoikums von Holms- und Amdrups-Land (Nordostgrönland). Medd. Grønland, **126**, No. 3, pp. 1—97, pls. 1—6. — Hills, J. M., 1942. Rhythm of Permian seas — a paleogeographic study. Bull. Amer. Assoc. Petrol. Geol., **26**, pp. 217—255. — Imlay, R. W., 1939. Paleogeographic studies in northeastern Sonora. Bull. geol. Soc. Amer., **50**, pp. 1723—2744, 4 pls. — Kay, M., 1951. North American geosynclines. Mem. geol. Soc. Amer., **48**, pp. 1—143, pls. 1—16. — King, P. B., 1948. Geology of the southern Guadalupe Mountain Texas. Prof. Pap. U.S. Geol. Surv., **215**, pp. 1—183, pls. 1—23. - 1951. The tectonics of middle North America. 203 pp. Princeton. — King, R. E., 1939. Geological reconnaissance in northern Sierra Madre Occidental of Mexico. Bull. geol. Soc. Amer., **50**, pp. 1625—1722, 9 pls. — King, R. E., Dunbar, C. O., Cloud, P. E., and Miller, A. K., 1944. Geology and Paleontology of the Permian area Northwest of Las Delicias, Southwestern Coahuila, Mexico. Sp. Pap. geol. Soc. Amer., **52**, pp. 1—172, pls. 1—45. — Lloyd, E. R., 1949. Pre-San Andres stratigraphy and oilproducing zones in Southeastern New Mexico. Bull. N. Mex. Bur. Min., **29**, pp. 1—87, pls. 1—10. — Lord, C. S., Hage, C. O., and Stewart, J. S., 1947. The Cordilleran region in Geology and economic minerals of Canada. 3rd Edit. Geol. Surv. Can., Econ. Geol., pp. 220—310. — Lovering, T. S., and Goddard, E. N., 1950. Geology and Ore deposits of the Front Range, Colorado. Prof. Pap. U.S. geol. Surv., 223, pp. 1—319, pls. 1—30. — McCABE, W. S., 1954. Williston Basin Paleozoic unconformities. Bull. Amer. Ass. Petrol. Geol., 38, pp. 1997—2010. — McKEE, E. D., 1951. Sedimentary basins of Arizona and adjoining areas. Bull. geol. Soc. Amer., 62, pp. 481—506, 3 pls. — McKELVEY, V. E., SWANSON, R. W., and SHELDON, R. P., 1953. The Permian phosphorite deposits of western United States. C. R. 19th Int. geol. Congr. Sec. XI, pp. 45—64. — McLEARN, F. H., and KINDLE, E. D., 1950. Geology of Northeastern British Columbia. Mem. geol. Surv. Can., 259, pp. 1—236, pls. 1—8, map. — McNAIR, A. H., 1951. Paleozoic stratigraphy of part of north western Arizonia. Bull. Amer. Assoc. Petrol. Geol., 35, pp. 503—541. — MAHER, J. C., 1954. Lithofacies and suggested depositional environment of Lyons Sandstone and Lykins Formation in Southeastern Colorado. Bull. Amer. Assoc. Petrol. Geol., 38, pp. 2233—2239. — MILLER, A. K., and YOUNGQUIST, W., 1949. American Permian nautiloids. Mem. geol. Soc. Amer., 41, pp. 1—218, pls. 1—59. — MOFFIT, F. H., 1954. Geology of the eastern part of the Alaska Range and adjacent areas. Bull. U.S. geol. Surv., 989-D, pp. 60—218, pls. 6, 7. — MOORE, R. C., FRYE, J. C., JEWETT, J. M., LEE, W., and O'CONNOR, H. G., 1951. The Kansas rock Column. Bull. geol. Surv. Kansas, 89, pp. 1—132. — MULLERRIED, F. K. G., MILLER, A. K., and FURNISH, W. M., 1941. The Middle Permian of Chiapas, southernmost Mexico, and its fauna. Amer. J. Sci., 239, pp. 397—406, 1 pl. — NEWELL, N. D., 1948. Key Permian section. Confusion Range, western Utah. Bull. geol. Soc. Amer., 59, pp. 1053—1058. — NOLAN, T. B., 1943. The Basin and Range Province in Utah, Nevada and California. Prof. Pap. U.S. geol. Surv., 197-D, pp. 141—196, pl. 40, 41. — ROOTS, E. F., 1954. Geology and mineral deposits of Aiken Lake Map-Area, British Columbia. Mem. geol. Surv. Can., 274, pp. 1—246, pl. i—x, maps. — SAPPER, K., 1937. Mittelamerika. Handb., reg. Geol. VIII, 4 a, pp. 1—160, pls. I—X. — SCHUCHERT, C., 1935. Historical Geology of the Antillean Caribbean Region. 811 pp. New York and London. — SMITH, P. S., 1939. Areal geology of Alaska. Prof. Pap. U.S. Geol. Surv., 192, pp. 1—100, pls. 1—18, 1 chart. — THOMPSON, M. L., 1954. American Wolfcampian fusulinids. Univ. Kansas, Paleont. Contrib., Protozoa. Art., 5, pp. 1—226, pls. 1—52. — THOMPSON, M. L., and MILLER, A. K., 1944. The Permian of southernmost Mexico and its fusulinid faunas. J. Palaeont., 18, pp. 481—504, pls. 79—84. — THOMPSON, M. L., WHEELER, H. E., and HAZZARD, J. C., 1946. Permian fusulinids of California. Mem. geol. Soc. Amer., 17, pp. 1—77, pls. 1—18. — TROELSEN, J. C., 1950. Contributions to the geology of Northwest Greenland, Ellesmere Island and Axel Heiberg Island. Medd. Grønland., 149, No. 7, pp. 1—86, 1 map. — WANLESS, H. R., et al., 1955. Paleozoic and Mesozoic rocks of Grosventre, Teton, Hoback, and Snake River Ranges, Wyoming. Mem. geol. Soc. Amer., 63, pp. 1—90, pls. 1—13. — WENGERD, S. A., and STRICKLAND, J. W., 1954. Pennsylvanian stratigraphy of Paradox Salt Basin, Four Corners Region, Colorado and Utah. Bull. Amer. Ass. Petrol. Geol., 38, pp. 2157—2199. — WHEELER, H. E., 1940. Permian volcanism in western North America. Proc. 6th Pan. Pacif. Sci. Congr., I. pp. 369—376.

South America

ADIE, R. J., 1952. Representatives of the Gondwana System in the Falkland Islands. Symposium sur les Séries de Gondwana. 19th Int. geol. Congr., pp. 385—392. — AHLFELD, F., 1946. Geologia de Bolivia. Rev. Mus. La Plata (NS), 3, Sec. Geol., pp. 11—370, 115 figs., map. — ALMEIDA, F. F. M. DE, 1952. État actuel des connaissances sur la formation de Gondwana, au Bresil. Symposium sur les Séries de Gondwana. 19th Int. geol. Congr., pp. 258—272. — BARBOSA,

O., 1952. Comparison between the Gondwana of Brazil, Bolivia, and Argentina. Symposium sur les Séries de Gondwana. 19th Int. geol. Congr., pp. 313—324. — BEURLEN, K., 1957. Das Gondwana-Inlandeis in Südbrasilien. Geol. Rdsch., 45, pp. 595—599. — CASTER, K. E., 1952. Stratigraphic and paleontologic data relevant to the problem of Afro-American ligation during the Paleozoic and Mesozoic. Bull. Amer. Mus. nat. Hist., 99, pp. 105—158, pls. 17—18, text figs. 1—17. — CASTER, K. E., and DRESSER, H., 1954. Contributions to knowledge of the Brazilian Paleozoic: No. 1. Bull. Amer. Pal., 35, No. 149, pp. 1—84, 8 pls. — DOLIANTE, E., 1952. La flore fossile du Gondwana au Brésil d'après sa position stratigraphique. Symposium sur les Séries de Gondwana. 19th Int. geol. Congr., pp. 285—301. — FOSSA-MANCINI, E., 1944. Las transgresiones marinas del Antracolitico en la America del Sur. Rev. Mus. La Plata (NS) 2, Sec. Geol., pp. 49—183. — FRENGUELLI, J., 1944. Apuntes acerca del Paleozoico Superior del Noroeste Argentino. Rev. Mus. La Plata (NS) 2, Sec. Geol., pp. 213—265. - 1946. Consideraciones acerca de la "Serie de Paganzo" en las provinces de San Juan y La Rioja. Rev. Mus. La Plata (NS) 2, Sec. Geol., pp. 313—376. — GERTH, H., and KRÄUSEL, R., 1931. Beiträge zur Kenntnis des Carbons in Südamerika, I. Neue Vorkommen von marinem Obercarbon in den nördlichen Anden. N. J. Min. Geol. Paläont. (B) Bd., 65, Abt. pp. 521—529, pl. 22. — HARRINGTON, H. J., 1950. Geologia del Paraguay oriental. Contr. Sci. Univ. Buenos Aires, Ser. E (Geol.), Vol. 1, pp. 1—82, pls. 1—3, 2 maps. - 1955. The Permian *Eurydesma* fauna of eastern Argentina. J. Paleont., 29, pp. 112—128, pls. 23—26. — HEIM, A., 1945. Observaciones tectonicas en Barreal, Precordillera de San Juan. Rev. Mus. La Plata (NS) 2, Sec. geol., pp. 267—286. — KEGEL, W., 1953. Das Paläozoikum des Parnaiba-Beckens (Piaui und Maranhao, Brasilien). 19th Int. geol. Congr. Sec. II, Fasc. II, pp. 165—169. — KOZLOWSKI, R., 1914. Les brachiopodes du Carbonifère supérieur de Bolivie. Ann. Paleont., 9, 102 pp., 11 pls. — MAACK, R., 1957. Über Vereisungsperioden und Vereisungsspuren in Brasilien. Geol. Rdsch., 45, pp. 547—595. — MENDES, J. C., 1952. Invertébrés du Système de Gondwana au Brésil. Symp. Séries de Gondwana. 19th Int. geol. Congr., pp. 302—307. - 1957. Das Karbon des Amazonas-Beckens. Geol. Rdsch., 45, pp. 540—546. — MILLER, A. K., and WILLIAMS, J. S., 1945. Permian cephalopods from northern Colombia. J. Paleont., 19, pp. 347—349, pl. 51. — MILLER, A. K., and GARNER, M. F., 1953. Upper Carboniferous goniatites from Argentina. J. Paleont., 27, pp. 821—823. — NEWELL, N. D., CHRONIC, J., and ROBERTS, T. G., 1953. Upper Paleozoic of Peru. Mem. geol. Soc. Amer., 58, pp. 1—276, pls. 1—44. — OLIVEIRA, A. I. DE, and LEONARDOS, O. H., 1943. Geologia do Brasil. 2nd Edit. Sér. didát. Serv. Inf. agric. Rio de J., No. 2, 813 pp., 37 pls. of fossils; map. — PUTZER, H., 1957. Beziehungen zwischen Inland-Vereisung und Kohlebildung in Oberkarbon von Südbrasilien. Geol. Rdsch., 45, pp. 599—608. — REED, F. R. C., 1927. Upper Carboniferous fossils from Argentina in Du Toit, A. L., A geological comparison of South America with South Africa. Publ. Carnegie Inst., 381, pp. 129—150, pls. 13—16. — SUERO, T., 1952. Las sucesiones sedimentarias suprapaleozoicas de la zona extraandina del Chubut (Patagonia Austral — Republica Argentina). Symposium sur les Séries de Gondwana. 19th Int. geol. Congr., pp. 373—384. — SUERO, T., and ROQUE, P. C., 1955. Descubrimiento de Paleozoico superior al oeste de Bahia Laura (Terr. Nac. de Santa Cruz) y su importancia paleogeográfica. Not. Mus. Univ. Nac. Eva Peron, XVIII, Geol. No. 68, pp. 157—168. — THOMPSON, M. L., and MILLER, A. K., 1949. Permian fusulinids and cephalopods from the vicinity of the Maracaibo basin in northern South America. J. Paleont., 23, pp. 1—24, pls. 1—8.

Africa

ASTRE, G., 1934. La Faune permienne des grès à *Productus* d'Ankitokazo dans le Nord de Madagascar. Ann. geol. Serv. Min. Madagascar, fasc. 4, 4 pls. — BESAIRIE, H., 1952. Les formations du Karroo à Madagascar. 19th Int. geol. Congr. Symposium sur les Séries de Gondwana, pp. 181—186. — BOND, G., 1952. The Karroo System in Southern Rhodesia. 19th Int. geol. Congr. Symposium sur les Séries de Gondwana, pp. 209—223. — BORGES, A., 1952. Le Systeme du Karroo au Moçambique. 19th Int. geol. Congr. Symposium sur les Séries de Gondwana, pp. 232—250. — BORGES, A., NUNES, A., and FREITAS, F., 1954. Contribution to the data on the Karroo of Portuguese Lake Nyasa. C. R. 19th Int. geol. Congr. Fasc. 21, pp. 83—92. — BRANDT, R. T., 1954. Notes on the Karroo System in Northern Rhodesia. C. R. 19th Int. geol. Congr. Fasc. 21, pp. 63—70. — CAHEN, L., 1954. Géologie du Congo belge, pp. 1—577, Liege. — COX, L. R., 1936. Karroo Lamellibranchia from Tanganyika Territory and Madagascar. Quart. J. geol. Soc., **92**, pp. 32—57, pls. 4, 5. — DAVIES, K. A., 1952. Karroo System in Uganda. 19th Int. geol. Congr. Symposium sur les Séries de Gondwana, pp. 191—194. — DIXEY, F., and WILLBOURN, E. S., 1951. The geology of the British African colonies. C. R. 18th Int. geol. Congr. Fasc. XIV, pp. 87—109. — EFREMOV, I. A., and VJUSCHKOV, B. I., 1955. [Catalogue of localities for Permian Triassic Reptiles in the Territories of the U.S.S.R.] Trav. Palaeont. Inst. Acad. Sci. U.R.S.S., **46**, pp. 1—185. — GREEN, D., and POLDERVAART, A., 1954. Coal in the Bechuanaland Protectorate. C. R. 19th Int. geol. Congr. Fasc., **21**, pp. 71—82. — GÜRICH, G., 1923. *Acrolepis Lotzi* und andere Ganoiden aus den Dwyka-Schichten von Ganikobis, Südwestafrika. Beitr. geol. Erforsch. dtsch. Schbeb. Heft **19**, Part II, pp. 26—74, pls. 1—3. — HARKIN, D. A., McKINLAY, A. C. M., and SPENCE, J., 1954. The Karroo System of Tanganyika. C. R. 19th Int. Geol. Congr. Fasc., **21**, pp. 93—102. — HAUGHTON, S. H., 1952. The Karroo System in the Union of South Africa. 19th Int. geol. Congr. Symposium sur les Series de Gondwana, pp. 254—255. - 1953. Gondwanaland and the distribution of early reptiles. Annex. to Vol. 56, Trans. geol. Soc. S. Afr., pp. 1—30. — HAUGHTON, S. H., BLIGNAUT, J. J. G., ROSSOUW, P. J., SPIES, J. J., and ZAGT, S., 1953. Results of an investigation into the possible presence of oil in Karroo rocks in parts of the Union of South Africa. Mem. geol. Surv. S. Afr., **45**, 130 pp. — HUENE, F. VON, 1940. Die Saurier der Karroo-, Gondwana- und verwandten Ablagerungen in faunistischer, biologischer und phylogenetischer Hinsicht. N. Jb. Min. Geol. Palaont. Abh. Abt. B, **83**, pp. 246—347. — JONGMANS, W. J., 1952. Some problems on Carboniferous stratigraphy. C. R. 3ième Congr. Strat. Geol. Carb. I, pp. 295—306. — MacGREGOR, A. M., 1947. An outline of the geological history of Southern Rhodesia. Bull. geol. Surv. S. Rhodesia, **38**, 73 pp. — MARTIN, H., 1950. Südwestafrika. Geol. Rdsch., **38**, pp. 6—14. - 1954. Notes on the Dwyka succession and on some pre-Dwyka valleys in S. W. Africa. Trans. geol. Soc. S. Afr., **56**, pp. 37—43, pl. 5. — MOUTA, F., and CAHEN, L., 1951. Le Karroo du Congo Belge et de l'Angola. C. R. 18th Int. geol. Congr. Fasc. XIV, pp. 270—281. — RILETT, M. H. P., 1955. Plant microfossils from the coal seams near Dannhauser, Natal. Trans. geol. Soc. S. Afr., **57**, pp. 27—38, pl. v. — ROCK, E., 1950. Histoire stratigraphique du Maroc. Notes Mem. Serv. geol. Maroc., **80**, pp. 1—435, pls. 1—22. — TEIXEIRA, C., 1951. Present state of our knowledge concerning the palaeontology of the Karroo of Portuguese Africa. C. R. 18th Int. geol. Congr. Fasc. XIV, pp. 214—217. — TEMPERLEY, B. N., 1952. A review of the Gondwana rocks of Kenya Colony. 19th Int. geol. Congr. Symposium sur les Séries de Gondwana, pp. 195—208. — WAGNER. P. A., 1916. The Geology and Mineral Industry of South-West Africa. Mem. geol. Surv. S. Afr., **7**, pp. 1—119, pls. 1—41, map. — WAYLAND, W. J., 1952. Karroo System of Bechuanaland Protectorate. 19th Int. geol. Congr. Symposium sur les Séries de Gondwana, pp. 251—253.

Antarctica

ADIE, R. J., 1952. Representatives of the Gondwana System in Antarctica. 19th Int. geol. Congr. Symposium sur les Séries de Gondwana. pp. 393—399. — SEWARD, A. C., 1914. Antarctic fossil plants. British Antarctic ("Terra Nova") Expedition, 1910. Geology Vol. 5, pp. 1—49, pls. 1—8, 3 maps.

Australia, New Guinea and New Zealand

CAMPANA, B., and WILSON, R. B., 1955. Tillites and related glacial topography of South Australia. Ecl. geol. Helv., **48**, pp. 1—30, pls. 1—4. — GLAESSNER, M. F., LLEWELLYN, K. M., and STANLEY, G. A. V., 1950. Fossiliferous rocks of Permian age from the Territory of New Guinea. Aust. J. Sci., **13**, pp. 24—25. — HILL, D., 1950. The Productinae of the Artinskian Cracow fauna of Queensland. Pap. Dept. Geol. Univ. Qd., **3**, (11), pp. 1—27, pls. 1—9. – (reviewer) 1955. Contributions to the correlation and fauna of the Permian in Australia and New Zealand. J. geol. Soc. Aust., **2**, pp. 83—107. — LUDBROOK, N. H., 1957. Permian Foraminifera in South Australia. Aust. J. Sci., **19**, pp. 161—162. — MAXWELL, W. G. H., 1954. *Strophalosia* in the Permian of Queensland. J. Paleont., **28**, pp. 533—559, pls. 54—57. — RICKWOOD, F. K., 1955. The geology of the Western Highlands of New Guinea. J. geol. Soc. Aust., **2**, pp. 63—82, pls. 1—4.*— THOMAS, G. A., 1954. Correlation and faunal affinities of the marine Permian of Western Australia. Proc. Pan-Ind. Ocean Sci. Congr. (C), pp. 5—6.

Manuscript completed March 1957.

Editor's Comments
on Papers 25 and 26

25 WALKOM
Gondwanaland: A Problem in Palaeogeography

26 RAVEN and AXELROD
Plate Tectonics and Australasian Paleobiogeography

Paper 25 by A. B. Walkom brings together four of the geological and paleobotanical hypotheses that were commonly used to explain the distribution of Recent biotas by neontologists during the 1930s and 1940s. After examining each hypothesis, he reaches several conclusions; the most significant is that without direct fossil evidence as old as the middle Mesozoic it is difficult, even impossible, to support or reject any of the four hypotheses. Walkom was a paleobotanist who became director of the Australian Museum in Sydney and included in his interests Australian Paleozoic and Mesozoic floras and their biogeographical and age relationships.

Paper 26 by Peter H. Raven and Daniel I. Axelrod adds a fifth hypothesis to those presented in Paper 25. The emphasis of Raven and Axelrod is toward tracing the history of the present Australasian vegetation as far as possible into the geologic history as viewed through a plate-tectonic model (see also Andrews, 1916). Peter H. Raven is director of the Missouri Botanical Garden and professor of biology at Washington University. He became interested in plate tectonics during a sabbatical leave in New Zealand in 1969–1970. Daniel I. Axelrod is professor of botany and geology at the University of California at Davis. He has been particularly interested in the ecological and evolutionary development of desert plants (see Paper 15); after initially taking a stand against the hypothesis of plate tectonics (Paper 16), he has adopted the ocean spreading-plate tectonic model to interpret botanical and geological distributions (Axelrod, 1972).

REFERENCES

Andrews, E. C. (1916). The geological history of the Australian flowering plants. Am. Jour. Sci., 4th ser. 42, 171–232.

Axelrod, D. I. (1972). Ocean-floor spreading in relation to ecosystematic problems, *in* R. T. Allen and F. C. James, eds., A symposium on eco-systematics. Univ. Arkansas Mus., Fayetteville, Ark., Occ. Pap. 4, 15–76.

25

Reprinted from *Rept. Hobart Meetings*, Aust. N.Z. Assoc. Adv. Sci., 1949, pp. 3–13

GONDWANALAND: A PROBLEM
IN PALAEOGEOGRAPHY

A. B. Walkom

A glance over the addresses delivered by Presidents of this Association reveals a wide range in subject and treatment. Choice of a subject is by no means easy, especially for one who for years has become immersed in administrative duties with little time for active research. My choice of a subject which might be of interest to a considerable proportion of the members was therefore somewhat limited. I have selected one which is associated with palaeobotanical research and which has aroused very much discussion over the years—Gondwanaland.

Gondwanaland is a name familiar to every student of geology and to many students of various branches of biology. It is a term often used loosely, and sometimes perhaps without an appreciation of the doubts that may exist as to its extent—or even its existence.

We are all familiar with the disposition of the continental land masses at the present day—but what of the past? Were the land-masses distributed essentially as they are to-day or were they widely different? This address deals with these questions at a period in the remote past.

4   A. B. Walkom

During the period of geological history known as the Permian—a period extending roughly from 250,000,000 years to 200,000,000 years ago—there flourished in the land masses of the southern hemisphere a remarkably uniform flora, very different from that of today. The most constant and characteristic member of this flora was the genus *Glossopteris*, and the whole assemblage of plants is widely spoken of as the Glossopteris flora.

Associated with this flora over a large part of the land in the Southern Hemisphere, especially in South Africa, there was a development of characteristic reptiles. In addition there was, at the same time, a widespread development of ice-age conditions, indicated by the glacial deposits associated with the occurrences of the Glossopteris flora. The distribution of this flora and the associated phenomena has set geologists—and biologists as well—a problem which has been much discussed over a long period of years and which has not yet been solved to the satisfaction of all.

The Glossopteris Flora occurs in abundance in South America (Brazil and Argentina), South Africa, India, Australia and Antarctica, and the problem before palaeogeographers has been to account satisfactorily for the spread of a land flora over land masses which are so widely separated today. I propose to place before you several of the explanations which have been suggested from time to time and which have received varying measures of support.

1. GONDWANALAND—A CONTINENT

The most widely known, and perhaps the most widely accepted, is the postulation of a huge continental mass, embracing all the land masses on which the Permian Glossopteris Flora is found. The name Gondwanaland was first given to part of this large continent by Edouard Suess in Volume I of his monumental work "Das Antlitz der Erde". In the original delineation by Suess this continent included part of Africa, Madagascar, and India, though in a later volume, the concept was extended to South America. In defining Gondwanaland (Fig. 1) he says (1904, p. 596), in speaking of the parts which have been welded together to form the mass of Asia, Africa, and Europe, "The first region comprises the southern and a great deal of the more central part of Africa, then Madagascar and the Indian Peninsula. The lofty table-lands of this region have never, so far as we know, been covered by the sea since primitive times, or the end of the Carboniferous period; it is only at the foot of the table-lands that marine sediments have been deposited, which followed the encroachment of the Indian Ocean, as this was formed by subsidence within the tabular mass. We call this mass Gondwana Land, after the ancient Gondwana flora which is common to all its parts". At a later stage (1909, p. 500) he says Gondwanaland comprises "South America from the Andes to the east coast between the Orinoco and cape Corrientes, the Falkland islands, Africa from the southern offshoots of the Great Atlas to the Cape Mountains, also Syria, Arabia, Madagascar, the Indian peninsular, and Ceylon. Here, as in Laurentia, there is no recent folding, except on the extreme western border. Just as the folding on the Mackenzie extends from the west, out of the Rocky Mountains, into the stratified series of the otherwise rigid Laurentian foreland, so the folding of the Andes likewise extends from the west into the stratified series of Bolivia and the Argentine, that is, parts of the otherwise rigid Gondwana Land", and still later (1909, p. 663), "This vast continent extended from the east side of India into the west of Brazil, and to the Argentine cordilleras . . . With the exception of

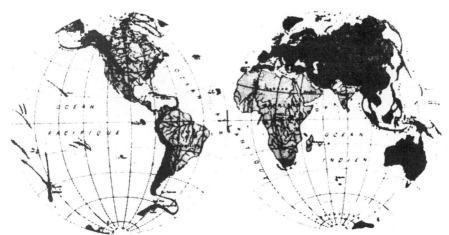

FIG. 1. Map of the world, showing the extent of Gondwanaland.   (after Suess.)

certain encroachments of the upper Cretaceous no sea has extended over this continent, now broken into fragments, since the Carboniferous period ".

Since Suess gave the name Gondwanaland to a hypothetical continent, this continent has been shown on many palaeogeographical maps and has become a familiar name to most students of geology. The continent, as depicted by later writers, has covered, to a varying degree, large areas of the South Atlantic and Indian Oceans and the Antarctic Sea, as well as the land areas mentioned by Suess.

Arber (1905) in his account of the Glossopteris flora included Australia and Tasmania in the southern continent—" The Glossopteris flora is found typically developed in four great provinces of Gondwana Land, viz., India,

originated in the Southern Hemisphere: and unless a very considerable area of what is now deep ocean was occupied by land in Mesozoic and Palaeozoic times, a change in favour of which there appears but slight evidence, it is far from improbable that the Antarctic continent was the original area of development '."

Schuchert (1928, p. 143) says " Long before the close of this era (Proterozoic) the sialsphere appears to have been welded into three zones of transverse or latitudinal lands of great extent, namely, Holarctis, Antarctis, and Equatoris. The latter embraced South America, a land-bridge across the present mid-Atlantic, Africa, Madagascar, and Lemuris including India. Antarctis had the Antarctic lands with extensions to South America and to Australia, which was then a part of this great polar land ".

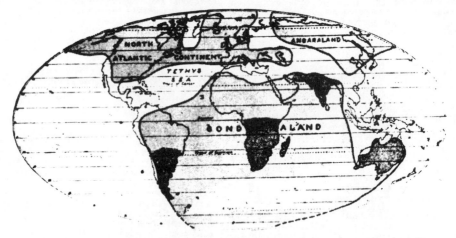

FIG. 2. Map of the world at the end of the Carboniferous period, based on a restoration by Arldt. (after Seward.)

Australia with Tasmania, Southern Africa, and South America ". His map shows the distribution of the Glossopteris flora in comparison with the distribution of the northern flora of the same period, and it emphasizes the latitudinal distribution of the floras.

In discussing the distribution Arber writes (p. xliii), " It would appear that climate as well as isolation must have had a determining influence on the distribution of Permo-Carboniferous vegetation. Probably the existence of widespread glacial conditions immediately preceding the deposition of the earlier *Glossopteris*-bearing sediments had a marked influence in this connection. Dr. Blanford has suggested there is ' some evidence in favour of the view that the transfer of the southern plants to the Northern Hemisphere was caused by a period of low temperature that drove a southern temperate flora northward to the equator '. He adds that ' it is highly probable that many other forms of terrestrial life besides the Mesozoic flora

" The Palaeozoic saw no marked changes in the major features of the earth's face, but the many readjustments in the continental masses toward the close of this era through their combined effect sounded the prophecy of doom for large parts of the sialsphere, since some land-bridges and most borderlands throughout the Mesozoic era were being absorbed by the periodically heated basaltic shell or simasphere. Cenozoic time completed what was begun in the Carboniferous ".

Thus he suggests that great changes commenced in the Carboniferous, and his map of the Palaeogeography of Early Permian time (see Fig. 6, below) shows Gondwana as a substantial part of the continental mass he calls Equatoris. This map emphasizes the fact that the conception of Gondwanaland as a continuous continental mass involves the recognition of an east-west arrangement of the continents in late Palaeozoic time as contrasted with the present-day north-south distribution.

Perhaps the extreme limit to the area envisaged for Gondwanaland is shown in Seward's (1941, p. 156) map of the world at the end of the Carboniferous period, based on a restoration by Arldt (Fig, 2). This includes not only most of South America, Africa, India and Australia, but also the areas of the South Atlantic, Indian and Southern Oceans, in one huge continent—covering practically the whole of the southern hemisphere, except that part occupied by the south Pacific Ocean.

With such a continental mass, the Glossopteris flora could have attained its known distribution in Permian rocks without difficulty from wherever on the southern continent it may have originated.

However, the disappearance of such a large continental mass as Gondwanaland raises problems of geophysics and climate which have been discussed by Bailey Willis (1932, p. 941), and I may be permitted to quote a selection of his remarks.

He says " The idea of extensive continents which have been replaced by oceanic basins was conceived before it was established that continents consist of relatively light rocks and ride high with reference to heavier sub-oceanic masses because of that lightness. Those relations being firmly established, however, we must suppose that Gondwana, if it ever had continental extent, also had continental lightness. If then through some failure of its foundations it sank 15,000 feet the resulting hollow in the earth's surface should show a strong defect of gravity, a negative anomaly. The facts, so far as they are known, are to the contrary: oceanic basins appear to be nearly in equilibrium, with a tendency over wide areas toward a plus anomaly. It is not conceivable that they are underlain by a sunken mass of granite of continental dimensions ".

And he concludes " the former existence of a Gondwana continent appears to be inconsistent with the facts of isostatic equilibrium as demonstrated by the broad relations of continents and ocean basins ".

Willis also discusses the climatic conditions that would result from the existence of such a continent as that shown on Arldt's map and concludes (1932, p. 943) " The heat and aridity of that Gondwanaland would have surpassed those of any known desert; yet the record demands that succulent floras shall have migrated there and ice-sheets shall have spread down to sealevel during one or more glacial epochs."

In a recent paper (1944) Willis has also criticized the conception of this continental mass thus " . . . Gondwana Land, the continent supposed to have extended from the East Indies westward to the Pacific, embracing India, Africa, and South America and occupying the sites of the Indian Ocean and South Atlantic oceans. It had no actual existence.

It was conceived to account for the transoceanic migrations of plants and terrestrial organisms, the identical line of reasoning of the argument for Continental Drift. In Suess's imagination it was a reality; but there is no reality, no geologic fact to demonstrate the one time existence of such a mass as it would need to have been."

2. Radial Distribution from Polar Lands

W. D. Matthew in 1906 and 1915 advocated a fundamentally different explanation of distribution. His arguments were based chiefly on his study of the development and distribution of mammals and on his adherence to the idea of the general permanency of the great ocean basins. He says (1915, p. 174): " The continents have been alternately partly overflowed, separated and insular, or raised to their greatest extent and united largely into a single mass. The great ocean basins have in the main been permanent. This principle is dependent upon the known facts in regard to isostasy."

He continues: " The present distribution of land and water shows the great land masses located mostly in the northern hemisphere. The land areas, extended to the borders of the continental shelf, form a single great irregular mass with three great projections, South America, Africa and Australasia, radiating out from it into the southern hemisphere," (see Fig. 3) and goes on to say that a rise of 600 feet would unite all the land into a single mass, with the doubtful exception of Australia.

Matthew's studies refer only to post-Cretaceous times, but my reason for referring to them in connection with the subject of Gondwanaland is that he himself says (1915, p. 191) " Perhaps the most widely accepted departure from the permanency of the ocean basins is the supposed Gondwana Land, invented to account for certain similarities in southern Palaeozoic floras, and since used to account for almost all cases of similarity among southern florae and faunae which were not demonstrably due to dispersal from the northern continent. This theory has in its original form gone so long uncontested that it is very generally regarded as incontestable. New discoveries have been interpreted in terms of it, the weakness of the original evidence, the possibility that it might be otherwise interpreted, has been forgotten, and like the Nebular Hypothesis, it has become almost impossible to dislodge it from its place in the affections of the average geologist. If the distribution of animals be interpreted along the lines here advocated, there is no occasion for a Gondwana Land even in the Palaeozoic."

Matthew's thesis is that distribution and dispersal took place along the longitudinal land masses which form the present continental lands with their shelves.

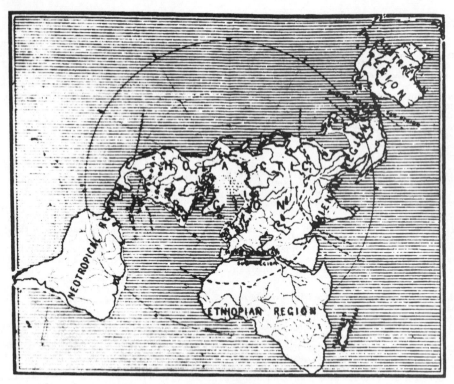

FIG. 3. Map on north polar projection showing Holarctic region and projecting masses. (after Matthew.)

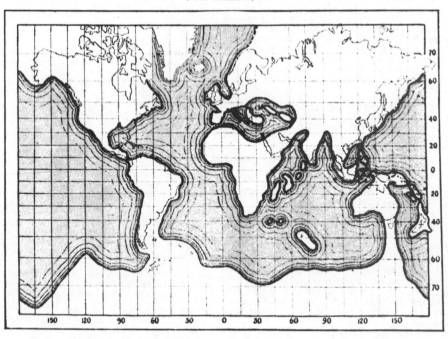

FIG. 4. Hypothetical continental outlines at end of the Cretaceous period. (after Matthew.)

350

In his 1906 paper, speaking of the land distribution at the close of the Cretaceous (Fig. 4) he says: "At this period I have placed the union of South America and Australia by way of the Antarctic continent, by which the Marsupials reached both continents, as did also many of the lower animals and plants which are common to these two continents, but not found in the rest of the world. This hypothesis has been advocated by a number of authorities of whom Professor Osborn and Dr. Ortmann are the most recent in date, and I have followed their views most nearly. I have omitted the doubtful connection with Africa which has also been proposed to account for a limited number of peculiar southern forms in that continent, as the evidence for it is not nearly as strong, and the difficulties are much greater (as Osborn has pointed out). If the connection occurred at all it may be supposed to have been at an earlier period".

In his 1915 paper, however, he reversed his ideas of direction of distribution and included in his thesis the following points: "The principal lines of migration in later geological epochs have been radial from Holarctic centres of dispersal".

"The geographic changes required to explain the present distribution of land vertebrates are not extensive and for the most part do not affect the permanence of the oceans as defined by the continental shelf."

"The numerous hypothetical land bridges in temperate, tropical and southern regions, connecting continents now separated by deep oceans, which have been advocated by various authors, are improbable and unnecessary to explain geographic distribution. On the contrary, the known facts point distinctly to a general permanency of continental outlines during the later epochs of geologic time, provided that due allowance be made for the known or probable gaps in our knowledge".

His map on north polar projection (see Fig. 3) shows the Holarctic region as a great land mass with three radiating projections—South America, Africa, and Australia.

A map of the southern continents on a south polar projection shows how South America, Africa, and Australia radiate from Antarctica and would be almost united to it by a rise of 1000 fathoms.

Matthew stresses the relative permanency of the North American continent, as indicated by Schuchert's series of palaeogeographic maps. He suggests too that available evidence points to similar permanence of the land platforms which we know as South America, Africa, and Australia.

It seems to me that these studies of Matthew are pertinent, though Schuchert (1932, p. 881) in reviewing them "admitted that Matthew's conclusion was sound when restricted to the Cenozoic, but to say there was no Gondwana in early Mesozoic time and especially none

in Permian time is to drag into this painstaking and most excellent study an unnecessary and unproved conclusion.'" Schucherts' own conception, as set out in the paper in which he makes the above criticism of Matthew, shows no continental mass differing much from present-day continents—nothing which could be called Gondwana in preference to Africa—which makes his criticism of Matthew seem unjustified. And after all, Matthew did not say there was no Gondwana in early Mesozoic or Permian time, but simply stated that if his interpretation of distribution were correct, then there was no occasion to postulate a continent such as Gondwana.

Mathew illustrated his paper with numerous maps, on the north polar projection, showing the distribution of the groups of vertebrates which formed the supporting evidence for his ideas. A recent botanical study of the genus *Crepis* (Babcock, 1947) shows four principal migration routes for the sections of this genus, from a probable centre of origin in Central Asia. Babcock's diagrammatic representation of these migration routes for *Crepis* is reproduced here (Fig. 5) to show how it agrees with many of the diagrams of vertebrate distribution and migration, used by Matthew, and gives some confirmation to Matthew's idea of distribution along the continental masses from a Holarctic centre of dispersal.

3. LAND BRIDGES CONNECTING CONTINENTAL MASSES

Schuchert was one of those geologists who held that "both continents and oceanic basins are, in the main, permanent features of the earth's surface; but that they have not necessarily always had their present shape and area" (1932, p. 876). His map of the world in early Permian time (Fig. 6), published in 1923, shows wide land bridges as well as a land mass covering a large area at present occupied by part of the Indian Ocean. He had long realized the difficulty of accounting for the foundering of such extensive land bridges and had often discussed this difficulty with the late J. Barrell before his death in 1919. Subsequently Schuchert took up the discussion with Bailey Willis and the results are embodied in two papers, one by each author, published in 1932.

The maps published with these papers show the distribution of the continental land masses in Permian time as not very different from that of today. They suggest, however, that there were a number of relatively narrow land bridges or isthmian links joining continents (Fig. 7) in particular, as far as concerns our subject, Antarctica to South America, Brazil to Guinea, and Madagascar to India. These land bridges, though narrow, were considered to have been sufficient to allow free migration—particularly of the Glossopteris flora and the late Palaeozoic reptiles. Schuchert concluded from his studies of faunal distribution that "the vast majority

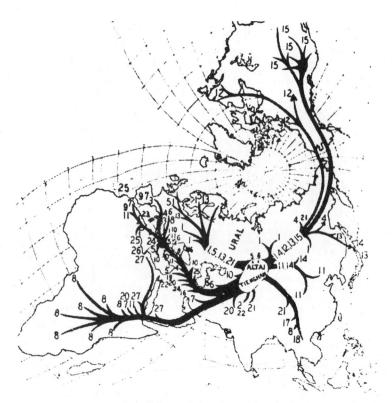

FIG. 5. Map showing distribution routes of species of *Crepis* (after Babcock.)

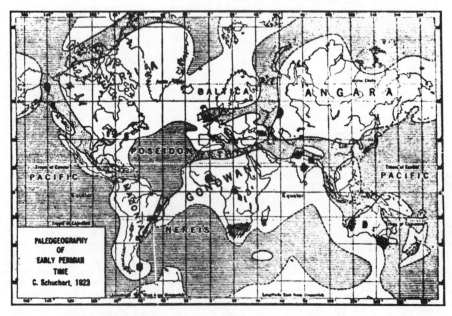

FIG. 6. Map of the world in early Permian time. (after Schuchert.)

of shallow-water life is dispersed along shore-lines, and only a small percentage in any fauna is accidently carried by the currents from one side of an ocean to the other ", from his entire study over a period of thirty years he considers that the evidence " is overwhelmingly in favour of the existence of the Gondwana Land bridge (connecting South America and Africa) from the pre-Cambrian until the end of Cretaceous time ".

Bailey Willis defines his conception of permanency of continents and ocean basins thus: " In recognizing the fact of permanency of large masses of the earth's crust in general form and relative position, there is no intention of maintaining that they have not changed in outline by loss of marginal sections or outlying parts. Neither could it be assumed that continental platforms or large areas of

Permian time—including the glaciation in India, north of the equator. Willis gives a detailed explanation of these phenomena, and adds his ideas of the climates and circulation to Schuchert's map of Permain geography.

Willis's conclusions (1932, p. 952) on the hypothesis of isthmian links are: (1) Reasoning regarding the orogenic process of their uplift is based on the facts of the Isthmus of Panama and the upthrusting of the framework of the Caribbean. (2) On being tried by the requirements of paleontologic facts the hypothesis is found fit as regards times and routes of migration, both for terrestrial and marine organisms. (3) Tested by the facts of the unusual climatic conditions it is found to provide a reasonable explanation of their peculiarities in accordance with known principles and effects of meteorology.

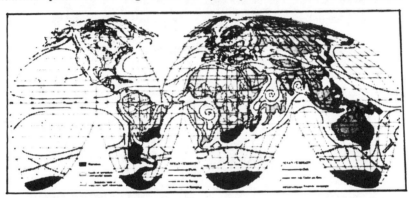

Fig. 7. Map of Permian land connections and climates. (after Schuchert and Willis.)

them had not been depressed so as to form basins and admit epicontinental seas. But even though submerged for a time and buried beneath sediments, the continental mass retains its identity and is permanent. No established fact of historical or structural geology can be denied because of, or can be in contradiction with the fact of permanency. Speculations which postulate impermanency, on the other hand, involve the theorist in insuperable difficulties ".

Willis bases his conception of " isthmian links " on studies of the Caribbean basin, and tests his hypothesis in terms of geophysics and climatology. He examines the effects on climate of the distribution of land and sea envisaged as a result of the isthmian links between continental areas. The subdivision of the oceanic areas of the South Atlantic and Indian Oceans by the land bridges suggested by Willis and Schuchert would profoundly modify the oceanic water circulation. Added to a general lowering of temperature which is generally admitted as a necessity for development of widespread glacial conditions, these changes in circulation and the resulting climatic changes could have produced the unusual distribution of glacial deposits in

4. CONTINENTAL DRIFT

Now we come to a hypothesis which involves a very different conception of Gondwanaland from those held under the three hypotheses discussed above. The hypothesis of continental displacement, or Wegener's Hypothesis, as it is perhaps most generally known, appeared in 1912 in *Petermann's Mitteilungen* and in the *Geologische Rundschau*. In 1915 Wegener published his work in book form and an English translation appeared in 1924 (The Origin of Continents and Oceans). This hypothesis, which provoked much discussion, involves very considerable lateral movement of the continental areas, and rejects the theory of permanency of continents and oceans which is one of the main ideas retained by most of those who have supported one or other of the explanations of distribution of Upper Palaeozoic land and sea mentioned above.

It is not intended here to enter into any detailed discussion of the Wegener hypothesis, but I might be permitted to quote the summary by Schuchert (1928, p. 105) who sets out the main points of the theory as follows:— " (1) The earth is not a shrinking mass. (2) The amount of water on the earth's surface has always been the same. (3) Early in the history of the earth there was a thin universal granitic shell which before the Silurian had been thrust and folded into a greatly thickened continent that he calls Pangaea. (4) The continental blocks (Pangaea) underwent great horizontal drifting movements in the course of geological time, and these presumably continue even to-day. The rifting probably began in Palaeozoic time, but it was not until the middle Jurassic that Australia-Antarctica began ·to separate and drift southeast. In the early Cretaceous the Americas began to move westward and finally in the Pleistocene Greenland-Newfoundland was separated from Norway and Great Britain. (5) The separated continents of today, when moved together, working on a globe, fit against one another as do the pieces of a jig-saw puzzle. (6) The poles of the earth have in the past slowly wandered about, and in Permian time they were as much as 2,500 miles from their present

Briefly, Wegener holds that for the greater part of geological time there was but one large continental land mass which drifted horizontally as a single mass until some time after the end of the Palaeozoic Era. During this long period the poles wandered about and in Permian time were some 2,500 miles away from their present position. After the end of the Palaeozoic the continental mass was rifted and various parts drifted away from one another until at the present time they are in the positions familiar to us.

For Gondwana Land, we look at Wegener's map of the distribution of continents and oceans in Upper Carboniferous time (Fig. 8). This shows his one large continental mass (Pangaea) in which all the present day continents—America, Europe, Asia, Africa, Australia, and Antarctica—are joined, some of the mass being covered by shallow epicontinental seas. Du Toit, who was a consistent supporter of the Wegener Hypothesis,

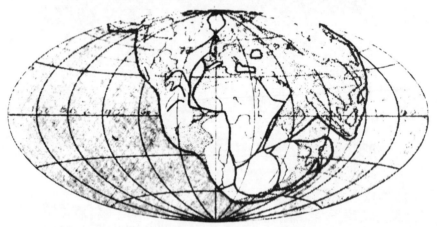

Fig. 8. Distribution of continents and oceans in upper Carboniferous time. (after Wegener.)

position. (7) The mountains of the earth are in the main not due to a shrinking earth but to the drifting of the continents, one set arising at the edge of the forward-moving granitic continents where they come against the resisting basaltic shell, as best exemplified by the Cordillera of North America and the Andes of South America; while those of the Euro-Asiatic continents are due to a 'striving toward the equator of the continental blocks', namely, the movement toward each other of the Euro-Asiatic mass and the African one. (8) Wegener holds firmly to the theory of isostasy, and accordingly believes that the land masses, large or small, cannot sink and vanish into the heavier basaltic layer. He does away with land bridges across the oceans by uniting all lands into a Pangaea. (9) Accordingly, Wegener rejects the theory of the permanency of oceans and continents as we see them to-day."

says (1921, p. 126) " That the continents might have originated by the actual tearing apart of one or more much larger masses is no new doctrine. Although generally dismissed as fantastic, it has been very ably championed recently by Wegener, and, when the hypothesis is studied in detail, the evidence in its support is found to mount up so remarkably as to become almost overwhelming."

Professor B. Sahni proceeds a step further and visualizes land masses not only drifting apart but also drifting together—as in his 1936 paper where he elaborates views he had published previously. He concludes, inter alia, " (1) We have not yet enough palaeobotanical data to prove the *drifting apart* of the different portions of Gondwanaland.

(2) But we at least seem compelled to agree that drift movements of large magnitude elsewhere have *brought into juxtaposition* continents once separated by wide oceans."

However, the great majority of the geologists who have published discussions of Wegener's Hypothesis have criticized it more or less severely. The comment, quoted above, of Du Toit, perhaps the most ardent geological supporter of Wegener over the years, may be contrasted with the comment of one of the severer critics published at about the same time thus, " Whatever Wegener's own attitude may have been originally, in his book he is not seeking truth; he is advocating a cause, and is blind to every fact and argument that tells against it. Much of his evidence is superficial. Nevertheless, he is a skillful advocate and presents an interesting case." (Lake, 1922.)

In view of the extremes represented by the two statements just quoted, the well-balanced account by Chester R. Longwell—in spite of the fact that Bailey Willis accuses him of sitting on the fence—is welcome. Longwell (1944) has analyzed the evidence for the hypothesis, and in the course of his analysis says (p. 220) " The Wegener hypothesis has been so stimulating and has such fundamental implications in geology as to merit respectful and sympathetic interest from every geologist. Some striking arguments in his favor have been advanced, and it would be foolhardy indeed to reject any concept that offers a possible key to solution of profound problems in the Earth's history ". Longwell adopts a quite impartial attitude in his analysis and points out weaknesses arising from lack of discrimination in presenting their case by some of the protagonists of the Wegener Hypothesis.

Further criticism of the hypothesis came from Bailey Willis (1944, p. 509), following on Longwell's statement, who says " I confess that my reason refuses to consider ' continental drift' possible. This position is not assumed on impulse. It is one established by 20 years of study of the problem of former continental connections as presented by Wegener, Taylor, Schuchert, du Toit, and others, with a definite purpose of giving due consideration to every hypothesis which may explain the proven facts ". He considers the theory a fairy tale, " a fascinating fancy which has captured imaginations "—one of a long line of hypotheses which " challenge by their stupendous appeal to the imagination and by the implication that there is no other explanation ".

It may not be unreasonable to say of the Wegener Hypothesis that many of the arguments put forward to support it are merely the facts of distribution which it was designed to explain.

In a recent paper, H. H. Nininger (1948), following consideration of the Arizona Meteorite Crater says (p. 105) " There has never been cited any reason why we should assume that the Arizona crater represents the largest impact our planet has been called upon to absorb. In fact there is good reason

to assume that larger bodies ranging up to the magnitude of the planetoids have from time to time collided with the earth." Then, after further discussion of the effects of large scale collisions, he continues (p. 107): " External rather than internal forces, or better stated, internal forces released by external impact will far more easily explain such phenomena as shifting of the poles, planetary drift, arcuate coast lines, the birth of continents, and curving mountain ranges ". It may be interesting to see what use is made of these ideas in explanation of past changes in distribution of land and sea.

SOME PROBLEMS OF PLANT DISTRIBUTION

My own palaeobotanical researches have dealt with fossil floras of Upper Palaeozoic and Mesozoic age, mostly in Eastern Australia. There are several problems of distribution of floras in the Gondwana regions which are of interest, especially in so far as they bear some relation to the possible distribution of land areas at the time they flourished.

The Lower Carboniferous flora, characterized by Rhacopteris—which is of common occurrence in Western European rocks of this age—is known from South America, Northern India, and Australia, but not from South Africa. If these four land masses were joined in one for the whole of the Palaeozoic Era, the absence of any trace of the Rhacopteris flora from Southern Africa needs some explanation.

The Glossopteris flora occurs throughout Gondwanaland and its persistence into the Lower Mesozoic in Tonkin presents no difficulty. But if New Zealand occupied the position shown on Wegener's maps well into the Mesozoic Era it seems strange that no trace of this flora has yet been found in that country.

European Upper Carboniferous plants are found associated with the Glossopteris flora in coal measures in Southern Rhodesia, presumably as a result of slow migration southward from Europe. This migration appears to have been limited to this one line from Matthew's Holarctic region since so far this Upper Carboniferous flora is not known to have penetrated to the Glossopteris bearing areas of South America or Australia.

The Permian flora of the Hermit Shale in Arizona includes plants, especially the genus Supaia, bearing a close resemblance to the Mesozoic Thinnfeldia flora which is of widespread occurrence in Gondwana lands. When I first saw David White's memoir on the Hermit Shale flora I felt that the plants must be of the same age as the Thinnfeldia flora. I had the opportunity of discussing this question with David White and his colleagues in 1933, and there is no doubt that the Hermit Shale is of Permian age, while the Thinnfeldia flora in the Gondwana beds is Lower

Mesozoic. If some of the Hermit Shale plants are the ancestors of the Thinnfeldias, their distribution could be explained either on the theory of isthmian links of Willis and Schuchert or on Wegener's theory.

That there are still these wide differences of opinion amongst geologists as to the distribution of the major land masses in the past is perhaps unfortunate, since individual biologists may appear to accept without question any one of the hypotheses. For example, I quote two recent statements which have come under my notice: (1) An entomologist says " According to the nearly unanimous opinion of the paleogeographers this Gondwana continent comprised all the recent continents of the southern hemisphere in Mesozoic time and was separated from the then existing continents of the northern hemisphere by a large ocean. It divided itself in Cretaceous time . . . ". (2) a botanist, in a revision of a recent Australian plant genus expresses the opinion that three recent allied genera " originated from a common ancestor, and that their present distribution can be satisfactorily explained on Wegener's Theory of Drifting Continents ", and later adds " It is suggested

that this genus reached Australia through Sub-Antarctica prior to late Jurassic times when New Zealand was still connected to the mainland of Australia . . . "

It seems to me that there is no sound foundation for the explanation of distribution of recent genera or species in terms of Gondwanaland or Wegener's Hypothesis unless there is some definite knowledge of their ancestors going as far back as about the Middle of the Mesozoic Era.

There I leave the problem. I have given some account of four hypotheses put forward to explain the distribution of continental land masses in Upper Palaeozoic time—(1) a huge continental mass known as Gondwanaland, (2) continental lands much as at the present day, but joined in north polar and/or south polar regions, (3) continental lands much as today, but joined latitudinally from time to time by isthmian links, and (4) one continental mass (Pangaea) which commenced to crack in Jurassic time and separate into fragments, which floated apart. Of these, in the light of my specialized branch of research, I would say that the isthmian links hypothesis is at present the most acceptable.

SELECTED LIST OF REFERENCES.

ARBER, E. A. N., 1905.--The Glossopteris Flora. *British Mus. Catalogue.*

BABCOCK, E. B., 1947.--The Genus *Crepis.* Univ. California Publ. in Botany. Vol. 21.

DU TOIT, A. L., 1921.--Land Connections between the other Continents and South Africa in the past. *S. Afr. J. Sci..* 18, 120-140.

————————, 1927.--A Geological Comparison of South America with South Africa. Carnegie Inst. Washington, Pub. 381.

LAKE, P., 1922.--Wegener's Displacement Theory. *Geol. Mag.,* 59, 338-346.

LONGWELL, C. R., 1944.--Some Thoughts on the Evidence for Continental Drift. *Amer. J. Sci.,* 242, 218-231.

MATTHEW, W. D., 1906.— Hypothetical Outlines of the Continents in Tertiary Times. *Bull. Amer. Mus. Nat. Hist.,* xxii, 353-383.

———————— ————, 1915.--Climate and Evolution. *Ann. N.Y. Acad. Sci.,* xxiv, 171-318.

NININGER, H. H., 1948.— Geological Significance of Meteorites. *Amer. J. Sci.,* 246, 101-108.

SAHNI, B., 1936.--Wegener's Theory of Continental Drift in the Light of Palaeobotanical Evidence. *J. Ind. Bot. Society,* xv, 319-332.

SCHUCHERT, C., 1923.--The Palaeography of Permian Time in relation to the Geography of earlier and later Periods. Proc. Pan-Pacific Sci. Congress, Australia, 1923, 1079-1091.

———————— , 1928.--The Hypothesis of Continental Drift. *The Theory of Continental Drift,* 104-144.

——— —— ——— , 1932.--Gondwana Land Bridges. *Bull. Geol. Soc. Amer.,* 43, 875-915.

SEWARD, A. C., 1941.--Plant Life through the Ages. *Camb. Univ. Press.*

SUESS, E., 1904.--The Face of the Earth (English Translation). Vol. I.

————————, 1909.--Id.. Vol. IV.

WEGENER, A., 1924.--The Origin of Continents and Oceans (English Translation).

WILLIS, BAILEY, 1932.--Isthmian Links. *Bull. Geol. Soc. Amer.,* 43, 917-952.

————————, 1944.--Continental Drift, Ein Marchen. *Amer. J. Sci.,* 242, 509-513.

Plate Tectonics and Australasian Paleobiogeography

The complex biogeographic relations of the region reflect its geologic history.

Peter H. Raven and Daniel I. Axelrod

The biogeographic relationships shown by the plants and animals of Australasia have long fascinated biologists. Because many widely different taxa (such as marsupials and *Nothofagus*) are common to temperate South America and to temperate Australia–Tasmania or New Zealand, some biologists, especially those in the Southern Hemisphere (*1, 2*), have espoused Wegner's (*3*) idea of continental drift to explain the relationships. In contrast, biologists in the Northern Hemisphere have remained largely skeptical (*4*) or have placed changes in the relative positions of land masses so far in the past that they would have had little or no effect on modern distributions (*5*).

During the past 5 years, however, the emergence of the theory of plate tectonics has provided a better understanding of the history of the earth's crust (*6*) and, hence, a more reasonable basis for explaining many problems of biogeography. As lavas form new ocean floor and grow laterally from ridges, the lighter, buoyant continents are rafted to new positions at rates of several centimeters per year through periods (and eras) of geologic time. Fully half the ocean floor is therefore of Cenozoic age, whereas the continents include the oldest rocks. Plate tectonics has subsumed earlier ideas of continental drift and is now generally accepted by geologists.

A reevaluation of the distribution of organisms in the light of plate tectonics is now called for, with a focus on the past positions of the continents and other lands. In this article we review the complex distributional problems in Australasia and the southwestern Pacific (Fig. 1) (*5, 7–9*). Our thesis is that the history of the biota in these widely scattered lands has been determined by (i) the Late Cretaceous position of the Australian plate, (ii) its northward movement during the Tertiary, (iii) the disruption of its eastern flank to form a series of microcontinents and archipelagos (Fig. 2), and (iv) its later Tertiary fusion with the Asian plate, which built island stepping-stones to Asia.

Geologic History of Australasia

In the middle Cretaceous, about 100 million years ago, Australia, New Zealand, and Antarctica were united with South America and with India–Madagascar–Africa, all forming a part of Gondwanaland (*10–19*). Australia was then in direct contact with lands now well removed from it—not only Antarctica, India, New Zealand, New Caledonia, Lord Howe Island, and Norfolk Island, but also rises now submerged (Lord Howe Rise, Norfolk Ridge, Queensland Plateau), composed of continental rocks (*17, 20–24*). The New Guinea part of the Australian

plate was largely submerged (*15, 25, 26*). The present regional geography of Australasia (Fig. 1) resulted from the breakup and spreading out of a geosynclinal belt of Mesozoic to Paleozoic age on the eastern (Pacific) margin of Gondwanaland. The fragments moved north and east into the Pacific by sea-floor spreading and transcurrent faulting (*17, 20–24*) as the north-moving Australian plate and the west-moving Pacific plate interacted.

The results are arcs spreading east from Australia, separated by basins that are now mostly inactive (*21, 27*). The scattered lands adjacent to the outer arc include New Zealand and Fiji along the "andesite line," marked by the Kermadec–Tonga Trench, which evidently has been a frontal arc since the early Tertiary (*21, 27*). By contrast, the basins of the inner arcs consist of linear ridges and troughs that are probably no older than Pliocene. That they were formed by extensional rifting within an older frontal arc seems probable, for the small, deep basins include younger volcanic archipelagos as well as young sea floor (*21, 27*).

Disruption of the Australian plate probably commenced from a Late Cretaceous (80 million years ago) arrangement like that assembled by Griffiths and a colleague (*17, 28*) (Fig. 2). Sea-floor magnetic records indicate that the Campbell Plateau (including New Zealand) separated from West Antarctica and began moving northwest as the Pacific-Antarctic rise was formed 80 million years ago (*11, 18, 29*). Somewhat later, northward movement of Australia resulted from activation of the Indian-Antarctic rise in the middle Eocene (45 million to 49 million years ago), the last time Australia was linked with South America via Antarctica (*12, 13, 30*). As Australia moved north it entered lower, warmer latitudes. Australia moved about 15° north from the Indian-Antarctic rise, Antarctica about 15° south (*11–13, 15–17, 31, 32*). Reconstructions that

Dr. Raven is director of the Missouri Botanical Garden, 2315 Tower Grove Avenue, St. Louis 63110, and professor of biology at Washington University, St. Louis. Dr. Axelrod is professor of geology and of botany at the University of California, Davis 95616.

place Australia 30° south of its present position are in conflict with biological evidence bearing on the position of Antarctica before the commencement of the movement (*33*). Current reconstructions of the positions of continents differ because plate movements are relative, and to indicate them, one area (such as Africa or Antarctica) has to be considered fixed.

By Miocene time, Australia had been rafted to about 10° south of its present position (*14, 15, 32*). The northward movement of New Zealand has been much less, and its isolation has been increased by the foundering of the Campbell Plateau.

These tectonic reconstructions are consistent with geologic evidence. New Zealand (*34*) and New Caledonia (*35*) consist of similar rocks of lower Mesozoic, Paleozoic, and probably greater age (*36*). Fiji is composed of ophiolites (slices of ocean floor, upper mantle rocks, and associated sedimentary rocks) that can be associated with the growth of island arcs. It was uplifted in the Eocene, probably as a result of the collision of the Pacific and Australian plates. However, the Solomon Islands and New Hebrides consist of rocks formed in inner arcs, and are Miocene archipelagos constructed behind the frontal arc system (*12, 27, 37*). The opening up of the Tasman basin, which contains con-

tinental rocks in Lord Howe Rise and Norfolk Ridge (*23, 36, 38*), has been related to the separation of Australia and Antarctica in the middle Eocene (45 million to 49 million years ago). A similar age has been postulated for the formation of the New Caledonia, South Fiji, and Coral Sea basins (*24, 27*). Topologically, reversing the movements would close the Tasman Sea and pull back the Campbell Plateau, New Caledonia, Fiji, Lord Howe Rise, and Norfolk Ridge close to the Australian mainland in the Paleocene (*15, 17, 22, 28*) (Fig. 2).

New Guinea, which defines the leading edge of the Australian plate, was largely submerged during its early northward movement (*15, 25, 26*). A period of moderate uplift that began in the late Oligocene and continued through the Miocene resulted from interaction between the Australian and Asian plates (*26, 39, 40*).

At the start of its northward movement 45 million to 49 million years ago, northern Australia, consisting of present northern Queensland and the now submerged Queensland Plateau (*41*), was all well south of the Tropic of Capricorn and mostly south of latitude 28°S, at approximately the present latitude of Brisbane. The emergence of large lowland areas of New Guinea above the sea during the Miocene ad-

vanced the land edge of the Australian plate about 10° of latitude, for the first time providing extensive land area within the tropics (*15*).

Many of the events in the area, including uplift of the mountains of New Guinea (*26, 39, 40*) and Malaysia (*42*) as well as Australia (*43*) and New Zealand (*44*), and the associated submergence of the Queensland Plateau (*41*), occurred in the past few million years. Much of this resulted from collision of the Australian with the Asian plate (*24, 26, 39, 40*).

Manifestly, (i) the northward rafting of the Australian plate and its fragments to new climates, (ii) the isolation of plate fragments by local sea-floor spreading, (iii) the appearance of new volcanic archipelagos and large islands and (iv) the subsidence or foundering of old plate fragments had profound consequences for the history of life in this region.

Austral Distribution Patterns

Many disjunct distribution patterns in the Southern Hemisphere reflect the geography of the region in the middle Cretaceous, when Africa, South America, and Australia were connected with Antarctica. The temperate parts of the austral landmass supported a mixed forest of austral gymnosperms and evergreen angiosperms, including members of the Podocarpaceae, Araucariaceae, Araliaceae, Myrsinaceae, Proteaceae, Winteraceae, Atherospermataceae (*45*), Lauraceae, Malvaceae, Loranthaceae (*46*), Sapindaceae, Casuarinaceae, and Myrtaceae, as well as *Nothofagus* and *Gunnera* (*47–51*). Associated with the forest were herbaceous and shrubby plants, such as Cruciferae (*49*), Pedaliaceae (*50*), Liliaceae (*49*), Epacridaceae, and possibly Umbelliferae subfamily Hydrocotyloideae; many groups of insects among them Peloridiidae, Idiostolidae (*52*), Belidae, Plecoptera (*53*) and Carabidae-Migadopinae; other invertebrates, such as Onychophora (*2, 5, 54–56*); and some bryophytes (*57*). The descendants of a number of these groups have become disjunct between South America and Australasia (*58, 59*). Direct overland migration between South America and Australia probably was feasible until approximately 45 million to 49 million years ago (*50, 58, 60*). Consequently, these two regions share more similar plants and animals

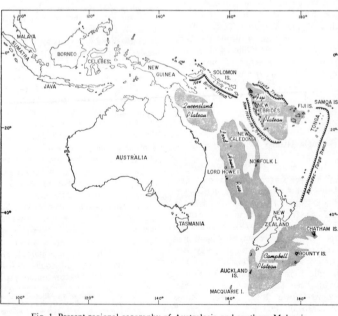

Fig. 1. Present regional geography of Australasia and southern Malaysia.

than do other southern lands, as stressed by Brundin (*61*).

Among the vertebrates, marsupials must have been present in association with this forest (*62*). They first appear in the fossil record in the Albian of North America (more than 100 million years ago) and may have originated there or in South America, where they are first known from the Upper Cretaceous (perhaps 80 million years ago). There was interchange between the landmasses of the Northern and Southern hemispheres during the Cretaceous, and marsupials certainly reached Australia via Antarctica (*63, 64*). Six endemic families had developed in Australia by the late Oligocene (*65*), when the group first appears in the fossil record. Marsupials are, and seem always to have been, absent from New Zealand (*51*), which implies that they reached Antarctica after New Zealand separated from it, about 80 million years ago (*11, 18, 29*).

Hylid frogs evidently originated in South America and had a history similar to that of the marsupials, as suggested by their absence from the Old World tropics and Africa; Ockham's razor rules out Darlington's elaborate alternative hypothesis based on radiation from the Old World tropics (*4*). Leptodactylid frogs (*66, 67*) and chelyid and meiolaniid turtles (*5*) evidently were present on this continuous southern land, but like marsupials, they seem to have reached Australia after New Zealand separated from Antarctica.

Ratite birds appear to have crossed Antarctica to reach their scattered stations in the Southern Hemisphere (*2, 7, 68, 69*). Their presence in Africa and New Zealand suggests that this occurred by the middle Cretaceous, more than 90 million years ago. Ceratodont lungfishes must also have reached Australia from the south during the Mesozoic, contrary to the arguments of Darlington (*5*). Most modern birds (*70*), lizards, and snakes (*4*), on the other hand, attained their present distributions after the lands of the southwestern Pacific reached approximately their present positions, in the late Tertiary and subsequently.

In general, it is the more generalized or primitive representatives of a particular group that display austral distribution patterns, as emphasized by Mackerras (*56*). This implies that in many groups the more advanced forms have evolved or radiated in post-Eocene time.

Austral Biota in Africa

The separation of Africa from South America took place in the early Albian (about 110 million years ago) (*12, 13, 16, 71*), while the separation of Africa from Antarctica seems to have taken place as much as 20 million years later. Even long after these lands had separated, access across narrow seas and via volcanic islands on the midocean ridges continued to be possible. Although Africa is now all north of latitude 35°S, and thus affords much less scope for cool-temperate plants and animals than South America, Australia, or New Zealand, it does have some

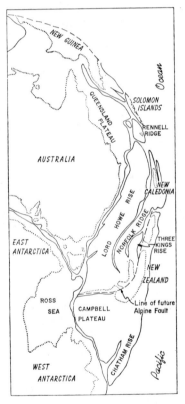

Fig. 2. Reconstruction of Australasia by Griffith and colleague (*17, 28*). A Senonian date (80 million years ago) appears reasonable. Proteaceae and *Nothofagus* were present in Australia, New Zealand, and doubtless New Caledonia at this time (*49, 50, 77*), but mammals do not seem to have arrived until later, since they are absent (except for bats) on the islands east of Australia. Monotremes are apparently of Triassic origin, but it is not known when or how they reached Australia.

taxa of austral affinity. These reflect the time when it was joined with Antarctica and was some 15° south of its present position, approximately 90 million years ago.

The austral gymnosperms *Araucaria* (now extinct in Africa) and *Podocarpus* (*48, 72*), some chironomid midges (*61*), and perhaps ratite birds (*2, 73*) seem to have reached Africa overland from the south. Very few angiosperm distribution patterns are comparable, however, which accords with a Tertiary origin for most genera. Pollen that can definitely be referred to living angiosperm genera is first known from the Senonian (*67, 74, 75*), that is, about 75 million years ago. It is problematic whether any living genera of angiosperms existed when Africa was directly connected by land with South America (110 million years ago) or Antarctica (about 90 million years ago), but some evidently evolved when there was fairly direct communication between these lands. There are numerous transtropical links involving Africa and South America (*71, 76*). Despite intensive searching, *Nothofagus*, which first appears in the fossil record of Australia, New Zealand, and Chile about 75 million years ago (*49, 50, 77*), has not yet been found in Africa (*78, 79*).

On the other hand, a few austral groups of angiosperms do seem to reflect former connections. The pattern of relationship in Proteaceae, best developed in Australia but with three groups in Africa [*Brabeium, Dilobeia*, and 13 of 19 genera of the tribe Proteeae (*80*)], is difficult to explain in any other way. The much closer relationships between the South American–Australian and Asian-Australian proteads than between the African-Australian ones is in accordance with the suggested relationships. Proteaceae first appear in the fossil record with *Nothofagus* in the Senonian, that is, 75 million years ago (*67, 74, 75*), when Africa and Antarctica were still relatively close. Possibly Restionaceae, with a very diverse assemblage of genera in Australia and one very distinct group of genera centering around *Restio* s. str. that radiated extensively in southern Africa (*81*), provide another example. In addition, the families Chloranthaceae and Winteraceae and the genera *Hibbertia* (Dilleniaceae), *Keraudrenia* and *Rulingia* (Sterculiaceae) are disjunct between the Australasian region and Madagascar (*19*), but do not occur on the mainland of Africa, where they may have become extinct. Analogous distributions, differing in detail, are found in *Adansonia, Cossignia, Cunonia,* Monimiaceae, and Brexiaceae (*82*). Most herbaceous plants disjunct between the two areas [for example, Compositae; but compare (*75, 83*)] probably attained their present distributions as a result of long-distance dispersal during the late Tertiary (*84*).

Relationships with the Northern Hemisphere

The ancestors of some southern plants and animals, like *Nothofagus* (*79, 85*), the marsupials, and hylid frogs, probably passed between the Northern and Southern hemispheres by way of Africa and Europe, since land connections were absent in Middle America (*64, 86*). The relatives of *Nothofagus* are all northern, and the evidence for its presence in the fossil record in Africa is weak (*78, 79*). Such a path is also implied by the existence of pollen of Proteaceae in the Late Cretaceous of North America, South America, and Nigeria; its simultaneous predominance (with *Nothofagus*) in Australia and Antarctica; and its absence in Borneo, where extensive samples have been studied (*67, 74, 75*). The occurrence of such present-day austral plants as *Hibbertia* (Dilleniaceae), *Leucopogon* (Epacridaceae), and Restionaceae in the Paleocene London Clay of southern England (*87*), *Araucaria* in the Mesozoic (*48*) and lower Tertiary (*87*) of Europe, and *Podocarpus* in the Tertiary of Europe (*88*) and North America (*89*) is evidence of Cretaceous links between the two hemispheres. This is not surprising for many plant families that are best developed today in the tropics and subtropics also reach into both the northern and southern mild-temperate regions (for example, Lauraceae, Palmae). The extinction of austral taxa in the north in response to more extreme climates has resulted in their present restriction to the Southern Hemisphere.

Other groups migrated into the Southern Hemisphere by long-distance dispersal across mountains uplifted in Malaysia and New Guinea in the late Pliocene and Pleistocene (*90, 91*); these presumably include *Veronica* (*59, 92*), *Euphrasia* (*75, 93*), *Poa, Carex,* the apioid Umbelliferae, and others that account for many bipolar distributions (*94–96*). The rich secondary radiation

of species of some of these taxa in the newly formed subalpine and alpine habitats of New Zealand (*95*) has at times led to the unwarranted assumption that they must possess great antiquity in Australasia, although it agrees in rate with other examples of adaptive radiation on islands (*97*). We suggest that all these groups reached New Zealand from Australia by means of the prevailing westerlies in the Pliocene or more recently (*98*).

The post-Eocene northward movement of Australia has been responsible for strengthening the prevailing westerlies (roaring forties, screaming fifties) into a dominant feature of world meteorology (*98*). This change in pattern may be related to the cooling trend in Antarctica that was evident by the late Miocene (*99*). With an ever-increasing thermal gradient between higher and lower latitudes, the winds would increase in velocity, and the opportunities for dispersal of seeds, spores, and small animals between the scattered southern lands would be enhanced (*58, 59*).

Australia

Many groups of archaic angiosperms occur today in the region from Australasia to Assam (*85, 100*). They became associated in this part of the world only in the Miocene, when the north-moving Australian plate collided with the Asian plate. Thus, the region as a whole could not have been an important early seat of evolution. Wallace's line (*4, 101*), which defines the region of mixing of Oriental and Australian plants and animals, came into existence when the plates collided. Many angiosperms crossed this line after the late Miocene [for example, *Drimys, Eucalyptus, Helicia,* Cunoniaceae, Philydraceae, Casuarinaceae (*67, 102*), Magnoliaceae, Dipterocarpaceae, Annonaceae], as have a great number of other plants and animals (*4, 101, 103*).

The northward movement of Australia across at least 15° of latitude in the past 45 million to 49 million years took place during a worldwide trend to decreased temperatures (*49, 104*). This movement into more moderate climates favored the survival of many archaic angiosperms, especially on islands. Although Australasia as a whole has become more accessible to immigration from Malaysia, islands such as New

Caledonia and Fiji have become more isolated, and relict biota have survived in their highly equable climates.

The Australian desert and semidesert developed in post-Eocene time as the continent moved into the horse latitude, the worldwide belt of reduced precipitation at the edge of the tropics. As stressed by Herbert (*105*), the taxa that made up the xeric vegetation of Australia can readily be linked with ancestral forms that occur in more mesic sites marginal to present desert and semidesert areas (*80, 106*). Communities of plants adapted to semiarid conditions may have existed locally along the northern margins of Australia even during Late Cretaceous time, as suggested by the presence of *Adansonia* and the great diversity of the bee family Colletidae (*107*). Pockets of sclerophyll vegetation almost certainly existed on infertile soils at this time and may have formed important source areas during the Tertiary. Descendants of the plants and animals found in these habitats became widespread as aridity achieved continental scope in the Tertiary. Thus evolution in response to expanding dry climate accounts for the rich flora and fauna of semiarid and arid Australia (*107*), a phenomenon typical of other dry regions as well.

As Australia continues to move north (*108*), the cool-temperate rain forest in its southeastern corner and in Tasmania will be replaced by subtropical and then by tropical rain forest, aside from the effects of human disturbance. The area of mediterranean climate in its southwestern and southern parts will become desert, and the thousands of species of plants and animals that typify its mediterranean ecosystem will become extinct. At the same time, the central area of *Acacia* shrublands and low woodlands will change to savanna and then to rain forest, and the present desert area will become more moist and suitable for open, xeric woodlands and savanna.

New Zealand

Much of the present lowland flora of New Zealand (*109, 110*) is similar to that of temperate Gondwanaland 80 million years ago. Relationships with other lands (*110, 111*) reflect the common origin of these lands and New Zealand in the breakup of the Australasian region. New Zealand has been forested throughout most of its history.

nd has been an archipelago since the ate Cretaceous (*51, 112*), at times much smaller than at present. It has lways had a highly equable oceanic limate that was maintained at a rela-ively stable temperature range as the rchipelago moved north. In the Creta-eous, New Zealand was joined to West Antarctica (Marie Byrd Land) and ronted the broad Pacific, whereas Aus-ralia was joined to East Antarctica Wilkes Land) and was situated in the ee of Africa-Madagascar-India. Con-equently, New Zealand would have had colder, cloudier, and rainier climate han Australia, even at that time. Hence, he absence of such characteristically Australian plants as *Acacia, Eucalyptus,* nd *Xanthorrhoea* may be explained artly on ecological grounds (*105*). Moreover, New Zealand has been iso-ated from Australia-Antarctica for ome 80 million years, which helps to xplain the distinctiveness of its plants nd animals (*61*).

The frog *Leiopelma,* the tuatara *Sphenodon*), and ratite birds almost ertainly reached New Zealand over-and when it was joined to West Antarc-ica (80 million years ago), as did many ymnosperms and ferns and some angio-perms. Biogeographic evidence suggests hat mammals had not reached Antarc-ca by this time, since they are absent rom New Zealand (except for bats). nakes are absent from New Zealand ecause they evolved in the Late Creta-eous; the geckos and skinks of New ealand presumably reached it across he sea in Neogene times (*4*). The bsence of large Mesozoic terrestrial eptiles in New Zealand (*4*) is presum-bly related to the scanty record of non-narine rocks (*51*), since such fossils ccur in both Australia and Antarctica *52*).

As the Campbell Plateau foundered nd as New Zealand moved farther way from possible source areas, some axa became extinct. The Proteaceae, ow represented in New Zealand by one pecies of *Knightia* and *Persoonia*, had dozen or more taxa in the Paleogene *113*). *Casuarina* (*51, 114*), *Araucaria* *48*), *Athrotaxis* (*48*), and many others *49*) also became extinct in New Zea-nd during the Tertiary, as did the irds *Malacorhynchus* and *Harpagornia* n the Pleistocene (*115*). The *brassii* roup of *Nothofagus*, now confined to Jew Guinea, New Caledonia, and outhern Argentina, disappeared from Australia and Tasmania in the Pliocene nd from New Zealand in the Pleisto-

cene (*79, 85*), presumably as a result of the lowered temperatures (*112*).

Such plants as *Rhopalostylis* (a palm), *Phormium* (Liliaceae), and *Coprosma* (Rubiaceae), which first appear in the Paleogene of New Zealand and occur today on Norfolk Island as well, sup-port an Eocene connection between a much enlarged Norfolk Island and New Zealand.

An endemic species of *Araucaria* per-sists on Norfolk Island, and there is an endemic genus of the archaic, austral hemipteran family Peloridiidae (*How-eria*) on Lord Howe Island, which in general has the more archaic flora and fauna. Numerous taxa must have be-come extinct as Lord Howe Rise, Nor-folk Ridge, and Campbell Plateau foundered in Paleogene and later times. The possibility of migration between New Zealand and New Caledonia was probably reduced by the same proc-esses.

New Zealand has become increas-ingly isolated during the past 80 mil-lion years, but has received immi-grants from many sources (*112*), no doubt at a declining rate (*116*). Sub-tropical climates appeared in northern North Island in the Miocene (*51*), probably as a consequence of its north-ward component of movement. The newly emerged volcanic archipelagos to the north served as source areas for invading subtropical taxa. The uplift of the mountains and spread of novel subalpine, alpine, and semiarid habitats in the late Pliocene and Pleistocene accommodated new immigrants such as *Dodonaea* (*49*) and *Veronica* (*59, 92*), some of which have radiated ex-tensively (*95*).

New Guinea

The distinctive plants and animals of Australia are mostly derived from tem-perate ancestors that evolved in re-sponse to increasing aridity as Australia moved north. When New Guinea, the leading edge of the Australian plate (*117*), was first elevated to form an extensive land area in the late Oligo-cene (*15, 25, 26*), it was colonized largely from the adjacent rich tropical lowlands of Malaysia (*7, 118–120*). Biogeographic affinities with Australia are poorly developed, but there has been a spectacular late Tertiary radia-tion of many groups. Among them are orchids (*121*), ferns, Sapotaceae (*122*), *Ficus* (*123*), *Elaeocarpus* and *Psycho-*

tria, the beetle families Cerambycidae and Chrysomelidae (*124*), and the birds, for which New Guinea has be-come an important secondary source area (*125*).

As the high mountains of New Guinea were uplifted in the late Plio-cene and Pleistocene (*26, 39, 40*), ex-tensive temperate montane areas ap-peared for the first time. *Nothofagus,* now richly represented there (*111, 126*), first appears in New Guinea in the upper Miocene (*79, 85, 127*), which suggests that the island may have been connected by land with the old, eroded mountains of the Cape York Peninsula and the now-submerged Queensland Plateau when it was first raised above the sea. Land connections between Australia and New Guinea are also suggested by the mid-Pliocene appearance of the marsupials (*65*). Other Australian groups, such as Win-teraceae, Proteaceae, *Araucaria,* and the monotremes, colonized New Guinea at various times (*103*). In the lee of the young mountains of New Guinea, an extensive rain shadow developed along the southern coast of Papua in the late Pliocene. Many Australian plants and animals, adapted to semi-arid conditions, occur only in this part of New Guinea, which they could not have invaded before this time.

Northern plants and animals like-wise reached the young mountains of New Guinea, where *Nothofagus* is as-sociated with *Castanopsis* and *Rhodo-dendron.* The island as a whole is a region of faunal and floral mixing, survival, and evolution in the middle to late Tertiary.

The Solomon Islands and New He-brides have floras (*8, 9*) and faunas [especially land snails (*119*)] heavily dominated by Indo-Malaysian taxa car-ried across water barriers, mainly by way of New Guinea (*9, 119, 128*). This is not surprising, for these islands are archipelagos of Miocene age (*37*) and would therefore not be expected to have numerous temperate austral links.

New Caledonia

The flora of New Caledonia is very rich in seed plants that represent ar-chaic taxa. This small island (about 17,000 square kilometers in area) is only one-seventh as large as North Island, New Zealand, yet it has about 40 endemic gymnosperms, several ves-selless angiosperms (Winteraceae, *Am-*

borella), and numerous other taxa (including Monimiaceae, Escalloniaceae, Cunoniaceae, Sapindaceae, *Nothofagus*, Araliaceae, Rutaceae, palms) that are relicts (*9, 129*). In strong contrast, it has a very poor representation of more advanced groups of Tertiary origin (*67, 74, 75*), such as Sympetalae in general, which probably reflects its isolation since the Late Cretaceous (perhaps 80 million years ago).

The New Caledonian flora appears to be a surviving, modified sample of the Late Cretaceous flora of eastern Australia. New Caledonia was then near latitude 35°S. The flora was preserved in isolation on serpentine (altered old sea floor) as the island moved north and east to its present position. Its survival was no doubt aided by late Cenozoic uplift (*35*), which provided mild temperatures in upland sites for taxa of basically temperate requirements as the island moved into warmer latitudes.

The archaic, austral land snails (*119*), some of the insects (*120, 130*), and possibly even galaxiid freshwater fish (*131*) and the endemic bird family Rhynochetidae presumably were derived from ancestors associated with the flora when the island was joined with Australia. The presence of terrestrial giant horned turtles of the family Meiolaniidae in southern South America in the early Tertiary, and in eastern Australia, Lord Howe Island, and on Walpole Island southeast of New Caledonia in the Pleistocene, provides further evidence of former land connections, as noted by Evans (*2*) and contrary to Darlington (*69*). Plants such as *Libocedrus* s. str., *Knightia*, and *Xeronema*, now restricted to New Zealand and New Caledonia, and other indications of biotic communality between these two lands (*9, 129*) can be related to the geography in Cretaceous and Paleogene time (Fig. 2).

Corner (*123*) has noted the highly endemic character of the fig (*Ficus*) flora of New Caledonia, which is consistent with its continental origin and subsequent rafting into greater isolation. In contrast, the more modern fig floras of New Guinea, the Solomons, and New Hebrides include many closely related species and are clearly derived from the Malaysian region. *Nothofagus* almost certainly has been present in New Caledonia since the Late Cretaceous; there is no biologic or geologic indication that it came via New Guinea in the upper Tertiary.

Fiji

This isolated island group has sometimes been regarded as having a continental flora and fauna, and sometimes an insular one (*120, 130*). Nearly 23 percent of the 445 indigenous plant genera, including *Dacrydium, Acmopyle, Agathis, Kermadecia* (Proteaceae), *Clematis*, and *Casuarina* div. *Gymnostomae*, and the families Annonaceae, Cunoniaceae, and Epacridaceae, among others, reach their eastern limits in the Pacific in Fiji (*9, 132*). *Balanops*, the only genus of Balanopaceae and one of very few austral Amentiferae, is confined to Queensland, New Caledonia, and Fiji. Other unusual taxa there include the primitive angiosperm *Degeneria*, as well as some land snakes and frogs. Among the plants of Fiji are many forest trees with large seeds (*132*), and it is unlikely that they or the animals mentioned dispersed over broad water gaps to reach Fiji: zoological evidence has been regarded as equivocal concerning this point (*4, 120, 130*).

We prefer to consider that Fiji arose from the sea close to the main Australian landmass and possibly connected with it at this time (Eocene). As it incorporated slices of sea floor (ophiolites) when it moved northeast to its present position, its climate was becoming more tropical. Hence, immigrants from across the water that played a progressively more important role in the flora were of Malaysian (*9, 132*) and Polynesian (*120, 130*) character. At the same time, numerous taxa with more temperate or marginally tropical requirements must have become extinct as the land area was rafted into warmer regions.

Summary

Complex tectonic events determined by plate movements had a guiding role in the biotic history of the Australasian region. These movements commenced in the middle Cretaceous, when Australia–Antarctica was still in proximity to Africa, India, and New Zealand, before the disruption of Gondwanaland. Ties between the Australian, Antarctic, and American plates provided routes of migration for southern temperate forests and associated land animals into later Eocene time. As the Australian plate was rafted north, it met the west-moving Pacific plate and underwent fragmentation, resulting in the develop-

ment of a complex series of basin plateaus, island arcs, and foredeeps. T isolated lands provided areas for t survival of ancient taxa, some of whi occur in distant regions (such as Mad gascar and Chile). Many new taxa al evolved over the lowlands as these lan were rafted north to warmer climate In Australia, the mixed gymnosperm evergreen dicot forest of Cretaceous Eocene time was replaced progressive by taxa adapted to mediterranean an desert climates, xeric shrublands an low woodlands, savanna, and tropic forest as the plate moved into low latitudes. The collision of the Australi with the Asian plate in the Mioce established Wallace's line, and the mi ing of Oriental and Australasian ta commenced. In the late Cenozoic, t elevation of high mountains from M laysia to New Guinea and Australi New Zealand provided dispersal rou for numerous herbs from cool-tempera parts of the Holarctic and new sites f their rapid evolution.

References and Notes

1. For example, L. Harrison, *Rep. Me Aust. N. Z. Ass. Advan. Sci.* **18**, 332 (192
2. J. W. Evans, in *Continental Drift. A Sy posium*, S. W. Carey, Ed. (Geology Depa ment, University of Tasmania, Hobart, 195 p. 134.
3. A. Wegener, *Die Entstehung der Kontine und Ozeane* (Vieweg, Braunschweig, 191
4. P. J. Darlington, Jr., *Zoogeography: T Geographical Distribution of Animals* (Wil New York, 1963).
5. ———, *Biogeography of the Southern E of the World* (Harvard Univ. Press, Ca bridge, Mass., 1965).
6. A. L. Hammond, *Science* **173**, 40, 133 (197
7. R. F. Thorne, in *Pacific Basin Biogeograp* J. L. Gressitt, Ed. (Bishop Museum Pre Honolulu, 1963), p. 311.
8. M. M. J. van Balgooy, *Blumea* **17**, (1969); *ibid.* (Suppl. 6), 1 (1971).
9. ———, *ibid.* **10**, 385 (1960).
10. W. P. Sproll and R. S. Dietz, *Nature* **2** 345 (1969); E. Irving and W. A. Roberts *J. Geophys. Res.* **74**, 1026 (1969); D. Hayes and W. C. Pitmann III, *Antarct.* **5**, 70 (1970); A. G. Smith and A. Halla *Nature* **225**, 139 (1970); I. W. D. Dal and D. H. Elliott, *ibid.* **233**, 246 (1971); Keast, *Quart. Rev. Biol.* **46**, 335 (1971).
11. X. Le Pichon and J. R. Heirtzler, *Geophys. Res.* **73**, 2101 (1968).
12. R. S. Dietz and J. C. Holden, *ibid.* **75**, 49 (1970).
13. D. H. Tarling, *Nature* **229**, 17 (1971).
14. M. W. McElhinny, *ibid.* **228**, 977 (1970).
15. J. G. Jones, *ibid.* **230**, 237 (1971).
16. J. J. Veevers, J. G. Jones, J. A. Tale *ibid.* **229**, 383 (1971).
17. J. R. Griffiths, *ibid.* **234**, 203 (1971).
18. J. R. Heirtzler, in *Research in the Antarc* L. O. Quam, Ed. (AAAS, Washington, D. 1971), p. 667.
19. ——— and R. H. Burroughs, *Science* **1** 488 (1971).
20. W. J. M. van der Linden, *Rec. Proc. 1 Pac. Sci. Congr.* **1**, 390 (1971).
21. D. E. Karig, *J. Geophys. Res.* **75**, 239 (197
22. D. J. Cullen, *N. Z. J. Geol. Geophys.* **13** (1970).
23. G. G. Shor, Jr., H. K. Kirk, H. W. Mena *J. Geophys. Res.* **76**, 2562 (1971).
24. C. G. Chase, *Geol. Soc. Amer. Bull.* **3087** (1971).
25. J. E. Thompson, *Aust. Petrol. Explor. Assoc.* **7**, 83 (1967); H. L. Davies and I. Smith, *Geol. Soc. Amer. Bull.* **82**, 3

(1971); J. G. Sclater, J. W. Hawkins, Jr., J. Mammerickx, C. G. Chase, *ibid.* **83**, 505 (1972).

26. J. J. Hermes, *Geol. Mijnbouw* **47**, 81 (1968).
27. D. E. Karig, *J. Geophys. Res.* **76**, 2542 (1971).
28. J. R. Griffiths and R. Varne, *Nature Phys. Sci.* **235**, 83 (1972).
29. W. C. Pitman III, E. M. Herron, J. R. Heirtzler, *J. Geophys. Res.* **73**, 2069 (1968).
30. Date of marine transgression: W. A. Berggren, *Nature* **224**, 1072 (1969); B. McGowran, "Paleontology Report 10/71" (Department of Mines, South Australia, unpublished, 1971); see also (*11, 18, 29*).
31. E. Irving, *Paleomagnetism* (Wiley, New York, 1964).
32. This would be in agreement with the calculation of an early Eocene (51.6 million years ago) latitude of 52°S for Canberra [P. Wellman, M. W. McElhinny, I. McDougall, *Geophys. J. Roy. Astron. Soc.* **18**, 371 (1969)].
33. D. I. Axelrod and P. H. Raven, in *Future Directions in the Life Sciences*, J. A. Behnke, Ed. (American Institute of Biological Sciences, Washington, D.C., in press).
34. C. A. Fleming, *Quart. J. Geol. Soc. London* **125**, 125 (1970).
35. A. R. Lillie and R. N. Brothers, *N. Z. J. Geol. Geophys.* **13**, 145 (1970).
36. J. L. Aronson and G. R. Tilton, *Geol. Soc. Amer. Bull.* **82**, 3449 (1971).
37. P. Quantin, *Rec. Proc. 12th Pac. Sci. Congr.* **1**, 5 (1971); B. D. Hackman, *ibid.*, p. 366; D. I. J. Mallick, *ibid.*, p. 368.
38. D. J. Woodward and T. M. Hunt, *N. Z. J. Geol. Geophys.* **14**, 39 (1971).
39. D. B. Dow, *Geol. Mijnbouw* **47**, 37 (1968).
40. T. W. E. David and W. R. Browne, *The Geology of the Commonwealth of Australia* (Arnold, London, 1950), vol. 1, p. 682.
41. M. Ewing, L. V. Hawkins, W. J. Ludwig, *J. Geophys. Res.* **75**, 1953 (1970).
42. For Mt. Kinabalu: B. Collenette, *Proc. Roy. Soc. London Ser. B* **161**, 56 (1964).
43. T. W. E. David and W. R. Browne, *The Geology of the Commonwealth of Australia* (Arnold, London, 1950).
44. R. P. Suggate, *N. Z. Geol. Surv. Bull.* **56** (1957); M. Gage, *Trans. Roy. Soc. N. Z.* **88**, 631 (1961); C. Cotton, *Tuatara* **10**, 5 (1962); D. A. Brown, K. S. W. Campbell, K. A. W. Crook, *Geological Evolution of Australia and New Zealand* (Pergamon, New York, 1968), p. 351.
45. R. Schodde, *Taxon* **19**, 324 (1970).
46. B. A. Barlow and D. Wiens, *ibid.* **20**, 291 (1971).
47. L. M. Cranwell, *Nature* **184**, 1782 (1959); D. J. McIntyre and G. J. Wilson, *N. Z. J. Bot.* **4**, 315 (1966).
48. R. Florin, *Acta Horti Bergiani* **20**, 121 (1963).
49. R. A. Couper, *N. Z. Geol. Surv. Paleontol. Bull.* **32** (1960).
50. L. M. Cranwell, in *Paleoecology of Africa*, E. M. van Zinderen Bakker, Ed. (Balkema, Cape Town, 1969), p. 1.
51. C. A. Fleming, *Tuatara* **10**, 53 (1962).
52. C. W. Schaefer and D. Wilcox, *Ann. Entomol. Soc. Amer.* **62**, 485 (1969).
53. J. Illies, *Verh. Deut. Zool. Ges. Bonn-Rhein 1960*, 384 (1960).
54. J. L. Gressitt, *Pac. Insects Monogr.* **23**, 295 (1970); E. E. Riek, in *The Insects of Australia* (Melbourne Univ. Press, Melbourne, 1970), p. 168.
55. W. Hennig, *Pac. Insects Monogr.* **9**, 1 (1966).
56. I. M. Mackerras, in *The Insects of Australia* (Melbourne Univ. Press, Melbourne, 1970), p. 187.
57. G. O. K. Sainsbury, *Roy. Soc. N. Z. Bull.* **5** (1955); R. M. Schuster, *Taxon* **18**, 46 (1969).
58. Many disjunct distributions between the southern lands were attained more recently. These involve plants with dustlike seeds, floating seeds, or fruits; fleshy fruits eaten by migratory birds; or adherent seeds or fruits. Examples are *Juncus, Colobanthus*, orchids, ferns, *Carpobrotus, Sophora, Tetragonia; Astelia, Pernettya, Fuchsia*, and *Acaena* and *Uncinia*. In the absence of a fossil record, all distributions must be interpreted with caution. W. R. B. Oliver, *Proc. 7th Pac. Sci. Congr.* **5**, 131 (1949); H. N. Barber, H. E. Dadswell, H. D. Ingle, *Nature* **184**, 204 (1959); E. J. Godley, *ibid.*

214, 74 (1967); W. R. Sykes and E. J. Godley, *ibid.* **218**, 495 (1968).
59. C. A. Fleming, in *Pacific Basin Biogeography*, J. L. Gressitt. Ed. (Bishop Museum Press, Honolulu, 1963), p. 369.
60. Forest vegetation persisted in Antarctica until at least the Oligocene [D. J. MacIntyre and G. J. Wilson, *N. Z. J. Bot.* **4**, 315 (1966); (*50*)]. On other grounds W. Hennig (*55*) has calculated that Australian–South American disjunct groups are at least Oligo-Miocene (26 million years) in age.
61. L. Brundin, *Evolution* **19**, 496 (1965); *Kgl. Sven. Vetensk. Handl.* **11**, 1 (1966); *Annu. Rev. Entomol.* **12**, 149 (1967); in *Antarctic Ecology*, M. W. Holgate, Ed. (Academic Press, New York, 1970), p. 41.
62. Unfortunately, no fossil vertebrates younger than the Triassic amphibians and reptiles discussed by E. H. Colbert [*Quart. Rev. Biol.* **46**, 250 (1971)] have yet been discovered in Antarctica [J. W. Kitching, J. W. Collinson, D. H. Elliot, E. H. Colbert. *Science* **175**, 524 (1972)].
63. R. Hoffstetter, *C. R. H. Acad. Sci. Ser. D* **271**, 388 (1970); C. B. Cox, *Nature* **226**, 767 (1970); W. A. Clemens, *Annu. Rev. Ecol. Syst.* **1**, 357 (1970); R. H. Tedford, *Geol. Soc. Amer. Abstr. Programs* **3**, part 7, 730 (1971); B. H. Slaughter, *J. Linn. Soc. London Zool.* **50**, 131 (1971); N. Jardine and D. McKenzie, *Nature* **235**, 20 (1972).
64. J. Fooden, *Science* **175**, 894 (1972).
65. E. D. Gill, *Trans. N.Y. Acad. Sci.* **28**, 5 (1965); R. A. Stirton, R. H. Tedford, M. O. Woodburne, *Univ. Calif. Publ. Geol. Sci.* **77**, 1 (1968).
66. J. D. Lynch, *Misc. Publ. Univ. Kans. Mus. Nat. Hist.* **53**, 1 (1971). The presence of a leptodactylid frog in the Eocene of India (*4*) is of special interest, because in the Cretaceous, India was part of Gondwanaland. The extinction of these frogs in India presumably occurred as it crossed the hotter equatorial zone. Proteaceae and *Casuarina* have also been reported from the Eocene of India, and other austral plants and animals should be sought (*67*).
67. Y. K. Mathur, *Quart. J. Geol. Mining Met. Soc. India* **38**, 33 (1966).
68. J. Gentili, *Emu* **49**, 85 (1949).
69. P. J. Darlington, Jr., *Quart. Rev. Biol.* **23**, 1 (1948).
70. E. Mayr, *Proc. 6th Pac. Sci. Congr.* **4**, 197 (1939).
71. D. I. Axelrod, *Bot. Rev.* **36**, 277 (1970).
72. J. B. Hair, in *Pacific Basin Biogeography*, J. L. Gressitt, Ed. (Bishop Museum Press, Honolulu, 1963), p. 401.
73. In evaluating the relationship between the ostriches of Africa and the rheas of South America, the close similarity of their external and internal parasites is significant [J. W. Evans (*2*)].
74. J. H. Germeraad, C. A. Hopping, J. Muller, *Rev. Palaeobot. Palynol.* **6**, 189 (1968). *Casuarina* is also known from the Eocene of India, which was then in south temperate latitudes [Y. K. Mathur (*67*)].
75. J. Muller, *Biol. Rev.* **45**, 417 (1970).
76. G. S. Myers, *Copeia* **1966**, 766 (1966); *Stud. Trop. Oceanogr. Inst. Mar. Sci. Univ. Miami* **5**, 614 (1967).
77. E. Dettmann and G. Playford, in *Stratigraphical and Palaeontological Essays in Honour of Dorothy Hill*, K. F. W. Campbell, Ed. (Australian National Univ. Press, Canberra, 1969), p. 174.
78. The report of *Nothofagus* pollen from Nigeria mentioned by S. K. Srivastava [*Bot. Rev.* **33**, 260 (1967)] may be valid; but see (*79*).
79. L. M. Cranwell, in *Pacific Basin Biogeography*, J. L. Gressitt, Ed. (Bishop Museum Press, Honolulu, 1963), p. 387. It cannot have reached the Southern Hemisphere via Malaysia as proposed by C. G. G. J. van Steenis (*85*).
80. L. A. S. Johnson and B. G. Briggs, *Aust. J. Bot.* **11**, 21 (1963).
81. D. F. Cutler, in *Taxonomy, Phytogeography and Evolution*, D. H. Valentine, Ed. (Academic Press, London, in press).
82. J. H. Willis, *Muelleria* **1**, 61 (1956); B. P. G. Hochreutiner, *Ann. Conserv. Jard. Bot. Genève* **11–12**, 136 (1908). One other genus of Cunoniaceae, the monotypic *Platylophus*, is found in South Africa.
83. Compositae are unknown in the fossil record prior to the lower Miocene, despite the

fact that they produce abundant, easily recognized pollen [J. Muller (*75*)]. As might have been expected, trans-Antarctic distributions are almost completely absent in this family, which probably arose in the Oligocene [G. L. Stebbins, Jr., *Proc. 6th Pac. Sci. Congr.* **3**, 649 (1940)].
84. Relationships between the biota of Madagascar and New Caledonia [R. Good, *Blumea* **6**, 470 (1950)], like those between the biota of Madagascar and the West Indies [W. T. Stearn, *Bull. Brit. Mus. Nat. Hist. Bot.* **4**, 261 (1971); (*4*)] came about because of extinction on the adjacent continents. Iguanid lizards, a dominant group in the New World, are represented in the Old World only in Madagascar and Fiji (*4*).
85. C. G. G. J. van Steenis, *Blumea* **19**, 65 (1971).
86. G. L. Freeland and R. S. Dietz, *Nature* **262**, 20 (1971). .
87. M. E. J. Chandler. "The Lower Tertiary Floras of Southern England," *Monogr. Brit. Mus. Nat. Hist. Geol.* (Suppl.), part 1 (1961); G. Erdtman, *Bot. Notis.* **113**, 46 (1960).
88. D. K. Ferguson, *Palaeogeogr. Palaeoclimatol. Palaeoecol.* **3**, 73 (1967).
89. D. L. Dilcher, *Science* **164**, 299 (1969).
90. The ancient continuous pathways envisioned by C. G. G. J. van Steenis [*Bull. Bot. Gard. Buitenzorg III* **13**, 139, 389 (1934); *ibid.* **14**, 56 (1936)] and G. E. DuRietz (*91*) are geologically impossible. All of the high mountain flora of Malaysia evolved locally or migrated into the area by long-distance dispersal during the late Pliocene and more recently, after the uplift of the mountains.
91. G. E. DuRietz, *Svensk. Bot. Tidskr.* **25**, 500 (1931); *ibid.* **42**, 99, 348 (1948).
92. Pollen of the entire *Veronica* complex, including *Hebe*, first appears in New Zealand in the Pleistocene, as stressed by C. A. Fleming (*59*) and W. F. Harris [in *The Natural History of Canterbury*, G. A. Knox, Ed. (Reed, Wellington, 1969), p. 334].
93. The austral distribution of this advanced sympetalous hemiparasite cannot be accounted for by postulating that it evolved into its modern form in the Cretaceous, as proposed by G. E. DuRietz (*91*). A Pliocene dispersal from Asia into Australasia, with subsequent dispersal to South America, is more consistent with the available facts. Differentiation of the order Scrophulariales into families seems to have been a Paleogene event [J. Muller (*75*)].
94. G. E. DuRietz, *Acta Phytogeogr. Suecica* **13**, 215 (1940). Many genera have reached the Southern Hemisphere independently in the Old World and the New World (*95*).
95. P. H. Raven, in *Taxonomy, Phytogeography and Evolution*, D. H. Valentine, Ed. (Academic Press, London, in press).
96. Other plant genera that may also have reached the Southern Hemisphere by way of South America in the Upper Cretaceous are *Geum* subg. *Oncostylus, Caltha*, and *Acaena*, although in the absence of a fossil record it cannot be determined that any one has been in the Southern Hemisphere so long [D. M. Moore, in *Taxonomy, Phytogeography and Evolution*, D. H. Valentine, Ed. (Academic Press, London, in press)].
97. W. J. Bock, *Evolution* **24**, 704 (1970).
98. The persistence of these winds is demonstrated by the flight of a balloon released from Christchurch, New Zealand, held at about 12 kilometers elevation, and tracked by satellite. It circled the world eight times in 102 days, reaching South America in just over 5 days [J. Mason, *Nature* **233**, 382 (1970); H. H. Lamb, *Quart. J. Roy. Meteorol. Soc.* **85** (No. 363), 1 (1959); R. Tyron, *Biotropica* **2**, 76 (1970)].
99. J. T. Tollin, in *Palaeoecology of Africa*, E. M. van Zinderen Bakker, Ed. (Balkema, Cape Town, 1969), p. 109.
100. A. Takhtajan, *Flowering Plants: Origin and Distribution* (Oliver & Boyd, Edinburg, 1969); A. C. Smith, *Univ. Hawaii Lyon Arbor. Lect. 1* (1970).
101. A. R. Wallace, *J. Linn. Soc. London* **4**, 172 (1860); E. Mayr, *Quart. Rev. Biol.* **19**, 1 (1944); R. Melville, *Nature* **207**, 48 (1965); *ibid.* **211**, 116 (1966); J. W. Dawson, *Tuatara* **18**, 94 (1970).
102. Of the two divisions of *Casuarina*, the more primitive Gymnostomae include 19 species ranging from the Philippines and Sumatra to Fiji, New Caledonia, and northern

Queensland; they are also known from the Miocene of Argentina [J. Frenguelli, *Notas La Plata Univ. Nac. Mus. Paleontol.* **8**, 349 (1943)]. The *Cryptostomae*, with 45 species, are primarily Australian but a few species have also crossed Wallace's line. Pollen of *Casuarina* is known from the Eocene of India, which was then in south temperate latitudes [Y. K. Mathur (67); L. A. S. Johnson, personal communication].

103. Such gymnosperms as *Agathis, Phyllocladus,* and *Dacrydium* presumably entered New Guinea in the upper Oligocene or Miocene and subsequently crossed Wallace's line.

104. C. A. Fleming (*51*); W. K. Harris, *Palaeontogr. Abt. B* **115**, 75 (1965); F. H. Dorman, *J. Geol.* **74**, 49 (1966); M. Schwarzbach, *Tuatara* **16**, 38 (1968); D. R. McQueen, D. C. Mildenhall, C. J. E. Bell, *ibid.*, p. 49; and V. Scheibnerová, *Rec. Geol. Surv. N. S. W.* **13**, 1 (1971) have shown that during the Cretaceous and Tertiary the temperature curves in Australia and New Zealand first rose and then fell much more gradually than those in North America and Europe. This reflects first the northward movement of the Australian plate to lower, warmer latitudes, and second the decline in temperature following the middle Miocene.

105. D. A. Herbert, *Victorian Natur.* **66**, 227 (1950).

106. E. C. Andrews, *Proc. Linn. Soc. N. S. W.* **38**, 529 (1913); E. C. Andrews, *J. Roy. Soc. N. S. W.* **48**, 333 (1914).

107. C. A. Gardner, *J. Roy. Soc. West. Aust.* **28**, x (1944); S. Smith-White, *Cold Spring Harbor Symp. Quant. Biol.* **24**, 278 (1959); N. T. Burbidge, *Aust. J. Bot.* **8**, 75 (1960); C. D. Michener, *Bull. Amer. Mus. Nat. Hist.* **130**, 1 (1965); B. A. Barlow, *Aust. J. Bot.* **19**, 295 (1971).

108. R. S. Dietz and J. C. Holden, *Sci. Amer.* **223**, 30 (Oct. 1970).

109. J. W. Dawson and B. V. Sneddon, *Pac. Sci.* **23**, 131 (1969).

110. J. W. Dawson, *Tuatara* **11**, 178 (1963).

111. R. G. Robbins [*ibid.* **8**, 121 (1961)] stresses the relationship between the lowland rain forest of New Zealand and the montane (not lowland) rain forest of New Guinea. J. W. Dawson (*110*) discusses relationships with New Caledonia.

112. P. Wardle, *N. Z. J. Bot.* **1**, 3 (1963).

113. L. Cranwell, cited by C. G. G. J. van Steenis [*Blumea* **11**, 294 (1962)].

114. If we accept the identification of *Triorites harrisii* Coup. as *Casuarina* [C. A. Fleming (*51*)], this genus appeared in New Zealand in the uppermost Cretaceous and became extinct with the onset of Pleistocene glaciation. Pollen of *Casuarina* is still blown from Australia to New Zealand, however, and there must be some uncertainty about the ancient records [N. T. Moar, *N. Z. J. Bot.* **7**, 424 (1969)].

115. R. A. Falla, *Emu* **53**, 36 (1953).

116. R. M. McDowall, *Tuatara* **17**, 1 (1969); R. H. MacArthur and E. O. Wilson, *Monogr. Pop. Biol. 1* (1967); E. O. Wilson and D. S. Simberloff, *Ecology* **50**, 267 (1969); D. S. Simberloff and E. O. Wilson, *ibid.*, p. 278; D. S. Simberloff, *ibid.*, p. 296.

117. This is contrary to the arguments of R. Good [*Aust. J. Sci.* **20**, 41 (1957); in *Pacific Basin Biogeography*, J. L. Gressitt, Ed. (Bishop Museum Press, Honolulu, 1963), p. 301].

118. H. J. Lam, *Blumea* **1**, 113 (1934).

119. A. Solem, *Nature* **181**, 1253 (1958).

120. J. L. Gressitt, *Syst. Zool.* **5**, 11 (1956).

121. R. Good, *Bull. Brit. Mus. Nat. Hist. Bot.* **2**, 205 (1960); *The Geography of Flowering Plants* (Longmans, Green, London, ed. 3, 1964).

122. H. J. Lam, *Proc. 6th Pac. Sci. Congr.* **4**, 673 (1939).

123. E. J. H. Corner, in *Pacific Basin Biogeography*, J. L. Gressitt, Ed. (Bishop Museum Press, Honolulu, 1963), p. 233; *Phil. Trans. Roy. Soc. London Ser. B* **255**, 567 (1969).

124. J. L. Gressitt, *Proc. 10th Int. Congr. Entomol.* **1**, 767 (1958); *Pac. Insects Monogr.* **2**, 1 (1961).

125. E. Mayr, *Proc. 7th Pac. Sci. Congr.* **4**, (1953).

126. C. G. G. J. van Steenis, *J. Arnold Arboretum Harvard Univ.* **34**, 301 (1953).

127. Upper Pliocene records: I. C. Cookson and K. M. Pike, *Aust. J. Bot.* **3**, 197 (1955); R. A. Couper, *Proc. Roy. Soc. London Ser.* **152**, 491 (1960); L. M. Cranwell (*79*). The late Miocene record was reported by C. G. J. van Steenis (*85*).

128. R. F. Thorne, *Phil. Trans. Roy. Soc. Ser.* **255**, 595 (1969).

129. R. F. Thorne, *Univ. Iowa Stud. Nat. Hist.* **20**, 1 (1965).

130. J. L. Gressitt, *Pac. Insects Monogr.* **2**, (1961).

131. G. S. Myers, *Proc. 7th Pac. Sci. Congr.* 38 (1953).

132. A. C. Smith, *J. Arnold Arboretum Harvard Univ.* **36**, 273 (1955).

133. We are most grateful to B. G. Briggs, M. Brundin, J. C. Crowell, J. W. Dawson, R. S. Dietz, C. A. Fleming, J. L. Gressitt, C. Holden, L. A. S. Johnson, D. E. Kar J. A. Lillegraven, A. R. H. Martin, H. A. Martin, B. McGowran, C. D. Michener, R. Moar, J. Muller, A. Ritchie, J. Sauer, V. Scheibnerová, A. G. Smith, W. R. Sykes, R. Taylor, R. H. Tedford, J. W. Walker, R. W. Warren, and G. Watson for useful discussions of some of the points considered in this article. This work has been supported by National Science Foundation grants to both authors. See also *Geotimes*, May 1972, p. 1, which appeared while this article was in press.

Editor's Comments
on Papers 27 and 28

27   SCHUCHERT
Climates of Geologic Time

28   DORF
Climatic Changes of the Past and Present

In the geological literature a number of articles and books have been published on past geological climates and the relationship that fluctuations and changes in these climates have had with respect to biota and their distribution. Many of these deal specifically with the effects of Pleistocene climatic fluctuations, with application also to earlier glaciations. Others are concerned with climatic indicators from a general viewpoint. Although the entire text of Paper 27 is too long to reproduce here, the biological and summary sections show the synthesis that Schuchert had prepared as early as 1914, in which he related climates to coal and limestone formation, sea-level changes, and mountain-building activity. At about the same time W. D. Matthew (1915) published "Climate and Evolution," which emphasized vertebrate evolution and climates. A few years later C. E. P. Brooks (1926, rev. ed. 1949) investigated the climatic evidence that pervaded Wegener's hypothesis of continental drift and the climatic interpretations of the hypothesis by Köppen and Wegener (1924).

Erling Dorf in Paper 28 examines the Cenozoic floral history of North America and draws climatic implications based on a combination of paleobotanical and geological data. He joined the faculty at Princeton University in 1926 and was curator of paleobotany from 1930 until his retirement in 1974. He is well known for his contributions to the solution of the Cretaceous–Tertiary boundary problem of western North America and the study of Tertiary plants from Yellowstone National Park. His spectrum has included plant fossils from the Early Devonian of Wyoming, Newfoundland,

and Maine to Gondwana floras of the Carboniferous, Permian, and Triassic. Paper 28 is the result of many of these interests as they apply to paleoclimates of the past 100 million years.

REFERENCES

Brooks, C. E. P. (1949). Climate through the ages: Ernest Benn. (Reprinted 1970, Dover, New York, 395 p.)

Köppen, W., and A. Wegener (1924). Die Klimate der geologischen Vorzeit. Berlin, 255 p.

Matthew, W. D. (1915). Climate and evolution. Ann. N.Y. Acad. Sci., 24, 171–318.

27

Reprinted from *Carnegie Inst. Washington, Publ. 192*, 1914, pp. 275–289, 297–298

CLIMATES OF GEOLOGIC TIME

Charles Schuchert

[*Editor's Summary of Pages 263-275:* Schuchert discusses the plane-tesimal hypothesis of Chamberlin and Moulton and the evidence for Pleistocene, Permian, Devonian, Cambrian, and Proterozoic glaciations. He considers the color of sediments an indicator of climate: black or gray indicates coolness, humidity, and glaciation, and red indicates warmth and aridity. This is followed by a brief review of volcanic dust as a contributor to climatic conditions.]

BIOLOGIC EVIDENCE.

In the previous pages there has been presented the evidence for cold climates during geologic time as furnished by the presence of the various tillites. This presentation has also been made from the standpoint of discovery of the tillites, which in general is in harmony with geologic chronology, *i. e.*, the youngest tillites were the first to be observed, while the most ancient one has been discovered recently.

Variability of climate is also to be observed in the succession of plants and animals as recorded in the fossils of the sedimentary rocks. In this study we are guided by the distribution of living organisms and the postulate that temperature conditions have always operated very much as they do now upon the living things of the land and waters. In presenting this biologic evidence we shall, however, begin at the beginning of geologic time and trace it to modern days, for the reason that life has constantly varied and evolved from the more simple to the more complex organisms.

Proterozoic.—The first era known to us with sedimentary formations that are not greatly altered is the Proterozoic, a time of enormous duration, so long indeed that some geologists do not hesitate to say that it endured as long as all subsequent time. These rocks are best known and occur most extensively over the southern half of the great area of 2,000,000 square miles covered by the Canadian shield. There were at least four cycles of rock-making, each one of which, in the area just north of the Great Lakes and the St. Lawrence River, was separated from the next by a period of mountain-making. These mountains were domed or batholithic masses of vertical uplift due to vast bodies of deep-seated granitic magmas rising beneath and into the sediments. In the Grenville area of Canada, Adams and Barlow (1910) tell us that the total thickness of the pre-Proterozoic rocks alone is 94,406 feet, or nearly 18 miles. Of this vast mass more than half (50,286 feet) is either pure limestone, magnesian limestone, or dolomite, and single beds are known with a thickness of 1,500 feet. Certainly so much limestone represents not only a vast duration of time but also warm waters teeming with life, almost nothing of which is as yet known. There is further evidence of life in the widely distributed graphites, carbon derived from plants and animals, which make up from 3 to 10 per cent by weight of the rocks of the Adirondacks (Bastin, 1910). The graphite occurs in beds up to 13 feet thick, and at Olonetz, Finland, there is an anthracite bed 7 feet thick.

It is also becoming plain that there was in the Proterozoic a very great amount of fresh-water and subaërial deposits, the so-called continental deposits, some of which indicate arid climates. Because of the apparent dominance of continental deposits and the great scarcity of organic remains throughout the Proterozoic, Walcott has called this time the Lipalian era (1910: 14).

We have seen that the Proterozoic began with a glacial period, as evidenced by the tillites of Canada, but that this frigid condition did not last long is attested by the younger Lower Huronian limestones of Steeprock Lake, Ontario, having a thickness of from 500 to 700 feet and replete with Archæocyathinæ, coral-like animals up to 15 inches in diameter, and forming reef limestones several feet thick, found there by Lawson and described by Walcott (1912). This discovery is of the greatest value, and opens out a new field for paleontologic endeavor in Proterozoic strata and for philosophic speculation as to the time and conditions when life originated.

We have also seen that the Proterozoic closed with a frigid climate, as is attested by the tillites of Australia, Tasmania, and possibly China, while the other glacial deposits of India, Africa, Norway, and Keweenaw certainly do in part indicate another and older period of cool to cold world climates.

Cambric.—Due to the researches of many paleontologists, but mainly to those of Charles D. Walcott, we now know that the shallow-water seas of Lower Cambric time abounded in a varied animal life that was fairly uniform the world over in its faunal development. It was essentially a world of medusæ, annelids, trilobites, and brachiopods, animals either devoid of skeletons or having thin and nitrogenous external skeletons with a limited amount of lime salts. The "lime habit" came in dominantly much later, in fact, not before the Upper Cambric. However, that the seas in Lower Cambric time had an abundance of usable lime salts in solution is attested by the presence of many Hyolithes, small gastropods and brachiopods, and more especially by the great number of Archæocyathinæ, which made reefs and limestones 200 feet thick and of wide distribution in Australia, Antarctica, California (thick limestones near the base of the Waucoba section), southern Labrador (reefs 50 feet thick), and to a smaller extent in Nevada, New York, Spain, Sardinia, northern Scotland, and Arctic Siberia.

With an abundance of limestone and reef-making animals of world-wide distribution in the Lower Cambric, we must conclude that the climate at that time was at least warm and fairly uniform in temperature the world over. We therefore see the force of a statement made to the writer by Walcott some years ago, in a letter, that "the Lower Cambrian fauna and sediments were those of a relatively mild climate uninfluenced by any considerable extent of glacial conditions," and also that "the glacial climate of late Proterozoic time had vanished before the appearance of earliest Cambrian time."

Toward the close of Lower Cambric time there was considerable mountain-making, without apparent volcanic activity, going on all along eastern North America and to a lesser extent in western Europe. These uplifts seemingly had much effect upon the marine life, for the Middle Cambric faunas became more and more provincial in character in comparison with the earlier, more cosmopolitan faunas of Lower Cambric time.

The Archæocyathinæ, which had endured since earliest Proterozoic time, now vanished, and their extinction is suggestive of cooler waters; there was, however, a greater variety of invertebrate forms, more lime-secreting invertebrates, and far more widespread limestone deposition in Middle Cambric time. In the Upper Cambric the brachiopods, gastropods, cephalopods, and bivalve crustaceans were abundantly represented by thick-shelled forms, and in most places throughout North America there was marked deposition of limestones, magnesian limestones, and dolomites, all of which is suggestive of warmer waters.

Ordovicic and Siluric.—The Ordovicic seas from Texas far into the Arctic regions were dominated by limestone deposits and a great profusion of marine life that was also more highly varied than that of any earlier time. The same species of graptolites, brachiopods, bryozoans, trilobites, and other invertebrate classes had a very wide distribution, all of which is evidence that at that time the earth had mild and uniform climates. In the Middle Ordovicic and again late in that period reef corals were common from Alaska to Oklahoma and Texas (Vaughan, 1911).

Toward the close of the Ordovicic, mountain-making was again in progress throughout eastern North America without significant volcanic activity, but in western Europe, where the movements were less marked, volcanoes were more plentiful. The seas were then almost completely withdrawn from the continents, and yet when the Siluric waters again transgressed the lands we find not only the same great profusion and variety of life as before, but as widely extended limestone deposition. The evidence is again that of mild and uniform climates. We can therefore say that the temperatures of a.˙ and water had been mild to warm throughout the world since the beginning of Cambric time; that there was a marked increase of warmth in the Upper Cambric; and that these conditions were maintained throughout the Ordovicic and the earlier half of the Siluric, since shallow-water corals, reef limestones, and very thick dolomites of Siluric time are as common in Arctic America as in the lower latitudes of the United States or Europe.

The Siluric closed with an epoch of sea withdrawal and North America was again arid, for now red shales, gypsum, thick beds of salt, and great flats of sun-cracked water-limestone were the dominant deposits of the vanishing seas. The marine faunas were as a rule scant and the individuals generally under the average size. In North America no marked mountain-making was in progress, but all along western Europe, from Ireland and Scotland across Norway into far Spitzbergen, the Caledonian Mountains were rising. In eastern Maine throughout Middle and Upper Siluric time there were active volcanoes of the explosive type, for here occur vast deposits of ash.

Devonic.—In the succeeding Lower Devonic time the Caledonian intermontane valleys of Scotland and north to at least southern Norway were filling with the Old Red sandstone deposits of a more or less arid climate. On the other hand, the invading seas of northern Europe were small indeed, and their deposits essentially sandstones or sandy shales, but in southern Europe and North America, where the invasions were also small and restricted to the margin of the continent, the deposits were either limestones or calcareous shales. The life of these waters was quite different from that of the earlier and Middle Siluric, and entire stocks had been blotted out in later Siluric time, as is seen best among the graptolites, crinids, brachiopods, and trilobites, while new ones appeared, as the goniatites, dipnoans or lung-fishes, sharks, and the terrible armored marine lung-fishes, the arthrodires.

From this evidence we may conclude that the early Paleozoic mild climates were considerably reduced in temperature toward the close of the Siluric and that even local glaciation may have been present. Refrigeration may have been greatest in the southern hemisphere, where the marine formations of Devonic time are coarse in character and, in Africa, of very limited extent. Corals were scarce or absent here, and in South Africa the glacial deposits of the Table Mountain series may be of late Siluric age; if so, they harmonize with the Caledonian period of mountain-making in the northern hemisphere. Warmer conditions again prevailed in the latter hemisphere early in Middle Devonic times, for coral reefs, limestones, and a highly varied marine life with pteropod accumulations were of wide distribution. On Bear Island workable coal beds were laid down in late Devonic time.

Throughout the Devonic, but more especially in the Lower and Middle Devonic, the entire area of the New England States and the Maritime Provinces of Canada was in the throes of mountain-making, combined with a great deal of volcanic activity. At the same time, many volcanoes were active throughout western Europe.

Carbonic.—The world-wide warm-water condition of the late Devonic seas of the northern hemisphere was continued into those of the Lower Carbonic. These latter seas were also replete with a varied marine life, among which the corals, crinids, blastids, echinids, bryozoans, brachiopods, and primitive sharks played the important rôles. Limestones were abundant and with the corals extended from the United States into Arctic Alaska. Reefs of Syringopora are reported in northern Finland at 67° 55' N., 46° 30' E., on Kanin Peninsula (Ramsay). Even several superposed coal beds, and up to 4 feet in thickness of pure coal, of early Lower Carbonic age, occur at Cape Lisburne, overlain by Lower Carbonic limestones with corals. It is generally held that the world climate at this time was uniformly mild and the many hundred kinds of primitive sharks lead to the same conclusion. There were in the American Devonic 39 species of these sharks, in the Lower Carbonic not less than 288, in the Coal Measures 55, and in the earliest Permic only 10. They had no enemies other than their own kind to fear, and as the same rise and decline occurred also in Europe, we must ask ourselves what was the cause for this rapid dying-out of the ancient sharks during and shortly after early Coal Measures time. With the sharks also vanished most of the crinids, but otherwise there was an abundance and variety of marine life (wide distribution of large foraminifers) with much limestone formation. The vanishing of the sharks does not appear therefore to have been due solely to a reduction of temperature, but may have been further helped by the oscillatory condition and retreat of the late Lower Carbonic seas.

Toward the close of the Lower Carbonic, or after the Culm and its coals of western Europe had been laid down, mountain movements on a great scale began to take place in central Europe, and then were born the Paleozoic Alps of that continent. These mountains, Kayser tells us, were in constant motion but with decreasing intensity throughout the Upper Carbonic, culminating in "a mighty chain of folded mountains." Toward the close of the Upper Carbonic began the rise of the Urals, which was finished in late Permic time when the Paleozoic Alps of Europe were again in motion. These movements are also traceable in Armenia and others are known in central and eastern Asia. Likewise, in America, the southern Appalachians were in movement at the close of the Lower Carbonic, but the greatest of all of the Upper Carbonic thrustings began to take place at the close of the period and culminated apparently in the earlier half of Permic time, when the entire Appalachian system from Newfoundland to Alabama, and the Ouachita Mountains, extending through Arkansas and Oklahoma, arose as majestic ranges anywhere from 3 to 4 miles high.

These mountain-making movements of long duration at first caused the oceans to oscillate frequently back and forth over parts of the continents, and great brackish-water marshes were developed, producing the greatest marsh floras and the greatest accumulations of good coals that the world has had. The paleobotanists White and Knowlton tell us that the climate of Upper Carbonic time was relatively uniform and mild, even subtropical in places, accompanied by high humidity extending to or into the polar circles. Plant associations were then "able to pass from one high latitude to the opposite without meeting an efficient climatic obstruction in the equatorial region" (1910: 760).

The marine faunas of Upper Carbonic time were fairly uniform in development, and many species had a wide distribution, although the biotas were still somewhat provincial in character. Limestones or calcareous shales predominated. The large Protozoa of the family Fusulinidæ occurred throughout the northern hemisphere and less widely in South America. They were also very common in Spitzbergen. Staff and Wedekind (1910) state that the Fusulinidæ occur here in a black asphaltic calcareous rock, i. e., a sapropel like those now forming in marine tropical regions, according to Potonié. The water, they state, was shallow, highly charged with calcium carbonate and of a tropical character, or at the very least not cooler than that of the present Mediterranean. The very large insects of the Coal Measures tell the same climatic story, for Handlirsch (1908: 1152) says that the cockroaches of that time were as long as a finger and the libellids as long as an arm. They were "brutal robbers" and scavengers living in a tropical and subtropical climate, or at the very least in a mild climate devoid of frosts. We therefore conclude that after Middle Devonic time the climate of the world was as a rule uniformly warm and more or less humid and that it remained so to the close of Upper Carbonic time.

During the time of these mild and humid climates vast accumulations of carbon extracted by the plants out of the atmosphere were being stored up in brackish and fresh-water swamps, and even greater quantities of this element were being locked up in the limestones and calcareous shales in the seas and oceans. According to the physico-chemist Arrhenius, and many geologists and paleontologists, so much loss of carbon dioxide and its associated water vapor from the air must have thinned the latter greatly and thus largely reduced the atmospheric blanket and retainer of the sun's heat rays. Therefore they hold that these factors alone were sufficient to have brought on a glacial climate. It may be that this theory will not stand the test of time, but even so we have learned that in Carbonic times there were earth movements on so grand a scale as to be but slightly inferior to those of the late Tertiary that were followed by the Pleistocene glacial climate.

Permic.—Very early in Permic time the mild climate of the past was greatly changed; the evidence is now overwhelming that throughout the southern hemisphere there was a glacial period seemingly of even greater extent than that of the northern hemisphere during the Pleistocene. This evidence is most easily seen in the wide distribution of the tillites and the scratched and polished grounds over which the land ice moved in Africa, Australia, Tasmania, India, and South America. In the northern hemisphere the evidence of ice work is far less marked; but tillites occur near Boston, Massachusetts, and in the Urals, and there is much evidence of thin and arid climates, seen in the widely distributed red formations. Then, too, the land life of this time clearly indicates that a great climatic change had taken place in the environment of the organic world.

The grand cosmopolitan swamp floras of the Upper Carbonic, consisting in the main of spore-bearing plants, such as the horse-tails (Equisetales), the running pines, and club-mosses (Lycopodiales), and the ferns, among which were also many broad-leaved evergreens (Cordaites) and seed-bearing ferns (Cycadofilices), were very largely exterminated in the southern hemisphere at the beginning of Permic time. In the northern hemisphere, however, the older flora maintained itself for a while longer, as best seen in North America, but finally the full effects of the cooled and glacial climates were felt everywhere. Then in later Permic time the old floras completely vanished, except the hardier pecopterids, cycads, and conifers of the northern hemisphere, and with these latter mingled the migrants from the hardy Gangamopteris flora originating in the glacial climate of the southern hemisphere (White, 1907). Some of the trees show distinct annual growth rings, and hence the presence of winters. It was these woody floras that gave rise to the cosmopolitan floras of early Mesozoic time.

With the vanishing of the cosmopolitan coal floras also went nearly all of the Paleozoic insect world of large size and direct development, for the insects of late Permic time were small and prophetic of modern forms. Then, too, they all passed through a metamorphic stage indicating, according to Handlirsch, that the insects of earlier Permic time had learned how to hibernate through the winters in the newly originated larval conditions.

Our knowledge of the land vertebrates of late Paleozoic time is increasing rapidly and it is becoming plainer that great changes were also in progress here. The vertebrates of the Coal Measures, either the armored amphibians (Stegocephalia) or the primitive reptiles, were still largely addicted to the "water habit" and lived in fresh waters or swamps, but this was much changed by the arid climates and vanishing swamps of later Permic times, and in the Triassic we meet with the first truly terrestrial reptilian faunas.

A climatic change naturally must affect the land life more quickly and profoundly than that of the marine waters, for the oceanic areas have stored in themselves a vast amount of warmth that is carried everywhere by the currents. The temperature of the ocean is more or less altered by the changes of climate, be they of latitude or of glaciation. The surface temperatures in the temperate and tropical regions, however, are the last to be affected, and only change when all of the oceanic deeps have been filled with the sinking cold waters brought there by the currents flowing from the glaciated area. We therefore find that the marine life of earlier Permic time was very much like that of the Coal Measures, and that it was not profoundly altered even in the temperate zones of Middle Permic time (Zechstein and Salt Range faunas). Our knowledge of Upper Permic marine life is as yet very limited and will probably always remain so because of the world-wide subtraction of the seas from the lands at that time. It was a period of continued arid climates, and the marginal shallow sea pans were, as a rule, depositing red formations with gypsum, and locally, as in northern Germany, alternations of salt with anhydrite or polyhalite in thicknesses up to 3,395 feet. In certain of these zones there were developed annual rings so regular in sequence as to lead to the inference that they were the depositions of warm summers and cold winters, enduring for at least 5,653 years (Görgey, 1911).

Triassic.—When we examine into the Triassic faunas we meet at once with a wholly new marine assemblage. The late Paleozoic world of fusulinids, tetracorals, crinids, brachiopods, nautilids, and trilobites had either vanished or was represented by a few small and rare forms. On the other side, in the Triassic, their places were taken by a rising marine world of small invertebrates, now hexacorals, regular echinids, modern bivalves (among them the oysters), siphonate gastropods, and more especially by a host of ammonites and a prophecy of the coming of squids and marine reptiles. Truly, there is no greater change recorded in all Historical Geology!

Plants are scarce in the rocks of Triassic time until near its close in the Rhætic, when we can again truly speak of Triassic floras. These are known from many parts of the world, and according to Knowlton there is nothing in the floras to suggest a "depauperate and pinched" condition, as has often been said. "In North Carolina, Virginia and Arizona, there are trunks of trees preserved, some of which are 8 feet in diameter and at least 120 feet long, while hundreds are from 2 to 4 feet in diameter. Many of the ferns [some are tree ferns] are of large size, indicating luxuriant growth, while Equisetum stems 4 to 5 inches in diameter are only approached by a single living South American species. * * * The complete, or nearly complete absence of rings in the tree trunks indicates that there were no, or but slight, seasonal changes due to alterations of hot and cold, or wet and dry periods." On the whole, the climate was "warm, probably at least subtropical" (1910A: 200–2).

Of insects, too few species (27) are known to be of value for climatic deductions. On the other hand, the reptilian life of the Triassic in America, Africa, and Europe was highly varied, and with the dinosaurs dominant and often of large size again gives evidence that appears to be indicative of uniform and mild climate.

The marine Triassic deposits consisted largely of thick limestones, and such are well developed in Arctic America and Arctic Siberia. One of the oldest faunas, known as the Meekoceras fauna, has a very great distribution from Spitzbergen to India and Madagascar, and from Siberia at Vladivostok to California and Idaho. In general, however, the Triassic assemblages were more provincial, and it was not until middle and late Triassic time that the faunas again had wide distribution. Limestones with thick coral reefs, of the same age, appear in the Alps (up to 1,000 meters thick), India, California, Nevada, Oregon, and Arctic Alaska. Smith, from whom most of these facts were taken, states that this shows there was during the Triassic "nearly uniform distribution of warm water over a great part of the globe" (1912A: 397–8).

We may therefore conclude that the rigid climate of the Permic had vanished even before the earliest of Triassic times, and that the climate of the latter period until near its close was again mild and fairly uniform though semiarid or even arid the world over.

Late Triassic-Lias.—Throughout much of late Triassic time there was renewed crustal instability, for we have the evidence of volcanism on a great scale all along the Pacific from central California into far Alaska, in eastern North America from Nova Scotia to Virginia, in Mexico, South America (in southern Brazil 600 meters thick), and New Zealand. The volcanoes of western North America were probably insular in position, for their lavas and ash beds are found interbedded with marine sediments. Just how important this movement was and what effect it had upon the climate is not yet clear, but there is important organic evidence leading to the belief that the temperature was considerably reduced during latest Triassic and earliest Jurassic time.

Pompeckj, Buckman, and Smith state that late Triassic time was a particularly critical one for the ammonites. Of the far more than 1,000 known species of Triassic ammonites, not one passed over into the Jurassic, and but a single family survived this time, the Phylloceratidæ. Pompeckj says that "out of Phylloceras has developed the abundance of Jurassic-Cretaceous ammonites" (1910: 64), while Buckman holds it was out of Nannites by way of the Liassic Cymbites that the later fullness of ammonite development came.

In the Liassic there are now known 415 species of insects that remind one much of modern forms. Nearly all were dwarf species, smaller than similar living insects of the same latitude and far smaller than Paleozoic or Upper Jurassic insects. Handlirsch (1910B) is positive that this uniform dwarfing of the Liassic insects was due to a general reduction of the climate and that the temperature was then cool and like that of present northern Europe between latitudes 46° and 55°. The climate, he states, was certainly cooler than either that of the Middle Triassic or Upper Jurassic.

In this connection we must not overlook the fact that the known Liassic insects are of wide distribution, for 172 species are known from England, 164 from Mecklenburg, northern Germany, 75 from Switzerland, and 2 from upper Austria. With this depauperating of the insects and the vanishing of the late Triassic ammonites, there is also to be noted a marked quantitative reduction and geographic restriction among the reef corals of Liassic time We therefore are seemingly warranted in concluding that the cooling of the climate in late Triassic and early Jurassic time was not local in character, but was rather of a general nature. Much workable coal was also laid down in Liassic time, not only in Hungary but also in many places eastward into China and Japan. In addition, the many black shales of this time furnished further evidence of cool and non-tropical climates; coal and black shales are so general in occurrence throughout the Liassic rocks that the time is often referred to as the Black Jura. Finally, certain Liassic conglomerates of Scotland have been thought by some to be of glacial origin (J. Geikie).

Jurassic.—The Jurassic formations of Europe are so rich in fossils that they have been the classic ground on which many paleontologists and stratigraphers were reared. From the studies of these faunas came the first clear ideas of climatic zones and world paleogeographic maps through the work of the great Neumayr of Vienna. As the result of a very long study of the ammonites and their geographic distribution, he came to the conclusion in 1883 that the earth in Jurassic time had clearly marked equatorial, temperate, and cool polar climates, agreeing in the main with the present occurrence of the same zones. He also said that "the equator and poles could not have very much altered their present position since Jurassic times." His conclusions were, however, assailed by many, and while no one has greatly altered his geographic belts of ammonite distribution, still the consensus of opinion to-day is that these are representative rather of faunal realms than of temperature belts. On the other hand, it is admitted that there were then clearly marked temperature zones--that is, a very wide medial warm-water area, embracing the present equatorial and temperate zones, with cooler but not cold water in the polar areas. That the oceanic waters of Middle and (somewhat less so) of Upper Jurassic times were warm throughout the greater part of the world is seen not only in the very great abundance of marine life—probably not less than 15,000 species are known in the Jurassic—but also in the far northern distribution of many ammonites, reef corals, and marine saurians. The Jurassic often abounds in reefs made by sponges, corals, and bryozoans. Jurassic corals occur 3,000 miles north of their present habitats.

The Jurassic floras were truly cosmopolitan, and Knowlton tells us that of the North American species, excluding the cycad trunks, about half are also found in Japan, Manchuria, Siberia, Spitzbergen, Scandinavia, or England. "What is even more remarkable, the plants found in Louis Philippe Land, 63° S., are practically the same [both generically and specifically] as those of Yorkshire, England. * * * The presence of luxuriant ferns, many of them tree ferns, equisetums of large size, conifers, the descendants of which are now found in southern lands, all point to a moist, warm, probably subtropical climate" (1910A: 204–5). The insects of this time were again large and abundant, indicating a warm climate—evidence in harmony with the plants.

At the close of the Jurassic the Sierra Nevadas of California and the Humboldt Ranges of Nevada were elevated; probably also the Cascade and Klamath Mountains farther north;

but this disturbance seemingly had no marked effect upon the world's climate, though there was a considerable retreat of the seas from the continents.

Cretacic.—The emergence of the continents at the close of the Upper Jurassic gave rise to extensive accumulations of fresh-water deposits, known in western Europe as the Wealden, and in the Rocky Mountain area of North America as the Morrison. These are now regarded as of Lower Cretacic (more accurately Comanchic) age. Along the Atlantic border of the United States occur other continental deposits, known as the Potomac formations, in the upper part of which the modern floras or Angiosperms make their first appearance. Before the close of the Lower Cretacic this early hardwood forest had spread to Alaska and Greenland, where elms, oaks, maples, and magnolias occurred. Knowlton concludes from this evidence that the climate "was certainly much milder than at the present time" and "was at least what we would now call warm temperate" (1910A: 205–6). It was therefore a climate somewhat cooler than that of the Jurassic. On the other hand, the Neocomian series of King Karl's Land has silicified wood, the trunks of which, according to Nathorst, are at least 80 cm. in diameter and show 210 annular rings. These rings are far better developed than in stems of the same age found in Europe, "which indicates that the trees lived in a region where the difference between the seasons was extremely pronounced" (1912: 339).

At this time, in the temperate and tropical belts, the world had the greatest of all land animals, the dinosaurs, reptiles attaining a length in North America of 75 feet or more and in equatorial German East Africa of probably 125 feet. Their bones range to 50° N. latitude, and the animals must have lived in a fairly warm and moist climate.

While the Lower Cretacic seas were prolific in life, the most characteristic shellfish of southern Europe, the Mediterranean countries, and Mexico, were the limestone-making rudistids, large ground-living foraminifers (Orbitolina), and reef corals. In northern Europe and in the United States from southern Texas to Kansas, nothing of these warm-water faunal elements is known. It is recognized that the north European seas had Arctic connections by way of Scandinavia and Russia, and along the west coast of North America are seen many other boreal migrants as far south as California and even Mexico. These waters, however, were not cold. The same geographic distribution prevailed in the Upper Cretacic of Europe. This distribution was first noted in Texas by Ferdinand Roemer in 1852, and he further observed that "in each case the European deposit is approximately 10° farther north than its American analogue," and concluded "that the differences between the northern and southern facies were due to climate and that the climatic relations between the two sides of the Atlantic were about the same in Cretaceous time as they are now" (Stanton, 1910: 67). Even though Roemer's conclusion as to climatic zones was founded on erroneous stratigraphic correlations, still his theory has long been looked upon favorably, but in 1908 Gothan showed that the fossil woods of the late Upper Cretacic of central Germany have distinct annual rings, while those of Egypt do not have a trace of them. The late Cretacic woods of Spitzbergen also have decided growth rings. Berry (1912) states that the climate of Upper Cretacic time was far more uniform than now and that there was an increase of warmth southward, Alabama having then a climate that was subtropical or even tropical. On the other hand, the early Upper Cretacic or Cenomanian flora of Atane in western Greenland, according to Nathorst, "is particularly rich in the leaves of Dicotyledonous trees, among which are found those of planes, tulip trees, and bread fruits, the last mentioned closely resembling those of the bread-fruit tree (*Artocarpus incisa*) of the islands of the southern seas" (1912: 340).

In Middle Cretacic times the oceans began again to spread over the continents and this transgression of the seas was one of the greatest of the geologic past. It is interesting to note that even though there was great opportunity for expansive evolution, but few new marine stocks appeared here, and it was rather a time of death to many characteristic stocks. This well-known fact is clearly brought out by Walther in his interesting book, "Geschichte der Erde und des Lebens" (1908), in Chapter 26, entitled "Cretaceous time and its great mortality." Entire stocks of specialized forms vanished, just as did other stocks at the close of the Paleozoic. In late Cretacic time it was the ammonites, belemnites, the rudistids that began to develop in great numbers in the Lower Cretacic, and the other thick-shelled large bivalves (Inoceramus) that perished. In addition, there was a great reduction among the reef corals, the replacing of the dominant ganoids by the teleosts or bony fishes, and, finally, the complete dying out of the various stocks of marine saurians.

On the land, with the further rise of the Angiosperm floras, we see the vanishing of the reptilian dragons known as pterodactyls, and, at the very close of the Cretacic, the last of the large and small dinosaurs and the birds with teeth. "We thus see the reptiles displaced from the seas by the fishes; on the land they are restricted by the rise of the mammals, in the air after a short struggle by the more finely organized birds—in short, the reptilian dominance is destroyed with the end of the Mesozoic era, in which entire time they were the characteristic feature" (Koken, 1893: 436).

The Upper Cretacic was therefore a time of great mortality among animals, "here sooner, there later; although numerous relict faunas are preserved for a time and last into the Cenozoic, still there never was so great a mortality as that taking place toward the close of the Cretacic" (Walther, 1908: 449).

During the Upper Cretacic, but more especially toward the close of the period, mountain-making on a vast scale went on, along with exceptional outpourings of lavas and ashes. These movements, though of less intensity, were repeated in early Tertiary times, and while they were equaled only by those of the closing period of the Paleozoic, they were exceeded by the crustal deformation of late Tertiary time; they form the Laramide revolution of Dana, embracing the mountains of western North and South America from Cape Horn to Alaska and the reëlevation of the Appalachian and Antillean Mountains. Throughout the Eocene in the Rocky Mountains there were many volcanoes throwing out immense quantities of ashes in which is entombed a remarkable vertebrate fauna. Then in late Cretacic time in peninsular India occurred the Deccan lava flows, the most stupendous eruptions known to geologists, covering an area of 200,000 square miles, in thickness anywhere up to a mile or more.

Although there were these great crustal movements toward the close of the Upper Cretacic, nevertheless they seem to have had no marked effect on the climates of the world, for nowhere has anyone shown the presence of unmistakable glacial tills of this age.[*] Then, too, the floras of early Tertiary times are said to be of about the same character as those of the late Cretacic and they indicate that the climates were warm with slight latitudinal variation, so slight that even in Greenland and Spitzbergen the early Tertiary floras were those of a moist and mild climate.

Tertiary.—We have seen that there was no marked climatic change in the time from the Cretacic to the Eocene, but that there was a reduction in temperature is admitted by paleobotanists and students of marine life. Berry states that the Middle Eocene floras of Europe "show many tropical characters absent in the earlier Eocene" (1910: 205). The Oligocene marine faunas were prolific in species, and the largest of all foraminifers, the nummulites, although still present at this time, had their widest distribution and largest species in the Middle Eocene and especially in the Tethyian Sea of the Old World, extending from 20° S. to 20° N. latitude (Stromer, 1909: 42).

In Miocene time on Spitzbergen (Cape Staratschin) lived the swamp cypress (*Taxodium distichum miocenum*), a leafy sequoia, pines and firs, besides various hardwood trees, such

---

[*] At the Princeton meeting of the Geological Society of America, December 29, 1913, Professor W. W. Atwood announced the discovery of a tillite about 90 feet thick in the San Juan Mountains of southwestern Colorado. The age of these glacial deposits is somewhere between late Cretacic and late Eocene. We therefore are now on the road to finding the physical evidence of a reduced climate during or following the close of the Laramide revolution.

as poplars, birches, beeches, oaks, elms, magnolias, limes, and maples.   The swamp cypress, Nathorst says, "formed forests, as in the swamps in the southern portion of the United States.   This conclusion is also confirmed by the occurrence of the remains of rather numerous insects" (1912: 341).   All of the plants mentioned then flourished as far north as 79° N. latitude, and even at nearly 82° in Grinnell Land.   This is evidence that in early Miocene time the climate was at least warm-temperate in Arctic America.

Again, Dall (1895) states that in Middle Miocene time considerable reduction of the climate appeared, for the Atlantic Chesapeake faunas were those of temperate waters and they spread southward as far as the eastern area of the Gulf of Mexico.   Similar conditions are noted by the same conchologist in the northern Pacific Ocean.   He says:

"The conditions indicated by the faunas of the post-Eocene Tertiary on the Pacific Coast from Oregon northward are a cool temperate climate in the early and Middle Miocene, a warming up toward the end of the Miocene culminating in a decidedly more warm-water fauna in the Pliocene, and a return to cold if not practically Arctic temperatures in the Pleistocene" (1907: 457–8).

The Tertiary was an era of extraordinary crustal movements, finally resulting in the greatest mountain chains of all geologic time.   These movements began in early Eocene time in the Rocky Mountains and at the close of this epoch further deformation took place in the Klamath and Coast Ranges of Oregon and the Santa Cruz Mountains of California. In Europe the elevations of Tertiary time started at the close of the Eocene in the Pyrenees, and in the Miocene the entire "Alpine system" was in elevation.   This unrest spread at the same time to the Caucasus, Asia, and to the entire Himalayan region of highest mountains and elevated plateaus, an area 22° of latitude in width.   It is probable that all of the world's great mountain chains were more or less reëlevated in Miocene and Pliocene times, resulting in the present abnormally high stand of the continents when contrasted with the oceanic mean level.

These elevations also altered the continental connections, for North and South America were reunited in Miocene times, and western Europe, Greenland, and America were severed late in the Tertiary era, the exact time being as yet not clearly established.   With these great changes also must have come about marked alterations in the oceanic currents and, as a consequence, in the distribution of heat and moisture over vast areas of the northern Atlantic lands.   It is admitted by all paleontologists that the marine waters of late Pliocene times in the Arctic region were cool, and the widespread glacial tills of the northern hemisphere are evidence of a glacial climate of varying intensity throughout Pleistocene time.

CONCLUSIONS.

Our studies of the paleometeorology* of the earth are summed up in figure 90.   We have seen that two marked glacial periods are clearly established.   The one best known was of Pleistocene time and the other, less well known in detail, of earliest Permic time.   Both were world-wide in their effects, reducing the mean temperatures sufficiently to allow of vast accumulations of snow and ice, not only at high altitudes, but even more markedly at low levels, with the glaciers in many places attaining the sea.   We also learn that the continental glaciers of Pleistocene time were dominant in the polar regions, while those of Permic time had their greatest spread from 20° to 40° south of the present equator, and to a far less extent between 20° and 40° in the other hemisphere.   There is also some evidence of glaciers in equatorial Africa in Permic time.   We may further state that, although Pleistocene glaciation was general in the Arctic region, there certainly was none at this pole in early Permic time, because of the widespread and abundant marine faunas that are not markedly unlike those of the Upper Carbonic; as for the south pole, our knowledge of pre-Pleistocene glaciation is as yet a blank.

* H. F. Osborn, Compte Rendu, Congrès Internat. Zool., Berne, 1904, 1905: 88.   For a review of the papers treating of paleometeorology, see M. Semper, Geol. Rundschau, I, 1910: 57–80.

A glacial period does not appear to remain constantly cold, but fluctuates between cold glacial climates and warmer interglacial times of varying duration. During the Pleistocene there were, according to the best glaciologists, at least three, if not four, such warmer intervals. The Permic glacial period also had its warmer times, while the interbedded red strata of the Proterozoic tillites seem to point to the same variability. It is this decided temperature fluctuation during the glacial periods that is so very difficult to explain.

In addition to the well-known Pleistocene and Permic glaciation, there is rapidly accumulating a great deal of evidence to the effect that there were at least two and probably three other periods of widespread glacial climates. All of these were geologically very ancient, earlier than the Paleozoic; in fact, one was at or near the close of Proterozoic time,

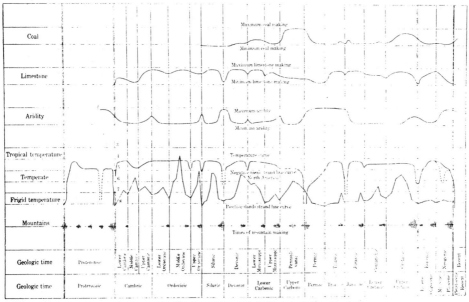

FIG. 90.—Chart of Geological Climates. Paleometeorology.

while another was at the very beginning of that era and almost at the beginning of earth history as known to geologists.

The oldest of all glacial materials occurs at the base of the Lower Huronian and is of great extent in Canada. Seemingly of the same time is the Torridonian glacial testimony of northwest Scotland. The Proterozoic tillites of China in latitude 31° N. may also be of this time. If these correlations are correct, then the oldest glacial evidence indicates that a greatly cooled climate prevailed near the very beginning of the known geologic record and that it was dominant in the northern hemisphere.

Toward or at the close of the Proterozoic there is other evidence of a glacial climate in Australia, Tasmania, and Norway. These occurrences of tillites lie immediately beneath Lower Cambrian fossiliferous marine strata and probably are of pre-Cambrian age.

In India there is also evidence of late Proterozoic tillites in two widely separated places, and it may be that the inadequately studied Keweenawan testimony of the Lake Superior region is of this time. If so, these occurrences record a distribution of glacial materials very similar to that of Permic time. Again, the Proterozoic tillites of Africa are clearly of another age, so that there is evidence of at least three periods of glaciation previous to the Paleozoic.

The physical evidence of former glacial climates is even yet not exhausted, for the Table Mountain tillites of South Africa point to a cold climate that apparently occurred, at least locally, late in Siluric time. Finally, there may have been a seventh cool period in early Jurassic time (Lias), but the biologic evidence so far at hand indicates that it was the least significant among the seven probable cool to cold climates so far discovered in the geologic record.

The data at hand show that the earth since the beginning of geologic history has periodically undergone more or less widespread glaciation and that the cold climates have been of short geologic duration. So far as known, there were seven periods of decided temperature changes and of these at least four were glacial climates. The greatest intensity of these reduced temperatures varied between the hemispheres, for in earliest Proterozoic and Pleistocene time it lay in the northern, while in late Proterozoic and Permic time it was more equatorial than boreal. The three other probable periods of cooled climates are as yet too little known to make out their centers of greatest intensity.

Of the four more or less well-determined glacial periods, at least three (the earliest Proterozoic, Permic, and Pleistocene) occurred during or directly after times of intensive mountain-making, while the fourth (late Proterozoic) apparently also followed a period of elevation. The Table Mountain tillites of South Africa, if correctly correlated, fall in with the time of the making of the great Caledonian Mountains in the northern hemisphere. On the other hand, the very marked and world-wide mountain-making period, with decided volcanic activity, during late Mesozoic and earliest Eocene times, was not accompanied by a glacial climate, but only by a cooled one. The cooled period of the Liassic also followed a mountain-making period, that of late Triassic time. We may therefore state that cooled and cold climates, as a rule, occur during or immediately follow periods of marked mountain-making—a conclusion also arrived at independently by Ramsay (1910: 27).

Geologists are beginning to see clearly that the lands have been periodically flooded by the oceans, and the times of maximum submergence and emergence of the continents since earliest Paleozoic time are fairly well known. The two marked glacial periods since Cambric time (Permic and Pleistocene) and the three other more or less cooled climates (late Siluric, Liassic, and late Cretacic) all fall in with the times when the continents were more or less extensively and highly emergent. There were no cold climates when the continents were flooded by the oceans, and it may be added that the periods of widespread limestone-making preceded and followed, but did not accompany, the reduced climates. On the other hand, the periods of greatest coal-making (Upper Carbonic and Upper Cretacic) accompanied the time of greatest continental flooding and preceded the appearance of cooled climates.

The more or less coarse red sediments seen at many horizons of the geologic column are interpreted as the deposits of variably arid climates, or those that are alternately wet and dry. In the Paleozoic they are seen more often at the close of the periods when the seas were temporarily withdrawn and the lands were most extensive. These red deposits alternate with formations that are either wholly marine or of brackish-water origin, and in the latter case of gray, green, blue, or black color.

Humphreys has shown that volcanic dust in the isothermal region of the earth's atmosphere does appreciably reduce the temperature at the surface of the globe. It is thought that if explosive volcanoes continued active through a more or less long geologic time, this factor alone would bring on, or largely assist in bringing on, a more reduced temperature or even a glacial climate. If then, we may further postulate that volcanic activity is most marked during times of mountain-making, *i. e.*, during the "critical periods" at the close of the eras and the less violent movements at the close of the periods, we should expect ice ages, or at least considerably cooled climates, occurring here also. Let us see how the facts agree with this hypothesis.

Of the "critical periods" at the close of the Paleozoic, Mesozoic, and Cenozoic eras, we know that the first and last were accompanied by glacial climates, but the Mesozoic, though a time of very extens've mountain-making and great and prolonged volcanic activity in North America, did not close with a glacial, but only with a slightly cooled climate. Not only this, but we find that volcanism was renewed in the Cordilleras of North America throughout much of the Eocene, and yet there was developed no glacial climate at this time.* In the same way the marked temperature reduction at the close of the Cenozoic in the Pleistocene was subsequent to the Miocene and Pliocene movements of this period and not coincident with them, while that of the Paleozoic appears to fall in with the rise of the Urals and Appalachians, though but little volcanism seems to have accompanied the movements in North America. It should also be said that equally extensive movements were going on in Europe in the rise of the European Alps during the geologic times before and after the Permic glaciation, and that the earlier movements did not appreciably affect the climate.

Again, there was decided mountain-making toward the close of the Siluric in the formation of the Caledonian Mountains all along western Europe from Spitzbergen to Scotland, with marked volcanic extrusions during the Siluric and early Devonic in Maine, the Maritime Provinces of Canada, and Europe. Yet we have no glacial climate at these times, certainly not in the northern hemisphere; rather it seems that the temperature was mild the world over. It is possible, however, that the Table Mountain tillites of South Africa may coincide with this time, and if so a colder temperature affected the southern hemisphere only locally.

On the other hand, the "life thermometer" indicates a cooled period at the close of the Triassic and the following Liassic, but this reduction of temperature, again, is geologically subsequent to, rather than coincident with the marked volcanic activity of the Triassic in many widely separated places.

Finally, there were earth movements of considerable magnitude at the close of the Lower Cambric, Ordovicic, and Jurassic that were not accompanied by glacial climates. At all of these times there appears, however, to have been a drop in temperature, slight for the two first-mentioned periods and more marked for the third one, for here we find in the austral region, during earliest Cretacic times, winters alternating with summers.

We may therefore conclude that volcanic dust in the isothermal region of the earth does not appear to be a primary factor in bringing on glacial climates. On the other hand, it can not be denied that such periodically formed blankets against the sun's radiation may have assisted in cooling the climates during some of the periods when the continents were highly emergent.

It has long been known that during times of intensive mountain-making and more or less cooled climates there was great destruction and alteration of life. The first effects of the environmental changes occurred among the organisms of the land, while the climax of alteration among the marine life appeared later. This is especially well seen in the Permic glaciation, which first blotted out the cosmopolitan Upper Carbonic flora and the insects, while the life of the sea continued without marked change into Middle Permic time. In the later Permic, in the northern equatorial waters of Tethys, occurred the final destruction of many stocks that had long dominated the Paleozoic seas. The explanation of these facts appears to be that on the lands the change of climate takes immediate effect on the organisms, while in the oceans a longer time is consumed in cooling down the warm and

---

* See footnote, page 283.

equable temperature and in filling all the basins with cold water. Accordingly the last regions in the oceans to come under the influence of glacial climates must be the shallow waters of the equatorial area. The proof of this conclusion is seen in that the last stand made by the marine Paleozoic world is recorded in the deposits of Tethys, the great Mediterranean sea of Permic time. It is also here that we find nearly all of the Paleozoic shallow-water hold-overs in the succeeding period, the Triassic.

The cooled but not frigid climate that followed the magnificent mountain-making at the close of the Cretacic also produced striking changes in the organic world. These changes were less marked than those of Permic time and more noticeable among the land animals than those of the marine waters, affecting especially the over-specialized, large, thick-shelled, and degenerate stocks.

Great changes were again produced among the large land animals of the world, as well as among those of the polar and temperate oceanic waters, by the glaciation of Pleistocene time. The present shallow waters of the equatorial region still maintain the late Tertiary faunas, and Africa is the asylum where the higher Pliocene land animals have been preserved into our time.

What the effects of the Proterozoic glacial climates were upon the living world of that time it is impossible to say, because we have as yet discovered but little of the organic record. The apparently sudden appearance of life at the base of the Cambric is partially explained by the widespread absence of the marine Proterozoic record, an era during which the nuclear portions of the continents appear to have been decidedly emergent for a very long time.

The marine "life thermometer" indicates vast stretches of time of mild to warm and equable temperatures, with but slight zonal differences between the equator and the poles. The great bulk of marine fossils are those of the shallow seas, and the evolutionary changes recorded in these "medals of creation" are slight throughout eternities of time that are punctuated by short but decisive periods of cooled waters and great mortality, followed by quick evolution, and the rise of new stocks. The times of less warmth are the *miotherm* and those of greater heat the *pliotherm* periods of Ramsay (1910: 15).

On the land the story of the climatic changes is different, but in general the equability of the temperature simulates that of the oceanic areas. In other words, the lands also had long-enduring times of mild to warm climates. Into the problem of land climates, however, enter other factors that are absent in the oceanic regions, and these have great influence upon the climates of the continents. Most important of these is the periodic warm-water inundation of the continents by the oceans, causing insular climates that are milder and moister. With the vanishing of the floods somewhat cooler and certainly drier climates are produced. The effects of these periodic floods must not be underestimated, for the North American continent was variably submerged at least seventeen times, and over an area of from 154,000 to 4,000,000 square miles (Schuchert, 1910: 601).

When to these factors is added the effect upon the climate caused by the periodic rising of mountain chains, it is at once apparent that the lands must have had constantly varying climates. In general the temperature fluctuations seem to have been slight, but geographically the climates varied between mild to warm pluvial, and mild to cool arid. The arid factor has been of the greatest import to the organic world of the lands. Further, when to all of these causes is added the fact that during emergent periods the formerly isolated lands were connected by land bridges, permitting intermigration of the land floras and faunas, with the introduction of their parasites and parasitic diseases,* we learn that while the climatic environment is of fundamental importance it is not the only cause for the more rapid evolution of terrestrial life. Unfortunately, the record of land life,

---

* This subject is fully discussed by R. T. Eccles, M.D., in the following papers: "Parasitism and Natural Selection," "Importance of Disease in Plant and Animal Evolution," "The Scope of Disease," and "Disease and Genetics." Medical Record for July 31, 1909; March 16, 1912; March 8, and August 2, 1913.

and especially of the animal world, is the most imperfect of all paleontologic records until we come to Tertiary time. The known mammal history is a vast one and, although very difficult to interpret from the climatic standpoint, we have in the work of Depéret (1909) and Osborn (1910) glimpses into the many temperature fluctuations, faunal isolations, and intercontinental radiations of Tertiary time. The history of the Tertiary is the last one of at least three previous and similar records (Mesozoic, later and earlier Paleozoic) of vastly longer eras, taking us back to a time when the lands were without visible life.

In conclusion, it is seemingly clear that the variability in the storage of solar radiation by the earth's atmospheric blanket and by oceanic waters, and the consequent climatic variations of the past and present are due in the main to topographic changes in the earth's crust. These telluric changes alter the configuration of the continents and oceans, the air currents (moist or dry), the oceanic currents (warm, mild, or cool), and the volcanic ash-content of the atmosphere.

On the other hand, a great deal has been written about the supply and consumption of the carbonic acid of the air as the primary cause for the storage of warmth by the atmospheric blanket. A greater supply of carbon dioxide is said to cause increase of temperature, and a marked subtraction of it will bring on a glacial climate. This aspect of the climatic problem is altogether too large and important to be entered upon here. It is permissible to state, however, that the glacial climates are irregular in their geologic appearance, are variable latitudinally, as is seen in the geographic distribution of the tillites between the poles and the equatorial region, and finally that they appear in geologic time as if suddenly introduced. These differences do not seem to the writer to be conditioned in the main by a greater or smaller amount of carbon dioxide in the atmosphere, for if this gas is so strong a controlling factor, it would seem that at least the glacial climates should not be of such quick development. On the other hand, an enormous amount of carbon dioxide was consumed in the vast limestones and coals of the Cretacic, with no glacial climate as a result; though it must be admitted that the great limestone and vaster coal accumulations of the Pennsylvanic were quickly followed by the Permic glaciation. Again it may be stated that the Pleistocene cold period was preceded in the Miocene and Pliocene by far smaller areas of known accumulations of limestone and coal than during either the Pennsylvanic or Cretacic, and yet a severe glacial climate followed.

Briefly, then, we may conclude that the markedly varying climates of the past seem to be due primarily to periodic changes in the topographic form of the earth's surface, plus variations in the amount of heat stored by the oceans. The causation for the warmer interglacial climates is the most difficult of all to explain, and it is here that factors other than those mentioned may enter.

Granting all this, there still seems to lie back of all these theories a greater question connected with the major changes in paleometeorology. This is: What is it that forces the earth's topography to change with varying intensity at irregularly rhythmic intervals? This difficult and elusive problem the older geologists solved with a great deal of assurance by saying that such change was due to a cooling earth, resulting in periodic shrinkage; but the amount of shrinkage that would necessarily have taken place to account for all the wrinklings and overthrustings of the earth's crust during geologic time would be far greater than that which has apparently occurred. Further, a cooling earth is yet to be demonstrated. Again, some paleogeographers seem to see a periodic heaping up of the oceanic waters in the equatorial region and a pulsatory flowing away later toward the poles. If these observations are not misleading, are we not forced to conclude that the earth's shape changes periodically in response to gravitative forces that alter the body-form?

[*Editor's Summary of Pages 290–296:* These pages are "Supplementary notes on glaciation before the Permic Period" and are mainly quoted passages from original references.]

BIBLIOGRAPHY.

PERMIC GLACIATION.

CHAMBERLIN, T. C., and SALISBURY, R. D., 1906. Geology. II: 632–639, 655–677.

COLEMAN, A. P., 1908. Glacial periods and their bearing on geological theories. Bull. Geol. Soc. America, 19: 347–366.

DAVID, T. W. E., 1896. Evidences of glacial action in Australia in Permo-Carboniferous time. Quart. Jour. Geol. Soc. London, 52: 289–301.

——, 1907. Conditions of climate at different geological epochs, with special reference to glacial epochs. Compte Rendu, Congrès Géol. Internat., Mexico: 449–482.

HATCH, F. H., and CORSTORPHINE, G. S., 1909. The geology of South Africa. 2d ed.: 219–243, 335–339.

KAYSER, E., 1911. Lehrbuch der Geologie. 4th ed., II: 263–318.

KOKEN, E., 1907. Indisches Perm und die permische Eiszeit. N. Jahrb. für Min., Geol., etc., Festband: 446–546.

RAMSAY, A. C., 1855. On the occurrence of angular, subangular, polished, and striated fragments and boulders in the Permian breccia of Shropshire, etc. Quart. Jour. Geol. Soc. London, 11: 185–205.

SAYLES, R. W., and LA FORGE, L., 1910. The glacial origin of the Roxbury conglomerate. Science, n.s., 32: 723–724.

SCHWARZ, E. H. L., 1906. The three Palæozoic ice-ages of South Africa. Jour. Geol., 14: 683–691.

WHITE, I. C., 1908. Final report of the chief of the Brazilian Coal Commission (Commissão de Estudos das Minas de Carvão de Pedra do Brazil): 11–14, 29–55, 227–233.

WOODWORTH, J. B., 1912A. Boulder beds of the Caney shales at Talihina, Oklahoma. Bull. Geol. Soc. America, 23: 457–462.

——, 1912B. Geological expedition to Brazil and Chile, 1908–1909. Bull. Mus. Comp. Zool., 56: 79–82.

DEVONIC GLACIATION.

GEIKIE, A., 1903. Text book of geology. 4th ed., II: 1001, 1011.

HATCH, F. H., and CORSTORPHINE, G. S., 1909. The geology of South Africa. 2d ed.: 62–78.

ROGERS, A. W., 1905. An introduction to the geology of Cape Colony: 94–121.

ROGERS, A. W., and SCHWARZ, E. H. L., 1901. Report on the geology of the Cederbergen and adjoining country. Ann. Rept. Geol. Comm. Cape Colony, 1898: 65–82.

SCHWARZ, E. H. L., 1906. The three Palæozoic ice-ages of South Africa. Jour. Geol., 14: 683–691.

PROTEROZOIC GLACIATION.

COLEMAN, A. P., 1907. A Lower Huronian ice-age. Amer. Jour. Sci. (4), 23: 187–192.

——, 1908A. Glacial periods and their bearing on geological theories. Bull. Geol. Soc. America, 19: 347–366.

——, 1908B. The Lower Huronian ice-age. Jour. Geol., 16: 149–158.

——, 1912. The Lower Huronian ice-age. Compte Rendu, Congrès Géol. Internat., Stockholm: 1069–1072.

DAVID, T. W. E., 1907. Glaciation in Lower Cambrian, possibly in pre-Cambrian time. Compte Rendu, Congrès Géol. Internat., Mexico: 271–274; Conditions of climate at different geological epochs, with special reference to glacial epochs, *Ibid.*: 440–446.

GEIKIE, A., 1880. A fragment of primeval Europe. Nature, 22: 400–403.

——, 1903. Text book of geology. 4th ed., II: 891, 899, 1309.

HOWCHIN, W., 1908. Glacial beds of Cambrian age in South Australia. Quart. Jour. Geol. Soc. London, 54: 234–259.

——, 1912. Australian glaciations. Jour. Geol., 20: 193–227.

MILLER, W. G., 1911. A geological trip in Scotland. Canadian Mining Jour., Feb. 15 and March 1.

PEACH, B. N., 1912. The relation between the Cambrian faunas of Scotland and North America. Nature, 90: 49–56.

✗REUSCH, H., 1891. Det nordlige Norges geologi. Norges geol. Undersögelse: 26–34.

SCHWARZ, E. H. L., 1906. The three Palæozoic ice-ages of South Africa. Jour. Geol., 14: 683–691.

STRAHAN, A., 1897. On glacial phenomena of Palæozoic age in the Varanger Fiord. Quart. Jour. Geol. Soc. London, 53: 137–146.

VREDENBURG, E. W., 1907. A summary of the geology of India: 19–23.

WILLIS, B., and BLACKWELDER, E., 1907, 1909. Research in China. Carn. Inst. Wash. Pub. No. 54, I, Pt. I: 264–269; II: 39–40.

GENERAL.

ADAMS, F. D., and BARLOW, A. E., 1910. Geology of the Haliburton and Bancroft areas (Ontario). Geol. Surv. Canada, Mem. 6.

BARRELL, J., 1907. Origin and significance of the Mauch Chunk shale. Bull. Geol. Soc. America, 18: 449–476.

———, 1908. Relations between climate and terrestrial deposits. Jour. Geol., 16: 159–190, 255–295, 363–384.

———, 1912. Criteria for the recognition of ancient delta deposits. Bull. Geol. Soc. America, 23: 377–446.

BASTIN, E. S., 1910. Origin of certain Adirondack graphite-deposits. Econ. Geol., 5: 134–157.

BERRY, E. W., 1910. An Eocene flora in Georgia and the indicated physical conditions. Bot. Gazette, 50: 202–208.

———, 1912. Contributions to the Mesozoic flora of the Atlantic coastal plain. VIII. Texas. Bull. Torrey Bot. Club, 39: 387–406.

CLARKE, F. W., 1911. The data of geochemistry. Bull. 491, U. S. Geol. Surv.

DALL, W. H., 1895. Contributions to the Tertiary fauna of Florida. Trans. Wagner Free Inst. Science, III, Pt. 6: 1547–1551.

———, 1904. The relations of the Miocene of Maryland to that of other regions and to the recent fauna. Maryland Geol. Surv., Miocene volume: CXXXIX–CLV.

———, 1907. On climatic conditions at Nome, Alaska, during the Pliocene. Amer. Jour. Sci., (4), 23: 457–458.

DEPÉRET, C., 1909. The transformations of the animal world.

GÖRGEY, R., 1911. Die Entwickelung der Lehre von den Salzlagerstätten. Geol. Rundschau, 2: 278–302.

HANDLIRSCH, A., 1908. Die fossilen Insekten.

———, 1910A. Das erste fossile Insekt aus dem Oberkarbon Westfalens. Verh. d. k. k. zool.-bot. Gesell. Wien, 60: 177–183.

———, 1910B. Die Bedeutung der fossilen Insekten für die Geologie. Mitt. Geol. Gesell. Wien, 3: 503–522.

KAYSER, E., 1911. Lehrbuch der Geologie. 4th ed., II: 471–472.

KNOWLTON, F. H., 1910A. In WILLIS and SALISBURY, Outlines of geologic history, chapter 10.

———, 1910B. The Jurassic age of the "Jurassic flora of Oregon." Amer. Jour. Sci. (4), 30: 33–64.

———, 1910C. Biologic principles of paleogeography. Pop. Sci. Monthly, 76: 601–603.

KOKEN, E., 1893. Die Vorwelt und ihre Entwickelungsgeschichte.

LAWSON, A. C., and WALCOTT, C. D., 1912. The geology of Steeprock Lake, Ontario; Notes on the fossils from limestone of Steeprock Lake, Ontario. Geol. Surv. Canada, Mem. 28.

NATHORST, A. G., 1912. On the value of the fossil floras of the Arctic regions as evidence of geological climates. Ann. Rep. Smithsonian Inst. for 1911: 335–344.

NEUMAYR, M., 1883. Ueber klimatische Zonen während der Jura- und Kreidezeit. Denk. d. k. Akad. d. Wiss., Math.-Nat. Classe, Wien, 47: 277–310.

OSBORN, H. F., 1910. The age of mammals in Europe, Asia, and North America.

POMPECKJ, J. F., 1910. Zur Rassenpersistenz der Ammoniten. Geol. Abth. d. Naturh. Gesell. zu Hannover: 63–83.

RAMSAY, W., 1910. Orogenesis und Klima. Oversigt af Finska Vet.-Soc. Forhandl., 52: 1–48.

ROEMER, F., 1852. Die Kreidebildungen von Texas.

SCHUCHERT, C., 1910. Paleogeography of North America. Bull. Geol. Soc. America, 20: 427–606.

SMITH, J. P., 1912A. Ancient portals of the earth. Pop. Sci. Monthly, 79: 393–399.

———, 1912B. The occurrence of coral reefs in the Triassic of North America. Amer. Jour. Sci. (4), 33: 92–96.

STAFF, H. VON, and WEDEKIND, R., 1910. Der Oberkarbone Foraminiferen-Sapropelit Spitzbergens. Bull. Geol. Inst. Upsala, 10: 81–123.

STANTON, T. W., 1910. Paleontologic evidences of climate. Pop. Sci. Monthly, 77: 67–70.

STROMER, E., 1909. Lehrbuch der Palaeozoologie.

VAUGHAN, T. W., 1910. A contribution to the geologic history of the Floridian plateau. Carn. Inst. Wash. Pub. No. 133: 99–195.

———, 1911. Physical conditions under which Paleozoic coral reefs were formed. Bull. Geol. Soc. America, 22: 238–252.

WALCOTT, C. D., 1910. Abrupt appearance of the Cambrian fauna on the North American continent. Smithsonian Misc. Coll., 57: 1–16.

WALTHER, J., 1908. Geschichte der Erde und des Lebens.

WHITE, D., 1907. Permo-Carboniferous climatic changes in South America. Jour. Geol., 15: 615–633.

———, and KNOWLTON, F. H., 1910. Evidences of paleobotany as to geological climate. Science, n.s., 31: 760.

28

Reprinted from *Contr. Mus. Paleont., Univ. Mich.*, **13**(8), 181–210 (1959)

CLIMATIC CHANGES OF THE PAST AND PRESENT*

BY

ERLING DORF[1]

CONTENTS

INTRODUCTION

INHABITANTS of the so-called temperate zone are quite familiar with both unusual weather and rapidly changing weather conditions. Lately, however, even the climate—that is, the composite weather conditions over a period of years—has seemed somewhat unusual and the "reality" of changing climatic conditions has become not only apparent, but even newsworthy. Some people may be old enough to have recollections of "the good old days" when the climate was different: the winters at least seemed to have been much colder and the snows much deeper than they are today. Although reliable meteorological records do not go back very far, they do seem to show that major climatic changes rather than minor fluctuations are taking place. Present conditions *are* warmer than they were in the latter half of the 19th century.

---

* Based on the 1957 Ermine Cowles Case Memorial Lecture delivered before the Society of Sigma Xi, University of Michigan, November 13, 1957.
[1] Department of Geology, Princeton University.

In terms of the remote geologic past, however, today's climate is actually unusually cold. In fact, we are still living in a "glacial age" compared to the much warmer conditions of 35 to 40 million years ago. An examination of Figure 1 will show that, since the beginning of the Cambrian Period, 500 million years ago, the earth's climate has been considerably warmer than at present about two-thirds of the time. The evidence indicates that major climatic changes from this warmer "non-glacial" to much colder "glacial" climate have occurred several times during geologic history. The world today is in one of these glacial episodes, which only a few thousand years ago was frigid enough to support an ice sheet over most of Canada and as far south as the northern United States. Furthermore, it may be noted (Fig. 1) that the cooling trend which culminated in the most recent glacial episode really started back in the Oligocene Epoch and not, as some people still believe, in the Late Pliocene.

In the present paper three major topics are discussed: (1) the climatic changes of the remote geologic past from about 50 million to about 12,000 years ago, that is, from about the beginning of the Eocene Epoch to the end of the Pleistocene "Ice Age"; (2) the climatic changes of the immediate past, from the end of the "Ice Age" to the present; and (3) a prediction of the possible climatic changes of the near future and of the next few thousand years.

CLIMATIC CHANGES OF THE GEOLOGIC PAST

Methods of Study

One may ask: how is it possible for geologists to reconstruct climates of thousands or even millions of years ago? Since climates are conditions and not, therefore, subject to fossilization, the evidence for depicting ancient climates must come entirely by inference from whatever clues are available in the geologic record. Throughout their work geologists rely at the outset on the doctrine that the present is the key to the past. It is supposed that fossil plants and animals lived under approximately the same climatic conditions as their most closely related living relatives. In this respect assemblages of fossils rather than individuals have proved to be the more reliable. Certain morphological features, as shape, size, and marginal character of fossil dicot leaves are useful. They confirm the inferences obtained through other methods. Even the rocks possess certain features which indicate that they originated under a particular kind of climate.

Fossil plant remains are, in general, proven to be the most widespread and dependable indicators of ancient climates of the earth's land areas available. This is true in large part because plants are more sensitive than animals to their environment. A comparison of any map of the world's vegetation zones with one of the world's climatic zones will show how nearly the two coincide. Plants, moreover, are stationary and they can, therefore, not migrate nor burrow underground to escape the rigors of an unfavorable season. No palms or breadfruit trees, for instance, will live in the parks of New York, Chicago, or Ann Arbor, simply because the climate is too cold or, more specifically, the winter season is too cold. Such plants represent types which can be used as climatic indicators of the past. For example, Plate I illustrates a specimen of a palm and one of a cycad that were collected from the Eocene rocks on Kupreanof Island in southeastern Alaska. These leaves were associated with the remains of laurels, magnolias, acacias, peppers, and other forms which clearly indicate conditions much warmer than exist in that region today, in fact, they were subtropical. A collection of poplars, maples, elms, and oaks would, on the contrary, indicate temperate conditions of growth; and a fossil occurrence of northern spruce, larch, alpine fir, and birch would point to subarctic (boreal) conditions.

Although the above is true, it must be emphasized that not all plants are equally valuable as climatic indicators. Some forms are too cosmopolitan, that is, they are too tolerant in their requirements. For example, pines occur in modern forests from sea level to the mountain tops and from the subtropics to the subarctic. The common brake, or bracken fern, is equally at home whether in the tropics or in temperate regions. Clearly neither fossil pines nor fossil bracken ferns would be reliable indicators of particular climates of the past. The best climatic indicators of the past are obviously those forms whose nearest living relatives have the narrowest climatic requirements in modern forests.

Generally, conclusions based on a few climatic indicators should be confirmed, whenever possible, by inferences based on the study of an entire assemblage of fossil plants. This eliminates the possibility of a few unrepresentative forms giving erroneous conclusions regarding past climates. A number of tropical families have temperate relatives which appear to be "foreigners," climatically speaking, in the temperate forests. Magnolia, for instance, which belongs to a tropical family, extends into the temperate forests as far north as Massachusetts and southern Ontario and the persimmon, and the sassafras, both members of tropical families, extend as far north as southern New England. Along the Pacific Coast the pepperwood (*Umbellularia*) is another typical temperate representative of a normally tropical family. If such forms as these were found in a fossil plant assemblage whose composition was predominantly temperate, they would be regarded as foreigners and eliminated from the list of valid climatic indicators.

In spite of the appearance of a few anomalous plants in a collection, the general facies of the total assemblage gives the only safe and reliable basis for reconstructing ancient climates. It is useful, however, to apply another, quite independent, method of determining past climatic conditions by means of certain anatomical features of plants. The majority of the deciduous leaves in forests in the humid subtropics and tropics are relatively large and smooth-margined, whereas those in temperate forests are dominantly smaller and variously lobed or toothed along their borders (Bailey and Sinnott, 1915). The arrangement and number of stomata on leaves are also helpful in determining the conditions under which the leaves developed. Since these morphological features can usually be observed in fossil leaves, their use in making paleoclimatic inferences is of great value.

Among animals used in the study of ancient climates, the marine corals, especially the reef corals, are the most reliable. Contemporary reef corals live only in the warm, clear seas of the subtropics and tropics. It is, therefore a fair assumption that closely related reef corals, when found as fossils, likewise indicate warm, clear seas at the time they lived. The fossilized remains of alligators and crocodiles, manatees, and tapirs are generally regarded as proof of subtropical to tropical conditions, whereas the recovery of fossil bones or skeletons of reindeer, muskoxen, walrus, or the boreal lemming points to cold, subarctic conditions. The discovery that the proportion of oxygen isotopes in the shells of marine shellfish depends upon the temperature of the sea water in which the animal built its shell has been found of value in reconstructing the changing oceanic temperatures of the past (Piggot and Urry, 1942). This last has been a particularly welcome method, because it is completely independent of the actual species, genus, or family to which the shells may belong.

Geologists have also found it possible to confirm climatic inferences by means of certain characters of associated sedimentary rocks. Lowland glacial deposits, for example, which are not too difficult to recognize, clearly indicate a past episode sufficiently cold to allow for the development of lowland ice sheets. On the other hand, reddish sedimentary rocks, generally owing their origin to the red soils called laterites, are known to form only under conditions having an annual temperature of at least 60° F. and 40 inches of annual rainfall (Krynine, 1949). Evaporite deposits, including salt and gypsum, usually indicate that conditions were semiarid to arid during their formation. Extensive coal deposits, from such unlikely places as Antarctica and Spitsbergen, are generally interpreted as having accumulated during a humid, temperate to subtropical period.

The known physical conditions of a particular period of the geologic past have also been used to reconstruct ancient climates. Inferences derived from this source are based on the relative heights of continents, the relative amount of land versus water, the inferred direction and temperature of ocean currents, and the amount of volcanic activity. Climatic curves based on such studies approximate very closely those derived from the study of fossil organisms (Brooks, 1951, p. 1016).

In the present discussion the reconstruction of past climatic conditions is limited to those of the Cenozoic Era, which began about 70 million years ago. The record of this portion of geologic time is more complete than that of more ancient periods. Furthermore, Cenozoic fossils show closer relationships to living plants and animals than is true of older fossils; hence, comparisons of the climatic requirements of Cenozoic with living assemblages are considered more reliable.

Tertiary Climates of North America

Late Eocene–Early Oligocene.—The inferred climatic zones[2] of the Late Eocene to Early Oligocene epochs, beginning about 40 million years ago and lasting until about 30 million years ago are illustrated in Map 1. The climatic zones of the preceding Paleocene Epoch, though not yet as well established, indicate somewhat cooler conditions. In both Europe and North America a general warming trend began before the end of the Paleocene Epoch and continued into the Eocene. As a consequence, by the end of the Eocene and continuing into the Oligocene the temperate forest belt in North America had shifted about 20 degrees of latitude farther north than its present position and, at the same time, the tropical forest belt extended about 10 to 12 degrees farther north than at present. In the Gulf Coast states, for example, there are numerous fossil remains of both Late Eocene and Early Oligocene forests whose nearest living relatives live in the tropical lowlands of northern South America and coastal Mexico. Fossil remains of these forests occur in widespread coastal plain deposits as far north as Tennessee and Missouri. Both mangrove swamp and beach jungle associations are represented. Some of the notable members of these floras include the date palm and the East Indian Nipa palm (Berry, 1937; Arnold, 1952). In the Cordilleran region fossils of subtropical vegetation are found in

---

[2] The terminology for the climatic zones here and on the remaining figures is a slight modification of that used by climatologists and geographers: the term "temperate" is used as a general term for the zone of the middle latitude continental climates, including both the humid, warm summer and cool summer places, the semiarid steppe, the arid desert, and the marine, cool summer phase (during most of the Tertiary Period only the warm and cool temperate phases can be distinguished); the term "subtropical" includes the climate which many botanists refer to as "warm temperate"; for example, southeastern United States, which the majority of geographers and climatologists call "humid, subtropical" (Köppen and Geiger, 1936; Lackey, 1944).

northern California, Oregon, Washington, and northern Wyoming. In western Oregon, for example, Dr. Ralph W. Chaney and his students collected a great many well preserved leaves of Eocene age, including figs, laurels, cinnamons, avocados, and magnolias. These forms are typical of a forest dominated by large, smooth-margined dicot leaves whose nearest living equivalents are found in the lowlands of Central America (Chaney and Sanborn, 1933). Remains of this same forest occur also in the Eocene rocks of the John Day Basin of eastern Oregon, where the living desert vegetation presents a striking contrast to the lush, subtropical forest recorded in the rocks. Fossilized remains of marine faunas along the Pacific Coast confirm the northward extension of warmer tropical climate during the Eocene (Durham, 1950).

Numerous fossil remains of a forest transitional between subtropical and warm temperate are found in Late Eocene-Early Oligocene deposits of British Columbia as well as on Kupreanof Island in southeastern Alaska. Modern equivalents of this transitional forest, which included palms and cycads (Pl. I), lie more than 20° in latitude south of their Eocene occurrences. Farther east among the spruces, firs, and quaking aspens of the boreal forest of Yellowstone Park, Wyoming, my own field parties have collected many Middle Eocene fossils of lowland, warmth-loving forms, such as breadfruit, laurels, figs, and magnolias in association with more temperate elements, such as true redwoods, hickories, maples, and oaks. Southeast of Yellowstone Park a Late Eocene flora is closely related to the subtropical forests of the same age in Oregon and California (Dorf, 1953). Fossil bones of subtropical alligators have also been found in rocks of this age as far north as central Wyoming. In the eastern states the Brandon fossil flora of Vermont—largely made up of seeds, fruits, and pollens—is interpreted as transitional between the warm temperate and subtropical forests of this age (Traverse, 1955, pp. 21–34). Further south along the Atlantic coast, the bryozoans of the Eocene deposits of New Jersey indicate subtropical marine waters. Truly warm temperate forests, dominated by such forms as dawn redwoods, maples, beeches, oaks, sycamores, and basswoods, are widely recorded in rocks of this age in the belt extending from central Alaska to west-central Greenland and eastward to Spitsbergen and Siberia (Chaney, 1947). In Greenland the fossil-leaf beds occur on the bleak Arctic wastes of Disco Island, almost within sight of the glacial mantle of ice covering the mainland. Today the stunted vegetation, with prostrate willows and dwarf birches, is striking witness to the hardships of the polar climate and contrasts vividly with the warm temperate aspect of the fossil forests. Fossil remains of subarctic boreal forests, including spruces, pines, willows, hazels, and birches, have been found in the rocks of Late Eocene

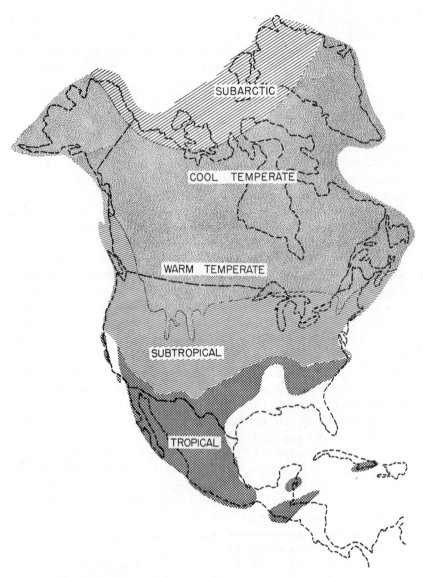

MAP 1. Generalized climatic zones of Late Eocene–Early Oligocene.

MAP 2. Generalized climatic zones of Late Oligocene–Early Miocene.

age as far north as Grinnell Land, only 8.5° from the North Pole (Berry, 1930, p. 10). There is no evidence of either a polar ice cap or a continental ice sheet on Greenland during this portion of geologic time. On the other hand, neither is there any evidence to support the view, unfortunately still often expressed, that the polar regions above the Arctic Circle supported subtropical or tropical forests during this time, or at any time, for that matter.

Late Oligocene – Early Miocene.—By the end of the Oligocene and the beginning of the Miocene Epoch, about 25 million years ago, the older subtropical forests of the Pacific Northwest had been replaced by warm temperate equivalents of the redwood forest association of the California Coast Ranges (Map 2). In eastern Oregon, for example, the Late Oligocene Bridge Creek flora from the John Day Basin contains numerous remains of a forest which included dawn redwoods, alders, maples, pepperwoods, dogwoods, tan oaks, hazels, and sycamores. This assemblage is essentially similar in its generic composition to the Late Eocene forest of central Alaska (Chaney, 1947). Its gradual southward shift in position during the course of 12 to 15 million years had apparently been accomplished without any major changes in its characteristic physiognomy. Eastward as far as the Dakotas the older subtropical forests appear to have moved to the south, with their places taken by warm temperate forests. This replacement was already under way earlier in the Oligocene Epoch, as is illustrated by the Middle Oligocene Ruby Basin flora of southwestern Montana which records a warm temperate forest dominated by dawn redwoods, oaks, beeches, maples, alders, and ash (Becker, 1956). Farther south in central Colorado the Florissant flora of the same age is an association of true redwoods and warm temperate hardwoods, in which there was a small lingering subtropical element as well as a few xeric forms (MacGinitie, 1953, pp. 36–42). Fossil records indicate that it was at this time that open woodland scrub and grasslands were beginning to develop in the lowland areas of the eastern Rockies and the Great Plains. In the southeastern United States the Oligocene Vicksburg flora, though small, is interpreted as a subtropical strand-line association (Berry, 1937). In marine deposits of Oligocene age, reef corals are known from as far north as 51.5° N. Lat. (at present their northern limit is about 32° N. Lat).

Middle Miocene – Late Miocene.—By the Middle to Late Miocene Epoch, about 18 to 12 millions years ago, the fossil plant record of western North America indicates a slight reversal of forest migrations (Map 3). A small subtropical element, which included palmettos, avocados, mahogany, and lancewood, returned north at least as far as Oregon and Washington (shown in the Mascall and Latah floras; Chaney, 1938, p. 387). Other

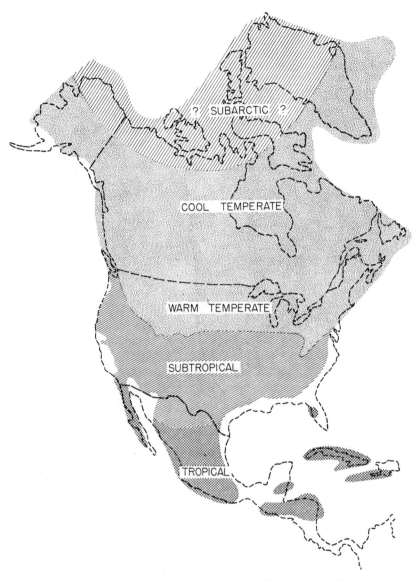

MAP 3. Generalized climatic zones of Middle Miocene–Late Miocene.

members of these two floras were of typically warm temperate aspect. From northern Mexico an element of hardy drought-resistant shrubs and chaparral moved north into southern California and the Great Basin regions, where there was a marked development of semiarid steppes (Axelrod, 1940). Fossil fish remains from the Miocene of southern California point to subtropical marine conditions that lasted well into the Late Miocene (David, 1943, pp. 47, 87, 96, 113, 174). Some time before the close of the Miocene Epoch, or early in the Pliocene, the climate began to change again and the cooling trend was resumed. In the Great Plains area the Miocene fossil record indicates a gradual development and spread of open grasslands and savannahs at the expense of true forests (Elias, 1942, pp. 15–18). Farther east the Miocene forests of the Virginia area were similar to the South Atlantic and Gulf Coast swamp forests of the present day. The Mid-Miocene deposits of Virginia have yielded bones of the southern manatee and crocodiles; neither of these warm-water animals occur north of Florida today. In the southern states the Miocene forests show a continuing replacement of tropical elements by subtropical and warm temperate forms (Berry, 1937).

Late Pliocene.—Near the end of the Pliocene Epoch, about 2 million years ago, the fossil records of forest distribution indicate an approach toward present climatic conditions (Map 4). Along the west coast of the United States, as a result of the cooling trend, the southward shifting of forest belts had continued since the Late Miocene trend. Major topographic changes brought about by the Cascadian mountain-building revolution produced a great variety of habitats and exerted a stronger influence than previously on the distribution of floral associations. Desert vegetation developed in the Great Basin and the interior of northern Mexico because of lower rainfall and greater seasonal temperature ranges (Chaney, 1944; Axelrod, 1944). Grasslands in the Great Plains appear to have reached essentially their modern aspect and distribution. In the southeastern states, Pliocene floras were wholly modern in general facies and were made up of species whose living forms are mostly native to the same region today (Berry, 1937).

Corroborative Evidence

The inferred trend from warmer conditions in the Eocene Epoch to much cooler conditions in the Pliocene has been amply substantiated by certain characters of the fossil leaves. In the Eocene floras of North American middle latitudes the dominantly large, smooth-margined leaves with elongate tips indicate humid subtropical to tropical conditions. But such leaf forms are gradually replaced in the succeeding Oligocene, Miocene, and Pliocene Epochs by the smaller, toothed or lobed leaves characteristic of more temperate conditions. This gradual cooling off has been further confirmed by study of the sequence of Tertiary marine faunas of the Pacific states (Durham, 1950). A drop in oceanic temperatures during the Tertiary has also been inferred from the gradual equatorward shifts of the northern limits of coral reefs and a gradual change in oxygen isotope ratios in shells of bottom-dwelling foraminifera from the mid-Pacific (Emiliani, 1954). On land, the record of the Tertiary mammals of western North America shows essentially the same general cooling of the climate (Colbert, 1953, pp. 266–70).

Tertiary Climates of Western Europe

In western Europe, for comparison, the fossil record of both plants and animals indicates the same climatic trend from essentially tropical conditions in the Eocene to progressively cooler and cooler conditions to the end of the Pliocene (Reid and Chandler, 1933, pp. 50–59) and, as in North America, the episode of maximum warmth occurred in the Eocene Epoch. The types of both fossil plants and invertebrates indicate that the preceding Paleocene Epoch began with a somewhat cooler climate. A gradually warming trend began before the Middle Paleocene and continued into the Eocene. The Eocene episode of maximum warmth is well illustrated by the tropical flora of the Eocene London Clay. This fossil flora, obtained from a region lying at about 50° N. Lat., finds its nearest living counterpart at about 10° N. Lat. in the lowland tropical rainforest of the Indo-Malayan region. The remains of alligators, crocodiles, the pearly nautilus, and large warm-water volutes in Eocene marine deposits of western Europe indicate subtropical-to-tropical marine waters far north of their present limits. Plant remains of the Eocene forests of Spitsbergen, collected about 11 to 13 degrees from the North Pole, include species of beech, sycamore, linden, oak, and water lilies—an assemblage whose modern equivalents are among the dominants of the temperate forests and lakes and now live 15 to 20 degrees farther south on the Continent (Heer, 1868, pp. 60–62). The fossil plant record of western Europe shows that a gradual change from the tropical conditions of the Eocene to subtropical in the Oligocene and warm temperate in the Miocene took place. This cooling trend apparently continued into the succeeding Pliocene Epoch, which ended with a climate believed to have been quite similar to the cool temperate conditions of western Europe at the present time. The annual mean temperature of the region is estimated to have dropped about 15° C. from the Eocene to the end of the Pliocene.

The gradual cooling off of western Europe during the Tertiary Period is further confirmed by the relative proportions of woody to herbaceous plants in the geologic sequence. The proportion of woody species shows a

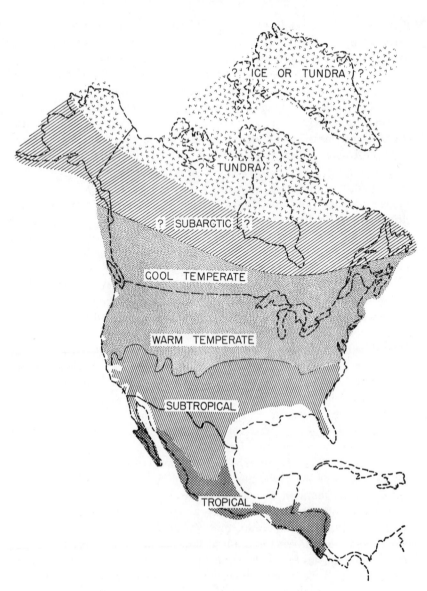

MAP 4. Generalized climatic zones of Late Pliocene.

downward progression from 97 per cent in the Eocene London Clay, to 57 in the Oligocene and to 22 in the Late Pliocene. Twenty-two per cent compares closely with the figure of 17 per cent for the woody species now in the region (Reid and Chandler, 1933, p. 53). This reduction in the course of time parallels the geographical reduction noted in living forests, from about 88 per cent woody species in the tropical lowlands of the Amazon Basin to about 10 per cent in the subarctic vegetation of Iceland.

Independent studies of Tertiary marine invertebrates and fishes, and of terrestrial insects, corroborate the climatic inferences based on fossil plants (Davies, 1934; Theobald, 1952; Schwarzbach, 1950). Evidence from the rocks themselves, the Tertiary laterites of Europe, has led to the same general conclusions regarding climatic changes as have the studies based on the fossil record (Harassowitz, 1926, pp. 544–88).

Quaternary Climates

Glacial.—In the Pleistocene, or "Glacial," Epoch the cooling trend, which had begun in the Oligocene Epoch, led to the formation and slow spread of lowland ice sheets in higher latitudes. Beneath the ice all plant and animal life was obviously obliterated. As the ice sheets spread southward from central Canada and extreme glacial climate caused a farther and more rapid southward shift of the forests south of the ice margins. Just how far south the forests were driven is a matter of debate. For many years both paleobotanists and botanists have believed that forest belts shifted only slightly (Berry, 1926, p. 99; Braun, 1950, pp. 458–72), but evidence for major forest migrations over wide distances has been accumulating during the past decade (Deevey, 1949, pp. 1360–66, 1375; Frey, 1953). The latter view is that supported here (cf. Map 5).

There are numerous Pleistocene records of both subarctic plants and animals far to the south in North America. Occurrences of northern spruce and fir are found in the lowlands of northern and central Florida, south-central Texas, northern Oklahoma, and southern Kansas (Deevey, 1949, pp. 1360–61; Davis, 1946; Potzger and Tharp, 1947). Cones of white spruce and Canadian larch, twigs of the northern arbor vitae, and the delicate remains of two species of northern mosses have been recorded in southeastern Louisiana, up to 1000 miles south of their present southern limits (Fisk, Richards, Brown, and Steere, 1938). Fir and northern pines occur on the coastal plain of North Carolina and larch in southeastern Georgia (Buell, 1945). Cold-temperate diatoms, whose living forms are characteritsic of New England and eastern Canada, are abundant in northern and central Florida (Hanna, 1933). In southern California there is a Pleistocene record of a kind of vegetation now found growing between 8 and 10 degrees farther north along the coast. A possible indirect effect of

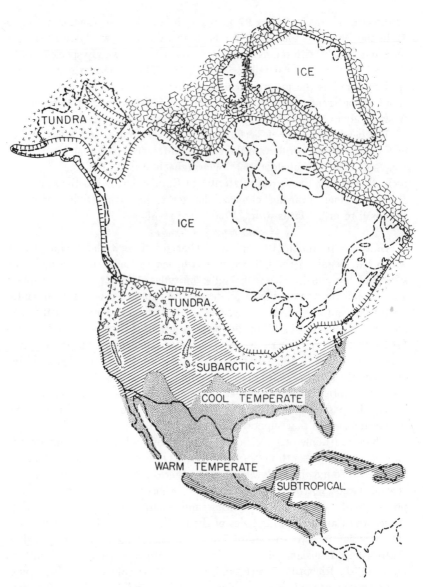

Map 5. Generalized climatic zones of a composite of glacial stages of Pleistocene.

the shifting of temperate forests as far south as shown in Map 5 is the presence today in the mountains and high plateaus of Mexico and Central America of remnants of the typical warm temperature to subtropical association normal to the lowlands farther north in the United States.

Among the vertebrates the records of musk oxen as far south as Mississippi, Texas, Oklahoma, and southern California indicate a considerable southward shift, as does the record of reindeer in southeastern Kentucky, Iowa, and central Nevada. Wooly mammoth remains are known from as far south as west-central Florida and southern Texas. Walrus bones have been found along the Atlantic coast as far as South Carolina and Georgia, over 1000 miles south of the southern limit of the modern walrus (Hay, 1923). Fossils of the northern moose are recorded from Oklahoma, Kentucky, and South Carolina. Conies, which live above timberline in the Rocky Mountains at the present time, occur as fossils in the Pleistocene deposits of Pennsylvania and Maryland. Fossils of western marmots are found from 2500 to 4500 feet below their present lower limit in central New Mexico (Deevey, 1949, pp. 1374–75; Stearns, 1942).

Among the invertebrates there is a record of a cold-water assemblage in southern California and of subarctic Foraminifera as far south as the Sigsbee Deep in the middle of the Gulf of Mexico (Trask, Phleger, and Stetson, 1947).

In western Europe extensive records of both faunas and floras of glacial age show that subarctic (boreal) conditions extended south as far as the shores and adjacent lowlands of the Mediterranean.

Interglacial.—Alternating with the glacial stages were the interglacial. During several of these the climates appear to have been somewhat warmer than today (Maps 6 and 7) and, as the ice sheets melted away, a rapid northward shift of the forests took place, extending them into the newly exposed glaciated wastelands of the north. At Toronto, for example, both the plants and invertebrates of a late interglacial age indicate conditions about 2 to 3° C. higher than today (Flint, 1957, p. 340); the plant remains include the pawpaw, red cedar, and osage orange, whose northern limits today are several hundred miles south of Toronto. On the Seward Peninsula of Alaska occurrence of fossil plants of an early interglacial age indicate a climate both warmer and more humid than in the same region today (Hopkins and Benninghoff, 1953). Cape Cod has a record of an interglacial forest similar to that of present-day Virginia and North Carolina. On Long Island an interglacial shellfish fauna points to water temperatures higher than now in the region, and in New Jersey the record of a manatee indicates conditions like those off the present coastline of Florida. In Pennsylvania there are interglacial records of tapirs of Central American aspect and of

MAP 6. Generalized climatic zones of a composite of interglacial stages of Pleistocene.

MAP 7. Generalized climatic zones of the Present.

peccaries whose nearest living relatives range only as far north as Arizona, New Mexico, and Texas. Farther west, in southwestern Kansas, interglacial records of a Gulf Coast box turtle and the rice rat point to conditions warmer than at present (Hibbard, 1955, p. 202). Sediment cores from the bottom of the Atlantic Ocean show layers of interglacial deposits with the remains of Foraminifera of warmer waters than those of the region today (Bradley, 1940).

CLIMATIC CHANGES OF THE NEAR PAST AND PRESENT

The knowledge of climatic changes since the last of the ice sheets (the Cochran readvance) depends in part upon the fossil record and in part on archeological and historical records. On the basis of the pollen record there was a period of somewhat higher temperatures than today (the so-called Thermal Maximum or Hypsithermal), which lasted from about 5000 to about 2000 B.C. (Fig. 1). This was apparently followed by a general cooling, which reached a minimum in about 500 B.C., and a subsequent rise in temperature. Archeological records indicate that the last major warm episode before the present occurred about 1000 to 1300 A.D. During this time a colony of about 3000 Norsemen grew crops and raised both cattle and sheep in southwestern Greenland (Ahlmann, 1949, p. 165) and vineyards were productive as far north as southern England. Beginning in about 1600, however, the climate began to change toward cooler conditions. Glaciers in the northern hemisphere began to readvance; in the Alps several valley settlements were completely overrun by advancing valley glaciers. This cooling episode led to the so-called "Little Ice Age," which lasted from about 1650 to about 1850. Since 1850 the general climatic trend has been toward warmer conditions.

In broad retrospect, then, where do we find ourselves today, near the middle of the twentieth century, in the everchanging pattern of climatic cycles? In the first place, it is quite clear that for the past million years or so (up to and including the present time) the earth has been subjected, geologically speaking, to an abnormally cold climate (see Fig. 1). The greater part of at least the last 500 million years, however, has had a warmer, nonglacial climate rather than the colder glacial climate of the past million years. In the second place, it is evident that the earth is not in one of the truly frigid glacial stages, but is rather in one of the slightly warmer interglacial stages, that it is perhaps about two-thirds of the way out of the last glacial stage, so to speak. By comparison with the long duration of past interglacial and glacial episodes in earth history, it is generally believed that we shall return to another glacial stage in about 10,000 to 15,000 years. Such a prospect, with its accompanying ice sheets devastating northern lands and settlements is not a happy one to contemplate in terms of physical, economic, or political consequences.

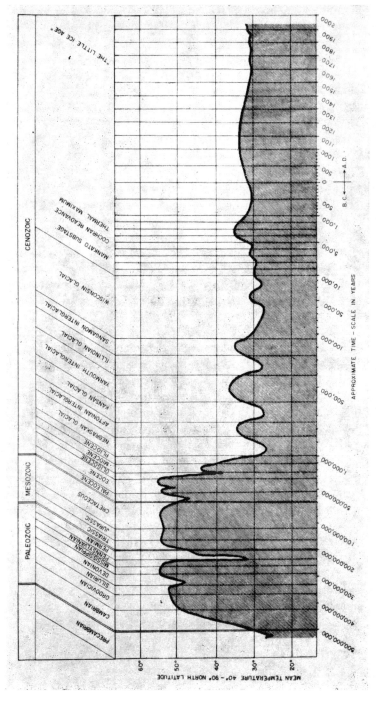

Fig. 1. Generalized temperature variations during the geologic past. After Brooks, 1951.

In the meantime we are apparently well along in a general world-wide warming trend (Fig. 2). In the United States the rise in mean annual temperature, since 1920, has been about 3.5° F., and the rise in winter temperatures has been about twice as much as that in summer temperatures (Visher, 1954, Part VII). In Boston winter temperatures are about 3.5° F. higher than they were a hundred years ago. In Philadelphia the temperature has risen more than 3° F. since 1880, and in Montreal more than 2.5° F. since 1850. In the years between 1931 and 1952 New Jersey's mean annual temperature has been above normal 18 years, below normal 3 years, and normal only 1 year; Michigan's temperature in the same interval has been above normal 15 years and below only 7 years. The U.S. Weather Bureau (1953) reported that during the same period all but 8 of the 48 states had annual temperatures which were above normal the majority of the years; those which were below normal during most of the years were the three Pacific coast states (California, Oregon, and Washington), four western states (Arizona, Idaho, New Mexico, and Montana), and one southern state (Texas). In both the Middle West and the Great Plains, the records of cities such as St. Louis and Kansas City indicate a similar rise in temperatures: 1.9° F. for the former since 1880, and 1.4° F. for the latter since 1894 (Oltman and Track, 1951, p. 17). It has been observed, however, that the greatest temperature increases during the last hundred years have been in the Arctic regions. In Spitsbergen, only about 10 to 12 degrees from the North Pole, the mean winter temperatures have risen about 14° F. since 1910 (Willett, 1950). Ice-free ports there are now open to navigation about 7 months of the year as compared with only 3 months fifty years ago (Ahlmann, 1953, p. 32). If the warming trend of the north polar region should continue at its present rate, it has been estimated that the entire Arctic Ocean would be navigable all year long within about a hundred years. At the opposite end of the world, according to recent reports from the Weather Bureau (Wexler, 1958), the Antarctic region has undergone a rise of about 5° F. in average temperature in the last fifty years. There has been no appreciable rise, however, in the mean annual temperatures in the tropical regions of the world.

What have been some of the notable results of this warming trend during the last hundred years? Glaciers throughout the world have been melting away at a rapidly increasing rate. Brooks (1949, p. 24), the eminent British paleoclimatologist, stated that "Since the beginning of the 20th Century glaciers have been wasting away rapidly, or even catastrophically." In the Juneau region of Alaska, all but one of the numerous glaciers began melting away as far back as 1765. Muir Glacier, for example, has retreated as much as two miles in 10 years. Baird and

Fig. 2. Changes in mean annual temperature, 1920 to 1940, in degrees Fahrenheit. After Willett, 1950.

Sharp (1954, p. 143) have referred to the "alarming retreat of glaciers" in the Alaskan region; along the Pacific Coast of North America and in Europe they believe the glacial melting "appears to be progressing violently." In the north polar region, measurements of melting of the ice islands in the Arctic Sea indicate an approach toward an open polar sea (Crary, Kulp, and Marshall, 1955). In only a few regions of the world, such as the Pacific Northwest, are there any records of glaciers advancing during the past century, and these have been mostly since 1950 (Hubley, 1956). The warmer temperatures have also caused a general rise of the snow line throughout the mountainous regions of the world, even in the tropics: in northern Peru it has risen about 2700 feet during the 60 years.

Believed in large part to be the result of the melting of the world's glaciers, sea level has been rising at a rapidly increasing rate, amounting to as much as a 6-inch rise from 1930 to 1948 (Marmar, 1948). This is about four times the average rate of sea level rise during the past 9000 years, as recorded by Shepard and Suess (1956). It should be noted that more than a six-fold increase in the rate of sea level rise occurred in the mid-1920's at the same time there was a striking change in the rate of glacial melting in the north (Ahlmann, 1953, Fig. 11).

Changes in vegetation brought about by the warmer temperatures include the encroachment of trees into the subpolar tundra as recorded in Alaska, Quebec, Laborador, and Siberia. In the Canadian prairies the agricultural crop line has shifted from 50 to 100 miles northward as a result of the lengthening of the growing season by as much as ten days. In parts of northern New England and eastern Canada the birch trees have been dying off over large areas, and the spruces and balsams have begun to suffer as a result of the rise in summer temperatures. In Sweden the timberline has moved up the mountain slopes as much as 65 feet since 1930 (Ahlmann, 1953, p. 35).

In the animal world many southern types of both birds and mammals have been extending their habitat ranges northward as a result of the warming trend. The cardinal, the turkey vulture, the tufted titmouse, and the blue-winged warbler, as well as the warmth-loving opossum, have slowly moved their ranges into the northern United States. A good many central European species of animals have been shifting their ranges northward into Scandinavia, Greenland, Iceland, and the Faero Islands. Twenty-five species of birds alone are reported to have invaded Greenland from the south since 1918 (Jensen and Fristrup, 1950). Codfish from the Atlantic have replaced the seals in the waters along the coast of Greenland. It is reported that compared to a shipment of 5 tons of codfish from Greenland in 1913, the 1946 shipment had risen to over 13,000 tons; the Greenland Eskimos have become cod fishermen instead of seal fishermen (Kimble, 1950). Farther south tunafish have moved northward into the waters off New England, and tropical flying fishes have become increasingly common off the coast of New Jersey.

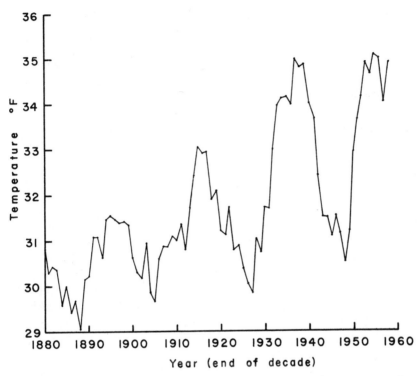

Fɪɢ. 3. January temperatures in New York City, 1871 to 1958, ten-year running means. After Spar, 1954.

In spite of its numerous and striking effects, the warming trend of the past hundred years has not been worldwide nor uniform in direction. Over most of the northern hemisphere there have been rather regular cyclic fluctuations from warmer spells to colder spells, each lasting a decade or so. Most of the eastern and central United States is at present in a decade that is somewhat colder than the one which ended in 1955. These shorter cyclic fluctuations are observable in both summer and winter temperatures. As shown in Figure 3, for example, the ten-year averages of

January temperatures in New York City show colder 10- to 11-year cycles followed quite regularly by warmer 10- to 11-year cycles since at least 1880 (Spar, 1954). Even here it may be noted that the trend is upward, that is, the average temperatures of the last few decades have been warmer than those of the decades prior to 1900. Certain regions of the world, moreover, have actually experienced a cooling trend during the time that most of the rest of the world was warming up. Such regions (Fig. 2) include the Hudson Bay Region, the west coast of North America, central South America, and the East Indies (Willett, 1950).

POSSIBLE CLIMATIC CHANGES OF THE NEAR FUTURE

The usefulness of observations from the past has been demonstrated in many fields to be of real value in the forecasting of events in the future. In the words of Byron, "The best prophet of the future is the past." Unfortunately, as one is reminded much too often, the forecasting of either weather or climate on the basis of past performance is notoriously even less reliable than predicting the future actions of human beings or race horses. Several climatologists expressed the opinion, in the late 1940's, that the 1930–1940 decade marked the culmination of the trend toward higher temperatures, and that a reversal of the trend had begun. In the middle 1950's, when it had become quite obvious that the reversal had not occurred and that the rising temperatures had continued after a short colder cycle, opinions were revised to the effect that it was the 1940–1950 decade which marked the maximum of the warming trend, which was due for a reversal. Other climatologists have stated their belief that the present warming episode will continue for at least a few centuries. Brooks (1949, p. 24), in his forecast of the future, stated that the melting of the world's glaciers "may be expected to continue until either the reduced ice-cap reaches a new position of stability, or until some meteorological 'accident' reverses the trend and ushers in a new period of re-advance."

Extrapolation of the known climatic curves into the future is one of the more reasonable approaches to the prediction of possible future trends. Such extrapolation suggests (1) that the earth is at present still in an interglacial stage but heading toward another glacial stage, perhaps 10,000 to 15,000 years hence; (2) that the winters are getting slightly colder again and should continue to average colder until about 1965; and (3) that though marked by minor alternating colder and warmer cycles, the general trend of increasing warmth should continue for at least two or three hundred years over most of the lowland regions of the northern hemisphere.

LITERATURE CITED

AHLMANN, H. W. 1949. The Present Climatic Fluctuation. Journ. Geog., Vol. 112, pp. 165–93.

——— 1953. Glacier Variations and Climatic Fluctuations. Amer. Geog. Soc., Bowman Mem. Lectures, Ser. 3, pp. 1–51.

ARNOLD, CHESTER A. 1952. Tertiary Plants from North America. Palaeobotanist, Vol. 1, pp. 73–78.

AXELROD, DANIEL I. 1940. Late Tertiary Floras of the Great Basin and Border Areas. Bull. Torrey Bot. Club, Vol. 67, pp. 477–87.

——— 1944. Pliocene Sequence in Central California. Chap. 8 *in:* Pliocene Floras of California, by R. W. Chaney. Carnegie Instit. Wash. Publ., No. 553, pp. 207–24.

BAILEY, I. W., and SINNOTT, E. W. 1915. A Botanical Index of Cretaceous and Tertiary Climates. Science, Vol. 41, pp. 831–34.

BAIRD, P. D., and SHARP, R. P. 1954. Glaciology. Arctic, Vol. 7, Nos. 3–4, pp. 141–52.

BECKER, HERMAN F. 1916. An Oligocene Flora from the Ruby River Basin in Southwestern Montana. Univ. Mich. Microfilm Publ., No. 1956. (Doctoral Dissertation.)

BERRY, E. W. 1956. Pleistocene Plants from North Carolina. U. S. Geol. Surv., Prof. Paper No. 140-C, pp. 97–120.

——— 1930. The Past Climate of the North Polar Region. Smithsonian Misc. Coll., Vol. 82, No. 6, pp. 1–29.

——— 1937. Tertiary Floras of Eastern North America. Bot. Rev., Vol. 3, pp. 31–46.

BRADLEY, WILMOT H. 1940. Geology and Climatology from the Ocean Abyss. Sci. Mon., Vol. 50, No. 2, pp. 97–109.

BRAUN, E. LUCY. 1950. Deciduous Forests of Eastern North America. Phila.: Blakiston. 533 pp.

BROOKS, C. E. P. 1949. Post-glacial Climatic Changes in the Light of Recent Glaciological Research. Geog. Ann., Vol. 31, pp. 21–24.

——— 1951. Geological and Historical Aspects of Climate Change. *In:* Compendium of Meteorology, Thomas F. Malone, ed. Boston: American Meteorological Society. Pp. 1004–1018.

BUELL, MURRAY F. 1945. Late Pleistocene Forests of Southeastern North Carolina. Torreya, Vol. 45, pp. 117–18.

CHANEY, RALPH W. 1938. Paleoecological Interpretations of Cenozoic Plants in Western North America. Bot. Rev., Vol. 4, pp. 371–96.

——— 1944. Pliocene Floras of California and Oregon. Carnegie Instit. Wash. Publ., No. 553, pp. 353–73, summary and conclusions.

——— 1947. Tertiary Centers and Migration Routes. Ecol. Monogr., Vol. 17, pp. 139–48.

——— and SANBORN, ETHEL I. 1933. The Goshen Flora of West Central Oregon. Carnegie Instit. Wash. Publ., No. 439.

COLBERT, EDWIN H. 1953. The Record of Climatic Changes as Revealed by Vertebrate Paleoecology. *In:* Climatic Change, H. Shapley, ed. Cambridge: Harvard Univ. Press. Pp. 249–71.

CRARY, A. P., KULP, J. L., and MARSHALL, E. W. 1955. Evidences of Climatic Change From Ice Island Studies. Science, Vol. 122, pp. 1171–73.

DAVID, L. R. 1943. Miocene Fishes of Southern California. Geol. Soc. Amer., Spec. Paper, No. 43. 193 pp.

DAVIES, A. MORLEY. 1934. Tertiary Fanuas. Vol. 2. London: Murby. 252 pp.

DAVIS, JOHN H., JR. 1946. The Peat Deposits of Florida, Their Occurrence, Development, and Uses. Florida Geol. Surv., Geol. Bull. No. 30, pp. 1–247.

DEEVEY, E. S., JR. 1949. Biogeography of the Pleistocene. Pt. I. Europe and North America. Bull. Geol. Soc. Amer., Vol. 60, pp. 1315–1416.

DORF, ERLING. 1953. Succession of Eocene Floras in Northwestern Wyoming. Bull. Geol. Soc. Amer., Vol. 64, p. 1413.

DURHAM, J. R. 1950. Cenozoic Marine Climates of the Pacific Coast. Bull. Geol. Soc. Amer., Vol. 61, pp. 1243–64.

ELIAS, MAXIM K. 1942. Tertiary Prairie Grasses and Other Herbs from the High Plains. Geol. Soc. Amer., Special Paper, No. 41, pp. 1–176.

EMILIANI, CESARE. 1954. Temperatures of Pacific Bottom Waters and Polar Superficial Waters during the Tertiary. Science, Vol. 119, pp. 853–55.

FISK, H. N., RICHARDS, H. F., BROWN, C. A., and STEERE, W. C. 1938. Contributions to the Pleistocene History of the Florida Parishes of Louisiana. Louisiana Dept. Conserv. Geol. Bull. No. 12, pp. 1–137.

FLINT, R. F. 1957. Glacial and Pleistocene Geology. New York: John Wiley & Sons. 553 pp.

FREY, DAVID G. 1953. Regional Aspects of the Late-glacial and Post-glacial Pollen Succession of Southeastern North Carolina. Ecol. Monogr., Vol. 23, pp. 289–313.

HANNA, G. DALLAS. 1933. Diatoms of the Florida Peat Deposits. Florida State Geol. Surv., 23rd–24th Ann. Rept. 1930–32, pp. 68–119.

HARASSOWITZ, HERMANN. 1926. Laterit. Fortschritte der Geol. und Palæont., Bd. 4, Heft 14, pp. 253–566.

HAY, O. P. 1923. The Pleistocene of North America and its Vertebrated Animals from the States East of the Mississippi River and from the Canadian Provinces East of Longitude 95°. Carnegie Instit. Wash. Publ., No. 322, pp. 1–499.

HEER, OSWALD. 1868. Die Fossile Flora der Polarländer. Flora Fossiles Arctica, Vol. 1, pp. 1–192.

HIBBARD, C. W. 1955. The Jinglebob Interglacial (Sangamon?) Fauna from Kansas and Its Climatic Significance. Contrib. Mus. Paleontol. Univ. Mich., Vol. 12, No. 10, pp. 179–228.

HOPKINS, D. M. and BENNINGHOFF, W. S. 1953. Evidence of a Very Warm Pleistocene Interglacial Interval on Seward Peninsula, Alaska. Abstract *in:* Bull. Geol. Soc. Amer., Vol. 64, pp. 1435–36.

HUBLEY, RICHARD C. 1956. Glaciers of the Washington Cascade and Olympic Mountains; their Present Activity and its Relation to Local Climatic Trends. Journ. Glaciol., Vol. 2, No. 19, pp. 669–74.

JENSEN, AD. S. and FRISTRUP, B. 1950. Den Arktiske Klimaforandring og dens Betydning, saerlig for Grönland. Geog. Tidskr., Vol. 50, pp. 20–47.

KIMBLE, C. H. T. 1950. The Changing Climate. Sci. Amer., Vol. 182, No. 4, pp. 48–53.

KÖPPEN, W., and GEIGER, R. 1936. Handbuch der Klimatologie, Vol. 2, Pt. J.

KRYNINE, PAUL. 1949. Origin of Red Beds. New York Acad. Sci., Ser. 2, Vol. 11, No. 3, pp. 60–68.

LACKEY, E. E. 1944. The Pattern of Climates. *In:* Global Geography, by George T. Renner and Associates. New York: Crowell. 714 pp.

MacGINITIE, HARRY D. 1953. Fossil Plants of the Florissant Beds, Colorado. Carnegie Instit. Wash. Publ. No. 599, pp. 1–188.

MARMAR, H. A. 1948. Is the Atlantic Coast Sinking? The Evidence from the Tide. Geog. Rev., Vol. 38, pp. 652–57.

OLTMANN, R. E., and TRACY, H. J.   Trends in Climate and in Precipitation—Runoff in Missouri River Basin. U. S. Geol. Surv., Circ. 98, pp. 1–113.

PIGGOT, C. S., and URRY, W. D.   1942.   Time Relations in Ocean Sediments. Bull. Geol. Soc. Amer., Vol. 53, pp. 1187–1210.

POTZGER, JOHN E., and THARP, B. C.   1947.   Pollen Profile from a Texas Bog. Ecology, Vol. 28, pp. 274–80.

REID, ELEANOR M., and CHANDLER, M. E. J.   1933.   The London Clay Flora. London: Brit. Mus. (Nat. Hist.). 561 pp.

SCHWARZBACH, M.   1950.   Das Klima der Vorzeit. Stuttgart: Ferdinand Enke. 211 pp.

SHEPARD, F. P., and SUESS, H. E.   1956.   Rate of Postglacial Rise of Sea Level. Science, Vol. 123, pp. 1082–83.

SPAR, JEROME.   1954.   Temperature Trends in New York City. Weatherwise, Vol. 7, No. 6, pp. 149–51.

STEARNS, CHARLES E.   1942.   A Fossil Marmot from New Mexico and Its Climatic Significance. Amer. Journ. Sci., Vol. 240, pp. 867–78.

THEOBALD, NICOLAS.   1952.   Les Climates de l'Europe Occidentale au cours des Temps Tertiares d'après l'Etude des Insectes Fossiles. Geol. Rundschau, Bd. 40, H. 1, pp. 89–92.

TRASK, PARKER D., PHLEGER, F. B JR., and STETSON, H. C.   1947.   Recent Changes in Sedimentation in the Gulf of Mexico. Science, Vol. 106, pp. 460–61.

TRAVERSE, ALFRED F., JR.   1955.   Pollen Analysis of the Brandon Lignite of Vermont. U. S. Bur. Mines, Rept. of Invest. No. 5151, pp. 1–107.

TREWARTHA, GLENN T.   1954.   An Introducton to Climate. New York: McGraw Hill. 395 pp.

U. S. WEATHER BUREAU.   1953.   Climatological Data, National Summary. 81 pp.

VISHER, STEPHEN S.   1954.   Climatic Atlas of the United States. Cambridge: Harvard Univ. Press. 403 pp.

WEXLER, HARRY.   1958.   (Quoted in New York Times, Saturday, May 31.)

WILLETT, H. C.   1950.   Temperature Trends of the Past Century. Centenary Proc. Royal Meteorol. Soc., pp. 195–206.

(Plate I is on the following page.)

PLATE I

EXPLANATION OF PLATE I

Fossil palm (left) and fossil cycad (right) from the Eocene of Kupreanof Island. U. S. Geological Survey.

AUTHOR CITATION INDEX

SUBJECT INDEX

About the Editor

CHARLES A. ROSS is Professor of Geology at Western Washingon State College and teaches courses in historical geology, paleontology, and paleoecology. Dr. Ross received his B.A. from the University of Colorado and his M.S. and Ph.D. from Yale University. He was a Research Associate at Peabody Museum of Natural History, Yale University, and Assistant Geologist and Associate Geologist at the Illinois State Geological Survey before going to Western Washington State College. He spent 1970–1971 in Australia at The University of Sydney studying the life habits of large calcareous Foraminifera (Protozoa) on the Great Barrier Reef in order to further interpret the significance and distribution of a group of extinct large Foraminifera of late Paleozoic age. In addition, he has traveled in many parts of Europe and the USSR and a large part of North America. Dr. Ross has published more than 70 articles and books and was organizer and editor of a symposium volume on Paleogeographic Provinces and Provinciality for the Society of Economic Paleontologists and Mineralogists. He is a member of the International Subcommission on Permian Stratigraphy.